Lecture Notes in Computer Science 9075

Commenced Publication in 1973
Founding and Former Series Editors:
Gerhard Goos, Juris Hartmanis, and Jan van Leeuwen

Editorial Board

More information about this series at http://www.springer.com/series/7407

Laurent Michel (Ed.)

Integration of AI and OR Techniques in Constraint Programming

12th International Conference, CPAIOR 2015
Barcelona, Spain, May 18–22, 2015
Proceedings

 Springer

Editor
Laurent Michel
University of Connecticut
Storrs, Connecticut
USA

ISSN 0302-9743   ISSN 1611-3349   (electronic)
Lecture Notes in Computer Science
ISBN 978-3-319-18007-6   ISBN 978-3-319-18008-3   (eBook)
DOI 10.1007/978-3-319-18008-3

Library of Congress Control Number: 2015936625

LNCS Sublibrary: SL1 – Theoretical Computer Science and General Issues

Springer Cham Heidelberg New York Dordrecht London

Printed on acid-free paper

Springer International Publishing AG Switzerland is part of Springer Science+Business Media
(www.springer.com)

Preface

The 12th International Conference on Integration of Artificial Intelligence (AI) and Operations Research (OR) techniques in Constraint Programming (CPAIOR 2015) was held during May 18–22, 2015 in Barcelona, Spain. The purpose of the conference series is to bring together researchers in the fields of Constraint Programming, Artificial Intelligence, and Operations Research to explore ways of solving hard and large-scale combinatorial optimization problems that emerge in various industrial domains. Pooling the skills and strengths of this diverse group of researchers has proved extremely effective and valuable during the past decade leading to improvements and cross-fertilization between the three fields as well as breakthrough for actual applications.

This year, the call for papers was opened to both long (15 pages) and short (8 pages) papers. A dedicated Program Committee of 29 members reviewed both types of submissions with the same stringent criteria. The conference received 90 submissions in total and the field narrowed down to 80 papers contending for publication. In the end, we accepted 35 papers for presentation and publication. This year, CPAIOR also featured a Journal Fast Track dedicated to the publication of selected papers with a level of maturity and details that warrants their immediate archival publication. Five papers among the 35 were identified for this process and are published in a special issue of the Journal.

In addition to the papers, the conference attendees enjoyed three invited talks. The speakers were Nikolaos Sahinidis, who is the John E. Swearingen Professor of Chemical Engineering at Carnegie Mellon University, Robert Nieuwenhuis who is a Professor in the Computer Science Department at the Technical University of Catalonia, and Jeffrey T. Linderoth who is a Professor in the Department of Industrial and Systems Engineering at the University of Wisconsin-Madison.

The traditional CPAIOR Master Class was delivered on the afternoon of May 18 and the entire day on the 19th. Pascal Van Hentenryck, who holds the Vice-Chancellor Chair in Data-Intensive Computing at the Australian National University and leads the NICTA Optimization Research Group organized the event whose theme was "Constraint Programming and Verification." The event featured no less than nine speakers all focused on verification. The morning of May 18 was devoted to three workshops titled "CPAIOR Meets CAV" organized by Justin Pearson and Michel Rueher, ISA: "Innovative Scheduling and Scheduling Applications using CPAIOR" organized by Pierre's Schaus and "Smart Cities" organized by Michele Lombardi.

I wish to extend my deepest gratitude to all the organizers. Carlos Ansotegui from Universitat de Lleida carried the torch as Conference Chair and was supported by a team consisting of Maria Bonet (UPC), Jordi Levy (IIIA-CSIC), and Mateu Villaret (UDG). David Bergman (UConn) was at the helm of the publicity effort and cannot be thanked enough. Pascal's enthusiasm delivered a stellar cast for this year's Master Class and I'm truly thankful for his energy and involvement. Finally, I wish to extend my thanks to the entire Program Committee for their efforts and participation throughout the entire process despite the tight schedule.

Naturally, I would be remiss if I did not express my gratitude to the sponsors of the conference which include, at the time of this writing, the ACP, Google Inc., National ICT Australia, The Catalan Association for Artificial Intelligence (ACIA), AIMMS, AMPL, ECCAI, Gurobi, and Inspires (The Polytechnic Institute of Research and Innovation in Sustainability). Last but not least, EasyChair deserves accolades for its flawless platform.

March 2015   Laurent Michel

Organization

Program Committee

Carlos Ansótegui	Universitat de Lleida, Spain
Claire Bagley	Oracle Corporation, USA
Chris Beck	University of Toronto, Canada
Nicolas Beldiceanu	TASC (CNRS/Inria), Mines Nantes, France
David Bergman	University of Connecticut, USA
Lucas Bordeaux	Microsoft Research, UK
Andre Cire	University of Toronto Scarborough, Canada
Philippe P. Codognet	CNRS/University of Tokyo, Japan
Andrew Davenport	IBM, USA
Tias Guns	Katholieke Universiteit Leuven, Belgium
Stefan Heinz	FICO Xpress Optimization, Germany
George Katsirelos	INRA, Toulouse, France
Philip Kilby	NICTA, Australia
Vitaly Lagoon	MathWorks, USA
Jeffrey T. Linderoth	University of Wisconsin-Madison, USA
Andrea Lodi	DEI, University of Bologna, Italy
Michele Lombardi	DISI, University of Bologna, Italy
Laurent Michel	University of Connecticut, USA
Michela Milano	DEIS, University of Bologna, Italy
Nina Narodytska	University of Toronto, Canada and University of New South Wales, Australia
Barry O'Sullivan	Insight Centre for Data Analytics, University College Cork, Ireland
Laurent Perron	Google, France
Louis-Martin Rousseau	École Polytechnique de Montréal, Canada
Michel Rueher	University of Nice Sophia, Antipolis, France
Paul Shaw	IBM, France
Pascal Van Hentenryck	NICTA, Australia
Willem-Jan van Hoeve	Carnegie Mellon University, USA
Petr Vilím	IBM, Czech Republic
Toby Walsh	NICTA and UNSW, Australia

Additional Reviewers

Abreu, Salvador
Babaki, Behrouz
Bal, Deepak

Berthold, Timo
Bettinelli, Andrea
Bonfietti, Alessio

Borghesi, Andrea
Cacchiani, Valentina
De Givry, Simon
Diaz, Daniel
Fontaine, Daniel
Frank, Michael
Gabàs, Joel
Gatti, Nicola
Gavanelli, Marco
Griffin, Joshua
Han, Zhi
Hashemi Doulabi, Hossein Seyed
Hendel, Gregor
Hurley, Barry
Iacovella, Sandro
Itzhakov, Avi
Jermann, Christophe
Kazachkov, Aleksandr
Kell, Brian
Kinable, Joris
Kiziltan, Zeynep
Ku, Wen-Yang
Lhomme, Olivier
Malitsky, Yuri
McCreesh, Ciaran
Michel, Claude
Miltenberger, Matthias

Monaci, Michele
Monfroy, Eric
Munera, Danny
Müller, Benjamin
Neveu, Bertrand
Nonato, Maddalena
Ponsini, Olivier
Rajeev, A.C.
Refalo, Philippe
Roli, Andrea
Rossi, Roberto
Rudi, Andrea
Sais, Lakhdar
Salvagnin, Domenico
Schaus, Pierre's
Schiex, Thomas
Sebbah, Samir
Steffy, Dan
Tjandraatmadja, Christian
Tran, Tony T.
Truchet, Charlotte
Tubertini, Paolo
Villaret, Mateu
Witzig, Jakob
Yahia, Lebbah
Zytnicki, Matthias

Invited Talks

Symmetry in Integer Programming

Jeff Linderoth

Department of Industrial and Systems Engineering,
University of Wisconsin-Madison, Madison, WI 53706-1572, USA

We will discuss mechanisms for dealing with integer programs that contain a great deal of symmetry [1,2]. The methods use information encoded in the symmetry group of the integer program to guide the branching decision and prune nodes of the search tree. These methods are adaptations of similar methods used in the constraint programming community, and they have been adapted into commercial integer programming software. We will discuss some recent work on using symmetry to augment the cutting plane procedures of branch-and-cut based solvers, and we will conclude with a brief discussion of using large-scale distributed computing platforms to solve difficult symmetric integer programs [3].

Joint work with Jim Ostrowski, Francois Margot, Fabrizio Rossi, Stefano Smriglio, and Greg Thain.

References

1. J. Ostrowski, J. Linderoth, F. Rossi, and S. Smriglio. Orbital branching. *Mathematical Programming*, 126(1):147–178, 2011.
2. J. Ostrowski, J. Linderoth, F. Rossi, and S. Smriglio. Constraint orbital branching. In *IPCO 2008: The Thirteenth Conference on Integer Programming and Combinatorial Optimization*, volume 5035 of *Lecture Notes in Computer Science*, pages 225–239. Springer, 2008.
3. J. Linderoth, F. Margot, and G. Thain. Improving bounds on the football pool problem via symmetry reduction and high-throughput computing. *INFORMS Journal on Computing*, 21:445–457, 2009.

IntSat: From SAT to Integer Linear Programming

Robert Nieuwenhuis

Barcelogic.com and Technical University of Catalonia (UPC), Barcelona, Spain

One of the most remarkable successes in Artificial Intelligence (AI) and Constraint Programming (CP) is probably the combination of activity-based heuristics and learning that allows Conflict-Driven Clause Learning (CDCL) propositional SAT solvers to solve, fully automatically, large and hard real-world industrial and scientific problems.

On the other hand, extremely powerful tools for Integer Linear Programming (ILP) and Mixed Integer Programming (MIP) exist, based on sophisticated techniques from Operations Research (OR), combining LP relaxations, simplex, branch-and-cut, including specialized cuts, heuristics and presolving methods, an extremely mature technology, with a 475000 times speedup between 1991 and 2012, according to [1].

Since SAT is the particular case of ILP where all variables are binary (0-1) and constraints have the form $x_1 + \ldots + x_m - y_1 \ldots - y_n > -n$, (usually written as *clauses* $x_1 \vee \ldots \vee x_m \vee \overline{y_1} \vee \ldots \vee \overline{y_n}$, i.e., disjunctions of *literals*), some natural questions arise. Why do OR solvers perform so poorly on pure SAT problems? Can CDCL-based techniques also beat OR ones for richer ILP problems than just deciding SAT?

Indeed, an extensive amount of work exists on CDCL-like techniques for Pseudo-Boolean optimization (aka. 0-1 ILP), see [4] for all background and references. IntSat [3] goes another step beyond, introducing a CDCL technique for full ILP (arbitrary integer variables, linear constraints and objective). IntSat extends CDCL by taking decisions on *bounds* (instead of literals), exhaustive *bound propagation*, and at each conflict it attempts to obtain by cuts a new ILP constraint, use it to *backjump*, and to *learn* it. This learning precludes "similar" future conflicts, which are ubiquitous in structured real-world problems (but not in randomly generated ones). IntSat appears to be the first method that is competitive, at least on certain ILP problem classes, with commercial OR tools such as CPLEX and Gurobi.

In this talk we will give an overview of CDCL-based SAT, the difficulties that arise when extending it, the key ideas behind IntSat, and several options between SAT and ILP, including Cutsat [2]. Along the way we will address further questions: How to exploit (and control) learned ILP constraints? Are CDCL-based techniques superior on problems with a combinatorial flavor? Are OR techniques better for the rather numerical ones? We also contribute to the theoretical background: completeness of cut-related inference rules, and infeasibility (or refutation) proof complexity.

References

1. Bob Bixby. Presentation: 1000X MIP Tricks, 12 June 2012, Bill Cunninghams 65th, 2012.
2. Dejan Jovanovic and Leonardo Mendonça de Moura. Cutting to the chase - solving linear integer arithmetic. *J. Autom. Reasoning*, 51(1):79–108, 2013.
3. Robert Nieuwenhuis. The IntSat method for integer linear programming. In *20th Principles and Practice of Constraint Programming, CP, LNCS 8656*, pages 574–589, 2014.
4. Olivier Roussel and Vasco M. Manquinho. Pseudo-boolean and cardinality constraints. Chapter in *Handbook of Satisfiability*, 695–733. IOS Press, 2009.

Constraint Programming
for Infeasibility Diagnosis with BARON

Yash Puranik and Nikolaos V. Sahinidis

Carnegie Mellon University, Pittsburgh, PA 15213, USA

Since its inception in the early 1990s, the BARON global optimization solver [1,2] was designed as a system that combined constraint programming and mathematical programming techniques for the global solution of algebraic nonlinear and mixed-integer nonlinear optimization problems (NLPs and MINLPs).

In this paper, we discuss recent developments in BARON, aiming to diagnose infeasibilities in nonconvex optimization models. Early work in the optimization literature to address the cause of infeasibilities has shown that the identification of Irreducible Inconsistent Sets (IIS) in a model can help speed up the process of correcting infeasible models [3]. An Irreducible Inconsistent Set is defined as an infeasible set of constraints with every proper subset being feasible. Identifying an IIS provides the modeler with a set of mutual inconsistencies that need to be diagnosed. Currently, efficient implementations for IIS isolation are only available for linear programs (LPs).

We propose a novel approach for IIS identification that is applicable to NLPs and MINLPs. This approach makes use of constraint programming techniques in a computationally inexpensive preprocessing stage to test for infeasibility in subparts of the infeasible model. This stage allows for rapid elimination of a large number of constraints in the model. Further, this preprocessing step itself could be sufficient to eliminate all constraints not part of an IIS for a large number of problems. The reduced model obtained can be filtered with any standard IIS isolation algorithm to obtain an IIS. The benefits of the approach lie in the efficient reduction in the model obtained by the preprocessing stage which leads to speedups in IIS identification. Extensive computational results are presented with an implementation of the proposed preprocessing algorithm in BARON along with four different filtering algorithms: deletion filter, addition filter, addition-deletion filter and depth-first binary search filter [4].

References

1. Sahinidis, N.V.: BARON: A general purpose global optimization software package. Journal of Global Optimization. **8** (1996) 201–205
2. Sahinidis, N.V.: Global optimization and constraint satisfaction: The branch-and-reduce approach. Lecture Notes in Computer Science. **2861** (2003) 1–16
3. Greenberg, H.J.: An empirical analysis of infeasibility diagnosis for instances of linear programming blending models. IMA Journal of Mathematics in Business and Industry. **3** (1992) 163–210
4. Chinneck, J. W.: Feasibility and infeasibility in optimization. Springer (2008)

Abstracts
of Fast Tracked Journal Papers

Lagrangian Bounds from Decision Diagrams

David Bergman[1], Andre A. Cire[2], and Willem-Jan van Hoeve[3]

[1] School of Business, University of Connecticut, USA
[2] Department of Management, University of Toronto Scarborough, Canada
[3] Tepper School of Business, Carnegie Mellon University, USA
david.bergman@business.uconn.edu, acire@utsc.utoronto.ca,
vanhoeve@andrew.cmu.edu

Decision diagrams are compact graphical representations of Boolean functions widely applied in circuit design and verification. Recently, *relaxed decision diagrams* were introduced as a new type of relaxation for discrete optimization problems [2]. In the context of constraint programming, relaxed decision diagrams have been successfully applied to improve constraint propagation and optimization reasoning [1,3]. One typically associates a decision diagram with a specific global constraint that is defined on a subset of variables (its scope), and the diagram is subsequently filtered and refined according to the other constraints of the problem. Additionally, if the objective function is evaluated on the same set of variables, the decision diagram can be used to obtain optimization bounds and for applying cost-based filtering.

In this work we propose a technique to strengthen a relaxed decision diagram of a problem by incorporating inference from constraints via a *Lagrangian relaxation* method. Namely, we associate penalties with the constraints that may be potentially violated by the solutions encoded in a relaxed decision diagram. These penalties are incorporated directly into the diagram as arc costs, which are taken into account in the diagram's objective function evaluation. We show that, with this generic approach, the resulting diagram may potentially yield stronger optimization bounds than the one obtained from the original relaxation, while the associated cost-based filtering allows for further refining the diagram and ultimately reducing the search space. If the incorporated constraints are linear, we also demonstrate that the optimal set of penalties are the duals of a shortest-path linear program derived from the decision diagram.

To evaluate our approach, we perform computational experiments on the traveling salesman problem with time windows. Relaxed decision diagrams are used in a global constraint enforcing that city visits must not overlap in time, and tests are performed with a state-of-the-art constraint-based scheduler. Results show that the diagram with improved Lagrangian bounds can drastically reduce solution times in comparison to the original relaxation.

References

1. Bergman, D., Cire, A.A., van Hoeve, W.J.: MDD propagation for sequence constraints. J. Artif. Intell. Res. (JAIR) 50, 697–722 (2014)
2. Bergman, D., Cire, A.A., van Hoeve, W.J., Hooker, J.N.: Optimization bounds from binary decision diagrams. INFORMS Journal on Computing 26(2), 253–268 (2014)
3. Cire, A.A., van Hoeve, W.J.: Multivalued Decision Diagrams for Sequencing Problems. Operations Research 61(6), 1411–1428 (2013)

A Constraint-Based Local Search Backend for MiniZinc (Summary)

Gustav Björdal, Jean-Noël Monette, Pierre Flener, and Justin Pearson

Uppsala University, Department of Information Technology, 752 37 Uppsala, Sweden
gustav.bjordal@gmail.com,
{Jean-Noel.Monette,Pierre.Flener,Justin.Pearson}
@it.uu.se

Solving combinatorial problems is a difficult task and no single solver can be universally better than all other solvers. Hence, when facing a problem, it is useful to be able to model it once and run several solvers to find the best one. MiniZinc [3] is a technology-independent modelling language for combinatorial problems, which can then be solved by a solver provided in a backend. There are many backends, based on various solving technologies. However, to the best of our knowledge, there is currently no constraint-based local search (CBLS, see [4]) backend. While most MiniZinc backends are just a parsing interface in front of the underlying solver, things are not as straightforward in the case of CBLS. In [1], we discuss the challenges to develop such a CBLS backend and give an overview of the design of a backend based on the OscaR/CBLS solver [2]. Our backend is called `fzn-oscar-cbls` and is publicly available from https://bitbucket.org/oscarlib/oscar/src/?at=fzn-oscar. The main *contributions* of [1] are:

- a description of a CBLS backend for MiniZinc;
- a heuristic to discover the structure of a model that can be used by a black-box local search procedure;
- a black-box local search procedure using constraint-specific neighbourhoods;

Experimental results show that, for some MiniZinc models, `fzn-oscar-cbls` is able to give good-quality results in competitive time. In [1], we focus on presenting a backend that works with existing MiniZinc models without modification, but we also briefly discuss how one can modify MiniZinc models or add annotations that would help a CBLS backend.

References

1. Björdal, G., Monette, J.N., Flener, P., Pearson, J.: A constraint-based local search backend for MiniZinc. Constraints, to appear (2015), http://dx.doi.org/10.1007/s10601-015-9184-z
2. De Landtsheer, R.: Oscar.cbls: a constraint-based local search engine (2012), https://bitbucket.org/oscarlib/oscar/downloads/Oscar.cbls.pdf
3. Nethercote, N., Stuckey, P.J., Becket, R., Brand, S., Duck, G.J., Tack, G.: MiniZinc: Towards a standard CP modelling language. In: Bessière, C. (ed.) CP 2007, LNCS, vol. 4741, pp. 529–543. Springer (2007), http://www.minizinc.org/
4. Van Hentenryck, P., Michel, L.: Constraint-Based Local Search. MIT Press (2009)

Supported by grants 2011-6133 and 2012-4908 of VR, the Swedish Research Council.

New Filtering for ATMOSTNVALUE and Its Weighted Variant: A Lagrangian Approach

Hadrien Cambazard and Jean-Guillaume Fages

Univ. Grenoble Alpes, G-SCOP, 38000 Grenoble, France
CNRS, G-SCOP, 38000 Grenoble, France
COSLING S.A.S., CS 70772, 44307 Nantes Cedex 3, France
hadrien.cambazard@grenoble-inp.fr, jg.fages@cosling.com

The ATMOSTNVALUE global constraint, which restricts the maximum number of distinct values taken by a set of variables, is a well known NP-Hard global constraint. The weighted version of the constraint, ATMOSTWVALUE, where each value is associated with a weight or cost, is a useful and natural extension. Both constraints occur in many industrial applications where the number and the cost of some resources have to be minimized. They have been initially introduced in [1,2]. Filtering these constraints has been proved to be NP-hard [4] but has been widely investigated by the Constraint Programming (CP) community [4] and remains an active topic [5,9,12].

This paper introduces a new filtering algorithm based on a Lagrangian relaxation for both constraints. The role of Lagrangian relaxation in constraint programming is an active area of research [11]. Its use for propagating NP-Hard global constraints is not new but has mostly been performed in very specific and applicative contexts [7,8,14,15] although propators for the Multi-cost-regular and Weighted-circuit global constraints have been designed with Lagrangian relaxation [3,8,10,13]. We believe that many global constraints (and in particuler NP-Hard global constraints involving costs) could be propagated using Lagrangian relaxation in a relatively generic manner [6]. This paper is investigating this idea for ATMOSTNVALUE and ATMOSTWVALUE .

The main contribution of this paper is a new filtering algorithm, based on Lagrangian relaxation, for both ATMOSTNVALUE and ATMOSTWVALUE. The algorithm proposed in this paper has several advantages : First, from a software engineering point of view, it is simple to implement and it does not require any connection with a linear solver. Second, it can provide a significantly stronger level of filtering compared to the state of the art algorithm for these constraints. Third, instead of the graph-based algorithm, it can be used directly to propagate the ATMOSTWVALUE global constraint for which no simple and efficient filtering algorithm exists. Thus, it is relevant to include it in a CP solver.

Several design options are discussed and empirically evaluated. The contribution is illustrated on problems related to facility location, which is a fundamental class of problems in operations research and management science. Results show that the Lagrangian propagator for both ATMOSTNVALUE and ATMOSTWVALUE provides significant improvement over a CP approach, up to being competitive with an Integer Linear Programming (ILP) approach. We believe it can help to bridge the gap between CP and ILP for a large class of problems related to facility location.

References

1. Nicolas Beldiceanu and Mats Carlsson. Pruning for the minimum constraint family and for the number of distinct values constraint family. In Toby Walsh, editor, *Principles and Practice of Constraint Programming – CP 2001*, volume 2239 of *LNCS*, pages 211–224. Springer Berlin Heidelberg, 2001.
2. Nicolas Beldiceanu, Mats Carlsson, and Sven Thiel. Cost-filtering algorithms for the two sides of the sum of weights of distinct values constraint. In *Technical report – T2002-14*. Swedish Institute of Computer Science, 2002.
3. Pascal Benchimol, Willem Jan van Hoeve, Jean-Charles Régin, Louis-Martin Rousseau, and Michel Rueher. Improved filtering for weighted circuit constraints. *Constraints*, 17(3):205–233, 2012.
4. Christian Bessiere, Emmanuel Hebrard, Brahim Hnich, Zeynep Kiziltan, and Toby Walsh. Filtering algorithms for the nvalue constraint. *Constraints*, 11(4):271–293, December 2006.
5. Christian Bessiere, George Katsirelos, Nina Narodytska, Claude-Guy Quimper, and Toby Walsh. Decomposition of the nvalue constraint. In David Cohen, editor, *Principles and Practice of Constraint Programming – CP 2010*, volume 6308 of *LNCS*, pages 114–128. Springer Berlin Heidelberg, 2010.
6. Hadrien Cambazard. Np-hard constraints involving costs: examples of applications and filtering. In *Dixièmes Journées Francophones de Programmation par Contraintes – JFPC*. 2014. Exposé invité.
7. Hadrien Cambazard, Eoin O'Mahony, and Barry O'Sullivan. A shortest path-based approach to the multileaf collimator sequencing problem. *Discrete Applied Mathematics*, 160(1-2):81–99, 2012.
8. Hadrien Cambazard and Bernard Penz. A constraint programming approach for the traveling purchaser problem. In Michela Milano, editor, *Principles and Practice of Constraint Programming - 18th International Conference, CP 2012, Québec City, QC, Canada, October 8-12, 2012. Proceedings*, volume 7514 of *LNCS*, pages 735–749. Springer, 2012.
9. Jean-Guillaume Fages and Tanguy Lapègue. Filtering atmostnvalue with difference constraints: Application to the shift minimisation personnel task scheduling problem. *Artificial Intelligence*, 212(0):116 – 133, 2014.
10. Jean-Guillaume Fages, Xavier Lorca, and Louis-Martin Rousseau. The salesman and the tree: the importance of search in cp. *Constraints*, pages 1–18, 2014.
11. Daniel Fontaine, Laurent D. Michel, and Pascal Van Hentenryck. Constraint-based lagrangian relaxation. In Barry O'Sullivan, editor, *Principles and Practice of Constraint Programming - 20th International Conference, CP 2014, Lyon, France, September 8-12, 2014. Proceedings*, volume 8656 of *LNCS*, pages 324–339. Springer, 2014.
12. Serge Gaspers and Stefan Szeider. Kernels for global constraints. *CoRR*, abs/1104.2541, 2011.
13. Julien Menana and Sophie Demassey. Sequencing and counting with the multicost-regular constraint. In Willem Jan van Hoeve and John N. Hooker, editors, *Integration of AI and OR Techniques in Constraint Programming for Combinatorial Optimization Problems, 6th International Conference, CPAIOR 2009, Pittsburgh, PA, USA, May 27-31, 2009, Proceedings*, volume 5547 of *LNCS*, pages 178–192. Springer, 2009.
14. Meinolf Sellmann and Torsten Fahle. Constraint programming based lagrangian relaxation for the automatic recording problem. *Annals OR*, 118(1-4):17–33, 2003.
15. Marla R. Slusky and Willem Jan van Hoeve. A lagrangian relaxation for golomb rulers. In Carla P. Gomes and Meinolf Sellmann, editors, *Integration of AI and OR Techniques in Constraint Programming for Combinatorial Optimization Problems, 10th International Conference, CPAIOR 2013, Yorktown Heights, NY, USA, May 18-22, 2013. Proceedings*, volume 7874 of *LNCS*, pages 251–267. Springer, 2013.

A Hybrid Exact Method for a Scheduling Problem with a Continuous Resource and Energy Constraints

Margaux Nattaf[1,2], Christian Artigues[1,2], and Pierre Lopez[1,2]

[1]CNRS, LAAS, 7 Avenue du colonel Roche, 31400 Toulouse, France
[2] Univ de Toulouse, LAAS, 31400 Toulouse, France
{nattaf,artigues,lopez}@laas.fr

Keywords: Continuous scheduling · Energy constraints · Energetic reasoning · Branching scheme · Mixed integer programming

We are studying a scheduling problem with a continuous resource and energy constraints, the Continuous Energy-Constrained Scheduling Problem (CECSP). In this problem, given a cumulative continuous resource of capacity B and a set of tasks, the goal is to find a schedule such that each task uses a variable resource quantity lying between a minimum and a maximum value, b_i^{min} and b_i^{max} respectively. Furthermore, each task needs to be executed during its time window $[r_i, d_i]$. Finally, each task has an energy requirement W_i and the energy used by a task is obtained by the integration of a function of the resource allocated to it, i.e. $\int f_i(b_i(t))dt \geq W_i$ (where $b_i(t) \leq B$, $\forall t$ is the resource quantity consumed by task i at time t). In this study, we focus on the case where function f_i is non-decreasing, continuous and linear.

For this NP-complete problem, we exhibit structural properties of the feasible solutions and we present a Mixed Integer Linear Program (MILP) based on an event-based formulation.

We also adapt the famous "left-shift/right-shift" satisfiability test (keystone of the so-called energetic reasoning) and the associated time-window adjustments to our specific problem. To achieve this test, we present three different ways for computing the relevant intervals.

Finally, we present a hybrid branch-and-bound method to solve the CECSP, which performs, at each node, the satisfiability test and time-window adjustments and, when the domains of all start and end times are small enough i.e. below a given parameter, the remaining solution space is searched via the event-based MILP.

Computational experiments on randomly generated instances are reported showing the interest of the hybrid method compared to pure MILP.

A Column-Generation Approach
for Joint Mobilization and Evacuation Planning

Victor Pillac[1], Manuel Cebrian[1], and Pascal Van Hentenryck[1,2]

[1] NICTA Optimisation Research Group, Australia
{Victor.Pillac,Manuel.Cebrian,Pascal.VanHentenryck}@nicta.com.au
http://org.nicta.com.au
[2] Australian National University, Australia

Large-scale evacuation planning is a critical part of the preparation and response to natural and man-made disasters. Surprisingly, local authorities still primarily rely on expert knowledge and simple heuristics to design and execute evacuation plans, and rarely integrate human behavioral models in the process. This is partly explained by the limited availability of algorithms and decision support systems able to produce evacuation plans that are compatible with operational constraints. Apart from a few exceptions, existing evacuation approaches rely on free-flow models which assume that evacuees can be dynamically routed in the transportation network. However, free-flow models violate a key operational constraint in actual evacuation plans, i.e., the fact that all evacuees in a given residential zone are instructed to follow the same evacuation route. In addition, only few studies have considered behavioral models and the mobilization process, and mainly from a simulation perspective.

This work addresses these issues and introduce the Joint Mobilization and Evacuation Planning Problem (JMEPP) which integrates evacuation planning and mobilization by incorporating the behavioral response of the evacuees. This methodological contribution is implemented through the integration of response curves into a column-generation algorithm that jointly decides the evacuation route, evacuation time, and the resource allocation for each evacuated area in order to maximize the number of evacuees reaching safety and minimize the total duration of the evacuation.

The column-generation algorithm for the JMEPP decomposes the problem as follows. The master problem selects *time-response evacuation plans*, which consists of an evacuation path, a response curve, and an evacuation time. The pricing subproblem generates columns of negative reduced costs, each representing a time-response evacuation plan associated with a single evacuated area. The approach leverages the response curve to solve the pricing subproblem efficiently by finding a shortest path for each evacuated node in a time-expanded graph.

Experimental results based on real instances demonstrate the practicability and benefits of the approach. Indeed, the case study shows that the quality of the resulting evacuation plans remains reasonably close to earlier approaches which assume full control of evacuation timing. In addition, the proposed approach produce evacuation schedules that ensure a continuous evacuation process, while previous approaches can lead to a schedule with numerous interruptions of the flow of vehicles — which would be difficult to enforce in a practical setting.

NICTA is funded by the Australian Government through the Department of Communications and the Australian Research Council through the ICT Centre of Excellence Program.

Contents

A Time-Dependent No-Overlap Constraint: Application to Urban Delivery Problems

Penélope Aguiar Melgarejo[1,2,3](✉), Philippe Laborie[3],
and Christine Solnon[1,2]

[1] Université de Lyon, CNRS, Lyon, France
[2] INSA-Lyon, LIRIS, UMR5205, 69621 Lyon, France
christine.solnon@insa-lyon.fr
[3] France Lab, IBM, Gentilly, France
{penelopeam,laborie}@fr.ibm.com

Abstract. The Time-Dependent Traveling Salesman Problem (TDTSP) is the extended version of the TSP where arc costs depend on the time when the arc is traveled. When we consider urban deliveries, travel times vary considerably during the day and optimizing a delivery tour comes down to solving an instance of the TDTSP. In this paper we propose a set of benchmarks for the TDTSP based on real traffic data and show the interest of handling time dependency in the problem. We then present a new global constraint (an extension of no-overlap) that integrates time-dependent transition times and show that this new constraint outperforms the classical CP approach.

1 Introduction

When we consider real world optimization problems, time is usually an important dimension to take into account. This is particularly the case for Delivery Problems for which time is typically present in different forms: travel times between consecutive deliveries; time windows where deliveries are allowed; precedence constraints. We are interested in the Time-Dependent Traveling Salesman Problem (TDTSP), an extended version of the Traveling Salesman Problem (TSP) where the travel time between deliveries (visits) depends on the date of the travel. The TDTSP is at the core of many real-world scheduling problems such as urban delivery problems, for example, since traffic conditions in urban areas usually vary a lot during the day. In these real-world problems, there are frequently additional constraints such as time-windows or precedence constraints.

Since the availability of extensive real-world data in this area is quite recent, the TDTSP has not been much studied in the literature and Constraint Programming (CP) approaches are even rarer. One reason for this is that CP is usually less efficient than Integer Linear Programming or Meta-heuristics (Local Search, Evolutionary Algorithms, Ant Colonies) for pure (non time-dependent) vehicle routing problems. On the other hand, Constraint-Based Scheduling [4], that is the application of CP to scheduling problems, is one of the biggest industrial success of CP and has shown that CP technologies can be very efficient for

© Springer International Publishing Switzerland 2015
L. Michel (Ed.): CPAIOR 2015, LNCS 9075, pp. 1–17, 2015.
DOI: 10.1007/978-3-319-18008-3_1

solving temporal problems. A variety of specialized variable types (interval variables, sequence variables) and related global constraints and search algorithms have been developed until recently [21–23] to improve the expressiveness and efficiency of CP-based models involving temporal domains.

In this paper, we start by giving a general definition of the TDTSP in Section 2. Then, we introduce a new benchmark for this problem in Section 3. This benchmark has been generated from real-world traffic data, coming from the city of Lyon, and we study the interest of handling time-dependent data on this benchmark. Section 4, describes related work. In Section 5, we introduce a basic CP model for the TDTSP and show some of its limitations. In Section 6, we describe a new global constraint (called TDnoOverlap) for efficiently tackling time-dependent data while ensuring that visits are not overlapping. This global constraint has been implemented on top of the CP Optimizer engine of IBM ILOG CPLEX Optimization Studio, and we experimentally evaluate it on our new benchmark in Section 7.

2   Definition of the TDTSP

The real world problem we address is the problem of scheduling a sequence of deliveries in a urban zone. The city of Lyon and other partners started a project called Optimod [1] with the goal to leverage city data to improve urban mobility. Traffic predictions are made from historic data and these predictions can be used to optimize moves in the city. In our case, we optimize delivery tours.

The theoretical problem involved is the TDTSP, an extension of the TSP where arc costs depend on the time when the arc is traveled.

In the TSP we are given a list of locations and the distances between every two of them. We are asked to find the tour minimizing the total traveled distance while visiting every location exactly once and coming back to the point of departure (depot).

In some cases though we are interested in minimizing the total travel time instead of distance or we simply need to schedule interventions or deliveries in certain time-windows predefined by the client. To do so we need to know the travel times between consecutive deliveries. To ensure a certain precision, since in urban zones travel times usually vary a lot during the day, we need time-dependent travel times. In order to take this variation into account we must know at which time the travel between two addresses starts so that we can take the time-dependent travel times into account.

We formally define the concepts of path, travel time function and timed-path in a graph and give a general formal definition of the TDTSP.

Definition 1 (Path). *A path $P = (v_1, ..., v_k)$ in a graph $G = (V, A)$ is a sequence of vertices such that $(v_i, v_{i+1}) \in A, \forall i \in \{1, ..., k-1\}, k \geq 2$.*

Definition 2 (Travel Time Function). *A travel time function $f : A \times \mathbb{R}^+ \to \mathbb{R}^+$ is a function such that for a given arc $(v_i, v_j) \in A$, $f(v_i, v_j, t)$ is the travel time from v_i to v_j when leaving v_i at time t.*

In the case of deliveries and specially of interventions, the duration of visits is generally not the same so we consider that each visit v_i is associated with a given duration $d(v_i)$. The notion of Timed-Path extends the notion of path in the context of TDTSP.

Definition 3 (Timed-Path). *Given a graph $G = (V, A)$, a starting time $\tau \in \mathbb{R}^+$, a travel time function $f : A \times \mathbb{R}^+ \to \mathbb{R}^+$ and a duration function $d : V \to \mathbb{R}^+$, a timed-path $P_{\tau,f}$ in G is a path $(v_1, \ldots, v_k)$ such that each vertex v_i has an associated start time $t(v_i, P_{\tau,f})$, corresponding to the time of arrival at v_i. Furthermore start times must respect:*

$$t(v_1, P_{\tau,f}) \geq \tau$$

$$t(v_{i+1}, P_{\tau,f}) \geq t(v_i, P_{\tau,f}) + d(v_i) + f(v_i, v_{i+1}, t(v_i, P_{\tau,f}) + d(v_i)), \forall i \in \{1, ..., k\}$$

Definition 4 (TDTSP). *Given a graph $G = (V, A)$, a depot $s \in V$, a starting time $\tau \in \mathbb{R}^+$ and a travel time function $f : A \times \mathbb{R}^+ \to \mathbb{R}^+$, the Time-Dependent Traveling Salesman Problem is the problem of finding the timed-path $P_{\tau,f} = (v_1, \ldots, v_k)$ which starts from the depot ($v_1 = s$) and visits each vertex exactly once ($\{v_1, \ldots, v_k\} = V$), and such that the returning time to the depot, $t(v_k, P_{\tau,f}) + d(v_k) + f(v_k, s, t(v_k, P_{\tau,f}) + d(v_k))$, is minimal.*

We could consider minimizing different objective functions but for our application's purpose we consider that minimizing the end time is a good objective.

3   A New Benchmark for the TDTSP

In the context of the Optimod project we had access to real traffic data measured from 630 sensors installed in the main axes of Lyon for 6 years. Those sensors measure vehicle's speed and estimations are made for the neighboring streets where there are no sensors. Given this data, a predictive model has been built, which gives predicted speeds on every street section, by 6-minute time steps.

We propose a benchmark generated using this model. The benchmark is built from 255 addresses randomly chosen from a list of delivery tours of transporters from Lyon, we can see their distribution in the city on Fig. 1.

Since the predictive model considers 6-minute time steps we produced a (stepwise) travel time function giving predicted travel times between every two addresses in the list for every 6-minute time step from 6h00 to 12h30 so that we have $m = 65$ time steps.

Travel times are computed using a time-dependent version of Dijkstra's point-to-point shortest path algorithm for every time step [27]. One limitation of our approach though is that we do not take into consideration the time spent in vertices in the path, where vertices are junctions between two or more street sections. We know from experience that the time it takes to traverse a crossroad or to turn left, for example, is an important factor in the augmentation of travel times during rush hours.

Fig. 1. 255 delivery addresses in Lyon

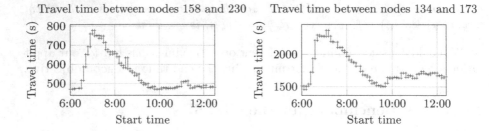

Fig. 2. Example of time-dependent travel time functions

The travel time functions given in the benchmark are represented as stepwise functions[1]. Examples of time-dependent travel time functions generated by this approach are depicted on Fig. 2.

We randomly generated 500 instances for each problem size n (10, 20 and 30 visits) by randomly selecting n locations among the 255 positions referenced in the travel time function. The duration of each visit is randomly selected in the interval [60s,300s]. Because the transition function tends to underestimate the travel time in congested areas, we generated two additional versions of the function with a dilatation of travel times of respectively 10 % and 20 % centered on the average travel time. So we end up with 3 functions: $T00$ (the original one) and $T10$, $T20$.

---

[1] This was a choice made to simplify the usage of the benchmark. We could generate piecewise linear functions from the same data by using the algorithm described in [19] in the travel time calculation section.

In a preliminary study, we want to evaluate the potential gain of using a *time-dependent* travel time model compared to a less precise TSP model that rules out time dependency. Given a stepwise travel time function T, we computed a TSP matrix $MedianTSP_T$ such that for every couple (i, j) of vertices, $MedianTSP_T(i, j)$ is the median value over the set of all time-dependent values associated with (i, j) in the stepwise travel time function T.

For each instance, we have solved the TSP defined by this median matrix $MedianTSP(T)$. Then, we have computed $cost_{MedianTSP}$, which is the cost of the TSP solution found previously when considering the time-dependent travel time function T as transition cost. We denote opt_{TDTSP} the cost of the optimal solution of the TDTSP. Obviously, $cost_{MedianTSP}$ gives an upper bound of opt_{TDTSP}. To evaluate the tightness of this bound, Fig. 3 gives the cactus plot of the relative gain defined as:

$$gain = \frac{cost_{MedianTSP} - opt_{TDTSP}}{opt_{TDTSP}}$$

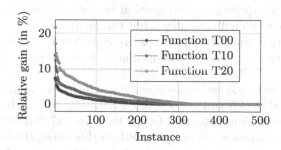

Fig. 3. Relative gain of TDTSP, instances ordered by decreasing gain

This figure shows that for more than 10 % of the instances, the gain is greater than 5 %, whereas for 40 % of the instances it is equal to 0 %. Note that a gain of 5 % is considered as very important in our context. Furthermore, real-world delivery problems usually have time-window constraints. In this case, it is mandatory to consider time-dependent data. As expected, the gain tends to increase (to more than 13 % and 21 %) when using functions $T10$ and $T20$ with larger amplitude. A similar behavior was observed for larger problems with 20 and 30 visits although the gain was slightly smaller, probably due to the fact that the peaks of traffic congestion occur between 06:00 and 09:00 which more or less corresponds to the time frame of a 10 visit problem so, for larger problems, part of the route is executed on less congested time windows.

For each problem size n (10, 20 and 30 visits) we selected a smaller set of 60 instances that is representative of the different types of gains between TSP and TDTSP (20 instances with the largest estimated gains, 20 instances with intermediate gains, 20 instances with negligible/zero gains). These instances are used in the experimental section to compare two models for solving the TDTSP problem. The benchmark is available on liris.cnrs.fr/christine.solnon/TDTSP.html.

4   Related Work

The name TSDTSP is used in the literature for two different problems. In 1978, [28] used the term TDTSP to describe the problem of scheduling jobs on a single machine with costs depending on the position of the job in the sequence. Since then, [2,8,16,17] also addressed this version of the problem.

We are interested in the actual time dependent problem introduced by [26] in 1992. They give several simple heuristics for the TDTSP and for the more general Time Dependent Vehicle Routing Problem (TDVRP) where a whole fleet must be routed instead of a single vehicle. Other heuristic approaches like ant colony systems [11], monte carlo [7], tabu search [19], simulated annealing [29] and others [13,18,24,25] are proposed for the TDTSP and TDVRP. Integer programming approaches are used in [8,9,30], some of these papers also take time-window constraints into account. To our knowledge, the only paper using a Constraint Programming approach for a time-dependent problem is [20], treating two scheduling problems with time-dependent task costs.

Aside from [14,25], the papers cited here consider instances which are randomly generated by applying some congestion rates during rush hours. The number of time steps considered is usually 3 but can go up to 16. In most cases, not more than 4 different congestion patterns are considered. The number of visits per tour varies between 10 and 65 but optimality is rarely proven for the largest instances.

Solutions used to tackle real time-dependent vehicle routing problems in Germany are presented in [14] and in the United Kingdom in [25]. In [25] they consider 96 time steps of 15 minutes and in [14] they use 217 time steps to model a whole day but test instances are not provided. In this sense, the benchmark provided by us offers a new testing parameter for the Time-Dependent VRP/TSP. In this new benchmark, we deal with a much larger number of time steps (i.e., 65), compared to what is usually considered in existing work.

5   Classical CP Model for the TDTSP and Its Limitations

5.1   CP Model

In this section we consider that f is a step function where each (time-)step has the same length l so that f is modeled with a cost matrix T. The input data is :

- A number $n > 0$ of visits, by convention the first (resp. last) visited vertex is 1 (resp. $n + 1$).
- A time horizon $H > 0$, a number of time steps $m > 0$ and a duration $l > 0$ of time steps so that $H = lm$.
- A cost matrix $T : [1, n + 1] \times [1, n + 1] \times [0, m] \to \mathbb{R}^+$ so that the travel time from vertex i to vertex j when leaving i at time t is given by $T[i][j][t/l]$.
- A visit duration vector $D : [1, n] \to \mathbb{R}^+$.

We present here a TDTSP model adapted from the classic CP model used to solve the TSP (see [6]). We added variables $time[i]$, which give the arrival time at each vertex i, and modified constraints to take into account the fact that a duration D_i is associated with every vertex i, and that travel durations are time-dependent.

intVar $position[1..n] \in 1..n$

          $next[1..n+1] \in 1..n+1, prev[1..n+1] \in 1..n+1$

          $time[1..n+1] \in 0..H$

minimize $time[n+1]$

subject to   alldifferent($position$), alldifferent($next$), alldifferent($prev$)

          inverse($prev, next$)

          $position[1] = 1, time[1] = 0, prev[1] = n+1, next[n+1] = 1$

          $\forall i \in 1..n+1 : next[i] \neq i, prev[i] \neq i$

          $\forall i \in 1..n : position[next[i]] = position[i] + 1$

          $\forall i \in 2..n+1 : position[prev[i]] + 1 = position[i]$

$$\forall i \in 1..n+1 : time[i] \geq time[prev[i]] + D[prev[i]] + T[prev[i]][i][time[prev[i]]/l] \quad (1)$$

$$\forall i \in 1..n+1 : time[next[i]] \geq time[i] + D[i] + T[i][next[i]][time[i]/l] \quad (2)$$

Note that the numbers of positions in a path is $n+1$ since we have to return to the depot. For each visit i: $next[i]$ and $prev[i]$ give the next and previous visits, $position[i]$ the position of the visit in the path and $time[i]$ the time of arrival at i. Beside constraints at the extremities of the tour to fix initial and end visits and the start time, an **alldifferent** constraint is posted on each group of variables whereas $prev/next$ variables are linked with **inverse** constraints. The relation between time and relative positions of visits is modeled with constraints (1) and (2). For a stronger propagation, the term T[...] in these constraints is modeled using a table constraint.

We also added the following redundant constraints to help improving the lower bound on the objective term $time[n+1]$. We noticed that these redundant constraints help reducing the number of branches by a factor close to 2 and the CPU time by a factor varying between 1 and 2.

$$time[n+1] \geq \sum_{i \in 1..n} D[i] + \sum_{i \in 1..n} T[i][next[i]][time[i]/l] \quad (3)$$

$$time[n+1] \geq \sum_{i \in 1..n} D[i] + \sum_{i \in 2..n+1} T[prev[i]][i][time[prev[i]]/l] \quad (4)$$

In the search branching scheme used to compare the performance of the propagation in Section 7, we use a search that builds the sequence of visits in a chronological order. For this reason, we added a new set of variables $atPosition[j]$ that represent the vertex at the jth position in the sequence. These variables are related with the rest of the model thanks to the following constraints:

intVar $atPosition[1..n] \in 1..n+1$

constraints   alldifferent($atPosition$)

          inverse($position, atPosition$)

          $atPosition[1] = 1, next[atPosition[n]] = n+1, atPosition[n] = prev[n+1]$

          $\forall j \in 1..n : next[atPosition[j]] = atPosition[j+1]$

          $\forall j \in 1..n : prev[atPosition[j+1]] = atPosition[j]$

5.2   Model Limitations

In the context of TDTSP, time variables $time[i]$ representing the dates of a visit are very important and their domain should be as tight as possible because the value of the travel time depends on the actual time value. An important limitation of the model presented above is the weakness of the propagation between temporal variables $time$ and sequencing variables (like $next$ and $prev$). For instance, it should be clear from their formulation that constraints like (1) and (2) would benefit from some more global reasoning over the travel time between i and $prev[i]$ (resp. between i and $next[i]$). Furthermore, reasoning only locally on direct successors of a visit ($next$, $prev$) may miss some important propagation as illustrated by the following example.

We call a visit a a *successor* of another visit b if a comes somewhere after b in the path, we call it $next$ of b if it is visited exactly after b. We can see from our variables that all the propagation is done reasoning with direct neighbors of a visit ($next$, $prev$).

We can show that reasoning with successors (besides prev/next variables) allows to obtain tighter bounds on the time of visits, as soon as the problem is asymmetric[2]. For simplification we will work here with a TSP example. Consider the following slightly asymmetric TSP problem where D is the depot, A, B, C are visits and the distance matrix T is as shown on Table 1. We suppose that the upper-bound of the objective is 100, therefore only two paths are feasible (D, B, C, A, D) and (D, C, B, A, D), each with a total length of 100. Given these two feasible solutions, the tightest possible domains of $prev$ and $next$ variables can be seen in Table 2.

In what follows we use $dom(a)$ to refer to the domain of a variable a, $\underline{a}$ for the smallest value in its domain and $\bar{a}$ for the biggest. If a is fixed then $\underline{a} = \bar{a}$ and $dom(a)$ is a singleton.

Table 1. Distance Matrix T

	D	A	B	C
D	0	9	46	8
A	8	0	46	8
B	46	46	0	38
C	8	8	38	0

Table 2. Next &Prev Domains

visit	dom(next)	dom(prev)
D	{B,C}	
A	{D}	{B,C}
B	{A,C}	{C,D}
C	{A,B}	{B,D}

If time bounds are computed using only $prev/next$ variables, the best we can do boils down to apply the following formulas to compute the minimum time bounds (smallest value in the domain) until a fix point is reached:

---

[2] Asymmetric in the sense that reversing a solution may change its total travel time or its feasibility. Some common causes of asymmetry are: asymmetric travel times (like time-dependent travel times), time windows constraints or precedences between visits.

$$time[A] = max(time[A], min(time[B] + T[B][A], time[C] + T[C][A]))$$
$$time[B] = max(time[B], min(time[C] + T[C][B], time[D] + T[D][B]))$$
$$time[C] = max(time[C], min(time[B] + T[B][C], time[D] + T[D][C]))$$

It gives $time[A] = 16$, $time[B] = 46$ and $time[C] = 8$. Let's now look at how we could propagate by also considering (indirect) successors. The distances in matrix T satisfy the triangle inequality[3] so we know that if A comes before B in the path the total length of any path starting from D, passing through A and B and returning back to D will be at least $T[D][A]+T[A][B]+T[B][D] = 101$, which is infeasible. Thus B must be visited before A and we can infer a precedence $B \rightarrow A$. This corresponds to the so-called disjunctive constraint in scheduling. With this, we know that A cannot start before $time[A] = T[D][B] + T[B][A] = 92$, which is a lot better than the value of $time[A] = 16$ found when reasoning only with $prev$ and $next$ variables.

If the TSP was purely symmetric we would not be able to deduce any successor links (indirect precedence) since any solution would be reversible and give the same cost. This type of reasoning is interesting as soon as solutions are asymmetric, which is usually the case for time-dependent travel times.

6 Time-Dependent No-Overlap Constraint

In order to integrate this kind of reasoning we use the concepts of *interval* and *sequence variables* in CP Optimizer [22,23]. Each visit i is modeled as an interval variable with start time denoted $time[i]$. The tour is modeled as a sequence variable over the set of visits. This variable maintains a precedence graph to propagate temporal relations between visits [15]. The vertices of this graph are the *time* variables associated to each visit. Two types of arcs are considered:

1. A **next arc** between two visits $i \Rightarrow j$ means that we pass through j directly after visiting i.
2. A **successor arc** between two visits $i \rightarrow j$ means that we visit j after i but we can go through other visits in between.

During the search, new **next** and **successor** arcs are added into the precedence graph because of search decisions (like when chronologically building a route) or as the result of constraint propagation (for instance by the extended disjunctive constraint sketched in subsection 6.3). The precedence graph incrementally maintains the transitive closure of the arcs.

In CP Optimizer the NoOverlap constraint allows to enforce a minimal transition time between vertices on the precedence graph. In this paper, we extended the NoOverlap into a TDNoOverlap constraint to take into account time-dependent

---

[3] If the triangle inequality is not satisfied, one can easily pre-compute a smaller transition time corresponding to the length of the shortest path (using Floyd-Warshall algorithm) to provide a lower bound on travel times. That is what the noOverlap constraint of CP Optimizer is doing internally.

transition times. The resulting model using the new `TDNoOverlap` constraint is sketched below.

$$
\begin{aligned}
&\text{intervalVar} \quad visit[i \in 1..n+1] \text{ size } D[i] \\
&\text{sequenceVar} \quad tour \text{ in all}(i \in 1..n) \ visit[i] \\
&\text{minimize} \quad startOf(visit[n+1]) \\
&\text{subject to} \quad first(tour, visit[1]) \\
&\qquad\qquad\quad\ last(tour, visit[n+1]) \\
&\qquad\qquad\quad\ tdnooverlap(tour, T)
\end{aligned}
$$

To propagate the bounds of time variable domains we need lower bound functions for the time-dependent transition time functions. Given two visits i and j, such that there exists a next arc or a successor arc going from i to j, we define two lower bound functions for each type of arc:

1. $f^{next}_{earliest}(i, j, t_d)$ and $f^{succ}_{earliest}(i, j, t_d)$, are the transition times giving the earliest arrival time at j if we leave i at time t_d or later.
2. $f^{next}_{latest}(i, j, t_a)$ and $f^{succ}_{latest}(i, j, t_a)$, are the transition times giving the latest departure time from i in order to arrive at j at time t_a or earlier.

With these functions, the `TDNoOverlap` constraint propagates the earliest time for j (5) and the latest time for i (6), by using the adequate lower bound function depending on whether we propagate a successor arc ($x = succ$) or a next arc ($x = next$):

$$
\underline{time[j]} \geq \underline{time[i]} + D[i] + f^{x}_{earliest}(i, j, \underline{time[i]} + D[i]) \tag{5}
$$

$$
\overline{time[i]} + D[i] \leq \overline{time[j]} - f^{x}_{latest}(i, j, \overline{time[j]}) \tag{6}
$$

Now we introduce the formal definitions of the bounding functions and explain how to calculate them.

6.1   Propagation of *Next* Arcs

Here we consider a next arc $i \Rightarrow j$ in the precedence graph. The earliest arrival time at j, if we leave i at time t_d, is propagated by formula (5), using the transition time function:

$$
f^{next}_{earliest}(i, j, t_d) = \min_{t \geq t_d}\{f(i, j, t) + t - t_d\} \tag{7}
$$

In $f^{next}_{earliest}$ we check if leaving from visit i later (waiting in place) allows to arrive at j sooner. In the case where waiting is never advantageous we say that the transition times satisfy the FIFO property.

Definition 5 (FIFO property). *A time-dependent transition time function f is said to satisfy the FIFO (First In First Out) property iff:*

$$
\forall i, j \in V, \forall t_1, t_2, (t_1 \leq t_2) \Rightarrow (t_1 + f(i, j, t_1) \leq t_2 + f(i, j, t_2))
$$

It follows from Def. 5 and from Eq. (7) that if f satisfies the FIFO property then $f^{next}_{earliest}$ and f are equal. Although the FIFO property generally holds in practice, our approach does not assume that f satisfies the FIFO property for two reasons: (1) stepwise functions do not satisfy it because of the discretization and (2) imprecision in data acquisition and time-dependent travel time calculations may introduce non-FIFO effects.

If f is a stepwise or a piecewise linear function, $f^{next}_{earliest}$ is a piecewise linear function. So, as we need in any case to handle piecewise linear functions in the propagation, in our implementation of the TDNoOverlap constraint we decided to treat the more general case where the input function f is a piecewise linear function.

In Algorithm 1 we describe the method used in pre-solve phase to calculate $f^{next}_{earliest}$ for each pair of vertices in the graph. We suppose a fixed arc (i, j) and f a piecewise linear function defined on the time domain $T = [t_{Min}, t_{Max}) \in \mathbb{R}$ and we simplify the notation to $f(t)$ instead of $f(i, j, t)$.

Each time interval $p_k = [t^k_{min}, t^k_{max})$ on which the function is linear is called a **piece**. Since p_k is open on t^k_{max}, by abuse of notation we write $f(t^k_{max})$ for $\lim_{x \to t^k_{max}} f(x)$. For instance, if f is not continuous on $t^k_{max} = t^{k+1}_{min}$ then $f(t^k_{max}) \neq f(t^{k+1}_{min})$. The notation $f \restriction_{p_k}$ means that f is restricted to interval p_k and therefore all operations are done only in this interval.

Finally, function $linear((t, v), (t', v'))$ denotes the linear function defined by the two points (t, v) and (t', v').

In the implementation, we used the class of piecewise linear function provided by CP Optimizer[4]. If ν is the number of pieces of the function, this class allows for a random access to a given piece with an average complexity of $O(log(\nu))$. Furthermore, when two consecutive pieces of the function are co-linear, these pieces are automatically merged so that the function is always represented with the minimal number of pieces.

The other type of propagation we do on next arcs in formula (6) depends on the estimation of the latest departure time from i in order to arrive at j at time t_a or before, given by:

$$f^{next}_{latest}(i, j, t_a) = \min_{t + f(i,j,t) \leq t_a} \{t_a - t\} \qquad (8)$$

Since $f^{next}_{earliest}$ already gives the minimum transition time from a given time it is clear that the minimum in Equation (8) is satisfied for the biggest t' such that $t' + f^{next}_{earliest}(i, j, t') \leq t_a$. Then, calculating f^{next}_{latest} comes down to finding this biggest t'.

Algorithm 2 describes the method used in a presolve phase to compute f^{next}_{latest} for each pair of vertices in the graph.

A similar procedure as the one described in Alg. 1 is used by [14], they also propose an algorithm like Alg. 2 but in their case, calculations are done for a single time t instead of calculating the whole function at once.

---

[4] Namely: IloNumToNumSegmentFunction.

Algorithm 1. Calculate $f_{earliest}^{next}$

Require: f, ν (the number of pieces of f)
1: $f_{earliest}^{next} \leftarrow f$
2: **for all** $k \in \{0, ..., \nu\}$ **do**
3:     $x_0 \leftarrow t_{min}^{k+1}$
4:     $v_0 \leftarrow f(x_0)$
5:     **if** $v_0 < f(t_{max}^k)$ **then**
6:         **for all** $j \in \{0, ..., \nu\} | p_j \subset [t_{Min}, x_0)$ **do**
7:             $f_{earliest}^{next} \lceil_{p_j} \leftarrow \min(f_{earliest}^{next}, linear((x_0, v_0), (v_0, x_0))) \lceil_{p_j}$
8:         **end for**
9:     **end if**
10: **end for**
11: **return** $f_{earliest}^{next}$

Algorithm 2. Calculate f_{latest}^{next}

Require: $f_{earliest}^{next}$, ν (the number of segments of $f_{earliest}^{next}$)
1: $arrivalTime \leftarrow linear((0,0), (1,1)) + f_{earliest}^{next}$
2: $f_{latest}^{next}(t) \leftarrow 0$
3: **for all** $k \in \{0, ..., \nu\}$ **do**
4:     $x_0 \leftarrow t_{min}^k$
5:     $x_1 \leftarrow t_{max}^k - 1$
6:     $v_0 \leftarrow arrivalTime(x_0)$
7:     $v_1 \leftarrow arrivalTime(x_1)$
8:     $slope_k(t) \leftarrow x_0 + \frac{x_1 - x_0}{v_1 - v_0} * (t - v_0)$
9:     **for all** $j \in \{0, ..., \nu\} | p_j \subset [v_0, v_1]$ **do**
10:         $f_{latest}^{next} \lceil_{p_j} \leftarrow \max(f_{latest}^{next}, linear((v_0, x_0), (v_1, x_1))) \lceil_{p_j}$
11:     **end for**
12:     **for all** $j \in \{0, ..., \nu\} | p_j \subset]v_1, arrivalTime(t_{Max})]$ **do**
13:         $f_{latest}^{next} \lceil_{p_j} \leftarrow \max(f_{latest}^{next}, x_1) \lceil_{p_j}$
14:     **end for**
15: **end for**
16: **return** f_{latest}^{next}

6.2 Propagation of *Successor* Arcs

Now we consider a successor arc $i \rightarrow j$ in the precedence graph. To estimate the earliest possible time of arrival at j if we leave i at time t_d or after we have to check if we can arrive faster at j by passing through other vertices. Let $\wp_{\tau, f}^{i,j}$ be the set of all timed-paths from i to j starting after time τ with travel times function f. We have:

$$f_{earliest}^{succ}(i, j, t_d) = \min_{p \in \wp_{t_d, f}^{i,j}} t(j, p) - t_d$$

where $t(j, p)$ is the start time of j in path p.

If there exists a shortest path from i to j, shorter than the direct arc, then the triangular inequality extended to the time-dependent case does not hold. It means that there is at least one vertex k such that passing through k allows to arrive faster at j.

Definition 6 (Time-dependent triangular inequality). *Function f is said to satisfy the triangular inequality property iff:*

$$\forall i, j, k \in V, \forall t \in \mathbb{R}^+, f(i, k, t) \leq f(i, j, t) + f(j, k, t + f(i, j, t))$$

To calculate $f^{succ}_{earliest}$ we use a time-dependent extension of the Floyd Warshall All Pairs Shortest Path algorithm [10]. We use $f^{next}_{earliest}$ as travel time function in the algorithm so that waiting at intermediate vertices to possibly go faster is already taken into account.

The second type of propagation on successor arcs is based on the estimation of the latest departure time from i in order to arrive at j at time t_a or before, given by:

$$f^{succ}_{latest}(i, j, t_a) = \min_{p \in \wp^{i,j}_{t,f}, t(j,p) \leq t_a} t_a - t$$

The reasoning for calculating f^{succ}_{latest} is exactly the same as the one we used for f^{next}_{latest} and the algorithm (2) is the same too, only this time we use $f^{succ}_{earliest}$ instead of $f^{next}_{earliest}$ as input.

6.3   Time-Dependent Disjunctive Propagation

Classical propagation algorithms used in constrained-based scheduling can be extended to time-dependent transition times. In our implementation of the TDNoOverlap constraint we extended the disjunctive reasoning [5]. As soon as two visits i and j are such that one of the conditions below is satisfied then it is clear that it is not possible to visit j before i and thus, we can add a successor arc $i \to j$ in the precedence graph:

$$\underline{time[j]} + D[j] + f^{succ}_{earliest}(j, i, \underline{time[j]} + D[j]) > \overline{time[i]}$$

$$\overline{time[i]} - f^{succ}_{latest}(j, i, \overline{time[i]}) - D[j] < \underline{time[j]}$$

This extended disjunctive reasoning helps discovering new arcs in the precedence graph that are themselves propagated as described in subsections 6.1 and 6.2.

6.4   Complexity

The complexity of the TDNoOverlap constraint is dominated by the complexity of maintaining the precedence graph and the disjunctive propagation. The worst-case complexity of the full-fledged propagation is quadratic with respect to the number of visits. We also implemented a slightly weaker but lighter propagation with linear complexity.

7 Experimental Evaluation

We compare the classical CP model presented in Section 5 with the model using the `TDNoOverlap` constraint[5] presented in Section 6.

In a first experiment we compare the filtering power of the two models. We use the same depth first search strategy for both models so that we can estimate the impact of constraint propagation on the number of branches of the complete search tree. We do a chronological scheduling of visits and choice of the nearest visit in terms of transition time first, given that the earliest date of the previous visit is known. For the classical CP model, this means that the search first fixes the variables $atPosition[i]$ for $i = 1, 2, ...n$. However, as the search strategy is not static, the search tree is different (one tree is not a sub-tree of the other). We could have tested on a static search strategy but this would have resulted in a more "artificial" type of search. We measured the number of branches and the CPU time of the two approaches on the 60 instances of size 10 on the 3 functions $T00, T10, T20$[6]. For those 180 tests (all solved to optimality with both models), the left side of Fig. 4 shows the comparison of the number of branches of the search tree explored by the two approaches while the right side compares the CPU times.

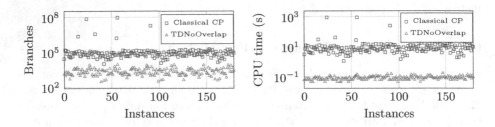

Fig. 4. Comparison of number of branches (left) and CPU time (right)

Not only the `TDNoOverlap` model propagates a lot more (about 50 times fewer branches) but it also finds better solutions faster than the classical CP model. For 10-sized instances the search is about 100 times faster on average.

On Fig. 5 we compare on the instances of size 20 and 30 the cost of the best solution found by the two approaches using the automatic search of CP Optimizer which is more sophisticated than depth first search. We used the same search heuristics as above and a time limit of 900s.

For instances of size 20, the `TDNoOverlap` model finds and proves the optimal solution for 165 instances out of 180 and, in average the solution found is more than 10 % better than the one of the classical CP model. For instances of size

---

[5] In these experiments we used the lighter version of the propagation but we noticed that there was not much difference with respect to the full-fledged version.

[6] Comparison was performed only on instances of size 10 as the classical CP model is not able to solve the larger problems to optimality.

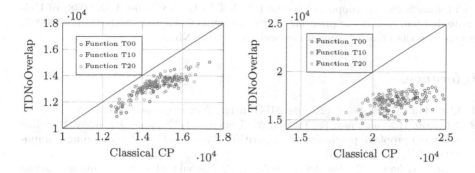

Fig. 5. Comparison of solution quality on problems of size 20 (left) and 30 (right)

30, both models are incapable of proving optimality in the time limit, but in average the solutions found by `TDNoOverlap` are more than 20 % better than the ones of the classical CP model.

We also compared the `TDNoOverlap` model with an intermediate model (described in [3]) using the `NoOverlap` constraint. In average on the 180 intances of size 10, the `TDNoOverlap` model proves optimality with 20 times less branches and is 40 times faster.

8 Discussion

In this paper we showed, in the context of scheduling a sequence of deliveries, the impact of reasoning with successors, other than just next/previous, and of taking time-dependent transition times into account directly into a global constraint. Reasoning on successors is crucial for problems involving time variables like the TDTSP. From an application perspective, the interest of the scheduling model we presented is that it is very easy to integrate additional constraints like precedence between visits or disjunctive time-windows. These constraints are in fact directly available in CP Optimizer and should work pretty well when added to the central TDTSP model presented in this paper.

Reasoning on successors is in fact complementary with reasoning on a prev/next graph. In future work we plan to improve our constraint propagation by calculating tighter bounds for the TDTSP by using Minimum Spanning Trees or Assignment Problem relaxations on the prev/next graph, extending the approaches described in [6,12,15]. We also want to see if the successor relations stored in the precedence graph can be exploited in this context. We plan to evaluate our new constraint on other benchmarks, and compare it with other approaches.

Acknowledgments. This work has been done in the context of the Optimod'Lyon project. We would like to give our special thanks to Thomas Baudel for his help in the obtention of traffic data.

Christine Solnon is supported by the LABEX IMU (ANR-10-LABX-0088) of Université de Lyon, within the program "Investissements d'Avenir" (ANR-11-IDEX-0007) operated by the French National Research Agency (ANR).

References

1. Optimod'Lyon website, (21 July 2014). http://www.optimodlyon.com/
2. Abeledo, H., Fukasawa, R., Pessoa, A., Uchoa, E.: The time dependent traveling salesman problem: polyhedra and algorithm. Mathematical Programming Computation **5**(1), 27–55 (2013)
3. Aguiar Melgarejo, P., Baudel, T., Solnon, C.: Global and reactive routing in urban context: first experiments and difficulty assessment. In: CP-2012 Workshop on Optimization and Smart Cities, October 2012
4. Baptiste, P., Laborie, P., Le Pape, C., Nuijten, W.: Constraint-based scheduling and planning. In: F. Rossi, P. van Beek, T.W. (eds.) Handbook of Constraint Programming, chap. 22, pp. 759–798. Elsevier (2006)
5. Baptiste, P., Le Pape, C.: Disjunctive constraints for manufacturing scheduling: principles and extensions. In: Proc. 3rd International Conference on Computer Integrated Manufacturing (1995)
6. Benchimol, P., van Hoeve, W.J., Regin, J.C., Rousseau, L.M., Rueher, M.: Improved filtering for weighted circuit constraints. Constraints, pp. 205–233 (2012)
7. Bentner, J., Bauer, G., Obermair, G.M., Morgenstern, I., Schneider, J.: Optimization of the time-dependent traveling salesman problem with Monte Carlo methods. Physical Review E **64**(3), August 2001
8. Bront, J.: Integer Programming approaches to the Time Dependent Travelling Salesman Problem. Ph.D. thesis, Universidad de Buenos Aires (2012)
9. Cordeau, J.F., Ghiani, G., Guerriero, E.: Properties and Branch-and-Cut Algorithm for the Time-Dependent Traveling Salesman Problem
10. Cormen, T.H., Leiserson, C.E., Rivest, R.L.: Introduction to Algorithms. MIT Press, McGraw-Hill edn. (1990)
11. Donati, A.V., Montemanni, R., Casagrande, N., Rizzoli, A.E., Gambardella, L.M.: Time dependent vehicle routing problem with a multi ant colony system. European Journal of Operational Research **185**(3), 1174–1191 (2008)
12. Fages, J.G., Lorca, X.: Improving the asymmetric tsp by considering graph structure. arxiv preprint arxiv:1206.3437 (2012)
13. Figliozzi, M.A.: The time dependent vehicle routing problem with time windows: Benchmark problems, an efficient solution algorithm, and solution characteristics. Transportation Research Part E Logistics and Transportation Review **48**, 616–636 (2012)
14. Fleischmann, B., Gietz, M., Gnutzmann, S.: Time-Varying Travel Times in Vehicle Routing. Transportation Science **38**(2), 160–173 (2004)
15. Focacci, F., Laborie, P., Nuijten, W.: Solving Scheduling Problems with Setup Times and Alternative Resources. AIPS Proceedings (2000)
16. Fox, K.R., Gavish, B., Graves, S.C.: An n-Constraint Formulation of the (Time-Dependent) Traveling Salesman Problem. Operations Research **28**(4), 1018–1021 (1980)
17. Gouveia, L., Voss, S.: A classification of formulations for the (time-dependent) traveling salesman problem. European Journal of Operational Research **83**(1), 69–82 (1995)

18. Hashimoto, H., Yagiura, M., Ibaraki, T.: An iterated local search algorithm for the time-dependent vehicle routing problem with time windows. Discrete Optimization **5**(2), 434–456 (2008)
19. Ichoua, S., Gendreau, M., Potvin, J.Y.: Vehicle Dispatching With Time-Dependent Travel Times. European Journal of Operations Research (2003)
20. Kelareva, E., Tierney, K., Kilby, P.: CP Methods for Scheduling and Routing with Time-Dependent Task Costs. EURO Journal on Computational Optimization **2**, 147–194 (2014)
21. Laborie, P., Godard, D.: Self-adapting large neighborhood search: application to single-mode scheduling problems. In: Proceedings of the 3rd Multidisciplinary International Conference on Scheduling: Theory and Applications (MISTA), pp. 276–284 (2007)
22. Laborie, P., Rogerie, J.: Reasoning with conditional time-intervals. In: FLAIRS Conference, pp. 555–560 (2008)
23. Laborie, P., Rogerie, J., Shaw, P., Vilím, P.: Reasoning with conditional time-intervals. Part II: an algebraical model for resources. In: FLAIRS Conference (2009)
24. Li, F., Golden, B., Wasil, E.: Solving the time dependent traveling salesman problem. In: Golden, B., Raghavan, S., Wasil, E. (eds.) The Next Wave in Computing, Optimization, and Decision Technologies, Operations Research/Computer Science Interfaces Series, vol. 29, pp. 163–182. Springer, US (2005)
25. Maden, W., Eglese, R., Black, D.: Vehicle Routing and Scheduling with Time Varying Data: A Case Study. Journal of the Operational Research Society **61**, 515–522 (2009)
26. Malandraki, C., Daskin, M.S.: Time Dependent Vehicle Routing Problems: Formulations. Properties and Heuristic Algorithms. Transportation Science **26**, 185–200 (1992)
27. Nannicini, G.: Point-to-Point Shortest Paths on Dynamic Time-Dependent Road Networks. Ph.D. thesis, Ecole Polytechnique, Palaiseau (2009)
28. Picard, J.C., Queyranne, M.: The Time-Dependent Traveling Salesman Problem and Its Application to the Tardiness Problem in One-Machine Scheduling (1978)
29. Schneider, J.: The time-dependent traveling salesman problem. Physica A: Statistical Mechanics and its Applications **314**(1–4), 151–155 (2002)
30. Stecco, G., Cordeau, J.F., Moretti, E.: A branch-and-cut algorithm for a production scheduling problem with sequence-dependent and time-dependent setup times. Computers & Operations Research **35**(8), 2635–2655 (2007)

Rectangle Placement for VLSI Testing

Merav Aharoni[1], Odellia Boni[1]([✉]), Ari Freund[2], Lidor Goren[3],
Wesam Ibraheem[1], and Tamir Segev[1]

[1] IBM Research, Haifa, Israel
odelliab@il.ibm.com
[2] work done while this author was at IBM Research, Haifa, Israel
[3] IBM STG - Fishkill, New York, USA

Abstract. We report our solution to the problem of designing test-
site chips. This is a specific variation of the VLSI floorplanning problem
where rectangular macros must be placed without overlap in a given area,
but no wiring between the macros exists. Typically, industrial problems
of this type require placing hundreds of macros of different sizes and
shapes and include additional constraints such as fixing or grouping some
of the macros. Many tools and techniques developed to solve similar
problems proved unsuitable for this specific variation. We used constraint
programming (CP) with additional heuristics, including sophisticated
variable and value orderings, to produce floorplans for real test-sites.
Our CP solution is successfully used in production by test-site designers.

Keywords: Floorplanning · Electronic design automation · Constraint
programming · Non overlapping rectangles placement

1 Introduction and Background

Floorplanning of a VLSI chip is the design phase in which placement of main
functional units is determined. General chip floorplanning is a highly complex
task, since the placement of the units must take into account not only spatial
considerations, but also wiring, pin placement, power planning and power grid
design.

Test-site chips are designed specifically for testing various aspects of the chip
manufacturing processes. These chips house design units called *macros* which
may be integrated circuits (IC) themselves. Floorplanning of test-site chips is
subject to considerations such as spatial arrangement, density, and precision
of manufacturing tools. However, this problem is significantly simpler than the
general floorplanning problem because the macros are not interconnected.

The actual placement of macros on a chip is limited by the precision of the
chip manufacturing tools. In order to account for that, a grid is defined on the
chip and macro placement is limited to the grid points. The distance between
any two adjacent points of the grid is the precision of the manufacturing tools.

Test-site chips are usually divided into several *chiplets* that are later diced
and can be tested as individual chips. This division into chiplets enables design

© Springer International Publishing Switzerland 2015
L. Michel (Ed.): CPAIOR 2015, LNCS 9075, pp. 18–30, 2015.
DOI: 10.1007/978-3-319-18008-3_2

and testing of different chiplets by separate design teams or even different companies.

When placing macros on a test-site, there are two fundamental constraints that must be satisfied. The first is that every macro must reside completely inside one of the chiplets. The second is that no two macros may overlap. Satisfying even just these two constraints may already prove challenging for real instances with high area utilization and a mix of macro sizes and shapes. Yet in many cases there are further requirements pertaining to the placement of individual macros or sets of macros.

Examples of requirements affecting a single macro:

Chiplet - Certain macros are confined to specific chiplets because these macros are to be designed, integrated, diced-out and tested separately.

Fixed-point - Some macros are pre-placed in specific locations.

Block - Some areas on the chip are reserved for later placement of special units and therefore macros cannot be placed there.

Perimeter - Certain macros may be confined to particular areas to accommodate various spatial considerations e.g., to minimize the efect of these macros on other macros.

Proximity-point - Some macros should reside as close as possible to a given point to enable chip-wide spatial analysis.

Examples of requirements affecting a set of macros:

Offset - A specific offset or distance may be defined between a given pair of macros. This enables local spatial measurements and analysis. A certain tolerance or deviation may be allowed in the specified offset.

Cluster - Manufacturing process-related spatial considerations may dictate that a certain group of macros form a single contiguous region. This requirement may be somewhat relaxed to require the macros to form as few contiguous regions as possible. In this case, we call this constraint *soft-cluster*.

In typical test-site placement problems, the width and height of the chip, when aligned to the precision grid, can be several tens of thousands of units. A single test-site is divided into around 10 chiplets and contains several hundred, up to 1500, macros. These macros are of 15-20 different sizes. Typical density, of area covered by macros, is between 85% to 95%. Figure 1 depicts a floorplan of a typical test-site. The black outlines indicate a perimeter constraint, and the diagonal lines over rectangles indicate that these rectangles participate in a cluster constraint. As can be seen in this figure, the high density dictates highly structured floorplans.

1.1 Placement Algorithms and Tools

Macro placement for test-site is currently mostly a manual process performed by experienced engineers. However, the increasing number of macros and shorter time to market requirement drive the need to automate this process.

Fig. 1. A typical test-site design containing 9 chiplets. Black outlines indicate a perimeter constraint. Diagonal lines over rectangles indicate cluster constraints.

In this section, we review existing solutions that aim to provide a convenient, fast and scalable tool for macro placement. We focus on tools that target the mixed size placement problem, as opposed to standard cell or block placement problems in VLSI [1]. The mixed-size placement problem requires placement of macro blocks of intermediate size intermixed with a large number of standard cells. These are more challenging problems, since moving a macro block involves the relocation of large numbers of standard cells. It should be noted that none of the reviewed tools and techniques is specifically designed to deal with the test-site placement problem.

To deal with large instances, most of the tools for mixed block placement use *hierarchical floorplanning* [4]. They divide the problem into multi-level placement problems by top-down divisioning or bottom-up clustering. For the placement problems at each level, most tools use a hybrid algorithm approach: one algorithm is used to reach an initial solution, and then another algorithm is used to improve or legalize the solution.

The algorithms used for placement fall into 4 categories:

- *recursive partitioning* - Min-cut [2] or bi-section [3] algorithms are used to recursively partition the area, until small enough regions are formed.
- *local search* - These algorithms exploit the fact that floorplans can be represented as graphs. For example, compact floorplans can be represented as B*-trees [5]. This representation allows easy introduction of perturbations in the floorplan, so that *simulated annealing* or *genetic algorithms* can be used to find a good placement. An academic tool that uses simulated annealing is Capo [6]. Local search algorithms usually converge slowly, require tuning of various input parameters and are not suitable when fixed macros exist.
- *analytical methods* - These methods are used for example in mPL6 [7], which models the macros as a set of masses and springs and uses force-directed

algorithms to find a placement. Another example is NTUplace [8] that formulates placement as density function control. Analytical methods usally require solving a quadratic program, or a linear relaxation.

- *greedy algorithms* - The floorplan is constructed by iteratively adding rectangles until all rectangles are placed. The location of the next rectangle depends on the current shape formed by the already placed rectangles. Hence rectangles order is important. With such algorithms, it is hard to accommodate requirements such as offsets between macros or proximity points. The *cluster growth* method [9], is an example of a greedy placement algorithm.

Existing commercial florplanning tools were found not suitable for test-site floorplanning by test-site designers. The main reason is that they try to account for power consumption and interconnectivity concerns which are not relevant for the test-site floorplanning case. In addition, they cannot accommodate some of the necessary constraints. Many of the algorithms mentioned above do not impose the no-overlap constraint while others treat this constraint inefficiently. Integer linear programming (ILP) based solvers, for example, model no-overlap constraint with two binary variables for every pair of macros, resulting in over a million binary variables for a typical test-site.

The remainder of the paper is organized as follows: in Section 2, we present a CP formulation of the test-site problem and in section 3 the heuristics used to solve it. In Section 4 we present our solution results for several industrial test-sites. Finally, in Section 5, we discuss the results.

2 Constraint Programming (CP) Formulation

We begin by introducing a few notations. We describe a *rectangular area* in the plane by the Cartesian product $[left, right] \times [bottom, top]$. For example $[-1, 3] \times [2, 4]$ is a rectangle with two diagonally opposite corners at positions $(-1, 2)$ and $(3, 4)$.

We denote a *placement* of a rectangle by a pair of coordinates (x, y) that indicate the position of its bottom-left corner. Note that when the rectangle with height H and width W is placed at $(\bar{x}, \bar{y})$, the rectangular area it occupies is $[\bar{x}, \bar{x} + W] \times [\bar{y}, \bar{y} + H]$.

Given a rectangular area $C = [L_C, R_C] \times [B_C, T_C]$ and a rectangle A with width W_A and height H_A placed at (x_A, y_A), we define the following relationships between A and C:

$$
\begin{aligned}
&A \ inside \ C && \Leftrightarrow \ L_C \leq x_A \leq R_C - W_A \ \wedge \ B_C \leq y_A \leq T_C - H_A \ (a) \\
&A \ overlaps \ C && \Leftrightarrow \ L_C - W_A < x_A < R_C \ \wedge \ B_C - H_A < y_A < T_C \ (b) \\
&A \ \neg overlap \ C && \Leftrightarrow \hspace{7.5cm} (c) \\
&\quad x_A \geq R_C \ \vee \ x_A + W_A \leq L_C \ \vee \ y_A \geq T_C \ \vee \ y_A + H_A \leq B_C \hspace{1cm} (1) \\
&A \ borders \ C && \Leftrightarrow \hspace{7.5cm} (d) \\
&\quad (L_C - W_A < x_A < R_C \ \wedge \ (B_C = y_A + H_A \ \vee \ y_A = T_C)) \\
&\quad \vee (B_C - H_A < y_A < T_C \ \wedge \ (x_A = R_C \ \vee \ x_A + W_A = L_C))
\end{aligned}
$$

In the test-site floorplanning problem, we are given a rectangular area of the entire chip (defined by $[L, R] \times [B, T]$), divided to K smaller, rectangular chiplets $C_k = [L_k, R_k] \times [B_k, T_k]$ $k = 1 \ldots K$.

Given N macros, represented as rectangles with respective heights H_i and widths W_i, where $i = 1, \ldots, N$, we have to find an on-grid placement (x_i, y_i) for each of these macros, such that no two macros overlap and each macro is inside a chiplet. If there are additional constraints on some of the macros, these must be satisfied as well.

2.1   Variables and Domains

We define for each macro two decision variables: x_i, y_i, $i = 1 \ldots N$, where (x_i, y_i) represent the placement coordinates (bottom-left corner) of macro i. The initial domain for variable x_i is $[L, R - Wi]$ and for variable y_i is $[B, T - H_i]$.

The actual placement of a macro on a chip is limited by the precision P of the chip manufacturing tools. Thus we define a grid on the chip such that macros can be placed only on grid points. Therefore, the coordinates are scaled by P and the variables domains are actually ranges of integers. For our convenience, we shall assume $P = 1$.

Note that this choice of decision variables may hinder prunning. Consider, for example, a macro that can be placed at only three locations: $(1, 1)$, $(2, 1)$ and $(1, 2)$. Using our choice of decision variables the initial domain of $x_i = [1, 2]$ and $y_i = [1, 2]$. If the location $(1, 1)$ proves to be infeasible for this macro, no prunning of variables x_i or y_i is possible. However, the alternative of having a single decision variable to represent the location of each macro entails handling very large domains, of magnitude $(R - L) * (T - B)$.

2.2   Constraints

Using the relationships given in Eq. 1(a),(b),(c) and (d), we can formulate the test-site floorplanning problem as the following set of constraints:

$$
\begin{array}{lll}
\exists k = 1, \ldots K : A_i \ inside \ C_k & \forall i = 1, \ldots, N & (a) \\
A_i \ \neg overlap \ A_j & \forall i, j = 1, \ldots, N, j \neq i & (b) \\
A_i \ inside \ C_k & i \in Chiplet(k) & (c) \\
A_i \ inside \ T & i \in Perimeter(T) & (d) \\
\forall i = 1, \ldots, N : A_i \ \neg overlap \ B & Blocked(B) & (e) \\
\exists j \in Cluster, j \neq i : A_i \ borders \ A_j & i \in Cluster & (f) \\
x_i = F_x \wedge y_i = F_y & i \in Fixed(F_x, F_y) & (g) \\
|x_i - x_j - X_o| \leq t_x \wedge |y_i - y_j - Y_o| \leq t_y & i, j \in Offset(X_o, Y_o, t_x, t_y) & (h) \\
A_i \ overlaps \ R(p_x, p_y, dist_x, dist_y) & i \in Proximity(p_x, p_y) & (i)
\end{array}
\qquad (2)
$$

where N is the number of rectangles to be placed, A_i, $i = 1, \ldots, N$ denotes the area occupied by rectangle i when placed at (x_i, y_i), i.e. $A_i = [x_i, x_i + W_i] \times [y_i, y_i + H_i]$, C_k $k = 1, \ldots, K$ denote the test-site chiplets and $R(p_x, p_y, dist_x, dist_y) = [p_x - dist_x, p_x + dist_x] \times [p_y - dist_x, p_y + dist_y]$ is a rectangular area around point (p_x, p_y).

Constraint (a) in Eq. 2 ensures that all placed rectangles fully reside inside the test-site chiplets, while constraint (b) ensures that no two placed rectangles overlap. These two constraints appear for every test-site design. The rest of the constraints listed in Eq. 2 describe types of possible additional constraints. These additional constraints are defined specifically per design. Thus, their number and parameters vary for different designs.

Chiplet constraints (type (c)) restrict macro i to reside within a specific chiplet (of given index k).

Perimeter constraints (type (d)) restrict certain macros to reside within a given rectangular area (denoted by T).

Block constraints (type (e)) define a rectangular area (denoted by B) which none of the placed rectangles can overlap.

Cluster constraints (type (f)) ensure that a given set of placed rectangles form a contiguous region. We use dedicated heuristics (described in Section 3.1) to deal with this type of constraints.

Fixed constraints (type(g)) place a certain rectangle i at a given location (F_x, F_y).

Offset constraints (type (h)) require that two particular macros i, j will be offset from each other by (Xo, Yo), perhaps with respective tolerance (t_x, t_y). Note that when no tolerance is allowed, the offset constraints become equalities.

Offset constraints can have 3 origins:

- Explicit - designer defines an offset constraint between two macros
- Transitive - If macros i, j have an offset constraint, and macros j, k also have an offset constraint, then an offset constraint is added between the macros i and k.
- Same cell - If a macro is an IC, it can contain several macros. In this case, the relative locations of the internal macros must be preserved in the solution. Therefore, the IC macro is translated into the set of its internal macros, with offset constraints between them, according to the internal placement in the IC macro.

Proximity constraints (type (i)) require that macro i should reside as close as possible to point $p = (p_x, p_y)$. Instead of resolving this exact constraint which involves an optimization problem, we define $(dist_x, dist_y)$ which determines the maximum allowed distance from p in each dimension. We set $dist$ to a predefined percentage of the total chip size in that dimension.

In our tool, we reformulated *Block* constraints as *Fixed* constraints: if area $T = [L_T, R_T] \times [B_T, T_T]$ is blocked, we introduce a new dummy macro d with $W_d = R_T - L_T$, $H_d = T_T - B_T$, fixed at location (L_T, B_T). Similarly, to account for chiplets, we added dummy fixed macros that cover all the chip area outside the chiplets. This reformulation eliminates constraints (a) and (d) appearing in Eq. 2, at the expense of adding some dummy fixed rectangles to the problem.

In addition, the *No-overlap* constraint (b) was modeled as the DIFF2 global constraint. The propagator we implemented is similar to that described in [10]. The main idea is that for every rectangle we maintain obligatory regions, that is regions which will be covered by any placement of this rectangle, and forbidden regions, which represent placements (lower-left corner locations) of this rectangle that lead to infeasibility. The obligatory and forbidden regions are updated for all rectangles during propagation by a single sweep through the xy plane. It is the most time-consuming constraint in the problem.

All unary constraints (*Fixed*, *Proximity*, *Chiplet*, *Perimeter*) can be handled by preprocessing the variable domains or by a single invocation of their propagators.

3 Heuristics for Test-site Placement

3.1 Cluster Handling

Cluster constraints (see Eq. 2 (f)) require that a group of macros (i.e., a subset of 1..N) form a single contiguous region after placement. We use two heuristics to ensure this type of constraints is satisfied. The first is named *bounding box* and the second *continuous placement*.

The continuous placement heuristic is similar to the cluster growth technique mentioned in Section 1.1. It is triggered when one of the cluster macros is placed during the solving process. Then, it selects the remaining macros that belong to the cluster one by one, and places them adjacent to any of the already placed cluster macros. While this heuristic ensures the creation of a contiguous region of the macros, it may be triggered too late in the solving process leaving insufficient space to place all cluster macros. In addition, this heuristic takes long, since after each placement, arc-consistency is performed. Therefore, the continuous placement heuristics is used only for soft clusters or for clusters that contain a fixed macro or a macro that has an offset from a macro outside the cluster.

The bounding box heuristic deals with the disadvantages presented by the continuous placement heuristics by trying to place all cluster macros at once. This is done by iterating two steps. First, we select a rectangular area on the chip whose area is large enough to accommodate the entire cluster. This area is called the bounding box. Then, a sub-problem of placing the cluster macros inside the bounding box is created. If the bounding box is small enough, the solution will form a contiguous region. In case the sub-problem created is not satisfiable, a new bounding box is selected, and so-on. When the sub-problem is satisfiable, we translate the macros in the sub-problem back to the original problem.

The bounding box heuristic is connected to the main problem by a mechanism that we name a *restrictor*. A *restrictor* can be viewed as a sophisticated type of constraint combined with variable and value ordering on a subset of the variables. Similarly to a propagator, a restrictor implements a constraint by removing all unsupported values, but unlike a propagator, it may leave only a subset of the supported values. At the extreme case, a restrictor may return a single tuple of valid values for all the variables participating in this constraint. Restrictors are integrated into the CP solving algorithm in place of an instantiation step (choice). Thus, if the instantiation by the restrictor fails in the main CSP, the algorithm will backtrack and the restrictor can make a new instantiation.

3.2 Merging IC Macros

Macros that are ICs can contain several macros. If these internal macros are spaced, their respective locations must be preserved. For packed floorplans, it is necessary to place macros in the spaces between these internal macros. As part of the pre-processing, we look for IC macros with spaced internal blocks that can be weaved together to form an (almost) contiguous region. When such macros are found, we merge them into a single macro.

3.3 Value Ordering

To comply with high area utilization, we must produce a compact placement. We achieve this by value ordering heuristics: we select a consistent direction to place all macros. For example, we try to position each macro in the minimum possible locations of x and y coordinates, effectively placing macros from bottom to top, left to right. At the beginning of the solving process, a direction is randomly selected from four possible directions: for each of x, y coordinates we can select the minimum or maximum. The chosen direction applies to all macros with the exception of macros inside a perimeter, where the direction may be reversed. This exception tries to avoid the collision between filling the perimeter area and filling the chiplet it resides in.

3.4 Variable Ordering

Our choice of variable ordering was derived from the typical characteristics of the macros to be placed. Most of the rectangles have an aspect ratio of between 20:1 to 80:1. There are very few nearly square or flat macros. Moreover, there are many macros sharing the same height. In addition, the test-site problem typically has few macros that are involved in more than 3 constraints. All these observations led us to the following variable ordering heuristics:

1. Macros with constraints that directly affect their placement (*Perimeter, Proximity, Cluster* and *Offset* constraints) are placed before macros with no such constraints.
2. Macros are ordered by height, so that macros of the same height will be adjacent.
3. Macros within a group with the same height are ordered by their width.
4. Randomly select some (a pre-defined percentage) macros and move them to a different place in the ordering. This can be done for individual macros and for sets of macros of the same height.

Although we describe order of placing macros, in fact the decision variables are x and y coordinates of their locations. Hence, we place every macro by first selecting its y coordinate, and then immediately its x coordinate.

3.5 Retries

We point out several observations which cause backtracks to be very ineffective on this problem. First, note that the variable domains are huge, of magnitude of 10,000 elements. Generally, small pertubations on the location of a particular macro will have a neglibible effect on the solutions. Furthermore, the effect of positioning a particular macro is not immediate, and we must perform a large number of choices, before we can identify a conflict.

This special nature of the problem may cause thrashing of the CP solver. In other words, if we use regular backtracks, the solver may get stuck on the same solution scheme for a very long time. One way to avoid thrashing is to direct the value ordering towards far/random values after each backtrack. In our case, such value ordering produces highly disordered placements that significantly reduce the chances of reaching a complete solution. Therefore, we believe that a better practice for the test-site problem is to refrain from any backtracks and perform a retry upon reaching a conflict. That is, reset the problem and run the solver again with a totally different randomization path. For this approach to be successful, we must ensure that different runs will produce different results. This is achieved by the randomness introduced in variable and value orderings that was described previously: in each retry the order in which macros are placed and the chosen placement direction are changed.

4 Experimental Results

Over the last two years, we used the CP-based tool described above to find placements for several test-site designs. Table 1 displays the properties of each test-site, including the number of chiplets and the number of macros. Note that some of the macros are ICs containing internal rectangles. The area utilization is calculated by dividing the sum of the areas of all the rectangles by that of the chiplets. The number of constrained macros is the number of macros that are subject to constraints defined by the designer (types (c)-(i) in Eq. 2). We do not include in this count same-cell offset constraints. Macros that appear in multiple designer constraints are counted only once.

Table 1. Characteristics of recent test-sites

Test-site	Chiplets	Macros	Area utilization	Constrained macros	Cluster sizes	Soft cluster sizes
B	7	909	95.56%	0	-	-
P	9	1006	91.96%	33	-	-
I	7	1064	92.44%	203	3,5	-
O	7	962	83.42%	715	2,3,7,7,44,44,47	3,5,7,8,9,10,41,90
S	8	1026	96.19%	177	3,4,4,6,6,6,7,10,23,40	3
G	8	1221	87.80%	0	-	-

We solved the floorplanning problems induced by these designs using the IBM CSP Solver [11], with the heuristics described in section 3. Table 2 summarizes the dimensions of the CP problem created for each design and the solver's performance on it. The number of variables appearing in this table is not exactly twice the number of macros appearing in Table 1, because some of the macros include internal macros and because of the merging pre-process described in 3.2.

For each design we performed 5 separate runs on a Linux machine with an Intel 1.9GHz processor and 2.1GB memory. The *Solved [%]* column shows the percentage of runs in which the solver managed to reach a solution within 7200 seconds (2 hours).

Under *Successful Runs* we show the average number of retries and the average run time of the solver until a solution was found.

Under *Timed Out Runs* we show data for the runs that were terminated due to time out. The seventh column shows the average number of retries in this case. In addition, the solver stores the placement of the best retry it had so far. If it times out, it returns this placement as the (partial) solution. The eighth column shows the average number of unplaced rectangles in the partial solutions of the timed out runs.

Table 2. Rectangles Placement Results

Problem				Successful Runs		Timed Out Runs	
Test-site	Variables	Constraints	Solved [%]	Retries	Solution Time [sec]	Retries	Unplaced Rectangles
B	1830	1	60	6	3101	14.5	6.5
P	2036	5	100	1	968	–	–
I	2158	44	80	12	4938	22	12
O	1834	178	100	1.8	1111	–	–
S	1982	263	0	–	–	6.6	16.4
G	2472	13	100	1.2	4182	–	–

The results in Table 2 show that the solver performed well on all test-site designs, solving most of them in under 2 hours. For those not solved completely, the solver produced an almost complete placement with very few macros left out. Such placements are also useful for test-site designers, and serve as good starting points to find a complete placement manually. Thus, the solver saves several days of manual placement by an experienced designer.

Examining the placements produced by our solver (as in Figure 1), we see a highly structured arrangement. Our choice of value and variable ordering results in rows of contiguous identical rectangles which are favored by the designers, and are similar to the type of solution an engineer would produce.

Figure 2 depicts the effectiveness of various retries. For every duration and number of placed rectangles, we measure how many retries over all designs achieved this result. For example, the height at a duration of 4000 [sec] and 200 placed rectangles is zero, indicating that no retry ended with such results.

Figure 2 demonstrates that at each retry, the solver either fails almost instantly (see the large peak at the origin), or succeeds in placing all or almost all the rectangles. Over all designs, there are very few retries that took a significant amount of time (a few minutes) and managed to place only a small or medium percentage of the rectangles.

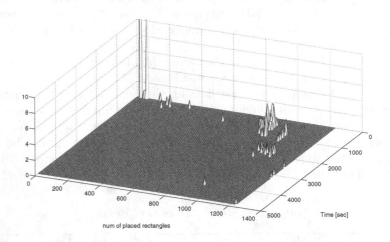

Fig. 2. Histogram showing the number of retries (over all test-sites) sharing the same duration and same success (num of placed rectangles)

4.1 Soft Retries

One of the main disadvantages of the retries heuristic is that we may give up on some good solution paths too early. However, as indicated in Section 3.5, we want to refrain from backtracking or backjumping. Therefore, after detecting a conflict, we perform an analysis on the already placed rectangles. If we identify chiplets that are almost full (occupied over some predefined percentage), we retain their placement for the next retry. We call this approach *Soft Retries*. To avoid getting stuck in some local neighborhood, we perform a full reset after a few such soft resets.

We re-ran the experiments described above on all designs with soft retries activated. A chiplet placement is retained if the chiplet was filled over 96%. A hard reset was performed on each fifth retry. Table 3 shows the results (The *Retries* column in this case counts both soft and regular retries).

Comparing the results in Table 3 with those in Table 2, we can conclude that the soft retries approach improved the percentage of solved designs. A fine grained examination of the single retry run times (not shown here) reveals that the soft retries took much less time on average and usually improved over the results of the initial retry. However, in some cases (especially when the number of retries required to solve the test-site was very small) soft retries

Table 3. Rectangles Placement Results with Soft Retries

Problem				Successful Runs		Timed Out Runs	
Test-site	Variables	Constraints	Solved [%]	Retries	Solution Time [sec]	Retries	Unplaced Rectangles
B	1830	1	80	14.5	4032	24	8
P	2036	5	100	1	975	–	–
I	2158	44	100	7.8	3450	–	–
O	1834	178	100	2.8	1739	–	–
S	1982	263	20	2	1351	8.5	14.75
G	2472	13	100	1.6	5538	–	–

caused a degradation in runtime. This may be a result of our small sample size and the large impact of being stuck on a bad path for 5 retries. Perhaps tuning of parameters of the soft retries heuristic can improve runtime as well.

5   Discussion

We transform the test-site florplanning problem into a rectangles placement problem within a fixed outline. The approach we took in addressing this placement problem is a combination of CP with dedicated heuristics.

Using CP for solving the test-site floorplanning problem provided several advantages:

1. The strength of propagation to continually prune the domains.
2. A controlled degree of randomness, giving unique solutions at every run.
3. The flexibility to introduce a range of heuristics.

We chose to use our own home-grown solver because it gave us the required flexibility. A strong feature of our solver is restrictors, described in Section 3.1. This allowed us to implement the cluster constraints as sub-problems that were solved separately, but integrated into the full flow of the solution. Other features we used are the ability to model variable and value ordering as C++ functions. We also implemented our own version of the DIFF2 constraint. As described before, the variables representing x and y coordinates of each rectangle have large, non-contiguous integer domains. Our solver supports the representation of integer domains as sets of ranges, which is very suitable for this problem.

The retries heuristic proves very suitable for this problem because of the large domains, late propagation and large number of choices made during the solving process. However, refraining from backtracking has two main disadvantages. The first is giving up on some good solution paths too early. This was handled by using soft retries, an approach that improved the percentage of solved designs and the average runtime of a single retry. We believe tuning of parameters of the soft retries heuristic may lead to further improvement.

The second disadvantage of refraining from backtracks is the inability to determine unsatisfiability. If the solver neither reaches a complete solution nor encounters a conflict during first arc-consistency, we cannot determine whether the given instance of the problem is unsatisfiable or just hard to solve. However, even with backtracking, the chances to prove unsatisfiability are very low due to domain sizes and number of choices. In either case, relaxing some of the constraints is advised.

In the time since this tool was introduced to test-site designers, we have been seeing a steady increase in the complexity of designs: the number of macros has been increasing, as has the number of constraints. New types of constraints have been introduced over time as well. This trend continues. The automatic solution enables designers to define more complex constraints they did not define before, which in turn drives ongoing improvement of the tool.

References

1. Roy, J.A., Ng, A.N., Aggarwal, R., Ramachandran, V., Markov, I.L.: Solving Modern Mixed-size Placement Instances. Integration, the VLSI Journal **42**(2), 262–275 (2006). Elsevier, Amsterdam
2. Lauther, U.: A Min-cut placement algorithm for general cell assemblies based on a graph representation. In: Papers on Twenty-five Years of Electronic Design Automation, pp. 182–191. ACM, New York (1988)
3. Khatkhate, A., Li, C., Agnihotri, A.R., Yildiz, M.C., Ono, S., Koh, C.K., Madden, P.H.: Recursive bisection based mixed block placement. In: Procceedings of ACM International Symposium on Physical Design, pp. 84–89. ACM (2004)
4. Gerez, S.H.: Algorithms for VLSI Design Automation. John Wiley & Sons(2000)
5. Chang, Y.C., Chang, Y.W., Wu, G.M., Wu, S.W. : B*-trees: a new representation for non-slicing floorplans. In: Procceedings of the Design Automation Conference, pp 458–463 (2000)
6. Adya, S.N., Chaturvedi, S., Roy, J.A., Papa, D.A., Markov, I.L.: Unification of partitioning, placement and floorplanning. In: Proceedings of the 2004 IEEE/ACM International Conference on Computer-aided Design, pp. 550–557. IEEE Computer Society, Washington DC (2004)
7. Chan, T.F., Cong, J., Shinnerl, J.R., Sze, K., Xie, M.: mPL6: enhanced multilevel mixed-size placement. In: Proceedings of the 2006 International Symposium on Physical Design, pp. 212–214. ACM, New York (2006)
8. Chen, T.C., Hsu, T.C., Jiang, Z.W., Chang, Y.M.: NTUplace: a ratio partitioning based placement algorithm for large-scale mixed-size designs. In: Proceedings of the 2005 International Symposium on Physical Design, pp. 236–238. ACM, New York (2005)
9. Batra, D.: Analysis of Floorplanning Algorithms in VLSI Physical Designs. International Journal of Advanced Technology & Engineering Research **2**(5), 62–71 (2012)
10. Beldiceanu, N., Carlsson, M.: Sweep as a generic pruning technique applied to the non-overlapping rectangles constraint. In: Walsh, T. (ed.) CP 2001. LNCS, vol. 2239, p. 377. Springer, Heidelberg (2001)
11. IBM Research - Haifa CP solver. http://www.research.ibm.com/haifa/dept/vst/csp_gec.shtml

A Constraint-Based Local Search for Edge Disjoint Rooted Distance-Constrained Minimum Spanning Tree Problem

Alejandro Arbelaez[✉], Deepak Mehta, Barry O'Sullivan, and Luis Quesada

Insight Centre for Data Analytics, University College Cork, Cork, Ireland
{alejandro.arbelaez,deepak.mehta,barry.osullivan,
luis.quesada}@insight-centre.org

Abstract. Many network design problems arising in areas as diverse as VLSI circuit design, QoS routing, traffic engineering, and computational sustainability require clients to be connected to a facility under path-length constraints and budget limits. These problems can be modelled as Rooted Distance-Constrained Minimum Spanning-Tree Problem (RDCMST), which is NP-hard. An inherent feature of these networks is that they are vulnerable to a failure. Therefore, it is often important to ensure that all clients are connected to two or more facilities via edge-disjoint paths. We call this problem the Edge-disjoint RDCMST (ERDCMST). Previous works on RDCMST have focused on dedicated algorithms which are hard to extend with side constraints, and therefore these algorithms cannot be extended for solving ERDCMST. We present a constraint-based local search algorithm for which we present two efficient local move operators and an incremental way of maintaining objective function. Our local search algorithm can easily be extended and it is able to solve both problems. The effectiveness of our approach is demonstrated by experimenting with a set of problem instances taken from real-world passive optical network deployments in Ireland, the UK, and Italy. We compare our approach with existing exact and heuristic approaches. Results show that our approach is superior to both of the latter in terms of scalability and its anytime behaviour.

1 Introduction

Many network design problems arising in areas as diverse as VLSI circuit design, QoS routing, traffic engineering, and computational sustainability require clients to be connected to a facility under path-length constraints and budget limits. Here the length of the path can be interpreted as distance, delay, signal loss, etc. For example, in a multicast communication setting where a single node is broadcasting to a set of clients, it is important to restrict the path delays from the server to each client. In Long-Reach Passive Optical Networks (LR-PON) a metro-node is connected to the set of exchange-sites via optical fibres, the length of the fibre between an exchange-site and its metro-node is bounded due to signal loss, and the goal is to minimise the cost resulting from the total length of fibres [1]. In VLSI circuit design path delay is a function of maximum interconnection path length

© Springer International Publishing Switzerland 2015
L. Michel (Ed.): CPAIOR 2015, LNCS 9075, pp. 31–46, 2015.
DOI: 10.1007/978-3-319-18008-3_3

while power consumption is a function of the total interconnection length [2]. In package shipment service guarantee constraints are expressed as restrictions on total travel time from an origin to a destination, and the organisation wants to minimise the transportation costs [3]. In wildlife conservation, which is an application from computational sustainability [4], the landscape connectivity is key to resilient wildlife populations in an increasingly fragmented habitat matrix where landscape connectivity is a function of the length of the path in terms of landscape resistance to animal movement.

Many of these network design problems can be modelled as Rooted Distance-Constrained Minimum Spanning-Tree Problem (RDCMST) [2] which is NP-hard. The objective is to find a minimum cost spanning tree with the additional constraint that the length of the path from a specified root-node (or facility) to any other node (client) must not exceed a given threshold. Many networks are complex systems that are vulnerable to a failure. A major fault occurrence would be a complete failure of the facility which would affect all the clients connected to the facility. Therefore it is important to provide network resilience.

In this paper we focus on the networks where all clients are required to be connected to two facilities via two edge-disjoint paths so that whenever a single facility fails or a single link fails all clients are still connected to at least one facility. We define this problem as the Edge-disjoint Rooted Distance-Constrained Minimum Spanning-Trees Problem (ERDCMST). Given a set of facilities and a set of clients such that each client is associated with two facilities, the problem is to find a set of distance-constrained spanning trees rooted from each facility with minimum total cost. Additionally, each client is connected to its two facilities via two edge-disjoint paths. Notice that if a same edge appears in two trees then it means there exists a client which is connected to its two facilities using the same edge. This would effectively mean that each pair of distance-bounded spanning trees would not be mutually disjoint in terms of edges. Certainly, ERDCMST is more complex than RDCMST as the former not only involves finding a set of distance-constrained spanning trees but also enforces that an edge between any pair of clients can only be used in at most one tree.

Previous works on RDCMST [5,6] have focused on dedicated algorithms which are hard to extend with side constraints, and therefore these algorithms cannot be extended for solving ERDCMST. We present a constraint-based local search algorithm which can easily be extended to apply widely. We present two efficient local move operators, an incremental way of maintaining information required to checking constraints efficiently and an incremental way of maintaining cost of the assignment which are key elements for efficient local search algorithms. Our local search algorithm is able to solve both RDCMST and ERDCMST problems. The effectiveness of our approach is demonstrated by experimenting with a set of problem instances taken from real-world passive optical network deployments in Ireland, the UK, and Italy. We compare our approach with existing exact and heuristic approaches. Our results show that our approach is superior to both of the latter in terms of scalability and its anytime behaviour.

2   Formal Specification and Complexity

Let M be the set of facilities. Let E be the set of clients. Let $E_i \subseteq E$ be the set of clients that are associated with facility $m_i \in M$. We use N to denote the set of nodes, which is equal to $M \cup E$. We use T_i to denote the tree network associated with facility i. We also use $N_i \subseteq N = E_i \cup \{m_i\}$ to denote the set of nodes in T_i. Let λ be the maximum path-length from a facility to any of its clients.

Rooted Distance-Constrained Minimum Spanning-Tree Problem (RDCMST)
Given a facility $m_i \in M$, the set of clients E_i, a set of feasible links $L_i \subseteq N_i^2$, two real numbers, a cost c_l and a distance d_l for each link $l \in L_i$, and a real number λ, the RDCMST is to find a spanning tree T_i with minimum total cost such that the length of the path from the facility m_i to any $e_j \in E_i$ is not greater than λ.

Edge-Disjoint Rooted Distance-Constrained Minimum Spanning-Trees Problem (ERDCMST). Given a set of facilities M, a set of clients E, a set of feasible links $L \subseteq N^2$, two real numbers, a cost c_l and a distance d_l for each link $l \in L$, an association of clients with two facilities $\pi : E \to M^2$, and a real number λ, the ERDCMST is to find a spanning tree T_i for each facility m_i such that:

1. The length of the unique path from the facility m_i to any of its clients is not greater than λ.
2. For each client e_k, the two paths connecting e_k to m_i and m_j, where $\pi(e_k) = \langle m_i, m_j \rangle$, are edge disjoint.
3. The sum of the costs of the edges in all the spanning trees is minimum.

Figure 1 shows an example with two facilities F_1 and F_2 and $N=\{$a, b, c, d, e, f$\}$, black (reps. gray) edge denote the set of edges used to reach F_1 (reap. F_2), the minimum valid value for λ is 12 and the total cost of the solution is 46 for this illustrative example. The indicated solution satisfies the length constraint (i.e., the distance from the facilities to any node is less or equal to λ=12), and the paths connecting the set of nodes to the facilities are edge disjoint. For instance a solution with lower cost would be replacing gray edge $\langle F_2, a \rangle$ and $\langle F_2, b \rangle$ for a gray edge $\langle a, b \rangle$, but this solution would not be edge disjoint, because edges $\langle a, b \rangle$ and $\langle b, c \rangle$ would be used for nodes b and c to reach the two facilities.

Complexity. ERDCMST involves finding a rooted distance-bounded spanning tree for every facility whose total cost is minimum. This problem is known to be NP-complete [2].

3   Constraint Optimisation Formulation

In this section we present a constraint optimisation formulation of ERDCMST. Without loss of generality, in the formulation of the problem we assume full connectivity in the graph, and non-existing links can be added in L by associating

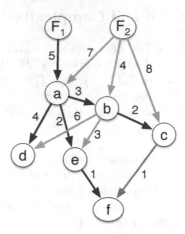

Fig. 1. Example of an instance of the ERDCMST problem. $\lambda=12$.

them with very large distances (w.r.t. to λ). The model only relies in integer variables and all constraints are linear, and therefore it can be encoded directly into existing MIP solvers.

Variables

- Let x^i_{jk} be a Boolean variable that denotes whether an arc between clients $j \in E_i$ and $k \in E_i$ of facility $i \in M$ is selected or not. Each arc (i,j) has an associated cost c_{ij}[1].
- Let f^i_j be a variable that denotes the length of the path from the facility i to its client j.
- Let b^i_j be a variable that denotes the maximum length of the path from the client j to any client in the tree of facility i that is acting as a leaf-node.

We remark that the partial order enforced by f or b help to rule out cycles in the solution.

Constraints. Each client associated with each facility has only one incoming arc:

$$\forall_{m_i \in M} \forall_{e_k \in E_i} : \quad \sum_{e_j \in N_i} x^i_{jk} = 1$$

Each facility is connected to at least one of its clients:

$$\forall_{m_i \in M} : \quad \sum_{e_j \in E_i} x^i_{ij} \geq 1$$

The total number of arcs in any tree T_i is equal to $|E_i|$:

$$\forall_{m_i \in M} : \quad \sum_{e_j \in N_i} \sum_{e_k \in E_i, e_j \neq e_k} x^i_{jk} = |E_i|$$

---

[1] In this paper we assume that the cost is symmetrical, i.e., the $c_{ij}=c_{ji}$.

If there is an arc from $e_j \in E_i$ to $e_k \in E_i$ then the length of the path from m_i to e_k is r equal to the sum of the lengths from m_i to e_j plus the length between e_j and e_k:

$$\forall_{m_i \in M} \forall_{\{e_j, e_k\} \in N_i} : \quad x_{jk}^i = 1 \Rightarrow f_k^i = f_j^i + d_{jk}$$

If there is an arc from $e_j \in E_i$ to $e_k \in E_i$ then the length of the path from e_j to any leaf-node through e_k is greater than or equal to the sum of the lengths from e_k to any of its leaf-node plus the length between e_j and e_k:

$$\forall_{m_i \in M} \forall_{\{e_j, e_k\} \in N_i} : \quad x_{jk}^i = 1 \Rightarrow b_j^i \geq b_k^i + d_{jk}$$

At any node in the tree the length of the path from a facility m_i to a client e_j and the length of the path from e_j to the farthest client on the same path should be less than λ:

$$\forall_{m_i \in M} \forall_{e_j \in N_i} : \quad f_j^i + b_j^i \leq \lambda$$

If m_i and $m_{i'}$ are the facilities of the client j, and if there exists any path in the subnetwork associated with facility i that includes the arc $\langle e_j, e_k \rangle$, then facility i' cannot use the same arc. Therefore, we enforce the following constraint:

$$\forall_{\{m_i, m_{i'}\} \in M} \forall_{\{e_j, e_k\} \in E_i \cap E_{i'}} : x_{jk}^i + x_{jk}^{i'} \leq 1$$

Objective. The objective is to minimize the total cost:

$$\min \sum_{m_i \in M} \sum_{\{e_j, e_k\} \in E_i} c_{jk} \cdot x_{jk}^i$$

4 Iterated Constraint-Based Local Search

The Iterated Constraint-based Local Search (ICBLS) [7,8] framework depicted in Algorithm 1 comprises two phases. First, in a local search phase, the algorithm improves the current solution, little by little, by performing small changes. Generally speaking, it employs a move operator in order to move from one solution to another in the hope of improving the value of the objective function. Second, in the perturbation phase, the algorithm perturbs the incumbent solution (s^*) in order to escape from difficult regions of the search (e.g., a local minima). Finally, the acceptance criterion decides whether to update s^* or not. To this end, with a probability 5% s'^* will be chosen, and the better one otherwise.

Our algorithm starts with a given initial solution where all clients are able to reach their facilities while satisfying all constraints (i.e., the upper bound in the length and disjointness). We use the trivial solution of connecting all clients directly to their respective facilities. We switch from the local search phase to perturbation when a local minima is observed; in the perturbation phase we perform a given number of random moves (20 in this paper). The stopping criteria is either a timeout or a given number of iterations.

Algorithm 1. Iterated Constraint-Based Local Search

1: $s_0 :=$ Initial Solution
2: $s^* :=$ ConstraintBasedLocalSearch(s_0)
3: **repeat**
4:      $s' :=$ Perturbation(s^*)
5:      $s'^* :=$ ConstraintBasedLocalSearch(s')
6:      $s^* :=$ AcceptanceCriterion(s^*, s'^*)
7: **until** No stopping criterion is met

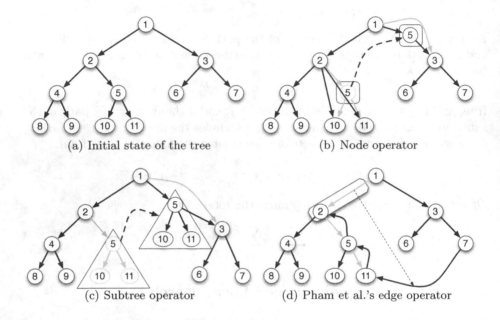

(a) Initial state of the tree   (b) Node operator

(c) Subtree operator   (d) Pham et al.'s edge operator

Fig. 2. Move operators

4.1   Move-Operators

In this section we propose *node* and *subtree* move operators. We use T_i to denote the tree associated with facility i. An edge between two clients e_p and e_q is denoted by $\langle e_p, e_q \rangle$.

Node operator (Figure 2(b)) moves a given node e_i from the current location to another in the tree. As a result of this, all successors of e_i will be directly connected to the predecessor node of e_i. e_i can be placed as a new successor for another node or in the middle of an existing arc in the tree.

Subtree operator (Figure 2(c)) moves a given node e_i and the subtree emanating from e_i from the current location to another in the tree. As a result of this, the predecessor of e_i is not connected to e_i, and all successors of e_i are still directly connected to e_i. e_i can be placed as a new successor for another node or in the middle of an existing arc.

Edge Operator (Figure2(d)). In this paper we limit our attention to moving a node or a complete subtree. [9] proposed to move edges in the context of the Constrained Optimum Path problems. Pham et al. move operator (Figure 2(d)) chooses an edge in the tree and finds another location for it without breaking the flow.

4.2   Operations and Complexities

We first present the complexities of node and subtree operators as they share similar features. For an efficient implementation of the move operators, it is necessary to maintain f_j^i (the length of the path from facility i to client j) and b_j^i (length of the path from e_j down to the farthest leaf associated with it in tree T_i) for each client e_i associated with each facility m_i. This information will be used to maintain the path-length constraint. Let e_{p_j} be the immediate predecessor of e_j and let S_j be the set of immediate successors of e_j in T_i. Table 1 summarises the complexities of the move operators. In this table n denotes the maximum number of clients associated with a single facility.

Table 1. Complexities of different operations

	Node	Subtree	Arc
Delete	$O(n)$	$O(n)$	$O(1)$
Feasibility	$O(1)$	$O(1)$	$O(n)$
Move	$O(n)$	$O(n)$	$O(n)$
Best move	$O(n)$	$O(n)$	$O(n^3)$

Delete. Deleting a node e_j from T_i requires a linear complexity with respect to the number of clients of facility m_i. For both operators, it is necessary to update $b_{j'}^i$ for all the nodes j' in the path from the facility m_i to client e_{p_j} in T_i. In addition, the node operator updates $f_{j'}^i$ for all nodes j' in the subtree emanating from e_i. After deleting a node e_j or a subtree emanating from e_j, the objective function is updated as follows:

$$obj = obj - c_{j,p_j}$$

Furthermore, the node operator needs to add to the objective function the cost of disconnecting each successor element of e_j and reconnecting them to e_{p_j}.

$$obj = obj + \sum_{k \in S_j} (c_{k,p_j} - c_{kj})$$

Feasibility. Checking feasibility for a move can be performed in constant time by using f_j^i and b_j^i. If e_j is inserted between an arc $\langle e_p, e_q \rangle$ then we check the following:

$$f_p^i + c_{pj} + c_{jq} + b_q^i < \lambda$$

If e_p is a leaf-node in the tree and e_j is placed as its successor then the following is checked:

$$f_p^i + c_{pj} + b_j^i < \lambda$$

Move. A move can be performed in linear time. We recall that this move operator might replace an existing arc $\langle e_p, e_q \rangle$ with two new arcs $\langle e_p, e_j \rangle$ and $\langle e_j, e_q \rangle$. This operation requires to update f_j^i for all nodes in the emanating tree of e_j, and b_j^i in all nodes in the path from the facility acting as a root node down to the new location of e_j. The objective function must be updated as follows:

$$obj = obj + c_{pj} + c_{jq} - c_{pq}$$

Best Move. Selecting the best move involves traversing all clients associated with the facility and selecting the one with the maximum reduction in the objective function.

Now we switch our attention to the edge operator. This operator does not benefit by using b_j^i. The reason is that moving a given edge from one location to another might require changing the direction of a certain number of edges in the tree as shown in Figure 2(d). Deleting an edge requires constant complexity, this operation generates two separated subtrees and no data structures need to be updated. Checking the feasibility of adding an edge $\langle e_{p'}, e_{q'} \rangle$ to connect two subtrees requires linear complexity. It is necessary to actually traverse the new tree to obtain the distance from $e_{q'}$ to the farthest leaf in tree of facility i. Performing a move requires a linear complexity, and it involves updating f_j^i for the new emanating tree of $e_{q'}$. And performing the best move requires a cubic time complexity, the number of possible moves is n^2 (total number of possible edges for connecting the two subtrees) and for each possible move it is necessary to check feasibility. Due to the high complexity ($O(n^3)$) of the edge operator to complete a move, hereafter we limit our attention to the node and subtree operators.

Disjointness. To ensure disjointness among spanning-trees we maintain a $2 \cdot |E|$ Matrix, where $|E|$ represents the number of clients. For every client, there are two integers indicating the predecessor in the primary and secondary facilities, these two numbers must always be different.

We also maintain a graph which we call *facility connectivity graph*, denoted by fcg, where the vertices represent the trees associated with the facilities and an edge between a pair of trees represent that a change in one tree can affect the change in another tree. An edge between two vertices in fcg is added if the facilities share at least 2 clients. Notice that if two facilities do not share at least 2 clients then they are independent from disjointness point of view.

[6] also proposed two move operators for the RDCMST. *Edge-Replace* is a special case of our subtree operator, which removes an arc in the tree, and finds the cheapest arc to reconnect the two sub-trees. *Component-Renew* removes an arc in the tree, nodes that are separated from the root node are sequentially added in the tree using Prim's algorithm, nodes that violate the length constraint are added in the tree using a pre-computed route from the root node to any node in the tree. It is worth noticing that the *Component-Renew* operator cannot be applied to DRDCMST as the pre-computed route from the root node to a give node might not be available due to the disjoint constraint.

4.3 Algorithm

The pseudo-code for constraint-based local search is depicted in Algorithm 2. It starts by selecting and removing a client e_j of a facility m_i randomly from Tree T_i (Lines 5 and 6) and performs a move by using one of the move operators as described before. Here the move-operator, which is itself a function, is passed as a parameter. In each iteration of the algorithm (Lines 9-19), we identify the best

---

Algorithm 2. ConstraintBasedLocalSearch (*move-op,sol*)

1: $\{T_1 \ldots T_n\} \leftarrow sol$
2: $list \leftarrow \{(m_i, e_j) | m_i \in M \wedge e_j \in E_i\}$
3: $fcg \leftarrow \{(m_i, m_j) || N_i \cap N_j| \geq 2\}$
4: **while** $list \neq \emptyset$ **do**
5:      Select (m_i, e_j) randomly from $list$
6:      Delete e_j from T_i and update T_i
7:      $Best \leftarrow \{(e_{p_j}, S_j)\}$
8:      $cost \leftarrow \infty$
9:      **for** (e_q, S) in $Locations(move\text{-}op, T_i)$ **do**
10:            **if** FeasibleMove(*move-op*, $(e_q, S), e_j$) **then**
11:                $cost' \leftarrow$ CostMove($(e_q, S), e_i$)
12:                **if** $cost' < cost$ **then**
13:                    $Best \leftarrow \{(e_q, S)\}$
14:                    $cost \leftarrow cost'$
15:                **else if** $cost' = cost$ **then**
16:                    $Best \leftarrow Best \cup \{(e_q, S)\}$
17:                **end if**
18:            **end if**
19:      **end for**
20:      Select $(e_{q'}, S')$ randomly from $Best$
21:      **if** $e_{q'} \neq e_{p_j} \vee S' \neq S_j$ **then**
22:            $list \leftarrow list \cup \{(m_k, e) | (m_i, m_k) \in fcg \wedge e \in N_k \cap (S_j \cup \{e_{p_j}\})\}$
23:      **end if**
24:      $list \leftarrow list - \{(m_i, e_j)\}$
25:      $T_i \leftarrow$ Move($T_i, move\text{-}op, (e_{q'}, S'), e_j$)
26: **end while**
27: **return** $\{T_1 \ldots T_n\}$

---

location for e_j. A location in a tree T_i is defined by (e_q, S) where e_q denotes the parent of e_j and S denotes the set of successors of e_j after the move is performed. Broadly speaking, there are two options for the new location: (1) Breaking an arc $\langle e_p, e_q \rangle$ and inserting e_j in between them such that the parent node of e_j would be e_p and the successor set would be singleton, i.e., $S = \{e_q\}$; (2) Adding a new arc $\langle e_p, e_j \rangle$ in the tree in which case the parent of e_j is e_p and $S = \emptyset$. *Locations* returns all the locations relevant w.r.t. a given move-operator (Line 9). Line 10 verifies that the new move is not breaking any constraint and *CostMove* returns the cost of applying such move using a given move operator. (Line 11).

If the cost of new location is the best so far then the best set of candidates is reiniliased to that location (Lines 12-14). If the cost is same as the best known cost so far then the best set of candidates is updated by adding that location (Lines 15-16). The new location for a given node is randomly selected from the best candidates if there is more than one (Lines 20).

Instead of verifying that the local minima is reached by exhaustively checking all moves for all clients of all facilities, we maintain a list of pairs of clients and facilities. The list is initialised in Line 2 with all pairs of facilities and clients. In each iteration a pair of facility and client, (m_i, e_j) is selected and removed from the list (Line 24). When *list* is empty the algorithm reaches a local minima. If an improvement is observed then one simple way to ensure the correctness of local minima is to populate *list* with all pairs of facilities and clients. However, this can be very expensive in terms of time. We instead exploit the *facility connectivity graph* (*fcg*) where an edge between two facilities is added (in Line 3) if they share at least two clients. If an edge between m_i and m_k exists in *fcg* then it means that if there is a change in T_i then it might be possible to make another change in T_k such that the total cost can be improved. Consequently, we only need to consider clients of affected facilities. This mechanism helps in reducing the time significantly by reducing the number of useless moves. We further strengthen the condition for populating *list* by observing the fact that the set of clients that can be affected in a facility m_k is a subset of $\{e_{p_j}\} \cup S_j$. The reason is all the other clients of m_k were not subject to any constraint from T_i (Line 22).

Algorithm 2 can tackle both RDCMST and ERDCMST. For the former, there would be only one facility. In addition, *FeasibleMove* will only check path-length constraint.

5   Long-Reach Passive Optical Networks

We now describe a real-world application whose instances are used for evaluating our approach. Long-Reach Passive Optical Networks (LR-PONs) provides an economically viable solution for fibre-to-the-home network architectures [1]. In LR-PON fibres are distributed from the Metro-Nodes (MNs) to the Exchange-Sites (ESs) through cables that forms a tree distribution network. A major fault occurrence in LR-PON would be a complete failure of the MN, which could affect tens of thousands of customers. The dual homing protection mechanism for LR-PON enables customers to be connected to two MNs, so that whenever

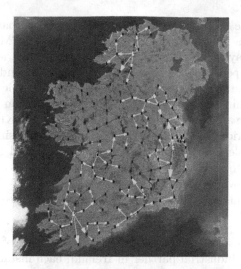

Fig. 3. Example of a LR-PON network for Ireland where each exchange-site is connected to two metro-nodes through disjoint paths. In the plot the subnetwork of each metro node is associated with a colour. Two subnetworks may have the same colour if they do not share nodes.

a single MN fails all customers are still connected to a back-up. Notice that the paths from an ES to its two MNs cannot contain the same link. Otherwise, this would void the purpose of having two MNs. Given as association of MNs with ESs the problem is to determine the routes of cables such that there are two edge-disjoint paths from an ES to its two MNs, the length of each path is below threshold and the total cable length required for connecting each ES to two MNs is minimised. Notice that here metro-nodes are facilities and exchange-sites are clients.

To evaluate the performance of the proposed local search algorithm and the move operators we use three datasets corresponding to real networks from three EU countries: Ireland with 1121 ESs, the UK with 5393 ESs, and Italy with 10708 ESs. For each dataset we use [10] to identify the position of the MNs and computed four instances for each country. Ireland with 18, 20, 22, and 24 MNs; the UK with 75, 80, 85, and 90 MNs; and Italy with 140, 150, 160, and 170 MNs. The LS algorithm starts with the trivial solution which consists in connecting all ESs (clients) to their MNs (facility). A solution obtained using our approach for Ireland is shown in Figure 3.

6   Empirical Evaluation

6.1   Experimental Protocol

All the experiments were performed on a 4-node cluster, each node features 2 Intel Xeon E5430 processors at 2.66Ghz and 12 GB of RAM memory.

In addition to the proposed MIP and local search approaches, we also considered BKRUS [2], a well known heuristic for solving RDCMST when the cost function and the delay function in RDCMST are equivalent and follow the Triangle inequality property (i.e., given three points a, b, and c then $d(a, b) + d(b, c) > d(a, c)$). The overall complexity of BKRUS is cubic in terms of number of nodes, which hinders its scalability when the bound on the path-length is tight. Although there are approaches that are better than BKRUS [5,6], our aim is not to claim superiority over the RDCMST approaches since we are solving a more general problem for which these approaches would not be applicable. Instead, our aim is to show that although our approach is not specialised for RDCMST, it still provides very good quality solutions.

In particular in this paper we consider the following three experimental scenarios.

- *RDCMST*: In this scenario we report results on a set of real-life instances coming from our industrial partner in Ireland, each instance contains $|E| \in \{200, 300, \ldots, 800\}$ nodes. In this dataset the minimum valid value for λ is 415.
- *ERDCMST*: In this case we consider the following two scenarios:
  - *Real-life*: We consider real-life instances coming from our industrial partners with real networks in Ireland, the UK, and Italy detailed in the previous section.
  - *Random*: We generated 10 random instances extracted from the previous real-life network in Ireland. Each instance is generated by using 18 facilities and for each facility we randomly select $|E| \in \{100, 200, \ldots, 1000\}$ nodes.

6.2   Experimental Results

Figures 4(a) and 4(b) report results of the RDCMST problem for the three approaches: the local search approach using the subtree move-operator (LS), CPLEX, BKRUS, and the lower bound of the solution computed using CPLEX (LB). For each experiment we increase λ from 415 to 623, and used a timelimit of one hour. When λ is not playing an important role (i.e., λ=623) the problem is close to the unbounded minimum spanning tree, and except for CPLEX with $|E|$=800, the three approaches report near-optimal values. However, for tight values of λ (i.e., 415) the problem becomes more challenging and BKRUS is unable to solve instances with $|E|$>700, and the gap for CPLEX (w.r.t. the lower bound) considerably increases as the number of nodes increases. On the other hand, for the difficult case (λ=415) LS reports better upper bounds than CPLEX and BKRUS in all 8 instances, and the quality of the solution w.r.t. the lower bound does not degrade with the problem size.

Table 2 reports results for the random set of instances of the ERDCMST problem, here we depict the median value across 11 independent executions of the node and subtree operators; the best solution obtained with CPLEX; the best solution obtained with CPLEX using the solution of the first execution of

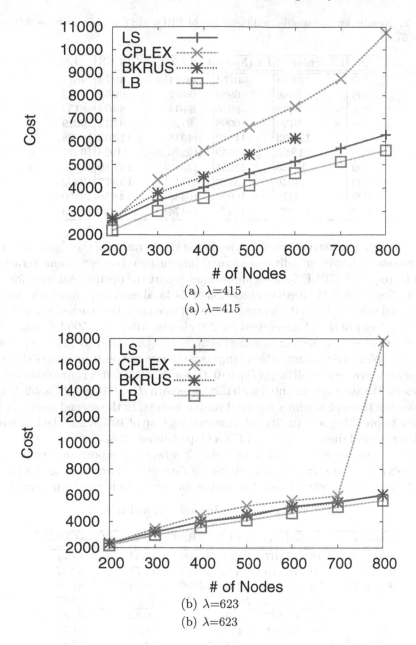

(a) $\lambda=415$

(b) $\lambda=623$

Fig. 4. RDCMST results with different limits on path-length

LS (subtree operator); and the best known LBs for each instance obtained with CPLEX using a larger time limit. The time limit for each local search experiment was set to 30 minutes, and 4 hours for CPLEX-based approaches.

In these experiments we observe that the subtree operator generally outperforms the node operator, we attribute this to the fact that moving a complete subtree

Table 2. Results for the small-sized instances of ERDCMST problem where $|M| = 18$ and $\lambda = 67$

$|E|$	LS (Subtree)	LS (Node)	CPLEX	LS+CPLEX	LB
100	**4674**	**4674**	**4674**	**4674**	4674
200	6966	6988	**6962**	**6962**	6962
300	8419	8575	**8404**	**8404**	8152
400	9728	10008	9728	**9721**	9329
500	**11203**	11672	11318	**11203**	10298
600	**11885**	12559	12276	11924	10517
700	13148	13981	13812	**13140**	11485
800	14040	15133	15118	**13977**	12402
900	14770	16098	16438	**14839**	12860
1000	**15962**	17479	18174	16009	13943

helps to maintain the structure of the tree in a single iteration of the algorithm. The node operator might eventually reconstruct the structure, however, more iterations would be required. CPLEX-based approaches report the optimal solution for 100 and 200 clients, while the median execution of the local search approaches reported the optimal solution for 100 clients, and the subtree operator reached the optimal solution in 5 out of the 11 executions for 200 clients. After $|E|=500$ LS dominates the performance for a margin ranging between 1% ($|E|=500$) to 12% ($|E|=900$). Moreover, cplex+LS was only able to improve the average performance of the subtree operator in a very small factor (up to 0.4% for $|E|=800$) after running the solver for 4 hours. We also experimented with instances with $|E|<100$ and $|E|>800$. In the first case the three algorithms reported similar results. In the second case only LS was able to provide good quality solutions with a Gap of 10% w.r.t. the LB, while BKRUS reported timeouts and CPLEX a Gap of about 59%.

Now, we move our attention to Table 3 where we report results for real ERDCMST instances. In this case, we used a time limit of one hour for LS (using the subtree move operator), and four hours for cplex. As it can be observed, LS

Table 3. Results for Ireland, UK and Italy

Country	$|M|$	Subtree	CPLEX	LB	Gap-Subtree	Gap-CPLEX
	18	**17107**	26787	14809	13.43	44.71
Ireland	20	**16819**	83746	14845	11.73	82.27
$|E|=1121$	22	**16711**	79919	14990	10.29	81.24
	24	**16163**	26918	14570	9.85	45.87
	75	**65377**	285014	54720	16.30	80.80
UK	80	**64565**	301190	54975	14.85	81.74
$|E|=5393$	85	**63517**	281546	55035	13.35	80.45
	90	**62163**	220041	55087	11.38	74.96
	140	**89418**	–	76457	14.49	–
Italy	150	**88255**	–	76479	13.34	–
$|E|=10708$	160	**88336**	–	76794	13.06	–
	170	**87405**	–	77013	11.88	–

dominates the performance in all these experiments, and once again the solution quality of LS does not degrade with the problem size. Indeed, the Gap w.r.t. to the LB for local search varies from 9% to 13% for Ireland, 11% to 16% for the UK, and 12% to 14% for Italy. CPLEX ran out of memory when solving instances from Italy, and we report '–' since no valid solutions were obtained. For the UK instances CPLEX also ran out of memory before the time limit.

7   Conclusions and Future Work

We have presented an efficient local search algorithm for solving Edge Disjoint Rooted Distance-Constrained Minimum Spanning-Trees problem. We presented two novel move operators along with their complexities and an incremental evaluation of the neighbourhood and the objective function. Any problem involving tree structures could benefit from these ideas and the techniques presented make sense for a constraint-based local search framework where this type of incrementally is needed for network design problems. The effectiveness of our approach is demonstrated by experimenting with a set of problem instances taken from real-world long-reach passive optical network deployments in Ireland, the UK, and Italy. We compare our approach with a MIP-based exact approach and a spanning tree-based heuristic approach. Results show that our approach is superior in terms of scalability and its anytime behaviour.

In future we would like to extend ERDCMST with the notion of optional nodes, since this extension is a common requirement in several applications of ERDCMST. Effectively this means that we would compute for every facility a Minimum Steiner Tree where all clients are covered but the path to them may follow some optional nodes. We also plan to make our constraint-based local search approach parallel in two different ways. First, we intend to study the performance of the portfolio approach [11] where multiple copies of the algorithm compete and cooperate to solve a given problem instance, and second we intend to exploit a decomposition of the problem into smaller sub-problems that can be solved in parallel. Preliminary results have been presented in [12]².

Acknowledgments. This work was supported by DISCUS (FP7 Grant Agreement 318137), and Science Foundation Ireland (SF) Grant No. 10/CE/I1853. The Insight Centre for Data Analytics is also supported by SFI under Grant Number SFI/12/RC/ 2289.

References

1. Payne, D.B.: FTTP deployment options and economic challenges. In: Proceedings of the 36th European Conference and Exhibition on Optical Communication (ECOC 2009) (2009)
2. Oh, J., Pyo, I., Pedram, M.: Constructing minimal spanning/steiner trees with bounded path length. Integration **22**(1–2), 137–163 (1997)

---

² We remark that [12] has been presented in a workshop without formal proceedings.

3. Ruthmair, M., Raidl, G.R.: A kruskal-based heuristic for the rooted delay-constrained minimum spanning tree problem. In: Moreno-Díaz, R., Pichler, F., Quesada-Arencibia, A. (eds.) EUROCAST 2009. LNCS, vol. 5717, pp. 713–720. Springer, Heidelberg (2009)

4. Gomes, C.P.: Computational sustainability: Computational methods for a sustainableenvironment, economy, and society. The Bridge **39**(4), 5–13 (2009)

5. Leitner, M., Ruthmair, M., Raidl, G.R.: Stabilized branch-and-price for the rooted delay-constrained steiner tree problem. In: Pahl, J., Reiners, T., Voß, S. (eds.) INOC 2011. LNCS, vol. 6701, pp. 124–138. Springer, Heidelberg (2011)

6. Ruthmair, M., Raidl, G.R.: Variable neighborhood search and ant colony optimization for the rooted delay-constrained minimum spanning tree problem. In: Schaefer, R., Cotta, C., Kołodziej, J., Rudolph, G. (eds.) PPSN XI. LNCS, vol. 6239, pp. 391–400. Springer, Heidelberg (2010)

7. Hoos, H., Stütze, T.: Stochastic Local Search: Foundations and Applications. Morgan Kaufmann (2005)

8. Hentenryck, P.V., Michel, L.: Constraint-based local search. The MIT Press (2009)

9. Pham, Q.D., Deville, Y., Hentenryck, P.V.: Ls(graph): a constraint-based local search for constraint optimization on trees and paths. Constraints **17**(4), 357–408 (2012)

10. Ruffini, M., Mehta, D., O'Sullivan, B., Quesada, L., Doyle, L., Payne, D.B.: Deployment strategies for protected long-reach pon. Journal of Optical Communications and Networking **4**, 118–129 (2012)

11. Arbelaez, A., Codognet, P.: From sequential to parallel local search for SAT. In: Middendorf, M., Blum, C. (eds.) EvoCOP 2013. LNCS, vol. 7832, pp. 157–168. Springer, Heidelberg (2013)

12. Arbelaez, A., Mehta, D., O'Sullivan, B., Quesada, L.: A constraint-based parallel local search for disjoint rooted distance-constrained minimum spanning tree problem. In: Workshop on Parallel Methods for Search & Optimization (2014)

A Benders Approach to the Minimum Chordal Completion Problem

David Bergman[1]([⊠]) and Arvind U. Raghunathan[2]

[1] University of Connecticut, 1 University Place, Stamford, CT, USA
david.bergman@business.uconn.edu
[2] Mitsubishi Electric Research Labs, 201 Broadway, Cambridge, MA, USA
raghunathan@merl.com

Abstract. This paper introduces an integer programming approach to the minimum chordal completion problem. This combinatorial optimization problem, although simple to pose, presents considerable computational difficulties and has been tackled mostly by heuristics. In this paper, an integer programming approach based on Benders decomposition is presented. Computational results show that the improvement in solution times over a simple branch-and-bound algorithm is substantial. The results also indicate that the value of the solutions obtained by a state-of-the-art heuristic can be in some cases significantly far away from the previously unknown optimal solutions obtained via the Benders approach.

1 Introduction

Given graph G, the *minimum chordal completion problem* (MCCP) asks for a minimum cardinality set of edges whose addition to G results in a *chordal graph* (a graph for which every cycle consisting of four or more vertices contains a *chord* - an edge connected vertices that do not appear consecutively in the cycle). The problem is also known as the *minimum fill-in problem* and the *minimum triangulation problem*.

Chordal completions find applications in a variety of fields. These include sparse matrix computation and semidefinite programming [18,23,27,31], database management [1,34], computer vision [12], many others (the interested reader may refer to a survey on the topic [19]). In addition to these applications, chordal completions are related to the *tree-width problem* [6], the *minimum interval completion problem* [25], and are a special case of graph sandwich problems [17].

Although the problem has applications in a variety of domains, computational approaches in the literature have been very limited, with the focus being on developing heuristics. The MCCP (in its decision version which ask whether or not a chordal completion containing fewer than k edges exists) was listed as one of the open problems in Garey and Johnson's classical book on computational complexity [15] and later proven NP-complete [36].

Surprisingly, the optimization community has largely overlooked computational approaches to the problem. Some algorithms are published which solve

© Springer International Publishing Switzerland 2015
L. Michel (Ed.): CPAIOR 2015, LNCS 9075, pp. 47–64, 2015.
DOI: 10.1007/978-3-319-18008-3_4

the MCCP exactly on particular classes of graphs [7,9,11,12,24,33]. For general graphs algorithms exists as well, although computational results have not been reported. The first fixed parameter tractable algorithm [22] was proven to have time complexity $O\left(2^{O(k)} + k^2 nm\right)$ (where n is the number of vertices in the graph and m is the number of edges in the graph). Algorithms have since been investigated [8] and recently the running time has been reduced to subexponential parameterized time complexity [13].

As opposed to exact algorithms, the literature is vast on algorithms designed to find *minimal* chordal completions (the interested reader can again refer to a survey written on the topic [19]). The objective of these algorithms are twofold. First, they seek to create chordal completion in the least possible computational time. Second, they search for chordal completions using as few edges as possible. A surprisingly simple algorithm based on ordering the vertices of the graph by their degree and running a *vertex elimination game* [16] runs in polynomial time [20] $(O(n^2 m))$ and produces very good solutions in terms of the number of edges added, compared to other heuristics [3–5,26,29].

This paper presents an integer programming approach to the MCCP based on Benders decomposition [2], and in particular on logic-based Benders decomposition [21]. The problem is modeled using only a quadratic (with respect to the number of vertices in the graph) number of variables (one per edge not in the graph), and cuts are added iteratively when they become violated. The problem is decomposed into a master problem and a subproblem. In the master problem, a solution is found which adds some subset of the complement edge set to the graph subject to Benders cuts that have been previously identified. This graph then defines a subproblem which returns either that the graph is chordal (and hence a minimum chordal completion) or finds chordal cuts that must be satisfied by any chordal completion of the graph. In the latter case, the cuts are added to the master problem, and any solution found by the master problem in subsequent iterations will never violate these cuts.

The remainder of the paper is organized as follows. In Section 2 graph notation is introduced which will be used throughout the paper. Section 3 formally introduces the MCCP. Section 4 introduces the Benders decomposition approach to the MCCP with Section 5 explaining in detail the solution to the subproblem and Section 6 describing the Benders cuts and proving that they are valid. Section 7 details the exact algorithm and heuristic algorithms used for comparison in the computational experiments presented in Section 8. The paper concludes in Section 9.

2   Graph Notation

Let S be any set. $\binom{S}{2}$ denotes the family of two-element subsets of S.

Let $G = (V, E)$ be a graph. For the remainder of the paper it is assumed that G is connected, undirected, has no self-loops or multi-edges.

Each edge $e \in E \subseteq \binom{V}{2}$ is a two-element subset of V. We denote by E^c the *complement edge set* of G: $E^c = \binom{V}{2} \backslash E$.

The *neighborhood* $N(v)$ of a vertex $v \in V$ is the set of vertices which share an edge with (are *adjacent* to) v: $N(v) = \{v' : (v, v') \in E\}$. The *closed neighborhood* $N[v]$ is the neighborhood of v together with v: $N[v] = N(v) \cup \{v\}$.

Let $V' \subseteq V$. The graph *induced* by V', $G[V'] = (V', E(V'))$ is the graph on vertex set V' with edge set $E(V') = E \cap \binom{V'}{2}$. Given $F \subseteq E^c$, a *completion* of G, denoted by $G + F$, is the addition of the sets in F to E: $G + F = (V, E \cup F)$.

A *cycle* C in $G = (V, E)$ is an ordered list of distinct vertices of G, $C = (v_1, \ldots, v_k)$, for which $\bigcup_{i=1,\ldots,k-1} \{v_i, v_{i+1}\} \cup \{v_k, v_1\} \subseteq E$. We denote by $V(C)$ the set of vertices that appear in the cycle, and $E(C)$ the edges connected to consecutive vertices in the cycle ($E(C) := \bigcup_{i=1,\ldots,k-1} \{v_i, v_{i+1}\} \cup \{v_k, v_1\}$). The *interior* of C, int(C), is the family of two-element subsets of the vertices in the cycle that do not coincide with the edges of the cycle: i.e., int$(C) = \binom{V(C)}{2} \backslash E(C)$. A cycle containing k vertices is called a *k-cycle*.

A cycle C for which the graph $G[V(C)]$ contains only those edges in the cycle is a *chordless* cycle. A graph is said to be *chordal* (or *chordless*) if the maximum size of any chordless cycle is three. A chordless cycle with k vertices is called a *k-chordless cycle*.

3   Problem Description

Let $G = (V, E)$ be a graph. A *chordal completion* of G is any subset of edges $F \subseteq E^c$ for which $G + F$ is chordal. A *minimal chordal completion* is a chordal completion F for which F' is not a chordal completion for any proper subset $F' \subset F$. A minimum chordal completion is a chordal completion of minimum cardinality. The minimum chordal completion problem (MCCP) is the problem of identifying such a subset of the complement edge set.

We refer to E^c as both the complement edge set and the set of candidate *fill edges*, since we think of filling G with edges in order to create chordless graphs. We will use both terms interchangeably.

Example 1. Consider the graph in Figure 1 (a). This graph has two chordless cycles, $C_1 = (1, 2, 3, 4)$ and $C_2 = (2, 3, 4, 5)$. Figure 1 (b) shows a minimal chordal completion, which adds edges $\{1, 3\}$ and $\{3, 5\}$. Removing either of these edges will result in a graph that is not chordal. Figure 1 (c) shows a smaller, minimum chordal completion consisting only of edge $\{2, 4\}$.

4   Integer Programming Approach

Benders decomposition is a general scheme proven to be useful for a variety of problems. Benders decomposition calls for the communication of *Benders cuts* between two models in order to communicate inferences.

In the scheme proposed in this paper, the decomposition is broken into an integer programming (IP) phase which identifies completions of G (the *master*

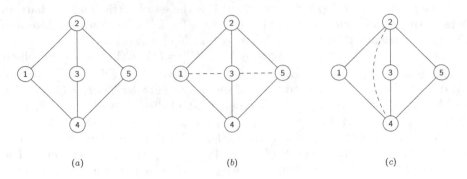

Fig. 1. (a) A graph. (b) A minimal chordal completion of the graph. (c) A minimum chordal completion of the graph.

problem) and a combinatorial optimization problem (the *subproblem*) of identifying cutting planes that will restrict the completions found by the subsequent IPs. Each time a completion is assigned and found to be inconsistent (i.e., leading to a graph which is not chordal), an inequality is produced at the end of the subproblem phase which will prohibit the IP from ever leading to this solution again. This passing of information is done iteratively until the subproblem no longer finds a Benders cut and certifies that the completion found by the master problem is not only feasible (a chordal completion), but a minimum such completion.

The master problem is the following:

$$\min \qquad \sum_{f \in E^c} z_f$$

subject to   [accumulated Benders cuts]

$$z_f \in \{0, 1\}, \qquad\qquad f \in E^c$$

(MP)

leading initially to the solution z^0 for which, $\forall f \in E^c, z_f^0 = 0$, Benders cuts are yet to be generated.

For a solution z' with elements in $\{0, 1\}$ for each $f \in E^c$, let $F(z') = \{f : z_f' = 1\}$. Each time a master problem is solved, the solution z^k encodes a graph $G(z^k) = (V, E \cup F(z^k))$. The subproblem has the goal of either determining that $G(z^k)$ is chordal, or producing some certificate of infeasibility which is translated into a linear inequality, restricting the master problem from ever producing this solution again (called *Benders cuts*). In the case of the MCCP, this certificate will be a k-cycle C with $k \geq 4$ (or a set of such cycles).

In a general iteration, the master problem will contain all of the elements in (MP), in addition to Benders cuts which are accumulated from earlier iterations.

The pseudocode for the algorithm is presented in Algorithm 1. We start with z^0 as defined above. This specifies that $G(z^0) = G$. Given solution z^k, the algorithm tests whether or not the graph $G(z^k)$ is chordal. If it is, the solution relates

---

Algorithm 1. Benders decomposition for the MCCP in graph $G = (V, E)$

1: $k \leftarrow 0$
2: let z^0 be the optimal solution to (MP), with no Benders cuts // $\forall f \in E^c, z_f^0 = 0$
3: **while** $G(z^k)$ is not chordal **do**
4:　　$k \leftarrow k + 1$
5:　　Let $\mathcal{C}^k$ be a set of chordless cycles in $G(z^{k-1})$
6:　　**for all** $C \in \mathcal{C}^k$ **do**
7:　　　　generate Benders cuts $B(C)$ and add them to (MP)
8:　　$z^k \leftarrow$ optimal solution to (MP) with all accumulated Benders cuts
9: **return** $F(z^k)$

---

to an optimal solution and $F(z^k)$ is returned. If not, the algorithm increases the iteration count by one, identifies a set of chordless cycles $\mathcal{C}^k$ in $G(z^k)$, generates a set of Benders cuts $B(C)$ for each $C \in \mathcal{C}^k$, and adds these cuts to the master problem. The master problem is then re-solved, to find a possible solution.

Several elements of Algorithm 1 need to be specified. Namely, line 3 which determines whether or not a given graph is chordal, line 5 which finds a set of chordless cycles in a graph which is not chordal, and line 7 which generates Benders cuts based on the cycles. In general, a relaxation is also added to the master problem in order guide the initial solution. One can view the first round of Benders cuts as such a relaxation since, as explained below, the first round of cuts is based entirely on the original graph. The following sections specify these particulars.

5　Finding Chordless Cycles

Determining whether a graph is chordal or not can be accomplished in linear time in the size of the graph [30]. This can be shown to be equivalent to finding a *perfect elimination order* of a graph, and even finding all perfect elimination orders has been investigated [10].

Papers have been published which seek to identify all chordless cycles in a graph too [32, 35], but the running time of these algorithms are exponential in the size of the graph. For the purpose of this paper, it is not necessary to list *all* chordless cycles; in any iteration of Algorithm 1, it is only necessary to find at least one chordless cycle, if one exists.

A simple strategy can be employed, based on searching through triples of vertices, that can be used to find a set of chordal cycles (or stopped prematurely to find a single chordal cycle) if one exists. This is presented in Algorithm 2.

The algorithm starts with no cycles in $\mathcal{C}$. For every ordered triple of vertices i, j, k for which i, j, k is an *induced path* (i.e., j is adjacent to both i and k, but i and k are not adjacent), the algorithm checks whether i and k are connected in the graph $G[(V \setminus N[j]) \cup \{i, k\}]$ induced by all vertices in G besides the closed neighborhood of j (include i and k). If so, the algorithm adds the set of vertices $\{i, j, k\}$ together with the shortest path between i and k in $G[(V \setminus N[j]) \cup \{i, k\}]$ to $\mathcal{C}$.

Theorem 1. *Algorithm 2 returns a set* $\mathcal{C}$ *for which every* $C \in \mathcal{C}$ *is a chordless cycle in* G, *and is empty at the end of the execution of the algorithm if and only if* G *is chordal.*

Proof. Suppose $\mathcal{C} \neq \emptyset$. Let C be any cycle in $\mathcal{C}$ and i, j, k the ordered triple of vertices that produced C. C is a cycle because i is adjacent to j, j is adjacent to k, and i, k are connected in G through the path $P = (i, v_1, \ldots, v_\ell, k)$ (determined during the algorithm) that connects i and k, does not contain j, and is connected in the subgraph $G[V \setminus N[j] \cup \{i, k\}]$ of G.

Furthermore, C must be a chordless cycle. By the construction of C, j is only adjacent to i and k (with $\{i, k\} \notin P$), so that j does not participate in any chord of C. In addition, since P is the shortest path in $G[V \setminus N[j] \cup \{i, k\}]$, there cannot be any edge $\{v_a, v_b\}$, for $a \leq b - 2$ (otherwise P would not be a shortest path) in the subgraph. Therefore, C is a chordless cycle in G.

What remains to be shown is that $\mathcal{C}$ is empty if and only if G is chordal. From the previous arguments, if $\mathcal{C}$ is non-empty, then every set $C \in \mathcal{C}$ is a cycle in G. Therefore in this case G is not chordal.

Finally, suppose G is not chordal. Let $C = (v_1, \ldots, v_\ell)$ be a smallest length chordless cycle in G ($\ell \geq 4$). Consider when the algorithm examines $i = v_1, j = v_2, k = v_3$. C is a chordless cycle, therefore v_2 is only adjacent to v_1 and v_3 in G. Therefore, the path $P = (v_1, v_\ell, \ldots, v_3)$ is in $G[V \setminus N[j] \cup \{i, k\}]$. Furthermore, this must be the shortest path in this subgraph, because otherwise C would not be a smallest length chordless cycle. Therefore, C is in $\mathcal{C}$ at the end of the execution of Algorithm 2, as desired, completing the proof.   □

Algorithm 2. Find a set of chordless cycles $\mathcal{C}$ in $G = (V, E)$ (or return $\mathcal{C} = \emptyset$ if G is chordal)

1: $\mathcal{C} \leftarrow \emptyset$
2: **for all** $i, j, k \in V$ // all ordered triples of vertices **do**
3:     **if** $\{i, j\}, \{j, k\} \in E$ and $\{i, k\} \notin E$ **then**
4:         **if** i, k are connected in $G[V \setminus N[j] \cup \{i, k\}]$ **then**
5:             $P \leftarrow$ shortest path from i to k in $G[V \setminus N[j] \cup \{i, k\}]$
6:             $\mathcal{C} \leftarrow \mathcal{C} \cup (P \cup \{j\})$
7: **return** $\mathcal{C}$

---

6   Benders Cuts

In this section a class of Benders cuts is presented, each generated by a chordless cycle in a graph. The main idea behind the Benders cuts developed here is that if a graph G contains a k-chordless cycle C then at least $k - 3$ of the edges interior to the cycle must be in any chordal completion of the graph.

Example 2. Consider the graph is Figure 2 (a). The graph is a cycle $C = \{1, 2, 3, 4, 5\}$ which is not chordal. Any chordal completion requires at least two

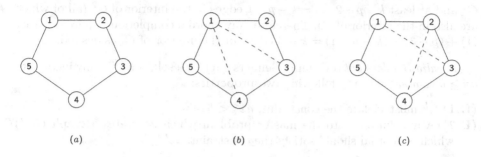

Fig. 2. (a) A graph. (b) A chordal completion with two fill edges. (c) A completion with two edges that is not chordal.

edges. One such completion is depicted in Figure 2 (b). Note only specific subsets of size two can be added to the graph to result in a chordal completion. Consider for example the graph depicted in Figure 2 (c), where the edges $\{1,3\}$ and $\{2,4\}$ are added to the graph although the graph is not chordal.

Lemma 1. *Let G be a graph containing a chordless cycle C of length $k \geq 4$. Then at least $k-3$ edges in $int(C)$ are in any chordal completion of G.*

Proof. We proceed by induction on k. For $k = 4$, if fewer that $4 - 3$ edges in the interior of the chordless cycle are added to G the graph will still contain this chordless cycle and hence not be chordal.

For the base case, let $k = 5$ and $C = (1, 2, 3, 4, 5)$. Suppose we add fewer than $5 - 3 = 2$ edges to the interior of C. Adding zero edges to the interior of C leaves G not chordal. Suppose we add only one edge to the interior of C. Since the cycle is symmetric with respect to the edges interior to the cycle, suppose $\{1, 3\}$ is added. This leaves chordless cycle $(1, 3, 4, 5)$.

Fix $k > 5$. Suppose that $\forall \ell, 4 \leq \ell \leq k - 1$, if a chordless cycle C' with length ℓ appears in G then at least $\ell - 3$ edges in the interior of C' must appear in any chordless completion of G.

Let C be a chordless cycle in G of length k. Let the vertices in C be the first k integers: $C = (1, 2, \ldots, k)$. There must be at least one edge in the interior of C that appears in any chordless completion of G. Since C is symmetric with respect to the vertices, let suppose that $\{1, p\}$ is added to G.

If $p = 3$, then $C' = (1, 3, 4, \ldots, k)$ is a chordless cycle of length $k - 1$. By the inductive hypothesis, any chordal completion requires at least $(k - 1) - 3 = k - 4$ edges in the interior of C' (which are also interior to C). Therefore any chordal completion of G requires at least $1 + k - 4 = k - 3$ edges in the interior of C.

If $p = k - 1$, the vertices can be renamed in opposite order, resulting in the same case as $p = 3$.

Let $4 \leq p \leq k - 2$. This chord cuts C and creates two, separate chordless cycles: $C^1 = (1, 2, \ldots, p)$ and $C^2 = (1, p, p + 1, \ldots, k)$. By the inductive hypothesis, any chordless completion will require at least $p - 3$ edges in the interior of

C^1 and at least $k - p + 2 - 3 = k - p - 1$ edges in the interior of C^2 (all of which are also in the interior of C). Therefore, any chordal completion requires at least $(1) + (p - 3) + (k - p - 1) = k - 3$ edges in the interior of C, as desired.   □

A *valid Benders cut* (or simply Benders cut) for a solution z^k is any inequality $az \geq b$ that satisfies the following two properties:

(B.1) z^k must violate the constraint: $az^k < b$
(B.2) Any solution z' to the master problem which generates a graph $G(z')$ which is chordal should satisfy the constraint: $az' \geq b$

Given a solution z^k in Algorithm 1 which yields graph $G(z^k)$ which is not chordal, after identifying a chordless cycle C (or set of chordless cycles) in line 5, the goal is to find a linear inequality which can be added to the master problem which prohibits z^k from appearing again and is satisfied by every solution z' which corresponds to a chordal completion $F(z^k)$. The simplest such cut, which is readily applicable in Benders schemes, is the following:

$$\sum_{f:z_f^k=1} (1 - z_f) + \sum_{f:z_f^k=0} (z_f) \geq 1 \tag{1}$$

This simple cut forbids only the given solution z^k

Example 3. Take for example z^0 with $z_f^0 = 0$ for all $f \in E^c$ for the graph in Figure 2. There are many possible Benders cuts associated with this solution. These include, for example:

$$z_{\{1,3\}} + z_{\{1,4\}} + z_{\{2,4\}} + z_{\{2,5\}} + z_{\{3,5\}} \geq 1 \tag{2}$$

(2) is the standard Benders cut (1).

The application of Benders decomposition necessitates the generation of problem specific valid Benders cuts so that the inequalities are tighter and eliminate additional infeasibility, for otherwise too many iterations are realized. The remainder of this section is devoted to proving that the following inequality, for a chordless cycle C identified in Algorithm 1, is a valid Benders cut:

$$\sum_{f \in \text{int}(C)} z_f \geq (|V(C)| - 3) \cdot \left(\sum_{f \in E(C) \cap E^c} z_f - |E(C) \cap E^c| + 1 \right) \tag{3}$$

Example 4. Let G be the graph in Figure 2. Suppose the solution z^k to the master problem is $z_{\{1,3\}}^k = z_{\{1,4\}}^k = z_{\{2,4\}}^k = z_{\{2,5\}}^k = z_{\{3,5\}}^k = 0$. (3) becomes

$$z_{\{1,3\}} + z_{\{1,4\}} + z_{\{2,4\}} + z_{\{2,5\}} + z_{\{3,5\}} \geq 2 \cdot 1 = 2$$

This is a valid Benders cut. The solution violates this constraint and in any chordal completion at least two of the edges in the interior of the cycle must be present.

Example 5. Take again the graph G in Figure 2. Suppose the solution z^k to the master problem is $z^k_{\{1,3\}} = 1$ and $z^k_{\{1,4\}} = z^k_{\{2,4\}} = z^k_{\{2,5\}} = z^k_{\{3,5\}} = 0$. Let $C = (v_1, v_3, v_4, v_5)$, which is a chordless cycle in the graph $G(z^k)$. In this case,

$$\begin{aligned}
V(C) &= \{1,3,4,5\} \\
E(C) &= \{\{1,3\}, \{3,4\}, \{4,5\}, \{1,5\}\} \\
\mathrm{int}(C) &= \{\{1,4\}, \{3,5\}\} \\
E(C) \cap E^c &= \{\{1,3\}\}
\end{aligned}$$

(3) now becomes

$$z_{\{1,4\}} + z_{\{3,5\}} \geq (4-3) \cdot \left(z_{\{1,3\}} - 1 + 1\right) = z_{\{1,3\}}$$

This inequality will force at least one of $z_{\{1,4\}}$ or $z_{\{3,5\}}$ to be equal to one if $z_{\{1,3\}} = 1$ in any subsequent master problem solution. This inequality is a valid Benders cut because **(B.1)** is satisfied (plugging in z^k into the converse of the inequality yields $0 < 1$) and **(B.2)** is satisfied (if a completion of G contains edge $\{1,3\}$, it *must* contain either edge $\{1,4\}$ or $\{3,5\}$, for otherwise cycle $C = (1,3,4,5)$ will be a chordless cycle in G).

Theorem 2. *Suppose C is a chordless cycle, identified in line 5, in the graph $G(z^k)$ derived from solution z^k obtained by solving the master problem in iteration k of Algorithm 1. (3) is a valid Benders cut.*

Proof. Let C be such a cycle.

(B.1): Because C is a chordless cycle, there cannot be any edges in the interior of C in $G(z^k)$ so that the left-hand side of (3) becomes $\sum_{f \in \mathrm{int}(C)} z^k_f = 0$.

Therefore it suffices to show that the right-hand side of (3) is strictly greater than 0. C is a chordless cycle, so $|V(C)| \geq 4$ and $(|V(C)| - 3) > 0$. In addition, each edge in $E(C) \cap E^c$ must be in C, for otherwise C would not be a cycle in $G(z^k)$, so that for the variables in this set $z^k_f = 1$. Therefore,

$$\sum_{f \in E(C) \cap E^c} z^k_f - |E(C) \cap E^c| + 1 = 1,$$

and the right-hand side of (3) evaluates to $(|V(C)| - 3)$ which is greater than 0.

(B.2): Let z' be any solution that generates a graph $G(z')$ that is chordal. First note that (3) is satisfied by *any* solution (not just feasible ones) if for any $f \in E(C) \cap E^c, z'_f = 0$. This is because the left-hand side of the inequality is always greater than or equal to 0, and if $z'_f = 0$ for some such z', the right-hand side of the inequality will be less than or equal to 0. Therefore consider only those z' for which the right hand side is greater than or equal to 1 i.e. $\forall f \in E(C) \cap E^c, z'_f = 1$.

For such z', the second of the two terms multiplied on the right-hand side of (3) evaluates to 0 at z'. It therefore suffices to show that

$$\sum_{f \in \mathrm{int}(C)} z'_f \geq |V(C)| - 3.$$

Because $\forall f \in E(C) \cap E^c, z'_f = 1$, $G(z')$ contains every edge in $E(C)$. C is a cycle in $G(z')$. Therefore, by Lemma 1, at least $|V(C)| - 3$ edges in the interior of C are in any chordal completion of $G(z')$. □

7   Minimum/Minimal Chordal Completion Algorithms

As discussed in Section 1, there is a vast literature on minimal chordal completions while the literature investigated computational approaches to the MCCP is surprisingly thin. In order to evaluate, computationally, Algorithm 1, it is necessary to compare with another MCCP algorithm. In the experimental results section, Section 8, the Benders approach is compared with a simple variant of an algorithm for the MCCP that has exponential running time [8]. This algorithm is explained below, along with a state-of-the-art heuristic.

Exact Algorithm. Developing any exact algorithm designed to solve the MCCP is non-trivial. Even formulating the problem with an IP model, a standard approach to solving combinatorial optimization problems, is difficult.

The fastest (in terms of time performance guarantee) algorithm for solving the MCCP [13], to the best knowledge of the authors, has time complexity $O\left(2^{O\left(\sqrt{k}\log(k)+k^2 nm\right)}\right)$, although it is not implemented nor have computational experiments on the algorithm been reported. The algorithm branches on chordless cycles of length four, and then on *moplexes*, keeping track of certain indicators which allow pruning. The interested reader can refer to the paper in the references for an explanation of the algorithm.

In place of implementing this complex algorithm, a simpler version is presented in Algorithm 3, which will be used for computational comparison. This algorithm searches on the complement edge set. It starts with a single search tree node s. Search tree nodes have two sets associated with them: $I(s)$ is the set of complement edges included in this search node, and $O(s)$ are the set of edges to be excluded. For any search tree node, if it is chordal, the upper bound is updated and the set $I(s)$ becomes the incumbent solution. If it isn't, two new nodes are created, only if they can lead to better solutions (it is required that $|I(s)| \leq z_{ub} - 2$ because at least one more additional edge has to be included and so it can only lead to a better solution if the number of edges is at least two less than the size of the incumbent solution). Then, a cycle is identified in $G' = (V, E \cup I(s))$ and for some edge in the interior of this cycle but not in $O(s)$, a search node is created which includes the edge and a search node is created which excludes this edge. The search concludes when there are no more nodes to explore. This algorithm can have a *warm start* where any heuristic can be ran prior to the algorithm for a starting incumbent solution F. The computational experiments in Section 8 assumes a depth-first search ordering.

Heuristic Algorithms. This section describes a heuristic that will be used for comparison with the Benders approach, to see the gap that results from running only a minimal chordal completion algorithm.

Algorithm 3. Branch-and-bound algorithm for the MCCP in graph $G = (V, E)$

1: $F \leftarrow E^c$
2: $z_{ub} \leftarrow |F|$
3: create search node s with $I(s) = O(s) = \emptyset$
4: $Q \leftarrow \{s\}$
5: **while** $Q \neq \emptyset$ **do**
6:      let s be some node in Q
7:      $Q \leftarrow Q \backslash \{s\}$
8:      **if** $G' = (V, E \cup I(s))$ is chordal **then**
9:          $F \leftarrow I(s)$
10:         $z_{ub} \leftarrow |F|$
11:     **else**
12:         **if** $|I(s)| \leq z_{ub} - 2$ **then**
13:             find chordless cycle C of length greater than 3 in G'
14:             let e be any edge in $\text{int}(C) \backslash O(s)$
15:             **if** e exists **then**
16:                 create search node s' with $I(s') = I(s) \cup \{e\}, O(s') = O(s)$
17:                 create search node s'' with $I(s'') = I(S), O(s'') = O(s) \cup \{e\}$
18:                 $Q \leftarrow Q \cup \{s', s''\}$
19: **return** F

A classical algorithm that can be used to find small minimal chordal completions of graphs is the elimination graph model: In this algorithm, the vertices are sorted by some permutation σ. The vertices are then relabeled according to σ. Then, for $i = 1, \ldots, n$, edges are added to G in order to make the neighbors of i in $\{i + 1, \ldots, n\}$ adjacent. The minimal chordal completions of G exactly coincide with the graphs that results from this process [14].

Finding the ordering that results in the smallest set of additional edges exactly corresponds to solving the MCCP. Finding this ordering is NP-hard, and many heuristics have been developed, most popular and highly implemented being the minimum degree ordering [16] which in general finds very good, nearly optimal solutions. This algorithm will henceforth be referred to as *MDOC* (for minimum degree ordering completion).

8   Computational Results

This section reports computational results on the Benders approach to the MCCP. All algorithms are implemented in C++ and ran on a 3.20 GHz Intel(R) Core(TM) i7-3930K CPU processor with 32 GB RAM. The IP solver used is Gurobi version 5.63. All settings were set to default, except that the solver was restricted to solving on one processor by setting Threads to 1.

The algorithms are tested on random graphs generated according to the Erdős-Réyni model. The graphs generated have n vertices and density d, with n ranging from 15 to 35, in increments of 5, and d ranging from 0.1 to 0.9, in increments of 0.1. 10 graphs are generated per configuration.

For the remainder of this section, the means reported are geometric means with shift 1.

Comparison with Simple Branch-and-bound Algorithm. Figure 3 depicts the time to solve the graphs generated with $n = 15$ vertices, displaying the geometric mean, shifted by 1, of the complete solution time for four different algorithms. BandB(no warm start) represents the simple branch-and-bound algorithm in Algorithm 3. BandB(no warm start) is the same algorithm, but initialized with a heuristic solution for bounding purposes, provided by MDOC. Benders(single cycle) represents Algorithm 1, where only one cycle is generated each time the subproblem is solved (Algorithm 2 is stopped once the first chordless cycle is identified), and Benders(multi cycle) is the same algorithm but with Algorithm 2 ran until completion.

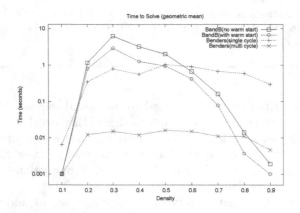

Fig. 3. Time to solve graphs with $n = 15$ vertices

As can be seen in Figure 3, the Benders approach outperforms the simple branch-and-bound algorithm except on those instances that are solved very quickly. Within 10,000,000 search nodes not all instances with $n = 20$ and above are solved by the simple branch-and-bound algorithm and so the plot is only generated for $n = 15$.

Even more elucidating of the difference in computational time is the number of instances that are solved within 30 seconds, depicted in Figure 4. For the same algorithms as above, this plot shows that within 30 seconds each of the instances with $n = 20$ are solved by Benders(multi cycle) (the maximum time on this set of instances for this technique is 1.13 seconds with mean of 0.053 seconds). In the alternative methods, there are many instances which remain unsolved in this time horizon, and several remaining unsolved even after 1800 seconds.

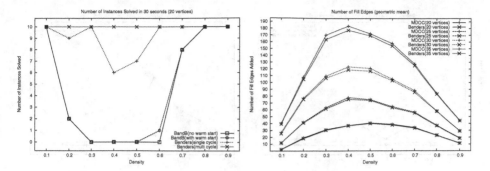

Fig. 4. Number of solved instances   **Fig. 5.** Size of chordal completions

Improvement Over MDOC. The heuristic, MDOC, works surprisingly well in practice and hence is heavily implemented in practice. In this section the difference between the solution provided by MDOC and the Benders approach (Benders(multi cycle)) are provided in order to evaluate the importance of searching for optimal solution, compared with a standard, simple, and powerful heuristic.

In general, as reported in the literature, the solutions outputted by MDOC are of high quality. Figure 5 shows the comparison between the optimal solutions found by the Benders approach with the solutions obtained by MCOD for $n = 20, 25, 30, 35$. As can be seen in this figure, the gap between the heuristic solution and the optimal solution is not too substantial, however the gap grows as the graphs increase in size.

To better see the differences in the chordal completions, consider Figure 6, where plots are generated for this data, for $n = 20, 25, 30, 35$, individually. In addition to the mean, the minimum and maximum difference per configuration is depicted. This plot more readily shows that as n grows the gap between the heuristic solution and the optimal solution grows.

This of course comes at the expense of extra computation time. The time necessary to run the heuristic is a fraction of a second, compared to the time to run the Benders algorithm, reported in Table 1, although solutions are provided with optimality guarantees. This table reports the numerical values depicted in Figure 6 along with the times to solve the MCCP using the Benders approach described in this paper. n and d are the number of vertices and density in the random graphs tested (10 instances per row in the table). MinFillDiff, MeanFillDiff, and MaxFillDiff report the minimum, mean, and maximum difference in the number of edges in the chordal completions calculated using MDOC and the Benders approach, respectively. MinBendersTime, MeanBenders Time, and MaxBendersTime report the minimum, mean, and maximum time to solve the instances using the Benders approach, respectively.

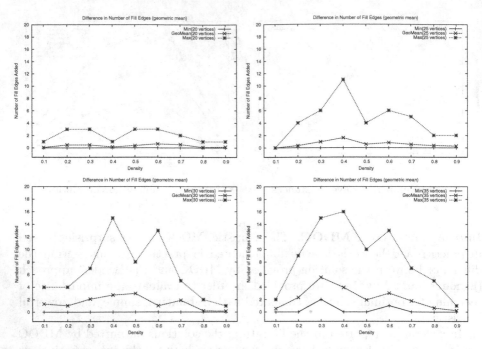

Fig. 6. Difference in number of fill edges added, $n = 20, 25, 30, 35$

Table 1. Difference in number of edges resulting from chordal completions using Benders versus MDOC ; Benders time

n	d	MinFillDiff	MeanFillDiff	MaxFillDiff	MinBendersTime	MeanBendersTime	MaxBendersTime
20	0.1	0	0.07	1	0.00	0.00	0.01
	0.2	0	0.47	3	0.02	0.04	0.13
	0.3	0	0.47	3	0.02	0.04	0.07
	0.4	0	0.15	1	0.02	0.15	1.13
	0.5	0	0.37	3	0.03	0.11	0.37
	0.6	0	0.64	3	0.02	0.05	0.10
	0.7	0	0.53	2	0.03	0.05	0.10
	0.8	0	0.07	1	0.02	0.03	0.04
	0.9	0	0.15	1	0.01	0.01	0.02
25	0.1	0	0.00	0	0.01	0.02	0.04
	0.2	0	0.35	4	0.03	0.87	24.04
	0.3	0	0.96	6	0.08	1.00	5.28
	0.4	0	1.61	11	0.17	1.05	2.72
	0.5	0	0.57	4	0.10	0.42	1.83
	0.6	0	0.83	6	0.05	0.11	0.33
	0.7	0	0.53	5	0.08	0.17	0.66
	0.8	0	0.37	2	0.04	0.06	0.12
	0.9	0	0.28	2	0.02	0.03	0.04
30	0.1	0	1.31	4	0.02	0.87	5.36
	0.2	0	1.03	4	0.28	8.05	181.68
	0.3	0	2.09	7	2.85	15.44	371.47
	0.4	0	2.73	15	0.23	6.39	38.70
	0.5	0	2.98	8	0.24	4.91	194.28
	0.6	0	1.02	13	0.24	0.91	2.17
	0.7	0	1.88	5	0.13	0.82	2.60
	0.8	0	0.20	2	0.08	0.23	0.75
	0.9	0	0.15	1	0.03	0.05	0.09
35	0.1	0	0.53	2	0.04	2.90	70.83
	0.2	0	2.32	9	1.31	86.49	1822.97
	0.3	2	5.54	15	2.92	635.38	11782.80
	0.4	0	3.93	16	4.51	91.90	2394.53
	0.5	0	1.94	10	0.75	17.24	624.04
	0.6	1	3.01	13	3.73	15.03	100.23
	0.7	0	1.76	7	0.52	5.39	19.57
	0.8	0	0.64	5	0.24	0.57	1.15
	0.9	0	0.23	1	0.07	0.10	0.19

Table 2. Chordal completion sizes on benchmark graphs ; * indicates that the value is a bound due to the Benders approach timing out in one hour

Instance	$|V|$	$|E|$	BendersTime	BendersValue	MDOCValue
1-FullIns_3	30	100	5.85	80	80
1-FullIns_4	93	593	-	623*	839
2-FullIns_3	52	201	-	206*	273
2-Insertions_3	37	72	-	68*	103
3-FullIns_3	80	346	-	379*	661
3-Insertions_3	56	110	-	102*	198
4-Insertions_3	79	156	-	148*	331
david	87	406	1.99	64	66
mug100_1	100	166	0.56	64	91
mug100_25	100	166	0.66	64	93
mug88_1	88	146	0.37	56	82
mug88_25	88	146	0.52	56	84
myciel3	11	20	0	10	10
myciel4	23	71	0.04	46	46
myciel5	47	236	35.71	196	197
myciel6	95	755	-	713*	753
queen10_10	100	1470	-	2045*	2671
queen5_5	25	160	262.57	93	94
queen6_6	36	290	-	218*	244
queen7_7	49	460	-	402*	444
queen8_12	96	1368	-	1863*	2401
queen8_8	64	728	-	772*	970
queen9_9	81	1056	-	1301*	1664

Benchmark Graphs. Table 2 reports results on benchmark graphs, appearing frequently in the literature in papers which test graph coloring algorithms and other problems as well [28]. The algorithm was tested only on those instances with 100 or fewer vertices that are connected.

The results indicate that there can be a significant gap between what the state-of-the-art minimal chordal completion algorithm returns and the optimal value of the MCCP. The table reports `Instance` (the name of the instance) the number of vertices and edges in the graph, followed by `BendersTime` (the amount of time, in seconds, to solve the instance), `BendersValue` (the optimal solution if the instance was solved within an hour and lower bound otherwise and indicated by an asterisk), and finally `MDOCValue` (the size of the chordal completion obtained by MDOC). Of particular note are the `mug` instances for which the gap between the heuristic solution value and the optimal value can be very substantial.

9 Conclusion

In conclusion, this paper introduces a Benders approach to the minimum chordal completion problem. Computational results indicate that the approach is promising as the algorithm significantly outperforms a simple branch-and-bound algorithm for the problem. In addition the paper reports that the gap between the value obtained using a state-of-the-art heuristic and the formally unknown optimal solutions to random graphs and benchmark instances can be significant.

References

1. Beeri, C., Fagin, R., Maier, D., Yannakakis, M.: On the desirability of acyclic database schemes. J. ACM **30**(3), 479–513 (1983). http://doi.acm.org/10.1145/2402.322389
2. Benders, J.: Partitioning procedures for solving mixed-variables programming problems. Numerische Mathematik **4**(1), 238–252 (1962). http://dx.doi.org/10.1007/BF01386316
3. Berry, A., Bordat, J.P., Heggernes, P., Simonet, G., Villanger, Y.: A wide-range algorithm for minimal triangulation from an arbitrary ordering. Journal of Algorithms **58**(1), 33–66 (2006). http://www.sciencedirect.com/science/article/pii/S0196677404001142
4. Berry, A., Heggernes, P., Simonet, G.: The minimum degree heuristic and the minimal triangulation process. In: Bodlaender, H.L. (ed.) WG 2003. LNCS, vol. 2880, pp. 58–70. Springer, Heidelberg (2003). http://www.dx.doi.org/10.1007/978-3-540-39890-5_6
5. Blair, J.R., Heggernes, P., Telle, J.A.: A practical algorithm for making filled graphs minimal. Theoretical Computer Science **250**(12), 125–141 (2001). http://www.sciencedirect.com/science/article/pii/S0304397599001267
6. Bodlaender, H.L.: A tourist guide through treewidth. Acta Cybernetica **11**, 1–23 (1993)
7. Bodlaender, H.L., Kloks, T., Kratsch, D., Mueller, H.: Treewidth and minimum fill-in on d-trapezoid graphs (1998)
8. Bodlaender, H.L., Heggernes, P., Villanger, Y.: Faster parameterized algorithms for MINIMUM FILL-IN. In: Hong, S.-H., Nagamochi, H., Fukunaga, T. (eds.) ISAAC 2008. LNCS, vol. 5369, pp. 282–293. Springer, Heidelberg (2008). http://dx.doi.org/10.1007/978-3-540-92182-0_27
9. Broersma, H., Dahlhaus, E., Kloks, T.: Algorithms for the treewidth and minimum fill-in of hhd-free graphs. In: Möhring, R.H. (ed.) WG 1997. LNCS, vol. 1335, pp. 109–117. Springer, Heidelberg (1997). http://dx.doi.org/10.1007/BFb0024492
10. Chandran, L.S., Ibarra, L., Ruskey, F., Sawada, J.: Generating and characterizing the perfect elimination orderings of a chordal graph. Theor. Comput. Sci. **307**(2), 303–317 (2003). http://dx.doi.org/10.1016/S0304-3975(03)00221-4
11. Chang, M.S.: Algorithms for maximum matching and minimum fill-in on chordal bipartite graphs. In: Nagamochi, H., Suri, S., Igarashi, Y., Miyano, S., Asano, T. (eds.) ISAAC 1996. LNCS, vol. 1178, pp. 146–155. Springer, Heidelberg (1996). http://dx.doi.org/10.1007/BFb0009490
12. Chung, F., Mumford, D.: Chordal completions of planar graphs. Journal of Combinatorial Theory, Series B **62**(1), 96–106 (1994). http://www.sciencedirect.com/science/article/pii/S0095895684710562
13. Fomin, F.V., Villanger, Y.: Subexponential parameterized algorithm for minimum fill-in. In: Proceedings of the Twenty-third Annual ACM-SIAM Symposium on Discrete Algorithms, SODA 2012, pp. 1737–1746. SIAM (2012), http://dl.acm.org/citation.cfm?id=2095116.2095254
14. Fulkerson, D.R., Gross, O.A.: Incidence matrices and interval graphs. Pacific Journal of Mathematics **15**(3), 835–855 (1965). http://projecteuclid.org/euclid.pjm/1102995572
15. Garey, M.R., Johnson, D.S.: Computers and Intractability: A Guide to the Theory of NP-Completeness. W. H. Freeman & Co., New York (1979)

16. George, A., Liu, W.H.: The evolution of the minimum degree ordering algorithm. SIAM Rev. **31**(1), 1–19 (1989). http://dx.doi.org/10.1137/1031001
17. Golumbic, M., Kaplan, H., Shamir, R.: Graph sandwich problems. Journal of Algorithms **19**(3), 449–473 (1995). http://www.sciencedirect.com/science/article/pii/S0196677485710474
18. Grone, R., Johnson, C.R., Sá, E.M., Wolkowicz, H.: Positive definite completions of partial hermitian matrices. Linear Algebra and its Applications **58**, 109–124 (1984). http://www.sciencedirect.com/science/article/pii/0024379584902076
19. Heggernes, P.: Minimal triangulations of graphs: A survey. Discrete Mathematics **306**(3), 297–317 (2006). http://www.sciencedirect.com/science/article/pii/S0012365X05006060
20. Heggernes, P., Eisenstat, S.C., Kumfert, G., Pothen, A.: The computational complexity of the minimum degree algorithm (2001)
21. Hooker, J., Ottosson, G.: Logic-based benders decomposition. Mathematical Programming **96**(1), 33–60 (2003). http://dx.doi.org/10.1007/s10107-003-0375-9
22. Kaplan, H., Shamir, R., Tarjan, R.E.: Tractability of parameterized completion problems on chordal and interval graphs: Minimum fill-in and physical mapping (extended abstract). SIAM J. Comput **28**, 780–791 (1994)
23. Kim, S., Kojima, M., Mevissen, M., Yamashita, M.: Exploiting sparsity in linear and nonlinear matrix inequalities via positive semidefinite matrix completion. Mathematical Programming **129**(1), 33–68 (2011). http://dx.doi.org/10.1007/s10107-010-0402-6
24. Kloks, T., Kratsch, D., Wong, C.: Minimum fill-in on circle and circular-arc graphs. Journal of Algorithms **28**(2), 272–289 (1998). http://www.sciencedirect.com/science/article/pii/S0196677498909361
25. Lekkeikerker, C., Boland, J.: Representation of a finite graph by a set of intervals on the real line. Fundamenta Mathematicae **51**, 45–64 (1962)
26. Mezzini, M., Moscarini, M.: Simple algorithms for minimal triangulation of a graph and backward selection of a decomposable markov network. Theoretical Computer Science **411**(79), 958–966 (2010). http://www.sciencedirect.com/science/article/pii/S030439750900735X
27. Nakata, K., Fujisawa, K., Fukuda, M., Kojima, M., Murota, K.: Exploiting sparsity in semidefinite programming via matrix completion ii: implementation and numerical results. Mathematical Programming **95**(2), 303–327 (2003). http://dx.doi.org/10.1007/s10107-002-0351-9
28. Nguyen, T.H., Bui, T.: Graph coloring benchmark instances. http://www.cs.hbg.psu.edu/txn131/graphcoloring.html (accessed: 14 July 2014)
29. Peyton, B.W.: Minimal orderings revisited. SIAM J. Matrix Anal. Appl. **23**(1), 271–294 (2001). http://dx.doi.org/10.1137/S089547989936443X
30. Rose, D., Tarjan, R., Lueker, G.: Algorithmic aspects of vertex elimination on graphs. SIAM Journal on Computing **5**(2), 266–283 (1976). http://dx.doi.org/10.1137/0205021
31. Rose, D.J.: A graph-theoretic study of the numerical solution of sparse positive definite systems of linear equations. In: Graph Theory and Computing, pp. 183–217. Academic Press (1972), http://www.sciencedirect.com/science/article/pii/B9781483231877500180
32. Sokhn, N., Baltensperger, R., Bersier, L.-F., Hennebert, J., Ultes-Nitsche, U.: Identification of chordless cycles in ecological networks. In: Glass, K., Colbaugh, R., Ormerod, P., Tsao, J. (eds.) Complex 2012. LNICST, vol. 126, pp. 316–324. Springer, Heidelberg (2013)

33. Spinrad, J., Brandstdt, A., Stewart, L.: Bipartite permutation graphs. Discrete Applied Mathematics **18**(3), 279–292 (1987). http://www.sciencedirect.com/science/article/pii/S0166218X87800033
34. Tarjan, R.E., Yannakakis, M.: Simple linear-time algorithms to test chordality of graphs, test acyclicity of hypergraphs, and selectively reduce acyclic hypergraphs. SIAM J. Comput. **13**(3), 566–579 (1984). http://dx.doi.org/10.1137/0213035
35. Uno, T., Satoh, H.: An efficient algorithm for enumerating chordless cycles and chordless paths. CoRR abs/1404.7610 (2014), http://arxiv.org/abs/1404.7610
36. Yannakakis, M.: Computing the minimum fill-in is np-complete. SIAM Journal on Algebraic Discrete Methods **2**(1), 77–79 (1981)

MaxSAT-Based Scheduling of B2B Meetings

Miquel Bofill[(✉)], Marc Garcia, Josep Suy, and Mateu Villaret

Departament d'Informàtica, Matemàtica Aplicada i Estadística,
Universitat de Girona, Girona, Spain
{mbofill,mgarciao,suy,villaret}@imae.udg.edu

Abstract. In this work we propose a MaxSAT formulation for the problem of scheduling business-to-business meetings. We identify some implied constraints and provide distinct encodings of the used cardinality constraints. The experimental results show that the proposed technique outperforms previous existing approaches on this problem.

1 Introduction

In recent years, solving combinatorial problems by encoding them into propositional formulae and proving their satisfiability has been proved to be a robust approach. Although the field of automated timetabling is dominated by local search heuristic methods, SAT and MaxSAT-based methods have been successfully applied to curriculum-based timetabling problems recently [2,5].

In this work we propose some MaxSAT encodings for the problem of scheduling business-to-business (B2B) bilateral meetings. Those meeting sessions occur in events from several fields like sports, research, etc. and aim to facilitate participants with similar interests meeting each other. We consider the generation of timetables for B2B meetings in the particular setting of the *Forum of the UdG's Science and Technology Park*.[1] The goal of this forum is to be a technological marketplace in Girona, by bringing the opportunity to companies, research groups, investors, etc., to find future business partnerships.

In the scheduling of B2B meetings it is desirable to avoid unnecessary idle time periods between meetings, and to be fair in such minimization, i.e., to avoid big differences in the number of idle time periods among participants. Experience shows that idle time periods may led some participants to leave the event, dismissing later scheduled meetings. We propose to face this optimization problem by encoding it as a partial MaxSAT formula [11], where some clauses are marked as hard whilst others are marked as soft, and the goal is to find an assignment to the variables that satisfies all hard clauses and falsifies the minimum number of soft clauses. In our case, the falsification of a soft clause will represent the existence of an idle time period for some participant.

M. Bofill, J. Suy, and M. Villaret—Supported by the Spanish Ministry of Science and Innovation (project TIN2012-33042).
[1] http://www.parcudg.com

© Springer International Publishing Switzerland 2015
L. Michel (Ed.): CPAIOR 2015, LNCS 9075, pp. 65–73, 2015.
DOI: 10.1007/978-3-319-18008-3_5

As far as we know, there are not many works dealing with this problem. In [7] we can find a system that is used by the company *piranha womex AG* for computing matchmaking schedules in several fairs. That system differs from ours in some aspects: e.g., it considers neither forbidden slots nor fairness. In [3] the authors proposed a Constraint Programming (CP) model and a pseudo-Boolean (PB) model, and provided performance results for basic configurations of several solvers. Among the model and solver combinations considered, the CP model with a SMT solver, and the PseudoBoolean model with a SAT-based PB solver, were shown to be the most robust. This fact motivates us to go one step further with SAT-based technology by providing a direct MaxSAT encoding for this problem in this paper. We report on experiments showing that a basic MaxSAT encoding outperforms previous results. We show that even better results can be obtained with state-of-the-art encodings of cardinality constraints and some improvements and extensions of the initial encoding. As test suite we use (and provide) the industrial instances of previous editions of the Forum and some crafted modifications of those.

2   The B2B Problem

Here we define the problem at hand. For more details on the nature of the problem and the instances from the Forum of the UdG's Science and Technology Park, see [3].

Definition 1 (B2BSOP-*d*). *Given a set of participants, a list of time slots, a set of available locations (tables) and a set of meetings between pairs of participants, where for each participant there can be forbidden time slots and meetings may be required to be held in morning or afternoon slots, the goal of the "B2B Scheduling Optimization Problem with homogeneity d" is to find a total mapping from the meetings to time slots and locations such that:*

- *At most one meeting involving the same participant is scheduled in each time slot.*
- *No meeting is scheduled in a forbidden time slot for any of its participants.*
- *Each meeting having a morning or afternoon slot requirement is scheduled in a time slot of the appropriate interval.*
- *The difference between the number of idle time periods of each pair of participants is at most d, where by an idle time period we refer to a group of idle time slots between two successive meetings involving the same participant.*
- *The total number of idle time periods is minimized.*

3   Encodings

3.1   MaxSAT Base Encoding for the B2BSOP-*d*

Parameters. Each instance is defined by the following parameters.

nMeetings: number of meetings
nTimeSlots: number of available time slots
nMorningSlots: number of morning time slots
nTables: number of available locations
nParticipants: number of participants
morningMeetings: subset of $\{1, \ldots, nMeetings\}$ to be scheduled in morning slots
afternoonMeetings: subset of $\{1, \ldots, nMeetings\}$ to be scheduled in afternoon slots
meetings, function from $\{1, \ldots, nParticipants\}$ to $2^{\{1, \ldots, nMeetings\}}$: set of meetings involving each participant
forbidden, function from $\{1, \ldots, nParticipants\}$ to $2^{\{1, \ldots, nTimeSlots\}}$: set of forbidden time slots for each participant

Variables. We define the following propositional variables.

$schedule_{i,j}$: meeting i is held in time slot j
$usedSlot_{p,j}$: participant p has a meeting scheduled in time slot j
$fromSlot_{p,j}$: participant p has a meeting scheduled at, or before, time slot j
$endHole_{p,j}$: participant p has an idle time period finishing at time slot j
$max_1, \ldots, max_{\lfloor (nTimeSlots-1)/2 \rfloor}$ and $min_1, \ldots, min_{\lfloor (nTimeSlots-1)/2 \rfloor}$:
unary representation of an upper bound and a lower bound of the maximum and minimum number of idle time periods among all participants, respectively. Note that there can be at most $\lfloor (nTimeSlots - 1)/2 \rfloor$ idle time periods per participant. By restricting the difference between these variables to be less than a certain value, we will enforce homogeneity of solutions (Constraints (16) to (21)).

We also use some auxiliary variables that will be introduced when needed.

Constraints. All constraints except (15) are hard. To help readability we define $M = \{1, \ldots, nMeetings\}$, $T = \{1, \ldots, nTimeSlots\}$, $P = \{1, \ldots, nParticipants\}$.

– *At most one meeting involving the same participant is scheduled in each time slot.*

$$atMost(1, \{schedule_{i,j} \mid i \in meetings(p)\}) \qquad \forall p \in P, j \in T \qquad (1)$$

– *No meeting is scheduled in a forbidden time slot for any of its participants.*

$$\bigwedge_{i \in meetings(p),\, j \in forbidden(p)} \neg schedule_{i,j} \qquad \forall p \in P \qquad (2)$$

- *Each meeting having a morning or afternoon slot requirement is scheduled in a time slot of the appropriate interval.*

$$exactly(1, \{schedule_{i,j} \mid j \in 1..nMorningSlots\}) \qquad \forall i \in morningMeetings \tag{3}$$

$$\neg schedule_{i,j} \qquad \begin{array}{l} \forall i \in morningMeetings \\ \forall j \in nMorningSlots + 1..nTimeSlots \end{array} \tag{4}$$

$$exactly(1, \{schedule_{i,j} \mid j \in nMorningSlots + 1..nTimeSlots\}) \\ \forall i \in afternoonMeetings \tag{5}$$

$$\neg schedule_{i,j} \qquad \begin{array}{l} \forall i \in afternoonMeetings \\ \forall j \in 1..nMorningSlots \end{array} \tag{6}$$

$$exactly(1, \{schedule_{i,j} \mid j \in T\}) \qquad \forall i \in M \setminus (morningMeetings \cup \\ afternoonMeetings) \tag{7}$$

- *At most one meeting is scheduled in a given time slot and location.*

$$atMost(nTables, \{schedule_{i,j} \mid i \in M\}) \qquad \forall j \in T \tag{8}$$

Note that with Constraints (3) to (8) we get a total mapping from the meetings to time slots and locations.

In order to be able to minimize the number of idle time periods we introduce channeling constraints between the variables *schedule*, *usedSlot* and *fromSlot*.

- *If a meeting is scheduled in a certain time slot, then that time slot is used by both participants of the meeting.*

$$schedule_{i,j} \rightarrow (usedSlot_{p_i^1,j} \wedge usedSlot_{p_i^2,j}) \qquad \forall i \in M, j \in T$$
$$\text{where } p_i^1 \text{ and } p_i^2 \text{ are the participants of meeting } i. \tag{9}$$

In the reverse direction, if a time slot is used by some participant, then one of the meetings of that participant is scheduled in that time slot.

$$usedSlot_{p,j} \rightarrow \bigvee_{i \in meetings(p)} schedule_{i,j} \qquad \forall p \in P, j \in T \tag{10}$$

- *For each participant p and time slot j, $fromSlot_{p,j}$ is true if and only if participant p has had a meeting at or before time slot j.*

$$\neg usedSlot_{p,1} \rightarrow \neg fromSlot_{p,1} \qquad \forall p \in P \tag{11}$$

$$(\neg fromSlot_{p,j-1} \wedge \neg usedSlot_{p,j}) \rightarrow \neg fromSlot_{p,j} \qquad \forall p \in P, j \in T \setminus \{1\} \tag{12}$$

$$usedSlot_{p,j} \rightarrow fromSlot_{p,j} \qquad \forall p \in P, j \in T \tag{13}$$

$$fromSlot_{p,j-1} \rightarrow fromSlot_{p,j} \qquad \forall p \in P, j \in T \setminus \{1\} \tag{14}$$

Optimization. Minimization of the number of idle time periods is achieved by means of soft constraints (except for the case of a cardinality network based encoding that we describe in Subsection 3.3).

– [Soft constraints] *If some participant does not have any meeting in a certain time slot, but it has had some meeting before, then she does not have any meeting in the following time slot.*

$$(\neg usedSlot_{p,j} \wedge fromSlot_{p,j}) \rightarrow \neg usedSlot_{p,j+1} \quad \forall p \in P, j \in T \backslash \{nTimeSlots\}$$
$$(15)$$

We claim that, with these constraints, an optimal solution will be one having the least number of idle time periods. Note that, for each participant, each meeting following some idle time period increases the cost by 1.

Remark 1. If we were just considering optimization, Constraints (11) and (12) would not be necessary, since minimization of the number of idle time periods would force the value of $fromSlot_{p,j}$ to be false for every participant p and time slot j previous to the first meeting of p. However, since we are also seeking for homogeneity, these constraints are mandatory. Without them, the value of $fromSlot_{p,j}$ could be set to true for time slots j previous to the first meeting of p, inducing a fake idle time period in order to satisfy the (hard) homogeneity constraints defined below. Constraints (11) and (12) were missing by mistake in [3].

Homogeneity. We reify the violation of soft constraints in order to count the number of idle time periods of each participant. This will allow us to find the maximum and minimum number of idle time periods among all participants, and to enforce homogeneity by bounding their difference.

– $endHole_{p,j}$ *is true if and only if participant p has an idle time period finishing at time slot j.*

$$endHole_{p,j} \leftrightarrow \neg\left((\neg usedSlot_{p,j} \wedge fromSlot_{p,j}) \rightarrow \neg usedSlot_{p,j+1}\right)$$
$$\forall p \in P, j \in T \backslash \{nTimeSlots\} \quad (16)$$

– $sortedHole_{p,1}, \ldots, sortedHole_{p,nTimeSlots}$ *are the unary representation of the number of idle time periods of each participant p.*

$$sortingNetwork([endHole_{p,j} \mid j \in T], [sortedHole_{p,j} \mid j \in T]) \quad \forall p \in P$$
$$(17)$$

– $max_1, \ldots, max_{\lfloor(nTimeSlots-1)/2\rfloor}$ *and* $min_1, \ldots, min_{\lfloor(nTimeSlots-1)/2\rfloor}$ *are (an approximation to) the unary representation of the maximum and minimum number of idle time periods among all participants, respectively.*

$$sortedHole_{p,j} \rightarrow max_j \quad \forall p \in P, j \in 1..\lfloor(nTimeSlots - 1)/2\rfloor \quad (18)$$
$$\neg sortedHole_{p,j} \rightarrow \neg min_j \quad \forall p \in P, j \in 1..\lfloor(nTimeSlots - 1)/2\rfloor \quad (19)$$

Constraints (18) and (19) are not enough to ensure that the *max* and *min* variables exactly represent the maximum and minimum number of idle time periods among all participants. However, together with Constraints (20) and (21), they suffice to soundly enforce the required homogeneity degree.

- *The difference between the maximum and minimum number of idle time periods can be at most d (in our setting the chosen number was 2).*

$$dif_j \leftrightarrow min_j \; XOR \; max_j \qquad \forall j \in 1..\lfloor (nTimeSlots - 1)/2 \rfloor \qquad (20)$$

$$atMost(d, \{dif_j \mid j \in 1..\lfloor (nTimeSlots - 1)/2 \rfloor\}) \qquad (21)$$

3.2   Extended Encoding

Implied Constraints. We have identified the following implied constraints.

- *The number of meetings of a participant p as derived from usedSlot$_{p,j}$ variables must match the total number of meetings of p.*

$$exactly(|meetings(p)|, \{usedSlot_{p,j} \mid j \in T\}) \qquad \forall p \in P \qquad (22)$$

- *The number of participants having a meeting in a given time slot is bounded by twice the number of available locations.*

$$atMost(2 \times nTables, \{usedSlot_{p,j} \mid p \in P\}) \qquad \forall j \in T \qquad (23)$$

Symmetry Breaking. With respect to symmetry breaking, the model implicitly eliminates possible table symmetries, since only the number of tables occupied is considered. Unfortunately, removing time symmetries in the presence of participants' forbidden time slots seems not to be feasible. However, we can break some time symmetries when there are no forbidden time slots and the meetings have neither morning nor afternoon slot requirements (there are several instances with these characteristics). Note that since we are minimizing the number of idle time periods, we cannot soundly break time symmetries by simply fixing a priori an ordering of meetings. Instead, what we do is to force some ordering in the "matrix" of *usedSlot* variables as follows, assuming an even number of time-slots and the existence of a participant with an odd number of meetings.[2]

- *For some participant p with an odd number of meetings we force the number of meetings of p taking place in the first half of time slots to be odd.*

$$((\ldots (usedSlot_{p,1} \; XOR \; usedSlot_{p,2}) \; \ldots) \; XOR \; usedSlot_{p,\lfloor nTimeSlots/2 \rfloor})$$
$$(24)$$

3.3   Encoding of Global Constraints

The cardinality constraints stating that at most (*atMost*) or exactly (*exactly*) k of a given set of variables must be true have been encoded in several ways. Similarly for the *sortingNetwork* constraint, which corresponds to a sorting network on a set of Boolean variables.

---

[2] All instances considered are like this.

Naïve Encoding.

- $atMost(1, \_)$: quadratic number of pairwise mutex clauses.
- $exactly(1, \_)$: $atMost(1, \_)$ plus a clause (disjunction) with all the involved variables, for the "at least" part of the constraint.
- $sortingNetwork$: odd-even sorting network.
- $atMost(k, \_)$: naïve sequential unary counter [12].

Cardinality Networks based Encoding.

- $atMost(1, \_)$: quadratic number of pairwise mutex clauses.
- $exactly(1, \_)$: commander-variable encoding [9].
- $exactly(k, \_)$, $sortingNetwork$ and $atMost(k, \_)$: cardinality networks [1].

By using cardinality networks we can deal with soft constraints in a more clever way: instead of soft constraints (15), we post as soft constraint the negation of each "output variable" $sortedHole_{p,j}$ of the $sortingNetwork$ corresponding to constraint (17). This way, knowing that each participant will have at most $\lfloor (nTimeSlots - 1)/2 \rfloor$ idle time periods, we can reduce the number of soft constraints, as well as the number of $sortedHole_{p,j}$ variables of each participant, to a half.

4   Experiments, Conclusions and Future Work

In this section we compare the performance of the state-of-the-art MaxSAT solver $QMaxSat14.04auto\text{-}g3$ [10] using the proposed base model and a naïve encoding of the global constraints, with the performance of the best known method and solver for each instance in [3]. We also show how using cardinality networks for the global constraints, and extending the model with implied constraints and symmetry breaking, we can significantly improve the solving time. We use the same nine instances that the authors used in [3], plus new eleven instances.[3] Among all there are five industrial instances (the ones without $craf$ annotation); the rest have been crafted from those by increasing the number of meetings, reducing the number of locations and removing the forbidden time slots.

All experiments have been run using the default options of each solver, on Intel® Xeon™CPU@3.1GHz machines, under CentOS release 6.3, kernel 2.6.32. Table 1 summarizes the results obtained. Only instances named tic do not contain forbidden time slots nor morning and afternoon preferences, hence symmetry related experiments are only reported for those. Column named **best known** shows the best results obtained in [3], where PB $clasp$ and PB $cplex$ refer to a pseudo-Boolean model solved with clasp 3.1.0 [8] and IBM ILOG CPLEX 12.6, respectively, and CP $sbdd$ refers to a CP like model solved with

---

[3] All instances can be found in http://imae.udg.edu/recerca/lap/simply/. Results from [3] have been updated according to Remark 1.

WSimply using shared BDD optimization [4] and Yices 1.0.33 [6] as SMT solver. Columns **naïve** and **cardinal** show the results using the naïve and cardinality network based encodings of our base model, respectively. Columns **imp1**, **imp2**, **imp1+2** and **imp1+2+sym** show the results for the cardinality networks based encoding using implied constraints 1, 2, both, and both with symmetry breaking, respectively. The three numbers below the names of each instance are: the ratio between the median of meetings per participant and *nTimeSlots*, the ratio between *nTables* and *nParticipants*, and the ratio between the number of meetings to schedule and the available slots (*nTables* × *nTimeSlots*).

Table 1. Solving time (in seconds) and optimum found (number of idle time periods) per instance and solver. TO stands for 2 hours timeout. For aborted executions we report the (sub)optimum found if the solver reported any.

instance	best known		naïve		cardinal		imp1		imp2		imp1+2		imp1+2+sym	
forum-13 (0.20, 0.40, 0.52)	153.1 (PB clasp)	0	24.8	0	20.5	0	13.4	0	25.2	0	18.3	0	-	-
forum-13crafb (0.24, 0.36, 0.66)	TO (CP sbdd)	12	1492.7	6	83.4	6	82.4	6	83.1	6	81.2	6	-	-
forum-13crafc (0.20, 0.34, 0.61)	TO (CP sbdd)	20	116.3	1	1872.4	1	1661.3	1	1800.5	1	1300.2	1	-	-
forum-14 (0.35, 0.56, 0.62)	TO (CP sbdd)	7	TO	-	431.2	2	349.1	2	409.2	2	240.2	2	-	-
forumt-14 (0.79, 0.90, 0.87)	-	-	21.1	5	8.0	5	8.5	5	11.9	5	10.2	5	-	-
forumt-14crafc (0.79, 0.83, 0.94)	-	-	148.9	5	32.7	5	28.8	5	33.1	5	31.5	5	-	-
forumt-14crafd (0.78, 0.83, 0.94)	-	-	84.9	4	32.4	4	26.6	4	37.1	4	35.5	4	-	-
forumt-14crafe (0.78, 0.80, 0.98)	-	-	TO	-	95.2	5	78.1	5	105.2	5	94.7	5	-	-
ticf-13crafa (0.21, 0.40, 0.52)	-	-	21.2	0	24.6	0	15.0	0	45.9	0	35.9	0	-	-
ticf-13crafb (0.51, 0.36, 0.66)	-	-	3866.1	3	118.3	3	117.3	3	111.3	3	114.2	3	-	-
ticf-13crafc (0.21, 0.34, 0.61)	-	-	309.4	1	574.2	1	562.3	1	416.9	1	432.3	1	-	-
ticf-14crafa (0.35, 0.56, 0.62)	-	-	TO	-	TO	-	1532.8	0	2044.1	0	1339.6	0	-	-
tic-12 (0.74, 1.00, 0.74)	0.2 (PB clasp)	0	0.2	0	0.2	0	0.3	0	0.2	0	0.2	0	0.4	0
tic-12crafc (0.74, 0.76, 0.97)	53.2 (PB cplex)	0	7.8	0	4.1	0	3.1	0	2.5	0	2.6	0	3.4	0
tic-13 (0.76, 0.89, 0.85)	3.0 (PB clasp)	0	18.4	0	5.9	0	4.1	0	4.6	0	4.2	0	5.7	0
tic-13crafb (0.80, 0.89, 0.87)	2.1 (PB clasp)	0	3.6	0	2.4	0	2.6	0	7.1	0	5.5	0	4.1	0
tic-13crafc (0.76, 0.80, 0.94)	TO (PB cplex)	4	TO	4	25.9	4	19.1	4	25.2	4	23.9	4	26.1	4
tic-14crafa (0.79, 0.90, 0.87)	-	-	30.0	0	16.3	0	10.2	0	24.4	0	16.4	0	14.2	0
tic-14crafc (0.79, 0.83, 0.94)	-	-	740.0	0	49.3	0	45.1	0	45.7	0	44.5	0	56.8	0
tic-14crafd (0.79, 0.83, 0.94)	-	-	190.7	0	35.2	0	47.9	0	32.5	0	34.9	0	53.2	0

From the results reported we can extract the following conclusions: a) our base MaxSAT model with a naïve encoding outperforms all approaches considered in [3]; b) the cardinality network based encoding of our base model outperforms the naïve encoding; c) using implied constraints is in general beneficial;

d) when the amount of information provided by the implied constraints is elevated is when really pays off to use them: in particular, for implied constraint 1, this happens when the ratio between the median of meetings per participant and $nTimeSlots$ is low; for implied constraint 2, this happens when the ratio between $nTables$ and $nParticipants$ is low; e) the use of symmetry breaking seems not to really help (in fact we think that we need some more hard instances to appreciate its possible benefits).

As future work we plan to find some more implied constraints and to improve symmetry breaking. We also plan to develop a portfolio with all these encodings, and to deeply compare with other solving techniques.

References

1. Abío, I., Nieuwenhuis, R., Oliveras, A., Rodríguez-Carbonell, E.: A parametric approach for smaller and better encodings of cardinality constraints. In: Schulte, C. (ed.) CP 2013. LNCS, vol. 8124, pp. 80–96. Springer, Heidelberg (2013)
2. Achá, R.J.A., Nieuwenhuis, R.: Curriculum-based course timetabling with SAT and MaxSAT. Annals OR **218**(1), 71–91 (2014)
3. Bofill, M., Espasa, J., Garcia, M., Palahí, M., Suy, J., Villaret, M.: Scheduling B2B meetings. In: O'Sullivan, B. (ed.) CP 2014. LNCS, vol. 8656, pp. 781–796. Springer, Heidelberg (2014)
4. Bofill, M., Palahí, M., Suy, J., Villaret, M.: Solving intensional weighted CSPs by incremental optimization with BDDs. In: O'Sullivan, B. (ed.) CP 2014. LNCS, vol. 8656, pp. 207–223. Springer, Heidelberg (2014)
5. Demirović, E., Musliu, N.: Modeling high school timetablingas partial weighted MaxSAT. In: LaSh: The 4th Workshop on Logic and Search (a SAT / ICLP workshop at FLoC 2014). Austria, Vienna (2014)
6. Dutertre, B., de Moura, L.: The Yices SMT solver. Tool paper, August 2006. http://yices.csl.sri.com/tool-paper.pdf (Accessed 23 November 2014)
7. Gebser, M., Glase, T., Sabuncu, O., Schaub, T.: Matchmaking with answer set programming. In: Cabalar, P., Son, T.C. (eds.) LPNMR 2013. LNCS, vol. 8148, pp. 342–347. Springer, Heidelberg (2013)
8. Gebser, M., Kaufmann, B., Neumann, A., Schaub, T.: *clasp*: A conflict-driven answer set solver. In: Baral, C., Brewka, G., Schlipf, J. (eds.) LPNMR 2007. LNCS (LNAI), vol. 4483, pp. 260–265. Springer, Heidelberg (2007)
9. Klieber, W., Kwon, G.: Efficient CNF encoding for selecting 1 from N objects. In: Fourth Workshop on Constraints in Formal Verification, CFV (2007)
10. Koshimura, M., Zhang, T., Fujita, H., Hasegawa, R.: QMaxSAT: A Partial MaxSAT Solver. Journal on Satisfiability, Boolean Modeling and Computation **8**(1/2), 95–100 (2012)
11. Li, C.M., Manyà, F.: Handbook of Satisfiability, chapter MaxSAT, Hard and Soft Constraints, pp. 613–631. IOS Press (2009)
12. Sinz, C.: Towards an optimal cnf encoding of boolean cardinality constraints. In: van Beek, P. (ed.) CP 2005. LNCS, vol. 3709, pp. 827–831. Springer, Heidelberg (2005)

Embedding Decision Trees and Random Forests in Constraint Programming

Alessio Bonfietti[✉], Michele Lombardi, and Michela Milano

DISI, University of Bologna, Bologna, Italy
{alessio.bonfietti,michele.lombardi2,michela.milano}@unibo.it

Abstract. In past papers, we have introduced Empirical Model Learning (EML) as a method to enable Combinatorial Optimization on real world systems that are impervious to classical modeling approaches. The core idea in EML consists in embedding a Machine Learning model in a traditional combinatorial model. So far, the method has been demonstrated by using Neural Networks and Constraint Programming (CP). In this paper we add one more technique to the EML arsenal, by devising methods to embed Decision Trees (DTs) in CP. In particular, we propose three approaches: 1) a simple encoding based on meta-constraints; 2) a method using attribute discretization and a global TABLE constraint; 3) an approach based on converting a DT into a Multi-valued Decision Diagram, which is then fed to an MDD constraint. We finally show how to embed in CP a Random Forest, a powerful type of ensemble classifier based on DTs. The proposed methods are compared in an experimental evaluation, highlighting their strengths and their weaknesses.

1 Introduction

Combinatorial Optimization methods have been successfully applied to a broad range of industrial problems. Many of such approaches rely on the availability of some declarative model describing decisions, constraints on these decisions, their cost and their impact on the considered system. In short, they rely on an accurate problem model. However, in some application domains, the model is either not fully known, or it is described in a way that is not useful for combinatorial optimization. As an example, for many domains there are predictive models to forecast the temporal dynamic of a target system via differential equations, but those are unfortunately impossible to insert into a combinatorial optimization model without incurring in computational issues.

In these cases, it is likely that the domain expert proposes some heuristic knowledge on the problem that is a (non measurable) approximation of the effect that some decisions have on the system dynamic. We propose here an alternative approach, which is an instantiation of a general method called Empirical Model Learning that we proposed in [2]. We aim at learning part of the combinatorial optimization model and to embed the learned piece of knowledge in the combinatorial model itself. In this way, we have two advantages: the first is that we

L. Michel (Ed.): CPAIOR 2015, LNCS 9075, pp. 74–90, 2015.
DOI: 10.1007/978-3-319-18008-3_6

can use this knowledge to reduce the search tree and the second is that we know the accuracy that we have obtained in the process.

In this paper, we consider the problem of embedding in a CP model two types of tree-based classifiers from Machine Learning, namely Decision Trees and Random Forests. Formally, a classifier is a function f mapping a tuple of values for a set of discrete or numeric *attributes* to a discrete *class*. We can embed a classifier in CP by introducing a vector of variables $\overline{x}$ to represent the attributes and a variable y to represent the class. Then we need to find a set of constraints such that they *guarantee* and *enforce some degree of consistency* on the relation:

$$\overline{x} = \overline{v} \wedge y = w \Leftrightarrow f(\overline{v}) = w \tag{1}$$

In other words, an effective embedding technique does not simply act as a function evaluator. Rather, it is capable to narrow the set of possible values for y given the current domain of $\overline{x}$ and vice-versa, i.e. it is capable of performing domain filtering.

Here, we show three CP encoding techniques for Decision Trees and Random Forests, in particular: 1) an approach based on rules and modeled via meta-constraints; 2) an encoding based the discretization of numeric attributes and a global TABLE constraint; 3) another approach relying on attribute discretization, but making use of an MDD constraint instead of TABLE. Each of the three approaches has its own merits: the rule-based encoding has the best scalability, but provides the weakest propagation. The TABLE and MDD approaches are both capable of enforcing GAC, but may suffer from scalability issues when dealing with large and complex trees.

We experiment our methods on a thermal-aware workload dispatching problem over an experimental multicore CPU by Intel, called Single-chip Cloud Computer (SCC, see [19]). Our goal is to map jobs so as to maximize the number of cores operating at high efficiency. The efficiency of each core depends on a number of complex factors, making it impossible to assess the effect of a job mapping via a traditional, expert designed, model. Hence, we obtained a model approximation by learning a set of Decision Trees (or Random Forests), each one trained to predict if a specific core will have high (class 1) or low (class 0) efficiency given a specific workload. We compare the behavior of the proposed techniques in a variety of conditions and we show how the EML approach is capable of providing improvements over a powerful Local Search method.

2 Background

In this section we discuss the basics of Decision Trees (DT) and Random Forests (RF), so as to establish the background to present our encoding techniques. A brief review of works that combine optimization and Machine Learning is provided in Section 5.

Decision Trees are a type of Machine Learning model typically employed for classification tasks. Each leaf of a Decision Tree (DT) is labeled with a *class*.

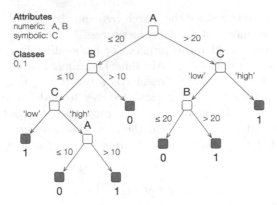

Fig. 1. Example of a simple Decision Tree

Each node is labeled with one of a set of numeric or symbolic *attributes* that are used to described the DT input. The outgoing branches of a node are labeled with conditions over its attribute. The conditions are such that they form a partition of the attribute domain: in particular, branches over symbolic attributes are labeled with a set of symbolic values, while branches over numeric attributes are labeled with a splitting condition such as $x_i \leq \theta$ or $x_i > \theta$. An example of a simple Decision Tree is depicted in Figure 1.

Trees can be learned in a greedy manner by procedures like the C4.5 algorithm [26]. The learning process starts from a set of *examples*, i.e. (tuples of attribute values, associated with a class). Then, this training set is recursively split into subsets, according to the attribute that makes the classes in the subsets most homogeneous (e.g. that achieves minimum Gini index maximum Information Entropy). The recursive process terminates when the remaining subsets are sufficiently pure to conclude with a classification, which is then associated with a leaf of the tree. A new example is classified by starting from the root node and traversing the tree, always taking the branches whose condition is satisfied by the values of the example attributes. For a fully specified example, this process will lead to a single leaf, corresponding to the predicted class.

Decision trees are quick to train and easy to understand, they can provide class probabilities and error bars, they can handle wrong or missing attribute values. On the downside, they require relatively large training sets in order to be effective and they do not always reach satisfactory accuracy levels.

The last drawback can be overcome by using DTs in an ensemble learning method, leading to Random Forests [10,18]. A Random Forest is a set of DTs and the forest output is the statistical mode of the classifications made by its components, i.e. the class predicted by the majority of the trees. Each DT is defined over a random subset of attributes and trained on a subset of the original examples (bagging). Additionally, randomization can also be employed for selecting the splitting value at each tree root. Such extensive use of randomization aims at breaking correlations between the trees, which greatly increases the prediction ability of the forest. Random Forests are widely considered among the most powerful Machine Learning models.

3 Embedding Decision Trees and Random Forests in CP

In the following paragraphs we describe several techniques to guarantee the satisfaction of and to enforce consistency on Equation (1) from Section 1, i.e.:

$$\overline{x} = \overline{v} \wedge y = w \Leftrightarrow f(\overline{v}) = w \tag{2}$$

assuming the f is a Decision Tree or a Random Forest.

On this purpose, it is useful to introduce some formal notation. For sake of uniformity with other graphical structures mentioned in this paper, we view a DT as a (tree structured) directed acyclic graph $T = \langle N, A \rangle$. N is the set of the nodes n_i and A is the set of arcs a_j. The source and sink endpoints of an arc are referred to as $src(a_j)$ and $snk(a_j)$. Each node is associated to an attribute $x(n_i)$ and we refer as $x(a_j)$ to $x(src(a_j))$. Each arc is associated to a set $\lambda(a_j)$, called the arc *label*. The labels correspond to the arc conditions: an arc associated to the condition $x(a_j) \in \{0, 2\}$ has the label $\{0, 2\}$ (symbolic values can always be associated to integers); an arc with condition $x(a_j) \leq 3$ has the label $]-\infty, 3]$. Arcs having the same source always have disjoint labels. The leaf nodes of the DT are associated to a class from a set of classes $\mathcal{C}$.

3.1 Rule Based Encoding

A DT can be converted to a set of classification rules by interpreting each path from root to leaf as a logical implication. A simple approach to encode a DT in CP consists in translating each implication into a boolean meta-constraint.

Formally, let $\mathcal{P}$ be the set of root-to-leaf paths π in the tree, each path π being a sequence of arcs indices $\pi(0), \ldots, \pi(k)$ such that $snk(a_{\pi(i)}) = src(a_{\pi(i+1)})$ for all $i = 0, \ldots k - 1$. Each path ends in a leaf node with a certain class, denoted here as $class(\pi)$. Then the DT can be encoded as a set of constraints:

$$\bigwedge_{\pi(i) \in \pi} \left[\left[x(a_{\pi(i)}) \in \lambda(a_{\pi(i)})\right]\right] \Rightarrow \left[\left[y = class(\pi)\right]\right] \qquad \forall \pi \in \mathcal{P} \tag{3}$$

where the notation $[[-]]$ refers to the reification of the constraint enclosed by the brackets. The notation $x(a_j)$ refers here to an attribute *variable* (attributes and attributes variables will often be considered interchangeable). If $x(a_{\pi(i)})$ is numeric, then the reified constraint $\left[\left[x(a_{\pi(i)}) \in \lambda(a_{\pi(i)})\right]\right]$ is defined as:

$$\left[\left[x(a_{\pi(i)}) \leq \theta\right]\right] \qquad \text{if } \lambda(a_{\pi(i)}) \text{ is in the form }]-\infty, \theta] \tag{4}$$

$$\left[\left[x(a_{\pi(i)}) > \theta\right]\right] \qquad \text{if } \lambda(a_{\pi(i)}) \text{ is in the form }]\theta, \infty[\tag{5}$$

while the constraint form for symbolic attributes is:

$$\bigvee_{\theta_j \in \lambda(a_{\pi(i)})} \left[\left[x(a_{\pi(i)}) = \theta_j\right]\right] \tag{6}$$

This encoding approach can be strengthened by observing that if the class variable takes a specific value, then one of the corresponding root-to-leaf paths in the DT must necessarily be true. This leads to this second, stronger, encoding:

$$[[y = w]] \Leftrightarrow \bigvee_{\substack{\pi \in \mathcal{P}\ : \\ class(\pi) = w}} \bigwedge_{\pi(i) \in \pi} [[x(a_{\pi(i)}) \in \lambda(a_{\pi(i)})]] \qquad \forall w \in \mathcal{C} \qquad (7)$$

which will serve as our reference rule-based encoding in the paper.

On solvers that do not provide support for logical constraints, sums can be used instead of $\vee$, multiplications instead of $\wedge$, and implications can be modeled as $\leq$ relations. A double implication such as the one in Equation (7) must be modeled using separate $\leq$ constraints for the two implication directions.

The rule-based encoding from Equation (7) is defined using a number of constraints that grows linearly with the number of leaves and logarithmically with the DT depth, yielding a space complexity of $O(|N| \log |N|)$. The encodings is simple, scalable, and easy to implement, but provides only a weak form of propagation. It makes therefore sense to investigate methods for embedding DTs in CP that strike a difference balance between propagation power and cost.

3.2   Table Based Encoding

The TABLE constraint in CP can be used to define a constraint in extensional form, i.e. by enumerating the allowed (or forbidden) tuples for its variables. The rules employed by our first encoding (corresponding to paths in the DT) are related to tuples in a TABLE constraint in that they specify conditions over the attribute and class variables. It makes therefore sense to investigate the possibility to embed a Decision Tree in CP using TABLE. For this to be possible, there are four important issues that should be addressed.

First, *(#1) DTs extracted via Machine Learning may feature numeric attributes*, for which enumerating the possible values is, strictly speaking, impossible. A straightforward solution consists in using integer variables to encode numerical attributes with finite precision. In fact, with many CP solver this is a mandatory step, since real valued variables are often not supported. However, with this form of discretization the variables corresponding to numeric attributes end up having a very large domain, dramatically reducing the scalability.

We address the issue via an interval-based discretization. Specifically, we traverse the DT and collect for each numeric attribute all the splitting thresholds. Let $\overline{\theta}_i$ be the sorted vector of the thresholds θ_i^k for a given numeric attribute x_i, to which we always add the values $\pm\infty$. For example, $\overline{\theta}_A = [-\infty, 10, 20, \infty]$ for DT from Figure 1. Then, we introduce a new integer variable $\delta(x_i)$ with domain $\{0, \dots |\overline{\theta}_i| - 2\}$ such that:

$$\delta(x_i) = k \Leftrightarrow \theta_i^k < x_i \leq \theta_i^{k+1} \qquad (8)$$

Equation (8) can be enforced via specific channeling constraints (if supported by the solver) or by using reification. Then, we update the numeric labels in the DT with the substitution:

$$\lambda(a_j) = \{k \in D(\delta(x_i)) : \;]\theta_i^k, \theta_i^{k+1}] \subseteq \lambda(a_j)\} \tag{9}$$

Once this is done, we can use $\delta(x_i)$ as a replacement for x_i in the encoding.

Then, *(#2) arcs in the DT are labeled with sets* and each path on the tree corresponds to a *set* of tuples rather than to a single one. This problem can be (inefficiently) addressed by generating all possible combination of values for the attribute and class variables, and removing those that violate Equation (1). A better approach would be to generate directly the set of feasible tuples, which is intuitively related to the cartesian product of the path labels.

However, this requires some care, because *(#3) a path in a DT obtained via Machine Learning can specify multiple conditions on the same attribute*, which makes a straightforward computation of the cartesian product $\prod_{\pi(i)} \lambda(a_\pi(i))$ meaningless. However, the problem can be easily fixed by replacing all the labels defined over a specific attribute with their intersection. Formally, we introduce the term *refined label* to refer to:

$$L(\pi, x_i) = \bigcap_{\substack{\pi(i) \in \pi : \\ x(a_{\pi(i)}) = x_i}} \lambda(a_{\pi(i)}) \tag{10}$$

Even with the refined labels, the set of allowed tuples cannot be formulated as a cartesian product, since *(#4) a path in the DT may specify no label for one or more attributes*. When this happens, it means that such attributes are irrelevant for the associated classification. Speaking in terms of tuples, this means that all possible completions obtained by assigning values to the missing attributes are feasible. In other words, the set of all allowed tuples corresponding to a path π is given by the following cartesian product:

$$\{class(\pi)\} \times \prod_{x_i \in \overline{x}} \begin{cases} L(\pi, x_i) & \text{if a label over } x_i \text{ exists in } \pi \\ D(x_i) & \text{otherwise} \end{cases} \tag{11}$$

where $D(x_i)$ is the domain of the attribute variable x_i. We have a polynomial number of products in the form of Equation (11), each one compactly representing a set of allowed tuples and corresponding to a c-tuple in the terminology of [23]. In that paper the authors introduce an algorithm to enforce GAC on a table constraint formulated by using c-tuples rather then regular tuples. We plan to test such method for future research, while in this work we focus on investigating a more classical approach, namely using Equation (11) to generate all the allowed tuples for a "traditional" TABLE constraint. This method may have limited scalability, but it is readily applicable on off-the-shelf available solvers.

3.3   MDD Based Encoding

Graphical structures closely related to decision trees are employed by propagators for several global constraints. In [13], the authors propose a filtering algorithm that achieves GAC on TABLE and is based on a *trie*. A trie is a particular

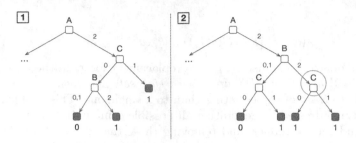

Fig. 2. Effect of attribute reordering

type of decision tree where the attributes are always 1) discrete, 2) considered exactly once, and 3) in the same order.

The approach from [13] is improved in [11] by converting the trie into a MDD, which is then used to define an efficient GAC propagator. From the purpose of this paper, an MDD can be considered as a trie generalization, where the requirement to have a tree structure is relaxed and multiple arcs are allowed to point to the same node. Thanks to this ability, an MDD can be considerably more compact than a trie, reducing the complexity of enforcing GAC. This raises interest in using MDDs to encode Decisions Trees and perform propagation.

Both MDDs and tries differ from the Decision Trees employed in Machine Learning (after the attribute discretization) in two important respects: 1) DTs can consider the same attribute multiple times and 3) in DTs attributes can be considered in a different order along different paths[1]. Such differences have deep consequences when trying to encode a DT. In fact, the encoding requires to re-order attributes, which causes the graph size to grow. Consider Figure 2.1, which shows a portion of the DT from Figure 1 after the attribute discretization. Assume that the selected attribute order is A, B, C. Moving the "B" split to the correct position requires to copy part of the graph, as shown in Figure 2.2. The phenomenon is multiplicative, possibly leading to an exponential growth. MDD reduction algorithms can mitigate the issue by sharing identical graph portions (e.g. the children of the circled node in Figure 2.2), but as of today we have no proof that exponential growth can be avoided, and some evidence that it actually occurs in practice.

Our approach for encoding a DT into an MDD is similar to the one from [11]: we construct a trie-like structure, which is then reduced using the *mddReduce* algorithm. The term "trie-like" is used because we allow multiple outgoing arcs of a node to point to the same child, as an (effective) measure for mitigating the graph expansion: strictly speaking, this make our structure already an MDD.

The algorithm starts from the set of all c-tuples obtained via Equation (11) and assumes that the attribute order has been pre-specified. We recall that in this formalization the class variable is considered an additional attribute. Algorithm 1

---

[1] Additionally, DTs can skip attributes that are irrelevant for a specific path, which may lead to an incorrect behavior of the MDD propagator from [11]. Despite this, skipped attributes are still easy to handle using MDDs.

Algorithm 1. build_trie(c-tuples, pos, [parent], [label])

1: *node* = a new MDD node
2: **if** *parent* is defined **then**
3:    **for** $v \in label$ **do** build an arc from *parent* to *node* with label v
4: **if** $pos = |\overline{x}|$ **then return** *node*
5: $L = \emptyset$, $D =$ the set of all values of x_{pos} in the c-tuples
6: **while** $|D| > 0$ **do** {*Cover the c-tuples values with a set L of disjoint labels*}
7:    $\lambda = \emptyset$
8:    **for** $v \in D$ **do**
9:       **if** if all c-tuples containing v at position *pos* contain also λ **then**
10:         $\lambda = \lambda \cup \{v\}$, $D = D \setminus \{v\}$
11:     $L = L \cup \lambda$
12: **for** $\lambda \in L$ **do**
13:    $T =$ set of c-tuples containing all values of λ at position *pos*
14:    build_trie(T, $pos + 1$, *node*, *label*)
15: **return** *node*

---

describes our *build_trie* function, which takes as input a set of c-tuples, an index over the sequence of attributes, plus an MDD node (to serve as a parent) and a label. The parent node and the label are left blank at the first invocation.

The function builds a new node and connects it to the parent (lines 2-3). Then the process stops if all x_i have been considered (line 4). Otherwise, we identify the set D of all values appearing at position *pos* in the c-tuples, and then we build a set L of labels such that: 1) the labels are disjoint; 2) the set of values for x_{pos} in any c-tuple can be expressed as a union of labels in L; 3) the cardinality of L is minimal. This is done via a loop with $O(|D||\text{c-tuples}|)$ iterations (polynomial in the size of the DT). This steps allows to limit the trie growth by grouping equivalent children of the current node. Finally, for each of the identified labels we build a set of compatible rules and we make a recursive call to *build_trie*. When the whole recursive process is over, the first call to the function returns the root of the trie/MDD.

Figure 3.1 shows the result of the conversion for the DT from Figure 1, using 'A, B, C, class' as attribute order. The inflating effect of attributes reordering is apparent. The graph size can be considerably reduced by feeding the trie to the *mddReduce* procedure from [11] (the output of the process is shown in Figure 3.2). For enforcing GAC over the MDD we use the same approach as [11].

3.4   Embedding Random Forests in CP

Embedding a Random Forest in CP requires to 1) embed in the CP model each DT from the forest and 2) define a constraint model for the mode computation, i.e. for aggregating the DT results and obtain the final classification. Since step (1) has already been throughly discussed, we now focus on step (2).

The statistical mode of a sample is the value that occurs most often in the collection. Formally, let y_F be the output of the forest and let y_j be the class

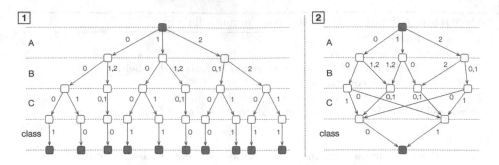

Fig. 3. [1] The trie-like structure for the DT from Figure 1. [2] The corresponding reduced MDD.

variable for the j-th DT. Then our goal is to enforce consistency on the following mathematical expression:

$$y_F = \operatorname{argmax}_{w \in \mathcal{C}} (cnt_{\overline{y}}(w)) \tag{12}$$

where $cnt_{\overline{y}}(w) = |\{y_j \in \overline{y} : y_j = w\}|$. Our approach to modeling the mode uses a global constraint to compute the cnt functions and one for $argmax$. In particular, we employ the following pair of constraints:

$$\text{GCC}(\overline{y}, \mathcal{C}, \overline{z}) \tag{13}$$

$$\text{ELEMENT}(\mathcal{C}, y_F, \max(\overline{z})) \tag{14}$$

where we recall that $\mathcal{C}$ is the set of the classes w, treated here as a vector; $\overline{z}$ is a vector of fresh variables z_w, each representing (thanks to the GCC constraint) the cardinality of the value w in the vector $\overline{y}$. The ELEMENT constraint ensures that y_F is the index of the value $w \in \mathcal{C}$ with the largest z_w, i.e. the mode.

This approach is based on classical global constraints and very easy to implement. As a drawback, the mode computation can be incomplete in case of ties among the z_w variables. If this happens, the ELEMENT constraint ensures that the domain of y_F contains the indices of all w values having maximal z_w. This situation can be resolved in the case of binary classifications by using an odd number of trees. In general, the user can force the solver to pick a value by adding y_F to the branching variables.

4   Experimentation

Our experimentation is divided in two parts. First, we compare all the presented techniques to embed DTs and RFs, in order to assess their effectiveness and scalability. Second, we evaluate the practical performance of our CP approach against a powerful solver based on Local Search.

For both the comparisons we consider a workload dispatching problem over an experimental 48-core CPU by Intel called Single-chip Cloud Computer (SCC, see [19]). The chip is designed to accept job batches and does not support process migration. We target a simulated version of the platform that has been augmented for research purpose with thermal controllers to slow down the cores in case they get too warm [1]. As a consequence, the efficiency of a core depends on its temperature. This in turns depends on many complex factors including the workload, the temperature of the other cores, the position on the silicon die, and the action of the thermal controllers. Our goal is to map jobs so as to maximize the number of cores operating at "high" efficiency ($\geq 97\%$).

Due to the complexity of the interactions determining the core behaviors, it is not possible to assess the effect of a job mapping on the efficiency via a traditional, expert designed, model. Hence, we obtained an approximation by generating a training set and then learning a Decision Tree (or Random Forest) for each core, to predict if it will have high (class 1) or low (class 0) efficiency given a specific workload. The input of each classifier is a set of four attributes (discussed later), whose values depend on the mapping decisions. We built the training set following a factorial design over a set of parameters identified in preliminary experiments. The training sets for each core range from 500 to 1000 tuples. Experiments were run on realistic sets of jobs.

Formally, there are n jobs that should be mapped to 48 cores. Each job i is characterized by its average Clocks Per Instruction (CPI) value cpi_i: jobs with low CPI are CPU-intensive and generate more heat, whereas jobs with high CPI are comparatively colder. All jobs are assumed to run indefinitely and hence, in order to avoid overloading, we require each core to run the same number of jobs. The mapping decisions are modeled via integer variables $core_i \in \{0, .., 47\}$ such that $core_i = k$ iff job i is mapped on core k. The overload prevention constraint is formulated using a single GCC with fixed cardinalities:

$$\text{GCC}(\overline{core}, [0, \ldots 47], [m, \ldots m]) \tag{15}$$

where $m = n/48$. The model objective is:

$$\max \sum_{k=0..47} \mathit{eff}_k \tag{16}$$

where eff_k is the integer class variable for the DT/RF associated to core k. The attributes for each DT are all numeric and correspond to:

- The average CPI of the jobs mapped on core k, i.e. $avgcpi_k$.
- The minimum CPI of the jobs mapped on core k, i.e. $mincpi_k$.
- The average $avgcpi_h$ for all the cores h that are within one tile from k.
- The average of $avgcpi_h$ for all the cores except k.

We model the attributes via integer variables (using a finite precision approximation). Attribute and mapping variables are connected by the following constraints:

$$avgcpi_k = \frac{1}{m} \sum_{i=0}^{n-1} cpi_i \cdot [[core_i = k]] \qquad \forall k = 0..47$$

$$mincpi_k = \min_{i=0..47} \left(\max(cpi_i) - [[core_i = k]] \left(\max(cpi_i) - cpi_i \right) \right) \qquad \forall k = 0..47$$

Table 1. Comparison of DT encodings

cores	attr.	Closed Inst. rls	tbl	mdd	Time rls	tbl	mdd	Sol.% T>R	M>R	T>M
3	4	100%	100%	100%	0.02	0.01	0.01	0.00%	0.00%	0.00%
	28	100%	100%	100%	0.06	0.04	0.06	0.00%	0.00%	0.00%
4	4	72%	98%	98%	19.27	2.22	2.51	1.67%	1.67%	0.00%
	28	57%	99%	99%	30.55	1.64	2.15	3.00%	3.00%	0.00%
5	4	13%	38%	34%	53.16	41.61	42.71	4.33%	4.00%	0.33%
	28	5%	38%	35%	57.47	40.60	42.05	8.33%	7.67%	0.67%
6	4	3%	14%	13%	58.30	52.40	52.59	9.67%	8.67%	1.00%
	28	1%	2%	2%	59.41	58.88	58.97	12.67%	11.67%	1.00%
7	4	0%	0%	0%	60.01	60.01	60.01	7.00%	7.00%	0.00%
	28	0%	1%	1%	60.01	59.47	92.54	10.00%	9.67%	0.67%

$$neighcpi_k = \frac{1}{|N(k)|} \sum_{h \in N(k)} avgcpi_h \qquad \forall k = 0..47$$

$$allcpi_k = \frac{1}{47} \sum_{h \neq k} avgcpi_h \qquad \forall k = 0..47$$

The constraints connecting the attribute and class variable eff_k for each core are obtained via the techniques discussed in Section 3.

4.1 Comparing the Different Encodings

For comparing the proposed DT and RF encodings, we generated benchmarks of instance with controlled size. This was done by 1) selecting random groups of cores and 2) generating a set of 6 jobs per core with realistic CPIs. Each benchmark contains 100 instances. We then solved each instance to optimality using a static search strategy and a time limit of 60 seconds. All experiments are run on a 2.3 GHz Intel Core i7. Our approaches have all been implemented over the Google or-tools solver [25].

Comparing DT Encodings: We tested our Decision Trees encodings, investigating the effect of the problem size and the number of attributes. This was done by building benchmarks with different numbers of cores and by augmenting the initial set of four attributes with up to 24 similarly computed features. The trees were learned from a training set using the implementation of C4.5 in Weka [16].

The results for this first evaluation are displayed in Table 1. Columns *cores* and *attr* report the number of cores and attributes for the benchmark. Then for each encoding approach (based on rules, TABLE, and MDD) we show the number of closed instances and the average solution time (time-outs at 60 sec of search are included). The last columns report the fraction of instances where each approach (R for rules, T for TABLE, and M for MDD) managed to find more

Table 2. Comparison of RF encodings, over different number of attributes

attr	Res. ok			closed			Mem (MB)			Time/10^6 brnc		
	rls	tbl	mdd	rls	tbl	mdd	rls	tbl	mdd	rls	tbl	mdd
4	100%	100%	100%	87%	92%	87%	38.5	39.3	40.3	25.17	13.59	30.29
8	100%	100%	100%	87%	92%	85%	40.9	49.5	53.4	34.13	18.36	42.29
12	100%	57%	55%	92%	57%	55%	46.0	351.4	197.4	40.59	28.38	51.60
16	100%	–	–	97%	–	–	47.6	–	–	45.43	–	–
20	100%	–	–	95%	–	–	53.2	–	–	49.72	–	–

solutions than another: since we use a static search strategy, this is an indication of the size of the explored search space.

As a general trend, both approaches capable of enforcing GAC considerably outperform the rule-based encoding. The trend stays the same when the number of attributes is inflated from 4 to 28: since more attributes tend to yield considerably bigger trees, this result show that the TABLE and MDD provide good scalability, despite the potential risk of combinatorial explosion of their main data structures. A hint of this risk is given by the average time for solving 7-cores, 28-attributes instances with the MDD approach, which is higher than 60 seconds. The catch is that the 60 seconds timeout is enforced only on search, while the total solution time takes into account also the model construction. Basically, in such case building the MDDs took a very long time. This is in part due to inefficiencies of our implementation, but is also symptomatic of possible scalability issues. In general the performance of the rule based approach seems to be the one most affected by the increased number of attributes.

Comparing RF Encodings: Next, we investigated the effectiveness of our encodings when applied to Random Forests. This was done by generating: 1) a first group of benchmarks defined over random groups of three cores, with several number of attributes and a fixed number of trees per forest (seven); 2) a second group of benchmarks defined again over core triplets, but with fixed number of attributes (four) and variable number of trees. The RFs were learned with the default algorithm provided by Weka.

The results of the first evaluation (variable number of attributes) are show in Table 2. The training algorithm for RFs yields trees that are radically different from those of C4.5 and translates to *much* bigger tuple-sets and MDDs. As a consequence, the TABLE and MDD based encodings have considerable scalability problems *at model construction time*, before the search process evens starts. In practice, we decided to stop some runs before the model construction started to take an impractical amount of resources (memory or time). The fraction of complete runs for all approaches is reported in the "Res. ok" columns. The memory usage (column "Mem") was found to be the main bottleneck for the TABLE based approach, whereas for the MDD method the biggest issue was the model construction time. The larger number of trees and their size had a negative effect on the branching speed: the TABLE was the most affected approach, despite

Table 3. Comparison of RF encodings, over different number of attributes

trees	closed			Time			Time/10^6 brnc		
	rls	tbl	mdd	rls	tbl	mdd	rls	tbl	mdd
3	95 %	97 %	97 %	10.71	8.58	12.22	14.58	10.01	17.05
5	92 %	97 %	90 %	14.08	9.86	15.95	18.07	10.81	22.39
7	87 %	92 %	87 %	17.07	11.83	18.39	24.51	13.58	29.60

being the one with the highest branching speed. Overall, the rule based approach managed to cope considerably better than any other with the scalability issues associated to RFs.

The effect of varying the number of trees can be observed in Table 3. The most striking finding is not reported in the table: basically, regardless of the number of trees, *all approaches reported exactly the same number of fails* in the instances they were able to close. This is probably due to the fact that the advantage of enforcing GAC on the individual trees gets lost at in the mode computation, performed by constraints (13) and (14). In such a situation, the best approach is the one with the lowest computation time. With this number of attributes (just 4) the TABLE encoding emerged as slightly faster, thanks also to the highly optimized implementation available in or-tools. Increasing the number of trees did not seem to have dramatic effects on the performance of the encodings.

4.2   Comparison with a State of the Art Local Search Approach

We performed a last set of experiments to evaluate our CP-based solution w.r.t. alternative approaches that can easily embed a Decision Tree in a model, but cannot benefit from constraint propagation. As a reference for the comparison, we used a model written and solved using Localsolver [6,12]. The choice was motivated by the simplicity of use of Localsolver, and its effectiveness in solving problems with non-trivial constraints and non-linear objective functions. Our Localsolver model is similar to the CP one, but the $core_i$ variables are missing, the reified $[[core_i = k]]$ are replaced by binary variables, and the GCC is replaced by a set of bounded sums.

For the CP model, we developed a solution approach that works by: 1) generating a first mapping via a heuristic; 2) using Large Neighborhood Search [27] to relax and re-map the jobs allocated to a subset of cores. In particular, we always select a few "bad" cores having $eff_k = 0$ at random, plus a few "good" cores with a probability based on their $avgcpi_k$ value. For exploring each neighborhood we use Depth First Search with random variable/value selection and restarts. We used the rules and table encoding for embedding Decision Trees.

We tested both approaches on a benchmark of 20 instances with 48 cores (modeled using DTs with four input attributes) and 288 jobs each. Since all methods are randomized, we performed 10 runs per instance, with a time limit of 90 seconds. A time limit of 2 seconds was enforced on each LNS iteration for CP. Table 4 reports the average value of the problem objective over the 10 runs

Table 4. Comparison with a state-of-the-art hybrid solver based on Local Search

ID	CP(rls) avg std	CP(tbl) avg std	LS avg std	ID	CP(rls) avg std	CP(tbl) avg std	LS avg std
0	29.60 0.98	30.50 0.50	27.70 0.46	10	35.60 0.98	36.80 0.87	33.20 0.88
1	33.60 0.80	35.60 0.80	32.40 0.67	11	24.30 0.78	24.20 0.75	24.00 0.00
2	28.70 1.01	30.00 0.78	28.10 0.70	12	33.10 0.83	34.20 1.17	31.30 0.78
3	25.10 0.70	25.60 1.02	25.10 0.54	13	24.00 1.00	25.30 0.78	24.00 0.00
4	23.30 0.46	23.80 1.08	23.50 0.50	14	29.60 0.66	30.90 0.70	28.60 0.92
5	30.50 1.12	31.80 0.60	29.00 0.63	15	32.90 0.83	35.00 0.78	30.80 0.60
6	39.10 0.70	40.00 0.63	36.90 0.70	16	25.10 0.70	25.90 0.70	24.70 0.64
7	28.60 0.80	30.30 0.64	27.90 1.04	17	38.90 0.54	40.10 0.54	36.60 1.02
8	32.90 1.30	34.10 0.54	30.70 0.64	18	26.40 1.02	26.80 0.60	25.20 0.40
9	37.30 0.64	38.70 0.90	34.90 1.04	19	32.90 1.14	34.80 0.98	31.80 0.75

and its standard deviation. The CP and LS approach proved to work very well[2], but CP managed to find the best solutions for the majority of the instances. This result was made possible by the use of domain knowledge for selecting the LNS fragments, *and by the propagation performed by our DT encodings.* Using the rule encoding was sufficient to quickly close many ill-chosen fragments and speed up the search. The additional propagation granted by enforcing GAC allowed the TABLE encoding to work even better.

5   Other Related Work

The integration of Machine learning and CP (and in general combinatorial optimization) has received increased attention in the last decade. Fertilizations in both directions have been studied: on one hand the machine learning community has studied how constraints can be used during mining and learning; on the other way round, machine learning may allow to automatically tune an optimization approach, or to acquire constraints and objective functions from data.

Along the first line of research, works such as [5, 28] have studied the core optimization problems in Machine Learning algorithms and proposed efficient methods for extracting knowledge from huge volumes of data. On the other way round, some researchers have considered the problem of learning optimization problem instances for testing new techniques [17]. Clustering methods have been employed for automatic algorithm selection in [22]. Several Machine Learning techniques have been used for predicting the run time of optimization algorithms (e.g. [20]), and in general for algorithm selection (e.g. [24]).

The approaches that are closest in spirit to this paper are those that focus on learning parts of the model from data. Along this line several papers [4, 7, 9] show how to learn a constraint network from a set positive and negative examples, while the QuAcq system [8] requires only partial queries on subsets

---

[2] w.r.t. other approaches that are not reported here due to lack of space.

of problem variables, with no need of positive examples. All such approaches have a focus on learning an unknown problem while simultaneously trying to solve it. Conversely, in our approach we focus on embedding in combinatorial optimization a well-defined, pre-extracted, Machine Learning model, which may however lack a straightforward encoding.

Many approaches for optimizing non-linear functions over continuous domains rely on on-line learning techniques to reduce the number of required function evaluations. The authors of [21] introduced methods to fit a response surface based on a few sampled points: the surface is then employed to guide the search process. The OptQuest [14] system integrates in a closed loop simulation and metaheuristics and relies on a simple form of learning (a neural network accelerator) to avoid trivially bad solutions. Only a few works (e.g. [15,29]) have resorted to using pre-extracted Machine Learning models to speed up the cost function evaluation. The LION book [3] proposes a similar approach, although in this case the goal is tackling problems where the cost function is difficult to model, rather than expensive to compute. The LION method focuses on extracting a function from available data, and then on obtaining solutions via model fitting. As a common feature, all such methods are designed for functions defined over an unconstrained (or loosely constrained) domain. Conversely, our method targets problems that mix a core combinatorial structure (typically having non-trivial constraints) with complex functions that we approximate via Machine Learning.

6   Concluding Remarks

The Empirical Model Learning (EML) method is aimed at learning part of the model from data or predictive tools. The learned component not only declaratively links decision variables with prediction/class variables, but contains an operational semantics enabling domain filtering and constraint propagation.

In this work, we have devised an additional component for EML: we have used Decision Trees and Random Forests as learning methods. We have provided three encodings of for DTs, respectively based on meta-constraints, on the global TABLE constraint, and on an MDD – the last two being able to enforce GAC. The experimentation on Decision Trees and Random Forests had quite different outcomes in terms of scalability, mainly because of differences between their learning algorithms. While for DTs the TABLE and MDD approaches are clearly the most effective, in RFs the TABLE constraint and MDD have serious scalability issues in terms of memory and model construction time respectively.

As part of future research, we plan to test the modified GAC-schema algorithm for c-tuples from [23], and to investigate methods to convert a DT into an MDD without the passing for an intermediate trie. We will also research the possibility to enforce GAC using the DT itself as the main data structure.

Finally, we are interested in finding ways to exploit the accuracy information provided by the Machine Learning models (e.g. class distributions in RFs). This is particularly important when combining different constraints, each representing an approximate relation between variables.

References

1. Bartolini, A., Cacciari, M., Tilli, A., Benini, L.: Thermal and energy management of high-performance multicores: Distributed and self-calibrating model-predictive controller. IEEE Trans. Parallel Distrib. Syst. **24**(1), 170–183 (2013)
2. Bartolini, A., Lombardi, M., Milano, M., Benini, L.: Neuron constraints to model complex real-world problems. In: Proc. of CP, pp. 115–129 (2011)
3. Battiti, R., Brunato, M.: The LION way: Machine Learning plus Intelligent Optimization. University of Trento, LIONlab (2014)
4. Beldiceanu, N., Simonis, H.: A model seeker: extracting global constraint models from positive examples. In: Proc. of CP, pp. 141–157 (2012)
5. Bennett, K.P., Parrado-Hernández, E.: The interplay of optimization and machine learning research. Journal of Machine Learning Research **7**, 1265–1281 (2006)
6. Benoist, T., Estellon, B., Gardi, F., Megel, R., Nouioua, K.: Localsolver 1.x: a black-box local-search solver for 0–1 programming. 4OR **9**(3), 299–316 (2011)
7. Bessière, C., Coletta, R., Freuder, E.C., O'Sullivan, B.: Leveraging the learning power of examples in automated constraint acquisition. In: Proc. of CP, pp. 123–137 (2004)
8. Bessière, C., Coletta, R., Hebrard, E., Katsirelos, G., Lazaar, N., Narodytska, N., Quimper, C., Walsh, T.: Constraint acquisition via partial queries. In: Proc. of IJCAI (2013)
9. Bessière, C., Coletta, R., O'Sullivan, B., Paulin, M.: Query-driven constraint acquisition. In: Proc. of IJCAI, pp. 50–55 (2007)
10. Breiman, L.: Random forests. Machine learning **45**(1), 5–32 (2001)
11. Cheng, K.C.K., Yap, R.H.C.: An mdd-based generalized arc consistency algorithm for positive and negative table constraints and some global constraints. Constraints **15**(2), 265–304 (2010)
12. Gardi, F., Benoist, T., Darlay, J., Estellon, B., Megel, R.: Mathematical Programming Solver Based on Local Search. John Wiley & Sons (2014)
13. Gent, I.P., Jefferson, C., Miguel, I., Nightingale, P.: Data structures for generalised arc consistency for extensional constraints. In: Proc. of AAAI, pp. 191–197 (2007)
14. Glover, F., Kelly, J.P., Laguna, M.: New Advances for Wedding optimization and simulation. In: Proc. of WSC, pp. 255–260 (1999)
15. Gopalakrishnan, K., Asce, A.M.: Neural Network - Swarm Intelligence Hybrid Nonlinear Optimization Algorithm for Pavement Moduli Back-Calculation. Journal of Transportation Engineering **136**(6), 528–536 (2009)
16. Hall, M., Frank, E., Holmes, G., Pfahringer, B., Reutemann, P., Witten, I.H.: The weka data mining software: an update. ACM SIGKDD explorations newsletter **11**(1), 10–18 (2009)
17. Hernando, Leticia, Mendiburu, Alexander, Lozano, Jose A.: Generating Customized Landscapes in Permutation-Based Combinatorial Optimization Problems. In: Nicosia, Giuseppe, Pardalos, Panos (eds.) LION 7. LNCS, vol. 7997, pp. 299–303. Springer, Heidelberg (2013)
18. Ho, T.K.: Random decision forests. In: Proc. of ICDAR, p. 278 (1995)
19. Howard, J., Dighe, S., et al.: A 48-Core IA-32 message-passing processor with DVFS in 45nm CMOS. In: Proc. of ISSCC, pp. 108–109, February 2010
20. Hutter, F., Xu, L., Hoos, H.H., Leyton-Brown, K.: Algorithm runtime prediction: Methods & evaluation. Artif. Intell. **206**, 79–111 (2014)
21. Jones, D.R., Schonlau, M., Welch, W.J.: Efficient global optimization of expensive black-box functions. Journal of Global optimization **13**(4), 455–492 (1998)

22. Kadioglu, S., Malitsky, Y., Sellmann, M., Tierney, K.: ISAC - instance-specific algorithm configuration. In: Proc. of ECAI, pp. 751–756 (2010)
23. Katsirelos, G., Walsh, T.: A compression algorithm for large arity extensional constraints. In: Proc. of CP, pp. 379–393 (2007)
24. Kotthoff, L., Gent, I.P., Miguel, I.: An evaluation of machine learning in algorithm selection for search problems. AI Commun. 25(3), 257–270 (2012)
25. Perron, L.: Operations Research and Constraint Programming at Google. In: Proc. of CP, p. 2 (2011)
26. Quinlan, J.R.: C4.5: Programs for Machine Learning. Morgan Kaufmann (1993)
27. Shaw, P.: Using constraint programming and local search methods to solve vehicle routing problems. In: Proc. of CP, pp. 417–431 (1998)
28. Sra, S., Nowozin, S., Wright, S.J.: Optimization for machine learning. MIT Press(2012)
29. Zaabab, A.H., Zhang, Q., Nakhla, M.: A neural network modeling approach to circuit optimization and statistical design. IEEE Transactions on Microwave Theory and Techniques 43(6), 1349–1358 (1995)

Scheduling with Fixed Maintenance, Shared Resources and Nonlinear Feedrate Constraints: A Mine Planning Case Study

Christina N. Burt[✉], Nir Lipovetzky, Adrian R. Pearce, and Peter J. Stuckey

Department of Computing and Information Systems,
The University of Melbourne, Parkville, Australia
cnburt@unimelb.edu.au

Abstract. Given a short term mining plan, the task for an operational mine planner is to determine how the equipment in the mine should be used each day. That is, how crushers, loaders and trucks should be used to realise the short term plan. It is important to achieve both grade targets (by blending) and maximise the utilisation (i.e., throughput) of the mine. The resulting problem is a non-linear scheduling problem with maintenance constraints, blending and shared resources. In this paper, we decompose this problem into two parts: the blending, and the utilisation problems. We then focus our attention on the utilisation problem. We examine how to model and solve it using alternative approaches: specifically, constraint programming, MIQP and MINLP. We provide a repair heuristic based on an outer-approximation, and empirically demonstrate its effectiveness for solving the real-world instances of operational mine planning obtained from our industry partner.

1 Introduction

In open-pit mines, a common form of materials handling is through truck and loader fleets [16], where the loaders excavate the material from blocks and the trucks haul it to dumpsites, stockpiles, run–of–mine (rom) stockpiles, or directly to the crusher. In this paper, we will consider a challenging scheduling problem that arises in the context of this form of materials handling. We denote the movement of material by a *movement*, which is a representation of the material, its source location (a block or stockpile), its destination location (a crusher or stockpile) and the grade of material. In Figure 1, we represent the *movements* of material by edges. Our task is to schedule these movements subject to constraints on the plants and equipment. At most one loader may excavate a movement. Since loader traversal is slow (≈5km/h), it is preferable to sequence the loader's tasks in such a way that loader traversal doesn't prohibit flow of material to the crusher. That is, at least one loader should always be working at any given time. Thus, we can think of the task of *sequencing the loaders* as sequencing the movements. The loaders transfer material to the trucks, which haul the material to one of various destinations. Importantly, the trucking resources are

© Springer International Publishing Switzerland 2015
L. Michel (Ed.): CPAIOR 2015, LNCS 9075, pp. 91–107, 2015.
DOI: 10.1007/978-3-319-18008-3_7

limited and *shared* between all loaders. The material flow to the crusher is key to measuring productivity of the mine, and therefore it is important to *maximise the feedrate* to the crusher at any moment in time.

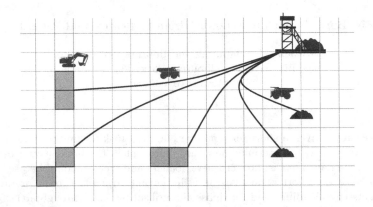

Fig. 1. An abstract representation of a mine. The squares are *movement* sources; the edges are *movements*; *loaders* excavate the movements at the source; *trucks* haul material from the loaders to the *crusher* or the various *stockpiles*. Icon images adapted from [4] and [8].

In open-pit mining operations, there are several levels of planning, each of which pass down restrictions in decisions. That is, long-term plans (strategic) are handed down to short-term planners, who in turn generate plans for operations planners (tactical). At each of these levels of planning, the task is to determine the order in which material *movements* should be mined and how they should be processed, such that blending and utilisation targets are met. Additionally, all of the equipment, including crushers, are subject to *maintenance tasks* which create periods of unavailability. The operations planners have the task of enacting the plans—physically, with trucks and loaders—such that the mine performance goals are met. At the core of our work is a nonlinear scheduling problem with shared resources, blending, and maintenance constraints. The goals of interest in our research are (a) correct blending of materials, and (b) utilisation of equipment. The output of a tactical mine plan is a sequence of *builds* (i.e., small, short-term stockpiles) with an allocation of partial movements to builds such that the builds have the correct blend and the crusher is maximally utilised.

In this paper, we investigate principled approaches for both modelling and solving a subproblem of tactical mine planning, which requires flexible partitioning of tasks to facilitate *fixed maintenance tasks*. We arrive at the subproblem by first decomposing the global problem into the blending and utilisation problems. We then derive *event-based* formulations for the latter subproblem: exploring formulations in constraint programming and mathematical programming. Motivated by finding a more computationally efficient solving approach, we derive an outer-approximation and repair heuristic which utilises aspects from each of our formulations.

The contributions of this paper are:

- a new modelling and solution approach for an operational mine scheduling problem with flexible partitioning, that allows for fixed maintenance tasks;
- a comparison of modelling approaches for this problem, using CP and OR techniques;
- a repair heuristic for obtaining efficient solutions to the subproblem; and
- a detailed set of experiments on real-world instances of the problem.

In previous work [12], we consider a different version of the build planning problem. The model in [12] does not consider maintenance, and does not include the crusher requirements. These considerations were considered crucial by our industry partner, and lead to a nonlinear problem. Another critical difference is that the modelling approach to the build planning problem in this paper uses CP and MIP/MINLP technology, whereas in [12] we use planning technology to account for state-dependent components that we do not include in this paper. Overall, this paper results from moving to a more realistic version of the problem we addressed in [12].

In the literature, scheduling problems with fixed maintenance have been addressed for up to 2 machines [1,11,15] and m machines [9,13]. None of these works consider the side constraints of shared resources, blending or feedrate constraints. The latter, [9], does consider a nonlinear objective function and provides a linearisation of the model. In [10], the authors consider an integrated scheduling problem, which is equivalent to the loader sequencing problem without maintenance tasks. However, it is clear from the scheduling with maintenance literature that the problem we consider here has not been addressed.

In the mining literature, the shared resources and blending constraints have been addressed. In particular, [2] develop an outer-approximation and repair heuristic for nonlinear blending constraints at stockpiles. This problem is the most similar to ours from the literature that we could find. However, as these authors consider a longer planning horizon, they do not consider the traversal of equipment or the feedrates at the crushers. The key difference between the nonlinear blending at stockpiles and the nonlinear feedrate at the crushers is that it is not possible to obtain tight bounds on the feedrate at the crushers. Apart from this, the nonlinear constraints that arise are both bilinear in form.

The remainder of the paper is organised as follows. We first describe the subproblem that addresses the problem of scheduling the equipment and sequencing the movements to keep the crushers busy. We formulate this problem using event-based models, allowing for mixed fleets, in Section 3. Then, we extend these formulations to account for maintenance constraints in Section 4. We provide a heuristic approach based on outer-approximation and feasibility repair in Section 5. We provide experiments across all models in Section 6. We conclude with a discussion of our experience in Section 7.

2   Problem Description

For our industry partner, operations planning and grade control—which together form the tactical mine scheduling problem—are performed separately. The grade controller (at the build) is interested in minimising deviation of the grade *blend*. Once the grade control plan is constructed we need to solve the utilisation problem to complete it as efficiently as possible.

In this sense, if there are multiple pits in a mine, each with their own crusher, then it is important that the scheduled tasks are aligned, i.e. complete at the same time. That is, there may be multiple crushers that create one build at the end of the supply chain. Therefore, these subproblems within pits (for example) are not completely independent.

Problem Description 1 (Tactical mine scheduling). *Given a set of movements to be mined, determine the mining sequence such that the movements form builds with correct blend bounds, the crushers contributing to a build are aligned, and the utilisation of the crusher is maximised.*

In practical applications, the sequence must respect *movement precedences*, which determine which movement is accessible first. For example, it may be required to clear all movements associated with a particular location before another movement (or set of movements) is possible. Additionally, all of the equipment— trucks, loaders and crushers—are subject to *maintenance* at pre-defined periods. The sequencing should therefore also account for maintenance of equipment.

In previous work (see [12]), we decomposed this problem into the blending and utilisation problems. In this paper, we will adopt the same scheme with the additional requirement that we align the crusher tasks contributing to the same build. We then focus completely on the utilisation problem in the remainder of this paper. For a general description of the decomposition approach, see [3].

In our decomposition approach, we partition the problem into two subproblems. We first model the blending subproblem as a mixed-integer program. We add new constraints that approximate the time required to mine a build, for each crusher, in the context of a mine with multiple crushers. To achieve this, we ensure that the contribution (in tonnes) of each crusher to the build, is proportional to the maximum feedrate of the crusher itself. That is:

$$\frac{C_i}{\bar{\Phi}_i} = \frac{C_j}{\bar{\Phi}_j},$$

where C_i is the total quantity of material contributed to the build from crusher i, and $\bar{\Phi}_i$ is the maximum feedrate of crusher i. To improve robustness, we introduce these alignment constraints in the objective function: i.e., as soft constraints. The output of this first partition is an allocation of movements to builds such that the blending constraints are met, and the builds are aligned with respect to maximum crusher utilisation.

The utilisation subproblem amounts to the question: how should the equipment be used so that our mining goals are met and the equipment is maximally utilised? The components of this problem include:

1. **sequencing the loaders**—this allows the loader traversal times between movements to be counted, and allows a feedrate to be allocated to each movement.
2. **allocating the trucks**—the cycle time (i.e., round trip travel time between pick-up and dumping locations and back) is accounted for.
3. **maximising feedrate at crusher**—the incoming ore feedrate is limited by the capacity of the crusher, and yet the crusher should not be scheduled to be underutilised.

And, key to this problem, we must allow for maintenance events. We assume that we have obtained a sub-set of the movements from a solution to the blending subproblem, such that the blending constraints are already met.

Problem Description 2 (Build Planning). *Given a set of movements that together meet blending constraints, determine the mining sequence that maximises the crusher utilisation, such that loading and truck fleet capacity constraints are met, and maintenance events are accounted for.*

The underlying challenges in this problem arise from the essential non-linearity of determining the mining rate (tonnes/minute) of a movement m. The feedrate of a movement, ϕ_m, is determined by the capacity, C_t^T, of each truck, t, assigned to the movement divided by the cycle time of the truck travelling from movement source to the crusher and back again, $T_{m,t}^T$. Conversely, the duration of mining the movement is the ratio of the size of the movement to the movement feedrate. Summing up the feedrates of the movements currently being sent to the crusher will determine the current crusher utilisation. Summing up the trucks assigned to all movements currently mined must remain below the trucking capacity limit. Hence we have a scheduling problem with variable durations and resource usages where we are trying to maximise utilisation of resources.

The fact that the movements themselves are discrete leads to an intuitive discretisation of the problem. That is, it is intuitive to assign a feedrate and truck allocation to each movement. However, this is actually an unnatural *restriction* on the problem—it might be better to, for example, change the truck allocation part way through mining the movement. This is not only important for optimising the solution (or finding feasible solutions in very tightly constrained instances), but also for accounting for maintenance.

With maintenance events, it is possible that the discretisation will lead directly to poor quality solutions in terms of crusher utilisation. Consider, for example, the case where the durations between maintenance events are so small that the given movement sizes cannot be mined, subject to operational constraints. In this case, it would be ideal to determine ways to partition the movements such that good solutions can be obtained.

3   Build Planning Models

We begin by addressing the problem of scheduling the crusher without maintenance tasks. In this section, we formalise our approaches as event-based models.

This is, in part, inspired by Automated Planning encodings, and in part by the logical representation of Constraint Programming.

3.1 Discretised Approach: Constraint Programming

We define the build planning problem by a set of movements, $m \in \mathcal{M}$, with a defined size (in tonnes), C_m^M, a source location, O_m, a destination location, D_m, and precedences between movements, $p \in \mathcal{P}$, defined by $p = (m', m) \in \mathcal{M} \times \mathcal{M}$ that require a specific movement, m', is completed before another, m, can begin. Additionally, the problem has a set of loaders, $l \in \mathcal{L}$, a set of crushers, $\kappa \in \mathcal{F}$, a set of truck types, $t \in \mathcal{T}$, and the number of trucks of each type, N_t^T.

For each crusher we have a maximum feed rate, $\bar{\Phi}_\kappa$. For each loader we have a maximum dig rate, $\bar{\Phi}_l^L$, and a time to traverse from the source location of movement m to the source location of movement m', $T_{m,m',l}^L$. For each truck type, we have a truck capacity, C_t^T, and a cycle time for each movement m, $T_{m,t}^T$—that is, how long it takes the truck to go from the source location O_m to the destination D_m and back to O_m. In addition, we assert a maximum number of movements per loader, N_l^M.

The principal decisions to be made are:

- for each loader l, the sequence of movements $d_{l,i} \in \mathcal{M} \cup \{\bot\}, i = 1, \ldots N_l^M$ it will complete (with $\bot$ representing dummy unused movements),
- for each movement, $\tau_{m,t}$ is the number of trucks of different types, $t \in \mathcal{T}$, assigned to the movement.

Auxiliary variables give:

- the start, s_m, duration, t_m, and end times, e_m, for each movement,
- the movement rate, ϕ_m, (tonnes/minute) for each movement,
- the loader assigned to each movement, λ_m,
- indicator variables, $z_{m,m'}^\wedge$, for which movements m are running when movement, m', is started.

The constraints are:

- Precedence constraints are satisfied:
$$e_m \leq s_{m'}, \qquad\qquad (m, m') \in \mathcal{P}. \qquad\qquad (1)$$

- Dummy movements $\bot$ are at the end
$$d_{l,i} = \bot \rightarrow d_{l,i+1} = \bot, \qquad l \in \mathcal{L}, i \in 1 \ldots N_l^M - 1. \qquad (2)$$

- The next task cannot begin until the loader has moved there:
$$e_{d_{l,i}} + T_{m,m',l}^L \leq s_{d_{l,i+1}}, \qquad l \in \mathcal{L}, i \in 1 \ldots N_l^M - 1, d_{l,i+1} \neq \bot. \qquad (3)$$

- Each movement is assigned to at most one loader (assuming $\bot = 0$):
$$\texttt{all\_different\_except\_0}([d_{l,i} \mid l \in \mathcal{L}, i \in 1 \ldots N_l^M]). \qquad (4)$$

- Ensure the λ_m and $d_{l,i}$ variables agree:

$$d_{l,i} = m \rightarrow \lambda_m = l, \qquad\qquad m \in \mathcal{M}, l \in \mathcal{L}, i \in 1 \ldots N_l^M, \qquad (5)$$

$$\lambda_m = l \rightarrow \exists_{i \in 1 \ldots N_l^M} d_{l,i} = m. \qquad (6)$$

- The feedrate for a movement is constrained by loader dig rate:

$$\phi_m \leq \bar{\Phi}_{\lambda_m}^L, \qquad\qquad m \in \mathcal{M}. \qquad (7)$$

- Movement rate is constrained by trucking capacity assigned:

$$\phi_m \leq \sum_{t \in \mathcal{T}} \tau_{m,t} \times C_t^T / T_{m,t}^T, \qquad\qquad m \in \mathcal{M}. \qquad (8)$$

- Duration of a movement is given by the tonnage divided by move rate:

$$t_m = \frac{C_m^M}{\phi_m}, \qquad\qquad m \in \mathcal{M}. \qquad (9)$$

- Start and end times are related by movement duration:

$$s_m + t_m = e_m, \qquad\qquad m \in \mathcal{M}. \qquad (10)$$

- Crusher is not overloaded:

$$\sum_{m \in \{\mathcal{M} | D_m = \kappa\}} z_{m,m'}^{\wedge} \phi_m + \phi_{m'} \leq \bar{\Phi}_\kappa, \quad \kappa \in \mathcal{F}, m' \in \{\mathcal{M} \mid D_{m'} = \kappa\}, \quad (11)$$

We only test overload at the start time of any movement, since that is the only time when more can be fed to a crusher.

- Trucking capacities are respected:

$$\texttt{cumulative}(s_m, t_m, [\tau_{m,t} | m \in \mathcal{M}], N_t^T), t \in \mathcal{T}. \qquad (12)$$

- Indicator variables for coinciding events are correct:

$$z_{m,m'}^{\wedge} \leftrightarrow (s_m \leq s_{m'} \wedge e_m > s_{m'} \wedge m \neq m'). \qquad (13)$$

The objective function is to minimise the makespan, i.e. minimise $\max_{m \in \mathcal{M}} e_m$. Our search strategy first assigns a movement to each loader, $d_{l,i} \in \mathcal{M}$, then assigns the earliest possible start times, s_m, and tries to assign the maximum feedrate, ϕ_m, while minimising the number of trucks assigned, $\tau_{m,t}$.

3.2   Discretised Approach: Mixed-integer Quadratic Programming

For a mathematical programming approach, we wish to keep the event-based representation and linearise the constraints as much as possible. The loader sequencing problem is represented by a graph where nodes are movement source locations, and edges are traversals of loaders from one movement source to another. Loader traversal decisions are represented by flow variables, $x_{m,m',l}$, which take a non-zero integer value if loader l performs movement m followed directly by m'. We extend the movement set $\mathcal{M}$ to include a dummy source, σ, and sink, σ', such that $\mathcal{M} \cup \{\sigma, \sigma'\} = \mathcal{M}'$.

We encode the loader sequences constraints with a node-disjoint multi-commodity flow formulation, which effectively allocates loaders to movements and derives their traversal sequence. That is,

$$\sum_{m' \in \mathcal{M}', l} x_{m',m,l} - \sum_{m' \in \mathcal{M}', l} x_{m,m',l} = \begin{cases} \min\{|\mathcal{M}|, |\mathcal{L}|\} & \text{if } m = \sigma, \\ -\min\{|\mathcal{M}|, |\mathcal{L}|\} & \text{if } m = \sigma', \quad \forall\, m \in \mathcal{M}, \quad (14) \\ 0 & \text{otherwise}, \end{cases}$$

$$\sum_{l, m' \in \mathcal{M}'} x_{m',m,l} = 1 \qquad\qquad \forall\, m \in \mathcal{M}', \quad (15)$$

$$e_{m'} + \sum_l T^L_{m',m,l} x_{m',m,l} - MS(1 - \sum_l x_{m',m,l}) \leq s_m \qquad \begin{array}{l} \forall\, m' \in \mathcal{M}' \quad (16) \\ m \in \mathcal{M} \\ m \neq m', \end{array}$$

where MS represents the maximum possible makespan. In constraint (14), we create sequences for each loader. However, if the number of blocks is less than the number of loaders, we must restrict the sequences to the smaller number. Constraint (15) ensures that all blocks are visited by exactly one loader. We use a big-M formulation for constraint (16) to ensure the travel time for each loader is accounted for, but only if that edge is traversed by that loader.

We can now use the flow variables to indicate which loader performs a task—this permits us to encode the loader maximum dig-rate bounding the movement feedrate:

$$\phi_m \leq \sum_{l, m'} \bar{\Phi}^L_l x_{m',m,l} \qquad\qquad \forall\, m \in \mathcal{M}. \qquad (17)$$

To encode the `cumulative` constraints we must determine whether two events coincide. To do this, we reason that, for any two events, if they both end after the other began, then the two events coincide. To formalise this, we introduce binary variables $z^\succ_{m,m'}$ to *indicate* that movement m finishes after m' starts. Recall that $z^\wedge_{m,m'}$ indicates that event m occurs at the same time as the start of event m'. The constraints to activate these variables are:

$$MS z^\succ_{m,m'} \geq e_m - s_{m'} \qquad\qquad \forall\, m \in \mathcal{M}, m' \in \mathcal{M}, \quad (18)$$

$$z^\wedge_{m,m'} \geq z^\succ_{m,m'} + z^\succ_{m',m} - 1 \qquad\qquad \forall\, m \in \mathcal{M}, m' \in \mathcal{M}, \quad (19)$$

$$z^\wedge_{m,m'} \leq \frac{z^\succ_{m,m'} + z^\succ_{m',m}}{2} \qquad\qquad \forall\, m \in \mathcal{M}, m' \in \mathcal{M}. \quad (20)$$

The `cumulative` trucking capacity constraints can now be encoded as

$$\sum_{m \in \mathcal{M} \setminus \{m'\}} \tau_{m,t} z^\wedge_{m,m'} + \tau_{m',t} \leq N^T_t \qquad\qquad \forall\, m' \in \mathcal{M}, t \in \mathcal{T}. \quad (21)$$

The crusher feed rate constraints are similarly encoded as

$$\sum_{m \in \{\mathcal{M} \setminus \{m'\} | D_m = \kappa\}} \phi_m z^\wedge_{m,m'} + \phi_{m'} \leq \bar{\Phi}_\kappa \qquad\qquad \forall\, m' \in \{\mathcal{M} | D_{m'} = \kappa\}. \quad (22)$$

With the exception of the nonlinear duration calculation, the remaining constraints as presented in the constraint programming model are linear, and can be used directly in the mathematical programming model. The constraints (22) and (21) can be linearised exactly, leaving us with a mixed-integer quadratic programming formulation with positive semi-definite form of constraint (9).

We linearise (21) and (22) by introducing two ancillary variables $\phi'_{m,m'}$ and $\tau'_{m,m',t}$, which will take on the value of ϕ_m if $z^\wedge_{m,m'}$ is 1, and zero otherwise. Let $\bar{\Phi}$ and $\bar{\tau}_m$ be the upper bounds on their respective variables. Then,

$$\phi'_{m,m'} \geq \phi_m - (1 - z^\wedge_{m,m'})\bar{\Phi} \quad \forall\, m, m' \in \{\mathcal{M} \mid D_m = \kappa, D_{m'} = \kappa\}, \quad (23)$$

$$\phi'_{m,m'} \leq \phi_m + (1 - z^\wedge_{m,m'})\bar{\Phi} \quad \forall\, m, m' \in \{\mathcal{M} \mid D_m = \kappa, D_{m'} = \kappa\}, \quad (24)$$

$$\tau'_{m,m',t} \geq \tau_{m,t} - (1 - z^\wedge_{m,m'})\bar{\tau}_m \quad \forall\, m, m', m \neq m', t \quad (25)$$

$$\tau'_{m,m',t} \leq \tau_{m,t} + (1 - z^\wedge_{m,m'})\bar{\tau}_m \quad \forall\, m, m', m \neq m', t. \quad (26)$$

Then, we alter constraints (21) and (22) as follows:

$$\sum_{m \in \mathcal{M} \setminus \{m'\}} \tau'_{m,m',t} + \tau_{m',t} \leq N^T_t \quad \forall\, m' \in \mathcal{M}, t \in \mathcal{T}, \quad (27)$$

$$\sum_{m \in \{\mathcal{M} \setminus \{m'\} | D_m = \kappa\}} \phi'_{m,m'} + \phi_{m'} \leq \bar{\Phi}_\kappa \quad \forall\, m' \in \{\mathcal{M} | D_{m'} = \kappa\}. \quad (28)$$

We improve computational performance with the following valid inequalities:

$$\sum_{m,m' \in \mathcal{M}'} x_{m',m,l} \leq 1 \qquad \forall\, l \in \mathcal{L}, \quad (29)$$

$$\sum_{m' \in \mathcal{M}',l} x_{m',m,l} = 1 \qquad \forall\, m \in \mathcal{M}, \quad (30)$$

$$\sum_{m' \in \mathcal{M}',l} x_{m,m',l} = 1 \qquad \forall\, m \in \mathcal{M}, \quad (31)$$

$$e_m \leq MS \sum_{m' \in \mathcal{M}',l} x_{m',m,l} \qquad \forall\, m \in \mathcal{M}, k \in \mathcal{K}, \quad (32)$$

$$z^\wedge_{m,m'} = z^\wedge_{m',m} \qquad \forall\, m \in \mathcal{M}, m' \in \mathcal{M}'. \quad (33)$$

In the objective function and constraint (34), we represent the makespan as a max function over all movement event end times. Thus we obtain a model of the discretised heterogeneous crusher scheduling problem as follows:

$$MIQPDisc: \qquad \min\ \omega$$

$$\text{s.t.} \quad \omega \geq e_m \qquad\qquad \forall\, m \in \mathcal{M}, \quad (34)$$

$$(1), (8)\text{--}(10), (14)\text{--}(20), (23)\text{--}(33),$$

$$z^\wedge_{m,m'}, z^\succ_{m,m'} \in \{0, 1\},$$

$$x_{m,m',l}, \phi'_{m,m'}, \tau_{m,t}, \tau'_{m,m',t}, \phi_m, t_m, \omega, s_m, e_m \in \mathcal{R}^+.$$

3.3 Overview: Discretised Approach

As we will see in the experiments section (Section 6), the models we presented solve easily in constraint programming, mixed-integer nonlinear programming and mixed-integer quadratic programming solvers for realistic sized instances. One key issue with this approach, however, is that the trucking fleet is allocated to one movement for the entirety of the mining event. This is equivalent to fixing the feedrate for the entire event. On one hand, this alone can lead to poor crusher utilisation. On the other hand, we wish to introduce maintenance tasks, which too can lead to poor crusher utilisation. For example, consider a scheduling problem with one loader and one movement. Suppose the minimum duration of the event is 1000, and the maintenance task occurs between $999 \le t \le 1099$. This will force the makespan to be 2099, while the crusher is doing nothing for the first 1099 time units.

This strongly motivates a need to be able to *partition* the movements on-the-fly. In the following section, we will extend our models to allow for this type of flexible partitioning. For simplicity, we present only the loader maintenance constraints, from which it is straightforward to extend the model to account for truck and crusher maintenance.

4 Build Planning Models with Flexible Partitioning

In this section, we extend the formulations from previous sections by introducing a flexible partitioning of each movement. We restrict our formulation to the case of two partitions. However, further partitions are an easy extension, but with particular attention paid to the symmetry constraints. The flow constraints are sufficient in the form of constraints (14)–(15). The variables s_m, e_m, $\tau_{m,t}$, $z^{\succ}_{m,m'}$, $z^{\wedge}_{m,m'}$, ϕ_m and t_m extend with an additional index representing the partition.

The new partitions have unknown size. Therefore, we require a variable, $c_{m,k}$, to represent the size of the partition (in tonnes), where the total size must equal the original size of the movement:

$$\sum_k c_{m,k} = C_m \qquad \forall\, m \in \mathcal{M}. \qquad (35)$$

Importantly, this leads to an expression for feedrate that is no longer positive semi-definite:

$$\phi_{m,k} = \frac{c_{m,k}}{t_{m,k}} \qquad \forall\, m \in \mathcal{M}, k \in \mathcal{K}. \qquad (36)$$

Symmetry is a big issue when we can partition a movement anywhere and then alternate the order of the partitions. To restrain this computational issue, we introduce the following constraint (in combination with a restriction on k and $k+1$ in constraint (38)):

$$s_{m,k+1} \geq e_{m,k} \qquad\qquad\qquad \forall\, m \in \mathcal{M} \quad (37)$$
$$k < |\mathcal{K}| - 1,$$

$$e_{m',k+1} + \sum_l T^L_{m',m,l} x_{m',m,l} - MS\Big(1 - \sum_l x_{m',m,l}\Big) \leq s_{m,k} \quad \forall\, m \in \mathcal{M} \quad (38)$$
$$m' \in \mathcal{M}'$$
$$m \neq m'$$
$$k < |\mathcal{K}| - 1.$$

We first ensure that the $k + 1$th partition follows the kth partition (of the same movement) with respect to start time—see constraint (37). Then, we ensure that only the first partition includes the loader traversal time.

We incorporate maintenance tasks for the loaders into the partition model as follows. Each maintenance task has a predefined start (s_l^H) and finish (e_l^H) time. We require that each maintenance task does not overlap with the scheduled tasks

$$e_{d_{l,i},k} \leq s_l^H \lor s_{d_{l,i},k} \geq e_l^H \lor d_{l,i} = \bot, \qquad \forall\, l \in \mathcal{L}, k \in \mathcal{K}, i \in 1 \dots N_l^M. \quad (39)$$

We can encode this for MIP models by introducing variables indicating a movement has finished before, $z_{m,k,l}^{H,\prec}$, or after, $z_{m,k,l}^{H,\succ}$, each maintenance task l, and use a big-M approach as follows:

$$e_{m,k} \leq s_l^H + (1 - z_{m,k,l}^{H,\prec})MS + \Big(1 - \sum_{m'} x_{m',m,l}\Big)MS \quad \forall\, m, k, l, \quad (40)$$

$$s_{m,k} \geq e_l^H - (1 - z_{m,k,l}^{H,\succ})MS - \Big(1 - \sum_{m'} x_{m',m,l}\Big)MS \quad \forall\, m, k, l, \quad (41)$$

$$\sum_{m'} x_{m',m,l} = z_{m,k,l}^{H,\prec} + z_{m,k,l}^{H,\succ} \qquad\qquad\qquad \forall\, m, k, l. \quad (42)$$

Thus we obtain the following mixed-integer nonlinear program:

$$MINLPexact:\qquad\qquad \min\quad \omega$$
s.t. $\quad (1)^*, (8)^*, (10)^*, (14)^*–(15)^*, (17)^*–(20)^*, (34)^*, (35)–(38), (40)–(42),$

$$z_{m,k,l}^{H,\prec}, z_{m,k,l}^{H,\succ}, z_{(m,k),(m',k')}^{\wedge}, z_{(m,k),(m',k')}^{\succ} \in \{0,1\},$$

$$\tau_{m,k,t}, x_{m,m',l}, \phi_{m,k}, \phi'_{(m,k),(m',k')} c_{m,k}, \tau'_{(m,k),(m',k')}, t_{m,k}, \omega, s_{m,k}, e_{m,k} \in \mathcal{R}^+.$$

Constraints marked with ($*$) are extended to account for partitions in the obvious way.

Since the partition is flexible, it seems plausible that the MINLP representation provides an optimal solution to the utilisation subproblem. However, this is not the case, as we show in the following theorem.

Theorem 1. *Let m_1 be a movement that can be flexibly partitioned, and let all remaining movements, $\{\mathcal{M} \backslash m_1\}$, be fixed such that they cannot be partitioned. Further, let maintenance tasks only exist for loaders. Then, the minimum number of partitions required for m_1 to guarantee a solution optimality for the overall problem is*

$$|\mathcal{K}_{m_1}| \geq \sum_l |Maint(l)| + |\mathcal{M}|.$$

Proof. Suppose there are $\mathcal{L} > 1$ loaders operating and there are no maintenance tasks. W.l.o.g, let l_1 mine only movement m_1 during the makespan of the build, and the remaining loaders mine the remaining $|\mathcal{M}| - 1$ movements. Since the remaining movements are fixed, if we consider $|\mathcal{M}|$ partitions, then we have considered all possible $|\mathcal{M}| - 1$ event end times, and therefore guarantee optimality. Now suppose each loader, l, has its own set of maintenance tasks, $Maint(l)$. Since each task may introduce a new event *partition* end time, we must consider a further $\sum_{i \in Maint(l)} i$ partitions in order to guarantee optimality.   □
$\forall l \in \mathcal{L}$

Furthermore, when we allow all movements to partition, then the number of partitions (per movement) to guarantee optimality depends on the number of partitions introduced in all the movements. That is, there is a recursive relationship. The take-home message here is that the number of partitions required cannot be determined *a priori*, and therefore should be determined during search.

5   Outer-Approximation and Repair Heuristic

We can obtain a completely linear outer-approximation by introducing McCormick inequalities to represent the bilinear term $\phi_{m,k} \, t_{m,k}$. While a solution to this model is not feasible for the original problem, it may provide us with useful information, such as a suggestion of the partitioning point for the movements. This gives rise to an Outer-Approximation and Repair Heuristic.

We approximate the bilinear term $\phi_{m,k} \, t_{m,k}$, using the upper $(\bar{\Phi}, \bar{T}_m^M)$ and lower $(\underline{\Phi}, \underline{T}_m^M)$ bounds on the variables, with constraints analogous to (23)–(26) to arrive at an ancillary variable, $\mu_{m,k}$. We substitute the ancillary variable into our feedrate constraint:

$$\mu_{m,k} \geq c_{m,k} \qquad\qquad \forall \, m, k. \qquad (43)$$

The outer-approximation model is therefore:

$$MIPouter: \qquad\qquad \min \quad \omega$$

s.t. $(1)^*, (8)^*, (10)^*(14)^*–(15)^*, (17)^*–(20)^*, (23)^*–(26)^*, (34)^*, (35)–(42), (43),$

$$z^{\wedge}_{(m,k),(m',k')}, z^{\smallfrown}_{(m,k),(m',k')} \in \{0, 1\},$$

$$\tau_{m,k,t}, \mu_{m,k}, \phi_{m,k}, \phi'_{(m,k),(m',k')}, x_{m,m',l}, c_{m,k}, \tau'_{(m,k),(m',k')},$$

$$t_{m,k}, \omega, s_{m,k}, e_{m,k} \in \mathcal{R}^+.$$

We use this model to obtain new partition breakpoints of existing movements. Specifically, we take the partitioned movement capacity, $c_{m,k}$, solution and fix this variable. This reduces the problem to a form that can be solved using $MIQPDisc$. This translation requires the following steps. We set the number of movements to be equal to the number of active partitions (i.e., movement partitions with capacity greater than zero). All movements that are split have their new indexes saved in a split movement set, $\mathcal{M}^S$. We perform a translation on all data indexed by movements.

Algorithm 1. The outer-approximation and repair heuristic for partitioning movements in the presence of maintenance tasks.

1: Solve $MIPouter$ model using a MIP solver.
2: **if** Optimal or Feasible **then**
3:    Partition movements according to $c_{m,k}$ solution.
4:    Fix new $C_m = c_{m,k}$.
5:    Solve $MIQPDisc$ model using MIQP solver.
6: **end if**

This repair heuristic is guaranteed to find a feasible partition and will never provide a solution worse than $MIQPDisc$, however it is not guaranteed to find an improved solution. In fact, when there are no maintenance constraints, the $c_{m,k}$ variables are not driven to find good solutions by any mechanism and we expect to obtain solutions equivalent to those found in $MIQPDisc$. However, once maintenance tasks are added, the $c_{m,k}$ variables have a strong bound and will snap to the maintenance tasks. Therefore, we expect much better quality solutions from the repair heuristic for problems with maintenance.

The quality of the repair heuristic is bounded from below by an optimal solution to $MIQPDisc$. Furthermore, if the repair heuristic is run with the same number of partitions as $MINLPexact$, then the quality of the repair heuristic is bounded from above by an optimal solution to $MINLPexact$.

6   Experiments

We validated all models by cross-checking the objective values on a set of validation instances. We created a set of test instances by extending a set of real instances provided by our industry partner. The base set included three weeks of movements, to be subdivided into five builds per week across three pits. In the context of our experiments, the equipment in the three pits are independent; they have their own crusher and truck and loader fleets. We therefore have 45 instances in the base set. On average there are 4.6 blocks and 1.4 stockpiles per pit—the biggest pit containing 9 blocks. The blocks and stockpiles range in size from 1.2kt up to 90kt. One pit has a mixed fleet of 5 loaders (4 types) and 13 trucks (1 type), while the other two pits have a mixed fleet of 6 loaders (2 types), and 17 trucks (2 types). We extend every pit instance in the following ways:

- *Symmetry*: we double the number of loaders from 5 and 6 to 10 and 12;
- *Truck constrained*: we decrease trucks from 13 and 17, to 6 and 8.
- *Maintenance*: for every pit, we schedule maintenance for the complete loader fleet in a cascade, each for 8 hours, decreasing the available loaders by 1 at any time. To make the instances more sensible to these events, we decreased the number of loaders available to 3 in total.

Table 1. When 5 or 300 is not followed by any tag, it refers to the original benchmark with their respective timeouts. Tags *sym, const, maint,* and *maint_e* stand for the symmetric, constrained, maintenance and extreme maintenance variations. *CPU* stands for average CPU time in seconds, *M [min]* stands for average Makespan in minutes, *M. D.* stands for the average build-crushers alignment difference in minutes.

Bench. Id	MIQdisc			MIPouter			MINLPexact			CP		
	CPU	M. [min]	M. D.	CPU	M. [min]	M. D.	CPU	M. [min]	M. D.	CPU	M. [min]	M. D.
‖5‖	1.3	**1699**	284	1.3	1709	288	4.9	2095	1557	3.7	1738	364
‖5‖sym‖	0.7	**1642**	157	1.2	1693	267	5.0	1771	972	3.9	1708	291
‖5‖const‖	1.4	**2405**	1412	1.7	**2405**	1420	5.2	2167	2169	4.2	2408	1853
‖300‖	23.0	1654	186	34.9	**1652**	174	235.0	1774	455	164.8	1713	308
‖300‖sym‖	4.6	**1642**	157	20.4	1653	178	240.8	1761	414	173.3	1699	272
‖300‖const‖	44.3	2369	1393	51.0	2368	1393	268.1	**2341**	1542	208.9	2357	1770
‖300‖maint‖	26.7	2011	750	45.2	2013	758	188.9	**1978**	779	–	–	–
‖300‖maint_e‖	10.7	2521	1263	62.0	2317	1139	152.0	**2277**	1049	–	–	–

- *Extreme Maintenance* as failure: for every pit, we schedule maintenance for the complete loader fleet for 1000 minutes, thus allowing the crusher to be fed only by stockpiles[1] if they are available.

We run experiments on the base set and all extension sets: thereby obtaining 225 instances. Each instance is tested with a time-out of 5 and 300 seconds. We solve the blending problem using Cplex version 12.6 with default settings— all instances solved within milliseconds, and therefore the solver required no intervention to improve computational efficiency. We solve *MIQPDisc* and *MIPouter* with Gurobi version 5.6.3, using tuned parameters GomoryPasses = 0 and PrePasses = 2. We tested the quadratic models using numerical stability settings (i.e., Presolve = 0, FeasTol = $1e - 9$, Quad = 1), but found these had no impact on the validity of solutions. We model *MINLPexact* using Pyomo version 3.5. We solve *MINLPexact* with the Scip Optimization Suite version 3.1.0, using IpOpt version 3.11.8, coinHSL version 2014.01.10 [7] and Cplex version 12.6. We model the constraint programming (*CP*) approach using MiniZinc [14], and solve it using Gecode 4.2.1 [6]. None of the solvers reach the maximum memory allowed of 4GB. *MIQPDisc* and *MIPouter* were able to find a solution for all tested instances, *CP* failed solving only 2 instances of the *constrained* benchmark, *MINLPexact* failed between 3 to 6 instances in all benchmarks when the time limit was set to 5 seconds.

In Table 1, we show the average (CPU) time for solving each instance, the average makespan of each build (M. [min]), and the average completion time difference of the slowest and fastest crusher within the same build (M. D.). *MIQPDisc* is consistently the fastest solver in terms of CPU time, and achieves the best (lowest) average makespan in 5 out of the 8 tested benchmarks. *MIPouter* have a similar performance in terms of makespan quality, outperforming *MIQPDisc* significantly in the extreme maintenance benchmark, which benefits from the flexible partitioning. In the other benchmarks, if *MIPouter* is able to solve the problem optimally, its makespans are as good or substantially

---

[1] Stockpiles have dedicated loaders and do not require a loader to be moved. In practice we extend our models to accommodate this restriction.

Table 2. When 5 or 300 is not followed by any tag, it refers to the original benchmark with their respective timeouts. Tags *sym, const, maint,* and *maint_e* stand for the symmetric, constrained, maintenance and extreme maintenance variations. *Cr. Util.* stands for crusher utilization in (%); *Tr. Util.* is truck utilization in (%).

Bench. Id	MIQdisc		MIPouter		MINLPexact		CP	
	Cr. Util.	Tr. Util.	Cr. Util.	Tr. Util.	Cr. Util.	Tr. Util.	Cr. Util.	Tr. Util.
||5||	**94.3**	53.5	94.0	52.2	79.8	46.5	93.4	55.8
||5||sym||	**96.9**	54.9	94.8	53.2	81.6	56.7	93.9	56.7
||5||const||	**74.3**	75.5	**74.3**	75.4	71.5	59.9	67.8	32.6
||300||	96.3	55.0	**96.4**	54.2	91.7	56.2	94.0	56.3
||300||sym||	**96.9**	55.3	96.2	55.3	91.8	59.5	94.2	56.8
||300||const||	75.1	76.7	**75.1**	76.6	72.4	70.9	68.8	32.8
||300||maint||	81.7	42.6	81.7	42.0	**83.7**	50.6	–	–
||300||maint_e||	66.3	34.2	72.1	36.5	**73.1**	46.2	–	–

better than *MIQPDisc*. This is not reflected on average, as *MIPouter* performs worse than *MIQPDisc* in those instances where it only finds a primal solution in the allotted time. Similarly, *MINLPexact* outperforms all other solvers in both maintenance and constrained benchmarks, when 300 CPU seconds are allowed. We remark that the only solver that substantially benefits from the increase of maximum computation time is *MINLPexact*. Therefore, we tested *MIPouter* and *MINLPexact* with a timeout of 900 seconds, but it did not result in the same quality improvement that we observed by changing from 5 to 300 seconds, so we omit those results. The *CP* approach has a competitive performance with respect to *MIP* solvers, but is not the best solver in any benchmark. *CP* failed solving the flexible partition model, and therefore we did not run this model on the maintenance sets. In terms of crusher alignment (M.D.), clearly the best results are achieved in the original benchmarks with 300 seconds, and even better in the symmetric version where more loaders were available. Crusher alignment is strongly correlated to the success of each solver on achieving 100% crusher utilisation, as we assumed the crushers operated at 100% efficiency in the blending model to help allocate the right proportion of tons for each pit. Whenever the average crusher feedrate decreases, the alignment is likely to be harmed. In Table 2, we provide the crusher and truck utilisation statistics. The best crusher utilisation is achieved by *MIQPDisc* in the 300 CPU time symmetric benchmark, which coincides with the best alignment achieved in Table 1.

MIQPDisc is the best model, achieving a crusher utilisation factor up to 96.9%. The constrained and maintenance benchmark variations harm the ability to fully utilise the crusher. Note that in those variations, it is not possible to achieve a 100% utilisation. The constrained version harms most significantly the *CP* approach, while the best models in the constrained version are *MIQPDisc* and *MIPouter*. The best models to handle maintenance tasks are *MIPouter* and *MINLPexact*, achieving an improvement of 6% and 7% respectively over *MIQPDisc*. Again, this highlights the benefits of flexible partitions.

Loader utilisation is not shown in the table, as this resource is unconstrained in practice. We observe that in the data given to us by our industry partner, truck fleet availability is not too constrained either. As they pointed out, truck resources can become scarce in other data sets. To confirm this statement, we observed in the original benchmark that the most extremely truck constrained pit solved by $MIQPDisc$ resulted in truck utilisation of 99.06% with 90% crusher utilisation, which highlights the ability of the solvers to push the truck utilisation to the maximum if it is required.

7   Discussion

An advantage of modelling the full problem using a high-level language, such as MiniZinc, is that it can lead to a more intuitive and natural representation of the problem. This simplified the extensions to a compact mathematical programming form, where the linearisations and logical constraints can sometimes be inelegant and cumbersome to de-bug. In this sense, it was beneficial to have multiple models from which we could validate the others results.

Beyond constraint programming and mathematical programming, the flexible partition scheduling problem could also be cast as a temporal planning problem over continuous variables with processes and events. We modelled this problem using the high-level *Planning Domain Description Language* (PDDL 2.1) [5], but no solver technology could handle the model. However, the flavour of planning encodings—event-driven modelling—is still present in our modelling approach.

Our industry partner solves this problem with an experienced planner and grade-controller, who develop plans by hand. They aim to achieve 80% utilisation at the crushers, after all practical constraints have been taken into account. While we cannot claim to have modelled all practical constraints, it it clear that these models will provide our partner with a useful decision-making tool with huge potential to push their crusher utilisation well beyond 80% where possible.

Acknowledgments. The authors wish to thank Mike Godfrey, Jon Lapwood, Vish Baht and John Usher from Rio Tinto for extensive discussions throughout our research. This research was co-funded by the Australian Research Council linkage grant LP11010015 "Making the Pilbara Blend: Agile Mine Scheduling through Contingent Planning" and industry partner Rio Tinto.

References

1. Aggoune, R.: Minimizing the makespan for the flow shop scheduling problem with availability constraints. European Journal of Operational Research **153**(3), 534–543 (2004)
2. Bley, A., Boland, N., Froyland, G., Zuckerberg, M.: Solving mixed integer nonlinear programming problems for mine production planning with stockpiling. Tech. rep., University of New South Wales (2012). http://web.maths.unsw.edu.au/froyland/bbfz.pdf
3. Burt, C.N., Lipovetzky, N., Pearce, A.R., Stuckey, P.J.: Approximate unidirectional Benders decomposition. In: Proceedings of PlanSOpt-15 Workshop on Planning, Search and Optimization AAAI-15 (2015)
4. Coal Shovel Clip Art: Accessed: 17/11/2014 (2014). gofreedownload.net/

5. Fox, M., Long, D.: PDDL2. 1: An extension to PDDL for expressing temporal planning domains. Journal of Artificial Intelligence Research **20**, 61–124 (2003)
6. Gecode Team: Gecode: Generic constraint development environment (2006). http://www.gecode.org
7. HSL: a collection of Fortran codes for large scale scientific computation (2013). http://www.hsl.rl.ac.uk
8. Immersive Technologies: Accessed: 17/11/2014 (2014). http://www.immersivetechnologies.com/
9. Jamshidi, R., Esfahani, M.M.S.: Reliability-based maintenance and job scheduling for identical parallel machines. International Journal of Production Research **53**(4), 1216–1227 (2015)
10. Khayat, G.E., Langevin, A., Riopel, D.: Integrated production and material handling scheduling using mathematical programming and constraint programming. In: Proceedings CPAIOR (2003)
11. Kubzin, M.A., Strusevich, V.A.: Planning machine maintenance in two-machine shop scheduling. Operations Research **54**(4), 789–800 (2006)
12. Lipovetzky, N., Burt, C.N., Pearce, A.R., Stuckey, P.J.: Planning for mining operations with time and resource constraints. In: Proceedings of the Twenty-Fourth International Conference on Automated Planning and Scheduling, ICAPS 2014 (2014)
13. Moradi, E., Ghoma, S.F., Zandieh, M.: Bi-objective optimization research on integrated fixed time interval preventive maintenance and production for scheduling flexible job-shop problem. Expert Systems with Applications **38**(6), 7169–7178 (2011)
14. Nethercote, N., Stuckey, P.J., Becket, R., Brand, S., Duck, G.J., Tack, G.R.: MiniZinc: towards a standard CP modelling language. In: Bessière, C. (ed.) CP 2007. LNCS, vol. 4741, pp. 529–543. Springer, Heidelberg (2007)
15. Sbihi, M., Varnier, C.: Single-machine scheduling with periodic and flexible periodic maintenance to minimize maximum tardiness. Computers & Industrial Engineering **55**, 830–840 (2008)
16. Ta, C., Kresta, J., Forbes, J., Marquez, H.: A stochastic optimization approach to mine truck allocation. International Journal of Surface Mining **19**, 162–175 (2005)

Learning Value Heuristics for Constraint Programming

Geoffrey Chu[(✉)] and Peter J. Stuckey

National ICT Australia, Victoria Laboratory, Department of Computing
and Information Systems, University of Melbourne, Melbourne, Australia
{geoffrey.chu,pstuckey}@unimelb.edu.au

Abstract. Search heuristics are of paramount importance for finding
good solutions to optimization problems quickly. Manually designing
problem specific search heuristics is a time consuming process and
requires expert knowledge from the user. Thus there is great interest
in developing autonomous search heuristics which work well for a wide
variety of problems. Various autonomous search heuristics already exist,
such as first fail, domwdeg and impact based search. However, such
heuristics are often more focused on the variable selection, i.e., picking important variables to branch on to make the search tree smaller,
rather than the value selection, i.e., ordering the subtrees so that the
good subtrees are explored first. In this paper, we define a framework
for learning value heuristics, by combining a scoring function, feature
selection, and machine learning algorithm. We demonstrate that we can
learn value heuristics that perform better than random value heuristics,
and for some problem classes, the learned heuristics are comparable in
performance to manually designed value heuristics. We also show that
value heuristics using features beyond a simple score can be valuable.

1 Introduction

Search heuristics are of paramount importance for finding good solutions to optimization problems quickly. Search heuristics can roughly be divided into two
parts: the variable selection heuristic, which selects which variable to branch on,
and the value heuristic, which determines which value is tried first. There has
been significant research on autonomous search heuristics including: first fail [1],
variable state independent decaying sum (VSIDS) [2], domain size divided by
weighted degree (domwdeg) [3], impact based search [4], solution counting based
search [5], and action[1] based search [6]. Most of these search heuristics concentrate on variable selection, as this is critical in reducing the size of the search
tree, although some, in particular impact and action based search also generate value heuristics. Phase saving [7] if a value-only heuristic which reuses
the last value of a Boolean variable (its phase) when it is reconsidered. In

---

[1] It was originally called activity-based search, we use the alternate name to distinguish
it from the long established activity-based search used in SAT, SMT and LCG
solvers.

© Springer International Publishing Switzerland 2015
L. Michel (Ed.): CPAIOR 2015, LNCS 9075, pp. 108–123, 2015.
DOI: 10.1007/978-3-319-18008-3_8

this paper, we focus on learning useful value heuristics for improving constraint programming search.

Given a current domain D and a variable x to branch on, assuming a maximization problem, the task of the value heuristic is to give the order in which the values in x's current domain should be explored. This is typically accomplished by defining a scoring function $g(D, x, v)$ which gives a score indicating how good assigning the value v to x is likely to be given the current domain D. The values are then sorted based on their scores and visited in decreasing score order. Ideally, g is a function such that $g(D, x, v_1) \geq g(D, x, v_2)$ iff the optimal value down the $x = v_1$ branch is greater than or equal to the optimal value down the $x = v_2$ branch. Such a value heuristic would immediately lead us to the optimal solution. In practice, a perfect scoring function is not likely to be feasible to compute and hence we will settle for a heuristic that is likely to have ordered the good/optimal branches near the front.

Many optimization problems have good, manually designed scoring function which allow the solver to find good solutions quickly. However, manually designing scoring functions can be a time consuming process and requires expert knowledge from the user. Thus there is significant value in developing autonomous search heuristics which work well for a wide variety of problems. One way to produce an autonomous value heuristic is to treat the design of the scoring function $g(D, x, v)$ as a machine learning problem. In order to do so, we have to characterize the current domain D using a set of appropriate features, generate a set of appropriate training instances along with their scores (i.e., values for D, x, v and the output value of the function for these arguments), and use an appropriate regression technique from machine learning to learn the function g.

Several autonomous search heuristics already exist, such as impact based search [4] and action based search [6]. The value heuristics suggested in these two methods can be seen as simple instances of machine learning. In both cases, the current domain is characterized by 0 features (i.e., both of these methods completely ignore the current domain when scoring a value), the training instances are collected during search or during an initial probing phase, and the score assigned to each training instance is the impact (i.e., proportional reduction in domain size) for impact based search, and the number of variables with reduced domains for action based search. In both cases, since there are no features used, the learning simply consists of taking the average score of all the training instances involving an assignment $x = v$ and assigning that average score as the value of $g(D, x, v)$.

There are several possible improvements to these methods. First, both of these methods do not use the current domain in the scoring function at all. This may be fine for problems where the merit of an assignment $x = v$ is largely independent of what else has been assigned. However, in problem classes where the merit of an assignment depends significantly on what else has been assigned, we should be able to learn a much better scoring function by taking into account the current domain D. Thus we are interested in finding features of D which help us to predict the merit of an assignment $x = v$ and using them in our machine learning algorithm. We claim that features of variables which are closer to the decision variable in the constraint graph are more likely to be predictive of the merit of its values. Thus we propose using the features of variables within a

k-neighborhood of the decision variable in the constraint graph as our features, where k is a parameter of our algorithm.

Second, the scores assigned to the training instances in the above two methods, i.e., impact and the number of domain changes, are only indirect measures of how good the subtree is, and there may be better ways to assign scores to the training instances. Indeed neither of these scores consider the objective function of the problem. We propose an alternative scoring method based on the pseudo-cost [8,9], i.e., the change in bound of the objective function after propagating the decision.

Note that the application of machine learning to Constraint Programming in this paper is significantly different from the large body of work using machine learning for solver/algorithm selection in portfolio based solvers (e.g. [10]). There, machine learning is used to predict how well existing solvers/algorithms may perform on a particular instance in order to select a solver/algorithm which works well for the instance. Here, we are using machine learning to predict how well a particular value assignment may do in order to generate new search heuristics. Clearly, these two uses of machine learning are complementary and it is possible to use the search heuristics we generate as the input to the algorithm selection problem.

The contributions of this paper are:

- A framework for learning value heuristics by defining scoring functions, and using linear regression over a restricted class of features of the problem
- A new scoring function, analogous to that used in pseudo-costs [8,9] we can use to define a value heuristics
- A new method of taking the objective function into account for constraint programming search.
- Experiments demonstrating that learnt value heuristics can be as effective as programmed value heuristics

The remainder of the paper is organized as follows. In Section 2, we go through our definitions and background. In Section 3, we describe how to generate training instances for the machine learning algorithm. In Section 4, we discuss feature selection and the machine learning algorithm. In Section 5, we present experimental results. In Section 7, we conclude and discuss future work.

2   Background

Constraint programming A constraint optimization problem (COP) P is a tuple (V, D, C, f) where V is a set of variables, D is a set of domains, C is a set of constraints, and f is an objective function. Let D_x be the domain of variable x. Without loss of generality, we assume the objective function f is to be maximized. A CP solver solves a COP P by interleaving search with inference. It starts with the original problem $P = (V, D, C, f)$ at the root of the search tree. At each node in the search tree, it repeatedly propagates the constraints $c \in C$ to try to infer variable/value pairs in the current domain D which cannot take part in any improving solution to the problem within that subtree. Such pairs are removed from the current domain D to create a new domain D'. The process is repeated until no more pairs can be removed. We denote this as $D' = \textbf{solv}(C, D)$. If the resulting domain D' is a *false domain*, i.e. $D'(v) = \emptyset$ for some $v \in V$, then the subproblem has no solution and the solver backtracks. Once the propagation fixed point is reached, if all the variables are assigned, then a solution θ has been found. The solver adds a branch and bound constraint constraining the solver to find only solutions with better objective value than θ, and then continues the search. If not all variables are fixed, then the solver further divides the problem into a number of more constrained subproblems and searches each of those in turn. The search heuristic determines how this division is performed. Typically, the search strategy consists of two parts, a variable selection heuristic which picks an unassigned variable x to branch on, and a value heuristic which pick a value v to try. The search will then explore $x = v$ down one branch and $x \neq v$ down the other branch.

Given a constraint problem $P \equiv (V, D, C, f)$, let its constraint graph G be the graph with the variables V as nodes, and with an edge between two variables $x, y \in V$ iff x and y appear together in at least one constraint $c \in C$. Given a graph G, let the k-neighborhood of a node x in graph G be the set of all nodes within a distance k of node x.

Impact Based Search. Impact based search was proposed in [4]. The impact of a decision $x = v$ can be defined as follows. Let D be the domain before the decision, and $D' = \textbf{solv}(C \cup \{x = v\}, D)$ be the domain after the decision has been propagated to fixed point. The impact of the decision is then: $1 - \prod_{x \in V} |D'_x| / |D_x|$. In impact based search, a running average $\bar{I}(x = v)$ of the impact of each assignment $x = v$ is maintained. The impact of a variable x given the current domain D is given by $\sum_{v \in D_x} 1 - \bar{I}(x = v)$. The variable heuristic picks the variable x with the highest impact and the value heuristic picks the value with the lowest impact.

Action Based Search. Action based search was proposed in [6]. At each decision in the search tree, the action $A(x)$ of each variable x is decayed by some factor α if x was not fixed before the decision, and $A(x)$ is increased by 1 if its domain was reduced after propagating the decision. The variable heuristic chooses the variable with the highest $A(x)/|D_x|$ value. The action of an assignment $x = v$ is the running average of the number of variables whose domains were reduced after propagating a decision $x = v$. The value heuristic chooses the value with the lowest action.

Linear Regression. In supervised learning, there is an underlying function $h : X \rightarrow Y$ which we wish to learn, and we are given a set of training instances $\{(\bar{x}_1, \bar{y}_1), \ldots, (\bar{x}_n, \bar{y}_n)\}$ such that $\bar{y}_i = h(\bar{x}_i)$ for each i. The goal is to learn an approximation h' of h which is as close to h as possible under some notion of error. The inputs $\bar{x}_i$ and the outputs $\bar{y}_i$ could be single values or could be a vector of values. In this paper, we are interested in the case where the inputs are a set of Boolean or numeric features $x_1, \ldots, x_m$, the output is a single numerical value y, and we are interested in learning a linear function $y = \sum a_i x_i + a_0$ which relates the inputs and the output, where Boolean features are considered 0-1 numeric variables. One common method for doing this is ordinary least squares regression (OLS) (see e.g.[11]). Unfortunately, OLS is insufficient for our purposes as it requires there to be more training instances than there are features, and the features must be linearly independent. An alternative is partial least squares regression (PLS) [12]. PLS attempts to project the input into a lower dimensional space represented by *latent variables* such that these latent variables explain as much of the variance in the output as possible. The number of latent variables to use is a parameter of the algorithm. PLS is able to handle cases where features may be linearly dependent or where there may be far more features than training instances.

3   Generating Training Instances

In this paper, we would like to treat the design of the scoring function $g(D, x, v)$ used in the value heuristic as a machine learning problem. However, unlike a typical machine learning problem where we are given a set of training instances, in this case, we need to generate our own. Furthermore, it is not obvious what the function $g(D, x, v)$ is supposed to output. One possibility is try to learn a function $g(D, x, v)$ which outputs the optimal value of the subproblem $(V, D, C \cup \{x = v\}, f)$. To do this, we could generate some training instances by picking some D, x and v values and solving the COP's $(V, D, C \cup \{x = v\}, f)$ exactly to get the correct output values. However, this is clearly highly impractical, because solving $(V, D, C \cup \{x = v\}, f)$ exactly is very expensive and we would have to do this for each training instance we want to generate. Alternatively, we could try to learn a function g which outputs an easier to calculate measure which is predictive of how high the optimal value of $(V, D, C \cup \{x = v\}, f)$ is. As long as this measure tends to have higher values for subproblems with higher optimal value, it can still be a good value heuristic.

We consider three different approximate measures for use in computing g: those used in impact and action based search which do not make use of the objective function f of the problem; and one other which attempts to take into account the objective. These measures are

score_impact Impact based search tries to learn a function g which predicts the impact of a particular assignment, with the assumption that lower impact tends to lead to better solutions. We will call this score_impact defined as $g_{score\_impact}(D, x, v) = 1 - \prod_{x \in V} |D'_x|/|D_x|$ where $D' = \mathbf{solv}(C \cup \{x = v\}, D)$.

score_num_red Action based search tries to learn a function g which predicts how many variables will have their domains reduced by a particular assignment, with the assumption that fewer domain reductions lead to better solutions. We'll call this score_num_red defined as $g_{score\_num\_red}(D, x, v) = |\{x \in V \mid D_x \neq D'_x\}|$ where $D' = \mathbf{solv}(C \cup \{x = v\}, D)$.

score_pseudo_cost Pseudo-cost branching [8,9] is an important variable selection strategy in mixed integer programming. Recall that we are assuming that the objective f is to be maximized. We try to learn a g which predicts how much the upper bound of the objective function f will decrease by when the assignment is made. Value choices for which the upper bound of f decreased less are likely to lead to better solutions. We'll call this score_pseudo_cost defined as $g_{score\_pseudo\_cost}(D, x, v) = \max D_f - \max D'_f$ where $D' = \mathbf{solv}(C \cup \{x = v\}, D)$.

Generating training instances to learn these three measures is much easier than generating instances to learn a function to predict the optimal value. Similar to impact based search and action based search, we propose to generate training instances with an initial probing phase followed by a normal search phase. In the probing phase, we use a random value heuristic, restart after every solution, and do not perform branch and bound. The aim of this phase is to get a good coverage of all the assignments. In the normal search phase, we use the learned value heuristic and perform branch and bound as normal. At each node during each of these two phases, when we get to the propagation fixed point and make a decision, we record those (D, x, v) values as a new training instance, and depending on which of the three scoring functions we are trying to learn, the score for this training instance will either be: the impact, the number of variables with reduced domains, or the change in the upper bound of f.

4   Feature Selection

In order to apply machine learning techniques to this problem, we need to define the set of features to be used in the model. Potentially, we could train a single model for $g(D, x, v)$ where x and v are considered features. However, we expect that the relevant features and the way that they affect the value could be very different for different values of x and v. Instead, we train a separate model for each possible assignment $x = v$, i.e., we learn a set of functions $g_{x_1,v_1}(D)$, $g_{x_1,v_2}(D), \ldots, g_{x_n,v_m}(D)$ s.t. $g(D, x, v) = g_{x,v}(D)$.

We need to extract from D a set of good features for predicting the value of the function we are trying to learn. We claim that the domains of the variables in the problem contain many of the features which are useful for predicting the value. Furthermore, we claim that it is typically the features of the variables which are close to the decision variable in the constraint graph which are the most useful. This is borne out by our analysis of the custom search heuristics for a variety of problems. In most of these custom search heuristics, the features used in the scoring function are simply the lower bounds, upper bounds or assignments of variables close to the decision variable in the constraint graph.

Example 1. Consider the minimization of open stacks problem [13]. We have a set of customers and a set of products. Each customer requires some subset of the products, and has a stack which must be opened when any product they require begins production. The customer's stack can be closed when all the products that the customer requires have finished production. We wish to find the order in which to produce the products such that the maximum number of open stacks at any time is minimized. It has been shown that rather than a model where we determine the order in which to produce products, it is better to determine the order in which we close the stacks of the customers [13]. In the model proposed in [13], we create a customer graph G where the nodes are customers and there is an edge between two customers iff there is a product that they both require. Closing a particular customer's stack means that all the products they require

```
 1 int: n;                                  % number of customers
 2 set of int: CUST = 1..n;
 3 set of int: TIME = 1..n;
 4 array[CUST,CUST] of bool: g;             % customer graph
 5
 6 array[TIME] of var CUST: x;              % which customer's stack is closed at time t
 7 array[CUST,TIME] of var bool: open_before;   % customer c open before time t
 8 array[CUST,TIME] of var bool: closed_before; % customer c closed before time t
 9 array[CUST,TIME] of var bool: open_during;   % customer c is open at time t
10 var CUST: stacks;                        % number of stacks required
11
12 constraint forall (c in CUST) (not closed_before[c, 1]);
13 constraint forall (c in CUST, t in 2..n)
14   ( (closed_before[c,t] = (closed_before[c,t-1] \/ x[t-1] = c)) /\
15     (closed_before[c,t] -> x[t] != c) );
16 constraint forall (c in CUST, t in TIME)
17   (open_before[c,t] = ((if t > 1 then open_before[c,t-1] else false endif)
18                        \/ exists (d in CUST where g[c,d]) (x[t] = d)) );
19 constraint forall (c in CUST, t in TIME)
20   (open_during[c,t] = (open_before[c,t] /\ not closed_before[c,t]));
21 constraint forall (t in TIME)
22   ( sum (c in CUST) (bool2int(open_during[c,t])) <= stacks );
23
24 solve minimize stacks;
```

Fig. 1. A MiniZinc [14] model for minimization of open stacks

must be produced before that time, which in turn means that the stacks of all its neighbors in the customer graph must be opened before that time. This leads to the model shown in Figure 1.

A good variable ordering is simply to label the x variables in order, as that produces the best propagation. The value heuristic proposed in [13] picks the customer which opens the fewest new stacks at each stage. In terms of the variables in this model, the score for the decision $x[t] = c$ can be written as: $\sum_{d \in CUST \text{ where } g[d,c]} (opened\_before[d,t] - 1)$, where we are simply giving a penalty of 1 to each stack that closing customer c's stack would force open and which is not already open. Clearly, this scoring function is simply a linear combination of the values of variables which already exist in the model.

We divide integer variables into two classes: value type integer variables and bound type integer variables.

- Value type integer variables typically have small domains. The value are unordered and each value means a completely different thing. They are typically involved in constraints like **alldifferent**, **element** or **table** where there is a lot of propagation based on values. For value type integer variables x for each value v in its original domain, we take the truth values of $D \Rightarrow x = v$ and $D \Rightarrow x \neq v$ as features where D is the current domain.
- Bound type integer variables on the other hand could have much larger domains, and the values are ordered, so values close together are closely related. They are typically involved in constraints like **cumulative** or linear constraints where there is only propagation based on bounds. For bounds type integer variables, we take their lower bound and upper bounds as features.

For a Boolean variable b, we take the truth values of $D \Rightarrow b$ and $D \Rightarrow \neg b$ as features. When used in a linear regression, integer features are kept as is, while Boolean features are converted to 0-1 integers.

Example 2. Suppose we have Boolean variables b_1, b_2 and b_3, with current domain $b_1 \in \{true\}$, $b_2 \in \{false\}$ and $b_3 \in \{false, true\}$. The two features for a Boolean variable b are the truth values of $D \Rightarrow b$ and $D \Rightarrow \neg b$. For b_1, they are 1 and 0 respectively. For b_2, they are 0 and 1 respectively. For b_3, they are 0 and 0 respectively. Suppose we have value type integer variables x_1 and x_2, both with original domain $\{1, 2, 3\}$ and current domains $x_1 \in \{1, 3\}$ and $x_2 \in \{2\}$. The features are the truth values of $D \Rightarrow x = v$ and $D \Rightarrow x \neq v$ for each v in the original domain. For x_1, this gives 0 and 0 for $v = 1$, 0 and 1 for $v = 2$, and 0 and 0 for $v = 3$. For x_2, this gives 0 and 1 for $v = 1$, 1 and 0 for $v = 2$ and 0 and 1 for $v = 3$. Suppose we have bound type integer variable x with current domain $\{2, \ldots, 153\}$. The two features are simply its lower and upper bound, i.e., 2 and 153.

In general, it is difficult to tell which variables have features which are useful for the function we wish to learn. We could of course, use the features of all the variables in the problem and use some standard feature selection algorithm to find a good subset of them. However, such methods are far too expensive in this context and are prone to over-fitting due to the large number of potential features and a limited number of training instances. Instead, we exploit our knowledge that variables closer to the decision variable in the constraint graph tend to be more useful and define a series of subsets of features to check. For each $k = 0, 1, 2, \ldots$, we pick the features of the variables in the k-neighborhood of the decision variable in the constraint graph as our features. Using a larger neighborhood may mean that useful features got included, improving the performance of the learned function, but it may also add irrelevant features and produce over-fitting as well as increase overhead. Note that using the 0-neighborhood with score_impact and score_num_red corresponds to the value heuristics used in impact based search and action based search respectively. However, here we have the potential to use higher k to learn that other assignments have an effect on the current decision.

After the training instances are generated and the features are selected, we can run our regression algorithm. We choose to use the partial least squares regression method. The reason is that the vast majority of custom scoring functions we analyzed were simple linear combinations of features, and thus we believe a linear function should do well. Secondly, we have to deal with co-linearity in the features as well as the possibility that there are more features than training instances. Partial least squares regression is able to handle all these and is therefore a good choice. We run the regression algorithm once when the probing phase is complete. After that, we re-run it every time we double the number of our training instances. In the special case where we are using a 0-neighborhood of features, there are actually no features at all, so we can simply keep a running average and update the scoring function whenever we get a new training instance.

Example 3. Consider the minimization of open stacks problem again. Suppose we are branching on $x[t]$. A 1-neighborhood will include the *open_before* and *closed_before* variables from time t and $t - 1$. A 2-neighborhood would include the *open_before* and *closed_before* variables from time $t - 2$ to $t + 1$, as well as the *open_during* variables from time t. Suppose we use a 1-neighborhood with the score_num_red scoring function. We pick a random decision from a random instance for illustrative purposes. In this instance, the custom scoring function for the assignment $x[3] = 4$ is: $1 * open\_before[3, 4] + 1 * open\_before[3, 15] + 1 * open\_before[3, 27] + 1 * open\_before[3, 29] - 4$.

The scoring function learned using partial least squares regression after the training instance has significantly more terms. However, the terms with the largest (absolute value of) coefficients are: $28.716 * open\_before[3, 4]$, $28.740 * open\_before[3, 15]$, $24.047 * open\_before[3, 27]$, $24.485 * open\_before[3, 29]$, $33.880 * closed\_before[3, 16]$, $-26.664 * closed\_before[3, 19]$, $-24.726 * closed\_before[3, 26]$, and it can be seen that the features considered important in the custom scoring function also have large coefficients in this learned scoring function. However, several terms not in the custom scoring function also have large coefficients here, possibly representing other useful features. In practice however, despite the differences, our experiments in Section 5 show that this learned value heuristic is almost identical in strength to the custom one.

5 Experiments

The experiments are run on Intel Xeon 2.40GHz processors using the CP solver Chuffed. We use the minimization of open stacks problem (see e.g. [13]) (MOSP), the talent scheduling problem (see e.g. [15]) (Talent), the resource constrained project scheduling problem (see e.g. [16]) (RCPSP), the nurse scheduling problem [17] (Nurse), the traveling salesman problem (TSP), and the soft car sequencing problem [18] (CarSeq). We select some hard instances from the J60 benchmark for RCPSP and generate 100 random instances for the other 5 problem classes. MiniZinc models and data for the problems can be found at: www.cs.mu.oz.au/~pjs/learn-value-heuristic/.

For each problem, we use a k-neighborhood for feature selection as described in Section 4 with $k = 0, 1, 2$. For RCPSP and TSP, $k = 1$ is identical to $k = 2$ since it already includes all the variables, so we only give results for $k = 1$. We try each of the three scoring methods for the training instances described in Section 3. We use a 10 second probing phase followed by a 590 second search phase for a total of 10 minutes per instance. We use a limit of 10 latent variables in the partial least squares regression method.

Since we are principally interested in the value heuristic part of the search heuristic in this paper, for the first experiment we use the same variable selection heuristic for all the different settings of the value selection heuristics so we can just compare the effect of the value heuristic.

We use an in-order variable heuristic for the minimization of open stacks problem, the talent scheduling problem, the nurse scheduling problem, and the soft car sequencing problem. We use a max-regret variable heuristic for the traveling salesman problem. And we use the earliest first variable heuristic (also called schedule generation [19]) for the resource constrained project scheduling problem. We also compare using a random value heuristic and a manually designed value heuristic. We use the following manually designed value heuristics. For the open stacks problem, we use the one described in [13], which tries to pick the customer which opens the fewest new stacks. For the talent scheduling problem, we pick the scene which minimizes the cost of new actors plus the cost

Table 1. Cost of partial least squares regression as a percentage of total run time

	1-neighborhood	2-neighborhood
MOSP	0.4%	3.3%
Talent	1.1%	3.2%
RCPSP	9.1%	–
Nurse	6.2%	67.5%
TSP	1.2%	–
CarSeq	0.4%	1.7%

Table 2. Solution quality at the end of 10 minutes

	random	custom	pseudo_cost-0	impact-0	num_red-0	pseudo_cost-1	impact-1	num_red-1	pseudo_cost-2	impact-2	num_red-2
MOSP	23.6	13.6	26.8	19.3	21.4	26.6	18.4	14.0	26.0	20.3	14.3
Talent	999	422	678	865	770	650	687	580	640	693	571
RCPSP	972	119	119	121	426	798	974	828	–	–	–
Nurse	66.5	136.9	127.6	70.3	86.5	93.8	68.2	78.4	98.9	66.5	77.1
TSP	1157	521	527	874	881	535	968	1036	–	–	–
CarSeq	32.4	8.4	24.0	35.4	35.4	31.6	26.6	27.0	29.0	32.7	31.1

of actors who are on-location but not in the scene. For the resource constrained project scheduling problem, we assign the start time to its current lower bound. For the nurse scheduling problem, we assign the nurse to the available shift they most prefer. For the traveling salesman problem, we pick the closest available city. For the car-sequencing problem, we pick the car type which utilizes the most heavily loaded available machine.

The average cost of the partial least squares regression as a percentage of run time is given in Table 1. The costs are generally quite small at just a few percent, however, for nurse scheduling with a 2-neighborhood, it grows to a rather massive 67.5%.

The solution quality at the end of 10 minutes is given in Table 2. The graph for average solution quality over time is given for each problem in Figures 2. The best learned heuristics are: for open stacks score_num_red-1, for talent scheduling score_num_red-2, for RCPSP score_pseudo-cost-0, for nurse scheduling score_pseudo-cost-0, for travelling salesman score_pseudo-cost-0, and for car sequencing score_pseudo-cost-0.

Note that these searches do not tend to find any good solutions during the initial 10 second probing phase where it is using a random value heuristic with no branch and bound. After that however, they may start finding much better solutions. It can be seen that in all the problems, there are some settings which allow the algorithm to learn a value heuristic which is significantly better than random. In some cases, the learned value heuristic is of comparable performance to the manually designed value heuristics. It can be seen that using a k-neighborhood with $k > 0$ is highly beneficial on problems like minimization of open stacks and talent scheduling, where whether a particular value is good or not depends significantly on what other decisions have been made. On other problems however, the extra features from using a larger neighborhood are not useful and only cause over-fitting, degrading the performance of the learned value heuristic. It can also be seen that using score_pseudo_cost to score the training instances is far better than using score_impact or score_num_red in many cases.

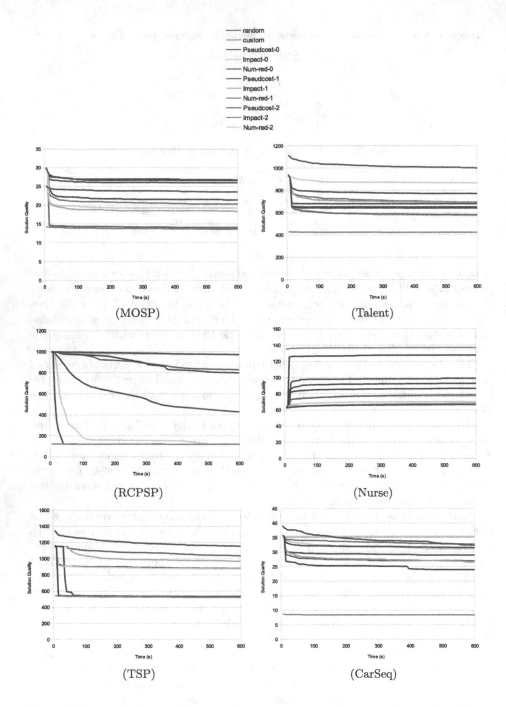

Fig. 2. Solution Quality vs Time graph for various value heuristics for the 6 problem classes. Smaller is better for all except Nurse where larger is better.

Table 3. Number of times each setting was best out of 1000 random samples of 5 instances

	pseudo_cost-0	impact-0	num_red-0	pseudo_cost-1	impact-1	num_red-1	pseudo_cost-2	impact-2	num_red-2
MOSP	0	0	0	0	0	635	0	0	365
Talent	0	0	0	0	1	382	5	1	611
RCPSP	1000	0	0	0	0	0	–	–	–
Nurse	1000	0	0	0	0	0	0	0	0
TSP	761	0	0	239	0	0	–	–	–
CarSeq	840	0	0	0	107	51	0	0	2

Table 4. Solution quality at the end of 10 minutes

	random	custom	impact	action	vsids	ml-5
MOSP	23.6	13.6	19.2	18.9	14.0	14.1
Talent	999	422	992	1068	1236	575
RCPSP	972	119	774	415	118	119
Nurse	66.5	136.9	69.0	76.1	140.5	127.6
TSP	1157	521	108	846	722	529
CarSeq	32.4	8.4	52.8	52.3	29.6	24.4

Although it may be difficult to know beforehand which settings will be best for a problem class, the relative performance of each setting is usually the same across all instances in a problem class, i.e., the good settings tend to do well on all instances and the bad settings tend to do badly on all instances. Hence if we need to solve a large number of instances from the same problem class, we can simply solve a few sample instances using the different settings, and use the setting which had the best average performance on the sample instances for the rest of the instances in the benchmark.

In Table 3, we show the number of times each setting had the best performance for 1000 different random samples of 5 instances from each of the benchmark. As can be seen, simply by trying the different settings on 5 instances, we will almost always end up picking the optimal or near optimal setting for the problem class.

In the third experiment we compare our method against various existing autonomous searches. From the first and second experiments, we can work out the expected solution quality over time curve of our method when we pick the setting by picking the best performing one on a random sample of 5 instances. That is, we take a weighted average of the curves in the first experiment, where they are weighted by the numbers in Table 3. We will call this ml-5. We compare against full impact based search impact, action based search action, the variable state independent decaying sum heuristic vsids, and the random and custom search heuristics from the first experiment.

The solution quality at the end of 10 minutes is shown in Table 4. The graph for average solution quality over time is given for each problem in Figures 3. It can be seen that our new value heuristic is generally much better at finding good solutions than the other autonomous searches, although for nurse scheduling, VSIDS is so good that it beats our heuristic and even beats the custom search.

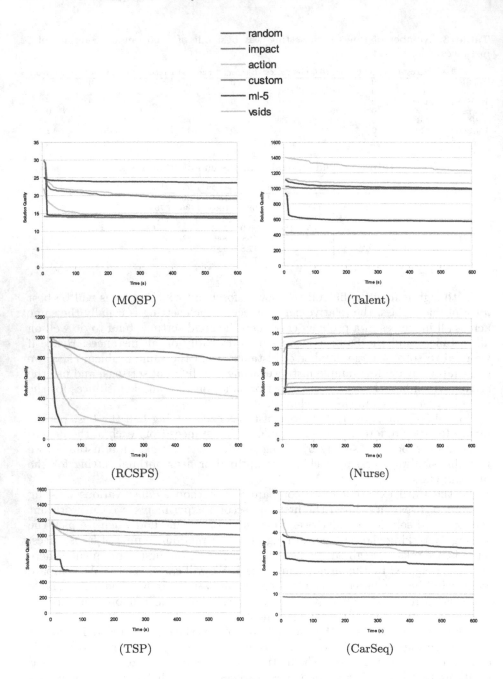

Fig. 3. Solution Quality vs Time graph for various search heuristics on the 6 problem classes. Smaller is better for all except Nurse where larger is better

6   Related Work

A closely related work is Bandit-based Search for Constraint Programming [20]. This method is based on Monte-Carlo tree search and uses reinforcement learning to learn which values should be explored first. They use a reward function based on whether the decision led to a failure depth which was above or below the average, so they do not really target optimization problems. This method differs from ours in that it uses reinforcement learning and does not attempt to predict the reward from features of the domain like our method.

Solution counting [5] is a powerful method for defining autonomous search, giving both variable and value heuristics. It relies on extending propagators to count or estimate the number of remaining solutions they have. It learns estimators for variable value pairs, similar to impact and action based search. Again it does not directly take into account the objective. It would be interesting to explore measures based on counting in our framework, where the current domain D can also be taken into account.

Regret [21] is a commonly used variable selection strategy for CP problems where the objective is the sum of a set of variables. The regret is equivalent to the difference in score_pseudo_cost for the two largest values that remain in the domain of a variable, for these problems.

Pseudo-cost branching [8,9] is an important MIP heuristic for variable selection. It is also used in the ToulBar2 [22] weighted CSP solver. It ranks variables by the expected gain per unit change in the variable, which is the difference between neighbouring values in score_pseudo_cost. Value heuristics are not common in MIP search, since node exploration is more commonly implemented by selecting from a frontier of open nodes, but it would be unsurprising if pseudo-cost had been used as a value heuristic in MIP.

Given that regret for CP and pseudo-costs for MIP are important search heuristics it is surprising that we are not aware of widespread use of pseudo-cost for CP search heuristics.

7   Conclusion

Autonomous search is an important topic for constraint programming, since its removes the burden from the modeller of deciding how best to search for solutions. The majority of work on autonomous search for CP has concentrated on variable selection heuristics since these can have a significant effect on the size of the search tree. But when we consider optimization problems, the value heuristic used can also substantially effect the size of the search tree. Similarly when we are considering optimization problems that are too hard to find/prove optimal solutions, value heuristics can make a significant difference on the quality of solutions found in a limited time. In this paper we define a framework for learning value heuristics by combining a score function, feature selection, and machine learning. We show that we can learn value heuristics that are comparable to programmed heuristics, and the cost of learning can be paid for during the search.

While we have investigated some choices for score functions, feature selection and machine learning each component of the framework could be replaced, leaving us wide scope for further exploration of the framework. Clearly we can imagine many other: score functions, e.g. the objective of the first solution found in a subtree; feature selections, e.g. a tighter definition of neighbouring variables using constraint activity; and learning methods, such as polynomial regression; which might be worth considering.

Acknowledgments. NICTA is funded by the Australian Government through the Department of Communications and the Australian Research Council through the ICT Centre of Excellence Program. This work was partially supported by Asian Office of Aerospace Research and Development grant 12-4056.

References

1. Haralick, R.M., Elliott, G.L.: Increasing tree search efficiency for constraint satisfaction problems. Artif. Intell. **14**, 263–313 (1980)
2. Moskewicz, M.W., Madigan, C.F., Zhao, Y., Zhang, L., Malik, S.: Chaff: engineering an efficient SAT solver. In: Proceedings of the 38th Design Automation Conference, DAC 2001, pp. 530–535. ACM, Las Vegas, June 18–22, 2001
3. Boussemart, F., Hemery, F., Lecoutre, C., Sais, L.: Boosting systematic search by weighting constraints. In: de Mántaras, R.L., Saitta, L. (eds.) Proceedings of the 16th Eureopean Conference on Artificial Intelligence, ECAI 2004, Including Prestigious Applicants of Intelligent Systems, PAIS 2004, pp. 146–150. IOS Press, Valencia (2004)
4. Refalo, P.: Impact-based search strategies for constraint programming. In: Wallace, M. (ed.) CP 2004. LNCS, vol. 3258, pp. 557–571. Springer, Heidelberg (2004)
5. Zanarini, A., Pesant, G.: Solution counting algorithms for constraint-centered search heuristics. Constraints **14**, 392–413 (2009)
6. Michel, L., Van Hentenryck, P.: Activity-based search for black-box constraint programming solvers. In: Beldiceanu, N., Jussien, N., Pinson, E. (eds.) CPAIOR 2012. LNCS, vol. 7298, pp. 228–243. Springer, Heidelberg (2012)
7. Pipatsrisawat, K., Darwiche, A.: A lightweight component caching scheme for satisfiability solvers. In: Marques-Silva, J., Sakallah, K.A. (eds.) SAT 2007. LNCS, vol. 4501, pp. 294–299. Springer, Heidelberg (2007)
8. Benichou, M., Gauthier, J., Girodet, P., Hentges, G., Ribiere, G., Vincent, O.: Experiments in mixed-integer programming. Mathematical Programming **1**, 76–94 (1971)
9. Linderoth, J., Savelsbergh, M.: A computational study of search strategies for mixed integer programming. INFORMS Journal of Computing **11** (1999)
10. Kotthoff, L.: Algorithm selection for combinatorial search problems: A survey. CoRR abs/1210.7959 (2012)
11. Amemiya, T.: Advanced Econometrics. Harvard University Press (1985)
12. Wold, H.: Estimation of principal components and related models by iterative least squares. In: Multivariate Analysis, pp. 391–420. Academic Press (1966)
13. Chu, G., Stuckey, P.J.: Minimizing the maximum number of open stacks by customer search. In: Gent, I.P. (ed.) CP 2009. LNCS, vol. 5732, pp. 242–257. Springer, Heidelberg (2009)
14. Nethercote, N., Stuckey, P.J., Becket, R., Brand, S., Duck, G.J., Tack, G.R.: MiniZinc: towards a standard CP modelling language. In: Bessière, C. (ed.) CP 2007. LNCS, vol. 4741, pp. 529–543. Springer, Heidelberg (2007)
15. Garcia de la Banda, M., Stuckey, P., Chu, G.: Solving talent scheduling with dynamic programming. INFORMS Journal of Computing **23**, 120–137 (2011)

16. Schutt, A., Feydy, T., Stuckey, P., Wallace, M.: Explaining the cumulative propagator. Constraints **16**, 250–282 (2011)
17. Miller, H., Pierskalla, W., Rath, G.: Nurse scheduling using mathematical programming. Operations Research, 857–870 (1976)
18. Dincbas, M., Simonis, H., Van Hentenryck, P.: Solving the car-sequencing problem in constraint logic programming. In: ECAI, vol. 88, pp. 290–295 (1988)
19. Hartmann, S., Kolisch, R.: Experimental evaluation of state-of-the-art heuristics for resource constrained project scheduling. European Journal of Operational Research **127**, 394–407 (2000)
20. Loth, M., Sebag, M., Hamadi, Y., Schoenauer, M.: Bandit-based search for constraint programming. In: Schulte, C. (ed.) CP 2013. LNCS, vol. 8124, pp. 464–480. Springer, Heidelberg (2013)
21. Savage, L.: The theory of statistical decision. Journal of the American Statistical Association **46** (1951)
22. Allouche, D., de Givry, S., Schiex, T.: Toulbar2, an open source exact cost function network solver. Technical report, INRIA (2010)

Derivative-Free Optimization: Lifting Single-Objective to Multi-Objective Algorithm

Cyrille Dejemeppe$^{(\boxtimes)}$, Pierre Schaus, and Yves Deville

ICTEAM, Université Catholique de Louvain (UCLouvain),
Louvain-la-Neuve, Belgium
{cyrille.dejemeppe,pierre.schaus,yves.deville}@uclouvain.be

Abstract. Most of the derivative-free optimization (DFO) algorithms rely on a comparison function able to compare any pair of points with respect to a black-box objective function. Recently, new dedicated derivative-free optimization algorithms have emerged to tackle multi-objective optimization problems and provide a Pareto front approximation to the user. This work aims at reusing single objective DFO algorithms (such as Nelder-Mead) in the context of multi-objective optimization. Therefore we introduce a comparison function able to compare a pair of points in the context of a set of non-dominated points. We describe an algorithm, MOGEN, which initializes a Pareto front approximation composed of a population of instances of single-objective DFO algorithms. These algorithms use the same introduced comparison function relying on a shared Pareto front approximation. The different instances of single-objective DFO algorithms are collaborating and competing to improve the Pareto front approximation. Our experiments comparing MOGEN with the state-of the-art Direct Multi-Search algorithm on a large set of benchmarks shows the practicality of the approach, allowing to obtain high quality Pareto fronts using a reasonably small amount of function evaluations.

1 Introduction

Continuous optimization aims at minimizing a function $f(x)$ with $x \in \mathbb{R}^n$. When some information is known about the derivatives of f, one generally uses gradient based methods. For some other problems the function is black-box which means it can only be evaluated (for instance the evaluation is the result of a complex simulation model). Original algorithms have been imagined to optimize f by only relying on its evaluation. This family of techniques is generally called *derivative-free optimization* (DFO). One differentiates further the applicability of the derivative-free algorithms depending whether or not the function evaluation is costly to evaluate. Genetic algorithms obtain very good results for DFO benchmarks but they generally require a prohibitive number of evaluations. Finally, in many practical applications, considering a single-objective is not sufficient. In many cases, objectives are conflicting, which means they do not share the same optimum. As such, dedicated multi-objective optimization

© Springer International Publishing Switzerland 2015
L. Michel (Ed.): CPAIOR 2015, LNCS 9075, pp. 124–140, 2015.
DOI: 10.1007/978-3-319-18008-3_9

methods aim at finding a set of solutions being tradeoffs between the different objectives. This set of tradeoffs is called the Pareto front. This is the context of this work: *We are interested at optimizing multi-objective black-box functions costly to evaluate providing to the user a set of non-dominated points.*

Current state-of-the-art multi-objective optimization algorithms use the Pareto dominance to determine if new points can be added to the current Pareto front approximation. Our first contribution is the definition of a comparison function allowing to compare points with regards to a current Pareto front estimation.

Our second contribution is the definition of a framework, MOGEN, making use of this comparison function to solve multi-objective optimization problems. This framework uses *derivative-free optimization single-objective algorithms* (such as the Nelder-Mead algorithm) in which we substitute our new comparison function to the classical one. With this comparison function, these DFO single-objective algorithms are able to identify directions to discover new points potentially improving the current Pareto front optimization. This framework can be instantiated with several algorithms and performs elitism such that algorithms bringing the most improvement to the Pareto front approximation will be favoured. The aim of MOGEN is to solve multi-objective optimization problems using a limited amount of evaluations; such behaviour is desired to solve problems for which the evaluation is expensive in terms of computation time. For example, problems where each evaluation requires a costly simulation could be solved using MOGEN.

2 Background

A generic multi-objective optimization problem can be expressed as follows:

$$\text{minimize } F(X) \equiv \{f_1(X), \ldots, f_m(X)\}$$
$$\text{such that } X \in \Omega \tag{1}$$

where $\Omega \subseteq \mathbb{R}^n$ is the feasible region and $f_i(X)$ are the objective functions. When $m = 1$, the problem is a single-objective optimization problem. For the rest of this paper, we consider the feasible region Ω defines upper and lower bounds on each dimension. In such case, $X \in \Omega$ can be translated as $\forall i \in \{1, \ldots, n\} : x_i \in [l_i, u_i]$ such that $\forall i \in \{1, \ldots, n\} : l_i < u_i$.

The *Pareto dominance* allows to evaluate if a point is better than another point with regards to several objective functions. Considering two points $x, y \in \Omega$, we say that the point x dominates the point y on functions $f_1, \ldots, f_m$, written $x \prec y$, if the two following conditions are satisfied:

$$x \prec y \equiv \begin{cases} \forall i \in \{1, \ldots, m\} : f_i(x) \leq f_i(y) \\ \exists i \in \{1, \ldots, m\} : f_i(x) < f_i(y) \end{cases}$$

Alternative dominance definitions exist, as those proposed in [11], [15] and [2], but they are not detailed in this article. These dominance definitions could also be used in the framework we define.

A point x is said to be Pareto optimal if it satisfies the following condition:

$$\nexists y \in \Omega : y \prec x$$

The Pareto optimal set is defined as the set of Pareto optimal points, i.e. the set of non-dominated points. Multi-objective optimization algorithms aim at finding an approximation of this Pareto optimal set.

3   Derivative-Free Optimization

In this section, we recall the concept of Derivative-Free optimization methods. Three popular DFO algorithms are described to illustrate this concept. These latter are used by the MOGEN framework described later in this article.

3.1   Derivative-Free Optimization Methods

Derivative-free optimization methods, as defined in [3] and [12], are optimization search techniques. These methods iteratively use comparisons between points to evaluate the search progress. This iterative design with comparisons is rather intuitive and many DFO algorithms rely on simple concepts and structures. The main advantage of these methods is that the only information they need is the comparison of evaluations of the objective functions.

DFO algorithms can be used to optimize *black-box* functions, i.e. functions for which the only information available is its evaluation. In many practical applications, it is desired to optimize black-box functions. There exists a huge range of single-objective DFO algorithms; several of them being described in [3]. In the following sections, we rapidly explain three single-objective DFO algorithms. These methods are designed to solve single-objective problems as defined in Equation 1 with $m = 1$. If new points x_i are discovered outside the box defined by Ω, they are replaced by the closest points in Ω to ensure the bound constraints are always respected.

The Directional Direct Search Algorithm. The Directional Direct Search Algorithm described in [3] converges to a local optimum by iteratively polling new points around the current iterate which is a point in Ω. This algorithm relies on a collection of unit vectors D and a step size α. At each iteration, for each direction $d_i \in D$, a new point around the current iterate, $x_{current}$ is created as follows: $x_i = x_{current} + \alpha \times d_i$. If a new point x_i is discovered such that it is better than $x_{current}$, i.e. $f(x_i) \leq f(x_{current})$, then x_i becomes the new iterate and a new iteration can begin. If no better point has been discovered at the end of an iteration, α is decreased. On the other hand, if the iteration was successful, α is either maintained or increased. By replacing the current iterate with better points, the algorithm eventually converges to a local minimum.

The Nelder-Mead Algorithm. The Nelder-Mead algorithm introduced in [13] is a popular single-objective DFO algorithm. This algorithm converges to a local optimum by iteratively applying transformations on a hypervolume (also called simplex). To solve a problem in $\Omega \subseteq \mathbb{R}^n$, the Nelder-Mead algorithm uses a hypervolume containing $n+1$ points. These points $y_0, \ldots, y_n$ are sorted such that $f(y_0) \leq f(y_1) \leq \ldots \leq f(y_{n-1}) \leq f(y_n)$. At each iteration, the worst point y_n is transformed into a new point y'_n such that $f(y'_n) \leq f(y_n)$. The transformations applied are, depending on the situation, reflection, expansion, inside and outside contraction. The Nelder-Mead transformations are applied around the centroid of all the hypervolume points but the worst. A 2D example of reflection of the worst point of a Nelder-Mead hypervolume is shown in Figure 1. In this example, the centroid of points y_0 and y_1 is y_c and the reflection of the worst point y_2 over the centroid is y_r.

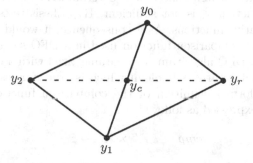

Fig. 1. Example of transformation of the worst point (y_2) from an hypervolume of the Nelder-Mead algorithm into its reflection (y_r) over the centroid (y_c)

If transformations fail to produce a new point better than y_n, the hypervolume is shrunk around y_0. By transforming the worst point of the hypervolume at each iteration, all points contained in the hypervolume are increasing in terms of quality eventually converging to a local minimum.

The MultiDirectional Search Algorithm. The MultiDirectional Search algorithm introduced in [7], similarly to the Nelder-Mead algorithm, applies transformation to a hypervolume structure to converge to a local optimum. This hypervolume contains $n + 1$ points to solve problems in $\Omega \subseteq \mathbb{R}^n$. These points $y_0, \ldots, y_n$ are sorted such that $f(y_0) \leq f(y_1) \leq \ldots \leq f(y_{n-1}) \leq f(y_n)$. At each iteration, all points but the best are transformed into new points y'_i by applying transformations: reflection, expansion (these are different from those of the Nelder-Mead algorithm since they are applied on all points but the best of the hypervolume while only the worst point is transformed in the Nelder-Mead algorithm). If at least one of the new point y'_i is better than the former best point y_0, then the iteration is a success. Otherwise, the hypervolume is shrunk around

y_0. These successive transformations of the hypervolume eventually converge to a local optimum.

3.2 The Comparison Step

DFO algorithms rely on comparisons of the objective evaluations to decide whether a point allows the algorithm to progress and is worth to be kept. DFO methods need a comparison function, cmp, to assess if a point is better than another one according to the considered problem. For example, in the case of a minimization problem, the comparison function used would be $cmp_<(x_1, x_2) \equiv f(x_1) < f(x_2)$. Indeed, if we have $f(x_1) < f(x_2)$ where f is the objective function, then x_1 is better than x_2 in order to minimize f.

These comparison functions could be replaced by any comparison function according to the type of problem considered. For example some industrial applications would require to minimize a non-deterministic function. In this case, the comparative function $<$ is not sufficient. Hypothesis tests would be more relevant as comparative functions. As a consequence, it would be interesting to be able to adapt the comparison function used in a DFO search to the type of problem considered. DFO algorithms are parametrized with a comparison function $cmp(x_1, x_2)$ that returns a boolean which is true if x_1 is better than x_2, false otherwise. A formal definition of the comparison function used in DFO algorithms can be expressed as follows:

$$cmp : \quad \mathbb{R}^n \times \mathbb{R}^n \to \mathbb{B}$$

A comparison function can be used in a DFO algorithm only if it is *transitive*. Using a non-transitive comparison function could lead to a non-productive looping search.

By using a comparison function as a parameter for a DFO algorithm, it is possible to solve different optimization problems declaratively. For example we could parametrize the Nelder-Mead algorithm as follows: $NM(f, cmp, x_0)$ where f is the objective to optimize, cmp is the comparison function to use (e.g. the function $cmp_<$ declared earlier) and x_0 is the starting point.

4   Multi-Objective Optimization Comparison Function

In this section we define a comparison function between two points in a multi-objective optimization context. The aim of this comparison function is to be used as a parameter in any DFO algorithm. Multi-objective optimization search techniques tend to iteratively improve their current approximation of the Pareto front. As stated in [5], the quality of a Pareto front approximation can be measured in terms of two criteria. First, it is desired that points in a Pareto front approximation are as close as possible to the optimal Pareto front. Then, it is desired that points in a Pareto front approximation span the whole optimal

Pareto front. The approximation of a Pareto front inside a multi-objective optimization algorithm is often referred to as an *archive*. An archive is a set of points such that no point is dominating (nor dominated by) any other point in the set.

The classical way of comparing two points with respect to several objectives is the Pareto dominance comparison function. In the context of an algorithm trying to improve its current archive, it is not sufficient. For example, comparing two points x_1 and x_2 with the Pareto dominance, could determine that no point is better than the other one (i.e. they don't dominate each other). On the other hand, one of these points could dominate more points in the archive. Therefore we introduce a new comparison function using the Pareto dominance between two points but applying it additionally to the whole archive. This comparison function compares two points x_1 and x_2 with respect to an archive A.

We define $dominates(x, A)$ as the set of points in A that x dominates. Similarly, $dominated(x, A)$ is the set of points in A by which x is dominated. From these two sets, we define the quantity $score(x, A)$ that represents the potential improvement brought by x in the archive A.

$$dominates(x, A) = \{y \in A \mid x \prec y\}$$
$$dominated(x, A) = \{y \in A \mid y \prec x\}$$
$$score(x, A) = |dominates(x, A)| - |dominated(x, A)|$$

The higher $score(x, A)$ is, the more promising x is to be included in the archive A. We base our comparison function $cmp_{\prec(A)}$ on this definition of $score$. It compares two points $x_1, x_2 \in \mathbb{R}^n$ with respect to a set of non-dominated points $A \subseteq \mathbb{R}^n$ and is defined as follows:

$$cmp_{\prec(A)}(x_1, x_2) = score(x_1, A') \geq score(x_2, A') \tag{2}$$

where $A' = A \cup \{x_1, x_2\}$. We use A' instead of A to guarantee that $cmp_{\prec(A)}$ is true if $x_1 \prec x_2$. Indeed, if the two points dominate or are dominated by the same number of points in the archive A, we would have $score(x_1, A) = score(x_2, A)$. In some cases, even if we obtain the same result for the evaluation of the $score$ function on x_1 and x_2 with respect to A, x_1 dominates x_2, or vice versa. Figure 2 illustrates an example where both x_1 and x_2 dominate all the points in A but x_1 dominates x_2. In this situation, using the $score$ function on A would not have been sufficient while using it on $A' = A \cup \{x_1, x_2\}$ would have shown that $x_1 \prec x_2$.

In [14], the transitivity of the Pareto dominance function is proven. It is then straightforward to prove that $cmp_{\prec(A)}$ defined in Equation (2) is transitive and could be used in DFO methods. $cmp_{\prec(A)}(x_1, x_2)$ has the behaviour desired to solve multi-objective optimization problems; it improves an archive A both in terms of distance to the real Pareto front and in terms of spreading of the points within A.

The next section defines a framework using single-objective DFO algorithms with $cmp_{\prec(A)}$ to perform multi-objective optimization.

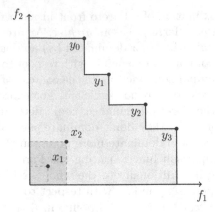

Fig. 2. The score values are equivalent: $score(x_1, A) = |A|$, $score(x_2, A) = |A|$ but $x_1 \prec x_2$. Need to use $score$ with $A' = A \cup \{x_1, x_2\}$.

5   The MOGEN Algorithm

As stated in [4], most current state-of-the-art multi-objective optimization algorithms are *evolutionary/genetic* algorithms. These algorithms provide Pareto front approximations of high quality both in terms of spreading and in terms of distance to the real Pareto front but at the cost of an important number of function evaluations [9]. In many real world applications, the cost of evaluating the objective functions is too high to make those techniques practicable. Alternative multi-objective optimization methods should allow to find high quality Pareto front approximations using a small number of evaluations of the objective functions. The MOGEN algorithm attempts to perform efficient multi-objective optimization search using a limited amount of evaluations of the objective functions.

According to [4], only few multi-objective optimization search techniques which are not evolutionary/genetic algorithms are efficient. In particular, the Direct Multi-Search (DMS) algorithm introduced in [5] can be considered state-of-the-art. The Direct Multi-Search algorithm has the particularity that its archive contains triplets (x, D, α) where x is a point in the input space, D is a set of direction vectors and α is a real number. At each iteration, a triplet (x, D, α) is selected and new points are computed for a subset of $d \in D$ with $x_{new} = x + \alpha \times d$. The new points are then inserted as triplets (x_{new}, D, α) in the archive if they are not dominated (and dominated points are removed from the archive).

The MOGEN algorithm described in Algorithm 1 follows a similar approach since it associates to each point in the archive an algorithm and its current state of execution. MOGEN aims at using several single-objective DFO algorithms to perform multi-objective optimization. In MOGEN, the single-objective DFO algorithms are used to discover new points that are potentially inserted into a Pareto front approximation shared by all the single-objective DFO algorithms used. It is possible to use these single-objective DFO algorithms by modifying the comparison function they usually use such that they favour points improving the shared Pareto front. All the data needed by an algorithm to continue its execution starting from where it has stopped is enclosed in its state of execution. This is illustrated on Figure 3. However MOGEN differs in two ways:

1. It may use several single-objective DFO algorithms. In Figure 3, three different DFO algorithms are used.
2. It uses the $cmp_{\prec(A)}$ comparison function, focusing on the possible improvements of the current archive.

To the best of our knowledge no other multi-objective optimization search method proposes to use single-objective DFO algorithms. In MOGEN, several single-objective DFO algorithms are in competition.

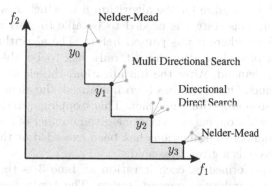

Fig. 3. An example of a MOGEN archive. The current state of execution is attached to each point of the archive. For example, y_1 is the best point of the simplex of a Nelder-Mead algorithm state.

Algorithm 1. MOGEN

Input: F A set of functions $f_i : \mathbb{R}^n \to \mathbb{R}$ to optimize
Input: Ω The feasible region ($\Omega \subseteq \mathbb{R}^n$)
Input: cmp A comparison function
Input: M A set of DFO algorithms
Output: $Arch$ A set of non-dominated points in $\mathbb{R}^n$ which is an
             approximation of the Pareto front

1 $Arch \leftarrow$ InitArchive(M, Ω)
2 **while** *stopping criterion not met* **do**
3 $\quad (x, a, s) \leftarrow$ SelectIterate($Arch$)
4 $\quad (Y_{new}, s_{new}) \leftarrow$ Apply(x, a, s, cmp, Ω)
5 $\quad Arch \leftarrow$ AddAndClean($Arch$, Y_{new}, a, s_{new})
6 **return** $Arch$

5.1   The Algorithm

The MOGEN algorithm described in Algorithm 1 takes several parameters as input. The first parameter, F, is the set of functions to optimize. The second parameter, Ω, is the feasible region of the considered problem. The third parameter, cmp, is the comparison function used by DFO algorithms to compare points. This comparison function must be adapted to the type of problem considered (e.g. deterministic maximization, stochastic minimization, ...). In this article, we assume MOGEN uses the comparison function $cmp_{\prec(A)}$. The last parameter, $M = \{a_1, \ldots, a_m\}$, is a set of single-objective derivative-free algorithms in which the comparison function cmp can be used. A simplified version of MOGEN could consider the use of a single algorithm which would be the same for every element in the archive.

The `InitArchive` function in Algorithm 1 at Line 1 initializes the archive as a set of triplets (x, a, s) where $x \in \Omega$, a is an algorithm from the set of algorithms M and s is a state for the algorithm a in which x is the best point (according to cmp). This state s is needed to be able to start a new iteration of the algorithm from where it was paused before. The algorithm a considered can still remain a black-box algorithm; it only has to be able to perform a single iteration on demand. After the initialization, the elements in $Arch$ are non-dominated points. Once $Arch$ has been initialized, the algorithm updates it iteratively until a stopping criterion is met. This stopping criterion is somewhat arbitrary. Examples of stopping criteria are: a given number of iteration has been reached, a given number of evaluations has been exceeded or the running time of the algorithm exceeds a given threshold.

The first step performed at each iteration at Line 3 is the selection of a triplet (x, a, s) in the archive as current iterate. The `Apply` function at Line 4 allows to perform a single iteration of a DFO algorithm from a given state[1]. It takes five elements as parameters. The first parameter is x, the best point of the state s. The second parameter, a is the algorithm on which the iteration has to be performed. The third parameter is s, the state from which the algorithm a begins its iteration. The fourth parameter, cmp, is the comparison function used in a. The last parameter is Ω, the feasible region. `Apply` returns two elements. The first one is Y_{new}, the set of new points discovered during the iteration of a (the algorithm sometimes discovers multiple interesting points, for example in the case of a Shrink of the Simplex in the Nelder-Mead algorithm). The second one is s_{new}, the new state reached after one iteration of a was performed starting from state s.

Once Y_{new} has been computed, the function `AddAndClean` at Line 5 updates the archive. It takes four arguments. The first argument is the current archive, $Arch$. The second argument, Y_{new}, is the set of new points found by using the `Apply` function with x, a, s, cmp and Ω. The third argument is the algorithm that was used to find the points in Y_{new} and the fourth argument s_{new} is the state of algorithm a in which new points from Y_{new} were found. `AddAndClean`

---

[1] For instance a reflection operation applied to the simplex of a Nelder-Mead instance.

compares every point y_i in Y_{new} with elements from the archive. If an element in the archive is dominated by a point y_i, it is removed from $Arch$. If a point y_i is not dominated by any point in the archive $Arch$, it is added in $Arch$ as a triplet (y_i, a, s_{new}). When an element (y_i, a, s_{new}) is added to the archive, the state of execution s_{new} associated to y_i is a *clone* of the state of execution from which it was obtained; as such, each instance of the same algorithm in the archive has its own state of execution that is not shared. When the stopping criterion is met, the algorithm returns the current archive, $Arch$.

If we consider an archive containing triplets with several different DFO algorithms a_i, then MOGEN performs elitism on these algorithms. Indeed, an algorithm a_{bad} performing poorly and failing to find a lot of new non-dominated points will generate fewer new point triplets (x_j, a_{bad}, s_j) in the archive. On the opposite, an algorithm a_{good} with good performances discovering many new non-dominated points will generate more new point triplets (x_k, a_{good}, s_k) in the archive. Furthermore, the AddAndClean function will clean points which are further from the optimal Pareto front. Triplets (x_j, a_{bad}, s_j) obtained with algorithms failing to get closer to the optimal Pareto Front will be eventually removed from the archive. Such algorithms will have less and less triplet representatives and potentially disappear from the archive. On the opposite, algorithms generating a lot of non-dominated points and getting closer to the optimal Pareto front will have more and more triplets representatives. As such, the MOGEN algorithm is elitist because it tends to use more and more often algorithms providing good results and to use less and less often (or even abandon) algorithms providing poor results.

5.2   Running Example

Let us consider a small running example of the MOGEN algorithm. This bi-objective problem has the following definition:

$$\text{minimize } F(x_1, x_2) \equiv \left\{ (x_1 - 1)^2 + (x_1 - x_2)^2 \right), (x_1 - x_2)^2 + (x_2 - 3)^2 \right\}$$
$$\text{such that } x_1, x_2 \in \Omega \equiv [-5, 5]$$

We consider an initial archive containing only two triplets: one associated to the Nelder-Mead algorithm and the other one to the Directional Direct Search algorithm. The first triplet is (x_0, NM, s_0) where $x_0 = (2.0, 2.0)$, NM is the Nelder-Mead algorithm and $s_0 = \{x_0 = (2.0, 2.0); x_{0,1} = (0.0, 2.0); x_{0,2} = (2.0, 0.0)\}$ is the hypervolume (also called simplex) structure (i.e. the *state*) of the Nelder-Mead algorithm. This hypervolume is well sorted according to the $cmp_{\prec(A)}$ since its evaluations are: $\{F(x_0) = (1.0, 1.0); F(x_{0,1}) = (5.0, 5.0); x_{0,2} = (5.0, 13.0)\}$. The second triplet is (x_1, DDS, s_1) where $x_1 = (0.75, 1.5)$, DDS is the Directional Direct Search algorithm and $s_1 = (D_1, alpha_1)$ is the state of the DDS algorithm where $D_1 = \{(1, 0); (-1, 0); (0, 1); (0, -1)\}$ is the collection of unit vectors and $\alpha_1 = 0.5$ is the step size. The initial archive is thus $Arch = \{(x_0, NM, s_0), (x_1, DDS, s_1)\}$.

The first iteration begins with SelectIterate($Arch$). For this example, we consider that SelectIterate considers the archive as a FIFO queue; it selects the first triplet in the archive. At the end of end of the iteration, the current iterate triplet is appended at the end of the archive with the new discovered points. The selected triplet is (x_0, NM, s_0). The iteration continues with $\texttt{Apply}(x_0, NM, s_0, cmp_{Arch}, \Omega)$ that applies an iteration of the Nelder-Mead algorithm starting from state s_0. The Nelder-Mead algorithm applies a reflection and an internal contraction discovering the new point $x_0^{new} = (1.5, 1.0)$ which evaluations are $F(x_0^{new}) = (0.5, 4.25)$. The $\texttt{Apply}$ function returns a pair (Y_0^{new}, s_0^{new}) where $Y_0^{new} = \{x_0^{new} = (1.5, 1.0)\}$ is the set of new points discovered and $s_0^{new} = \{x_0 = (2.0, 2.0); x_0^{new} = (1.5, 0.5); x_{0,1} = (0.0, 2.0)\}$ is the new hypervolume for the Nelder-Mead algorithm associated to x_0. Then the iteration ends with $\texttt{AddAndClean}(Arch, Y_0^{new}, NM, s_0^{new})$ which inserts points from Y_0^{new} in $Arch$ such that they are associated to NM and s_0^{new}. The new archive is thus $Arch = \{(x_1, DDS, s_1), (x_0^{new}, NM, s_0^{new}), (x_0, NM, s_0^{new})\}$ at the end of the first iteration.

The second iteration begins with $\texttt{SelectIterate}(Arch)$ which returns the triplet (x_1, DDS, s_1). Then, $\texttt{Apply}(x_1, DDS, s_1, cmp_{Arch_0}, \Omega)$ applies an iteration of the Directional Direct Search algorithm starting from state s_1. Polling discovers a new point $x_1^{new} = (1.25, 1.5)$ which evaluation is $F(x_1^{new}) = (0.125, 2.3125)$. The $\texttt{Apply}$ function returns a pair (Y_1^{new}, s_1^{new}) where $Y_1^{new} = \{x_1^{new} = (1.25, 1.5)\}$ and $s_1^{new} = (D_1, \alpha_1 = 1.0)$. Finally, $\texttt{AddAndClean}(Arch, Y_1^{new}, DDS, s_1^{new})$ inserts points from Y_1^{new} in $Arch$ to obtain $Arch = \{(x_0, NM, s_0^{new}), (x_1^{new}, DDS, s_1^{new})\}$. Note that x_0^{new} and x_1 have been removed from the archive since they were dominated by x_1^{new}.

6   Performance Assessment and Benchmarks

Intuitively it is desirable for a Pareto front estimation to contain points close to the optimal Pareto front and representing a diversified subset of the optimal Pareto front. Several measures exist to evaluate these two criteria (see in [17] and [16]). We present only the purity and delta metrics both used in [5] to evaluate the state-of-the-art Direct Multi-Search algorithm.

6.1   The Purity Metric

The *Purity* metric defined in [1] allows to compare several Pareto front approximations and define which one is the closest to the optimal Pareto front. Let us consider several different multi-objective DFO algorithms. Let $\mathcal{A}$ be the set of archives A_i produced by each algorithm i, and let A_{global} be the union of these archives A_i with dominated points removed:

$$A_{global} = \left\{ u \in \bigcup_{\mathcal{A}} A_i \,\middle|\, \forall v \neq u \in \bigcup_{\mathcal{A}} A_i : u \nprec v \right\}$$

The Purity metric of an archive $A_i \in \mathcal{A}$ is then defined as the ratio of the number of points in both A_i and A_{global} and the number of points in A_{global}:

$$Purity(A_i) = \frac{|A_i \cap A_{global}|}{|A_{global}|}$$

The higher the Purity of an archive A_i is, the less points from A_i are dominated by other archives and the closer A_i is to the optimal Pareto front. As mentioned in [5], two similar solvers produce similar archives, which can decrease their performances for the Purity metric since many points from these approximations will dominate each other. Therefore a third solver could benefit from this effect and obtain a higher Purity metric performance than the other two. To avoid this phenomenon, we only compare solvers in pairs for the Purity metric.

6.2   The Delta Metric

The Delta metric, Δ, proposed in [6] and extended in [5], is a spreading metric like the Gamma metric (Γ) defined in [5]. The Γ metric is however ambiguous for problems with more than two objectives. For a given archive, F_i is an objective space containing k dimensions, the Delta metric $\Delta(A_i)$ is defined as follows:

$$\Delta(A_i) = \max_{j=1,\ldots,n} \left(\frac{\delta_{0,j} + \delta_{k,j} + \sum_{i=1}^{k-1} |\delta_{i,j} - \bar{\delta}_j|}{\delta_{0,j} + \delta_{k,j} + (k-1)\delta_j} \right)$$

where $\bar{\delta}_j$ for $j = 1, \ldots, n$ is the average of the distances $\delta_{i,j} = f_{i+1,j} - f_{i,j}$ with $i = 1, \ldots, k - 1$ (assuming the objective function values have been sorted by increasing value for each objective j). This metric allows to measure how well an archive is spread along several objective dimensions.

6.3   Benchmark Problems

Our goal is to assess the performances of the MOGEN algorithm on a wide range of multi-objective problems reported in the literature. We consider problems involving bound constraints on each dimension, i.e. problems where the input space is contained in a hyper-volume: $\Omega = [l, u]$ with $l, u \in \mathbb{R}^n$ and $l < u$.

In [5], a collection of 100 problems with bound constraints from the literature was modelled in AMPL (A Modelling Language for Mathematical Programming) [10]. This collection of problems, available at http://www.mat.uc.pt/dms, contains a wide range of problems with different dimensions and properties. We used this collection to assess the performances of our algorithm.

The results obtained on these benchmarks for the Purity and the Δ metrics are reported graphically using performance profiles as in [8] and [5]. Let $t_{p,s}$ represent the performance of the solver $s \in S$ on the problem p such that lower values of $t_{p,s}$ indicate better performance. The ratio of the performance is defined as follows: $r_{p,s} = \frac{t_{p,s}}{\min\{t_{p,s^*}|s^* \in S\}}$. A ratio $r_{p,s} = 1$ means that solver s obtained the best value on problem p. The performance profile $\rho_s(\tau) = \frac{1}{|P|} \times |\{p \in P | r_{p,s} \le \tau\}|$ is a cumulative distribution of the performance of s compared to other solvers.

7   Results

Our results are divided in two parts. We first compare different MOGEN variants, then we compare MOGEN to the Direct Multi-Search algorithm.

7.1   Comparison of MOGEN Variants

The MOGEN algorithm can be instantiated in several ways since it involves several parameters and heuristics. We restrict our experimental comparison to different algorithm sets in the initial archive; the other (meta-)parameters are fixed and described as follows.

We decided to initialize all MOGEN variants with the *Line Multiple Point Initialization*. It selects n equidistant points on the line connecting l and u, respectively the lower and the upper bound of the input space $\Omega \subseteq \mathbb{R}^n$. The initial archive is defined as follows: $A_0 = \{l + (\frac{i}{n-1})(u-l)\}$ where $i = 0, \ldots, n-1$. The results obtained in this paper were all performed for $n = 10$ to ensure that all MOGEN variants start from the same initial archive. To select the iterate at the beginning of each iteration, we apply the (fair) *First In First Out Queue* selection heuristics. This means every iteration, a point is popped from the queue, an iteration is performed and the point (and possibly the new points found) is inserted back into the queue.

Finally, MOGEN variants may differ according the DFO algorithm associated with each point in the initial archive. We consider four different versions of MOGEN with the following algorithm proportions:

- **MOGEN(NM)** uses only the Nelder-Mead algorithm introduced in [13].
- **MOGEN(DDS)** uses only the Directional Direct Search algorithm as described in [3].
- **MOGEN(MDS)** uses only the MultiDirectional Search algorithm introduced in [7].
- **MOGEN(ALL)** uses the three algorithms described above; namely Nelder-Mead, Directional Direct Search and MultiDirectional Search. These algorithms are represented in the initial archive in identical proportions.

In Figure 4 we see the evolution of the average Purity metric on the benchmarks for the four MOGEN variants. As can be observed, MOGEN(NM) outperforms the other three variants. Even after a very small number of evaluations, MOGEN(NM) already has a high average Purity metric. It remains the case after a larger number of evaluations.

In Figure 5 we see the performance profiles of the Purity metric for the four MOGEN variants with a maximum budget of 20,000 evaluations. The performance profiles graph have to be read as follow: for a given solver s, a point on this graph in $(\tau, \rho(\tau))$ means that for a proportion $\rho(\tau)$ of the tested instances, the solver s was at worst τ times worse than the best performance obtained on all solvers represented in the graph. As such, the height $\rho(\tau)$ reached by a solver s in $\tau = 1$ represents the proportion $\rho(\tau)$ of instances on which solver s obtained the best performance among all represented solvers. If the curve of a given solver s never reaches $\rho(\tau) = 1$ but stops at $\rho(\tau) = r$, it means that the solver obtained an infinite performance ratio for a proportion $(1 - r)$ of the instances.

As can be observed, MOGEN(NM) outperforms the three other variants. These curves even show that MOGEN(NM) is very efficient. Indeed, for $\tau = 1$, MOGEN(NM) obtains the best metric value for more than 85% of the problems. MOGEN(NM) is also very robust since it is able to find at least one non-dominated point for more than 95% of the problems (i.e. there were less than 5% of the instances for which MOGEN(NM) was not able to discover a single point in the global archive). MOGEN(ALL) is only the second best. This could be explained by the fact that, eventually, points in the archive associated to the MDS and DDS algorithms are dominated by points associated to NM, leading to an increasing proportion of points associated to NM, performing better.

In Figure 6 we see the evolution of the average Δ metric on the benchmarks for the four MOGEN variants. As can be observed, MOGEN(NM) outperforms the other three variants. Even after a very small number of evaluations, MOGEN(NM) already has a low average Δ metric. It remains the case after a larger number of evaluations as this metric stabilizes after around 1,000 evaluations.

Figure 7 shows performance profiles of the Delta metric for the four MOGEN variants. For the Δ metric, MOGEN(NM) seems to outperform the other variants. We can also see that it is again the MOGEN(DDS) variants that seems to have the worst performance for the Δ metric while the MOGEN(MDS) and the MOGEN(ALL) variants have similar performances. We can conclude that, according to the Purity and the Δ metrics, MOGEN(NM) is the MOGEN variant showing the best performance.

7.2   Comparison of MOGEN and Direct Multi-Search

In [5], the authors show that the Direct Multi-Search (DMS) algorithm outperformed state-of-the-art solvers. We compare the best MOGEN variant, MOGEN(NM), to DMS. The parameters used for MOGEN(NM) are the same that those used to compare MOGEN variants. In Figure 8, we compare the purity metrics for MOGEN(NM) and the DMS algorithm. MOGEN(NM) performs globally better than the DMS algorithm. Indeed, the performance profile reveals that MOGEN(NM) has the best purity metric value for more than 65% of the instances while the DMS algorithm only has the best purity metric value for less than 40% of the instances. The fact that the DMS algorithm seems to perform better than MOGEN(NM) for 8 instances between $\tau = 6$ and $\tau = 100$ only means that when the archive of DMS is largely dominated by another archive, it tends to have a few more non-dominated points than MOGEN(NM).

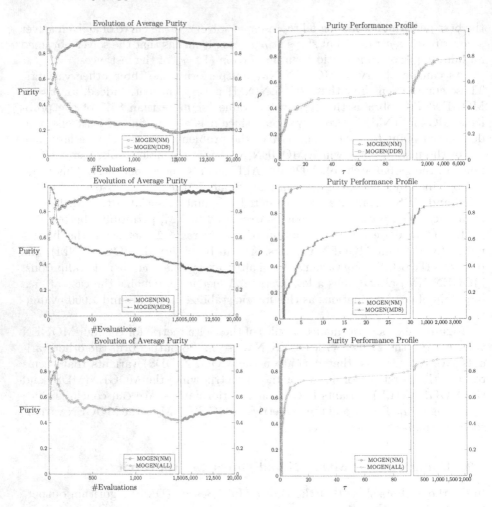

Fig. 4. Average Purity evolution - MOGEN

Fig. 5. Purity performance profiles - MOGEN

Figure 7 shows performance profiles of the Δ metric for the four MOGEN variants and DMS. For the Δ metric, DMS seems to outperform the MOGEN variants. However, a solver can have a very good Δ metric performance while its archive is completely dominated by the archives of other solvers. As such, DMS seems to diversify more than MOGEN(NM) but tends to produce archives dominated by those produced by MOGEN(NM). Similarly to what is done in DMS, a *search step* could be added at the beginning of each iteration of MOGEN to perform more diversification.

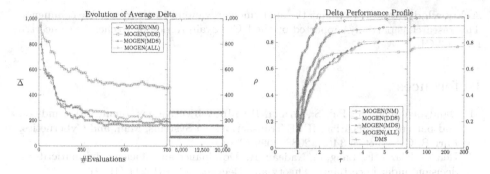

Fig. 6. Average Δ metric evolution - MOGEN

Fig. 7. Δ metric performance profiles

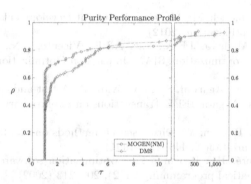

Fig. 8. Purity performance profiles - MOGEN(NM) & DMS

8   Conclusion

In this paper, we introduced a new comparison function, $cmp_{\prec(A)}$, allowing to compare points in a multi-objective optimization context with regards to an existing archive A. We defined a new generic multi-objective optimization framework, MOGEN, that uses single-objective DFO algorithms with $cmp_{\prec(A)}$. Several MOGEN variants using different sets of DFO algorithms have been compared and the one with the Nelder-Mead algorithm has obtained the best performances. The comparison between DMS and MOGEN revealed that the latter produces archives closer to the optimal Pareto front but tends to be less diversified.

Several research directions could be explored as future work. It would be interesting to use MOGEN with different comparison functions. Other algorithms could be used in the algorithm as well and it is possible that other single-objective DFO algorithms would improve greatly the performances obtained by MOGEN. The heuristics used in this algorithm should also be studied to reveal how much they impact the produced archive.

Acknowledgments. We thank the anonymous reviewers for their valuable comments. This research is partially supported by the UCLouvain Action de Recherche Concert?e ICTM22C1.

References

1. Bandyopadhyay, S., Pal, S., Aruna, B.: Multiobjective gas, quantitative indices, and pattern classification. IEEE Transactions on Systems, Man, and Cybernetics, Part B: Cybernetics **34**(5), 2088–2099 (2004)
2. Ben Abdelaziz, F., Lang, P., Nadeau, R.: Dominance and efficiency in multicriteria decision under uncertainty. Theory and Decision **47**, 191–211 (1999)
3. Conn, A., Scheinberg, K., Vicente, L.: Introduction to Derivative-Free Optimization. Society for Industrial and Applied Mathematics, Mps-siam Series on Optimization (2009)
4. Custódio, A., Emmerich, M., Madeira, J.: Recent Developments in Derivative-Free Multiobjective Optimization (2012)
5. Custódio, A.L., Madeira, J.F.A., Vaz, A.I.F., Vicente, L.N.: Direct multisearch for multiobjective optimization. SIAM Journal on Optimization **21**(3), 1109–1140 (2011)
6. Deb, K., Pratap, A., Agarwal, S., Meyarivan, T.: A fast and elitist multiobjective genetic algorithm: Nsga-ii. IEEE Transactions on Evolutionary Computation **6**(2), 182–197 (2002)
7. Dennis Jr., J.E., Torczon, V.: Direct search methods on parallel machines. SIAM Journal on Optimization **1**, 448–474 (1991)
8. Dolan, E.D., Moré, J.J.: Benchmarking optimization software with performance profiles. Mathematical programming **91**(2), 201–213 (2002)
9. Echagüe, E., Delbos, F., Dumas, L.: A global derivative-free optimization method for expensive functions with bound constraints. In: Proceedings of Global Optimization Workshop, pp. 65–68 (2012)
10. Fourer, R., Gay, D.M., Kernighan, B.W.: A modeling language for mathematical programming. Management Sci. **36**, 519–554 (1990)
11. Kollat, J., Reed, P., Kasprzyk, J.: A new epsilon-dominance hierarchical bayesian optimization algorithm for large multiobjective monitoring network design problems. Advances in Water Resources **31**(5), 828–845 (2008)
12. Mor, J., Wild, S.: Benchmarking derivative-free optimization algorithms. SIAM Journal on Optimization **20**(1), 172–191 (2009)
13. Nelder, J.A., Mead, R.: A simplex method for function minimization. In Comput. J. **7** (1965)
14. Voorneveld, M.: Characterization of pareto dominance. Operations Research Letters **31**(1), 7–11 (2003)
15. Zitzler, E., Brockhoff, D., Thiele, L.: The hypervolume indicator revisited: On the design of pareto-compliant indicators via weighted integration. In: Obayashi, S., Deb, K., Poloni, C., Hiroyasu, T., Murata, T. (eds.) EMO 2007. LNCS, vol. 4403, pp. 862–876. Springer, Heidelberg (2007)
16. Zitzler, E., Knowles, J., Thiele, L.: Quality assessment of pareto set approximations. In: Branke, J., Deb, K., Miettinen, K., Słowiński, R. (eds.) Multiobjective Optimization. LNCS, vol. 5252, pp. 373–404. Springer, Heidelberg (2008)
17. Zitzler, E., Thiele, L., Laumanns, M., Fonseca, C., da Fonseca, V.: Performance assessment of multiobjective optimizers: an analysis and review. IEEE Transactions on Evolutionary Computation **7**(2), 117–132 (2003)

Branching on Multi-aggregated Variables

Gerald Gamrath[1]($\boxtimes$), Anna Melchiori[2], Timo Berthold[3], Ambros M. Gleixner[1], and Domenico Salvagnin[4]

[1] Zuse Institute Berlin, Takustr. 7, 14195 Berlin, Germany
{gamrath,gleixner}@zib.de
[2] University of Padua, Via Trieste 63, 35121 Padua, Italy
melch.anna@gmail.com
[3] Fair Isaac Europe Ltd, c/o ZIB, Takustr. 7, 14195 Berlin, Germany
timoberthold@fico.com
[4] DEI, University of Padua, Via Gradenigo 6/b, 35121 Padua, Italy
salvagni@dei.unipd.it

Abstract. In mixed-integer programming, the branching rule is a key component to a fast convergence of the branch-and-bound algorithm. The most common strategy is to branch on simple disjunctions that split the domain of a single integer variable into two disjoint intervals. Multi-aggregation is a presolving step that replaces variables by an affine linear sum of other variables, thereby reducing the problem size. While this simplification typically improves the performance of MIP solvers, it also restricts the degree of freedom in variable-based branching rules.

We present a novel branching scheme that tries to overcome the above drawback by considering general disjunctions defined by multi-aggregated variables in addition to the standard disjunctions based on single variables. This natural idea results in a hybrid between variable- and constraint-based branching rules. Our implementation within the constraint integer programming framework SCIP incorporates this into a full strong branching rule and reduces the number of branch-and-bound nodes on a general test set of publicly available benchmark instances. For a specific class of problems, we show that the solving time decreases significantly.

1 Introduction

Since the invention of the branch-and-bound method for solving mixed-integer linear programming in the 1960s [1,2], branching rules have been an important field of research, being one of its core components. For surveys, see [3–5]. In this paper we address branching strategies for mixed-integer linear programs (MIPs) of the form

$$\min\{c^T x : Ax \leq b, \ell \leq x \leq u, x_i \in \mathbb{Z} \text{ for all } i \in \mathcal{I}\} \qquad (1)$$

with $c \in \mathbb{R}^n$, $A \in \mathbb{R}^{m \times n}$, $b \in \mathbb{R}^m$, $\ell, u \in \bar{\mathbb{R}}^n$ where $\bar{\mathbb{R}} := \mathbb{R} \cup \{\pm\infty\}$, and $\mathcal{I} \subseteq \mathcal{N} = \{1, \ldots, n\}$ being the index set of integer variables. When removing the integrality restrictions, we obtain the *linear programming (LP) relaxation* of the problem.

© Springer International Publishing Switzerland 2015
L. Michel (Ed.): CPAIOR 2015, LNCS 9075, pp. 141–156, 2015.
DOI: 10.1007/978-3-319-18008-3_10

If the solution $\tilde{x}$ to the LP relaxation of (1) is fractional, i.e., if the index set $\tilde{\mathcal{I}} := \{i \in \mathcal{I} : \tilde{x}_i \notin \mathbb{Z}\}$ of *fractional variables* is not empty, the task of a branching rule is to split the problem into two or more subproblems. The strategy is typically to exclude the LP solution from all subproblems while keeping the feasible integer solutions, each being present in exactly one subproblem.

The choice of which subproblems to create is crucial for the performance of the algorithm. The approach most widely used by MIP solvers is to branch on *simple disjunctions*

$$x_k \leq \lfloor \tilde{x}_k \rfloor \quad \bigvee \quad x_k \geq \lceil \tilde{x}_k \rceil. \tag{2}$$

each side being enforced in one subproblem. As this procedure splits the domain of a single variable at a time, it is also called *branching on variables*. Alternatively, branching can be performed on a *general disjunction*

$$\pi^T x \leq \pi_0 \quad \bigvee \quad \pi^T x \geq \pi_0 + 1. \tag{3}$$

where $(\pi, \pi_0) \in \mathbb{Z}^n \times \mathbb{Z}$, and $\pi_i = 0$ for all $i \notin \mathcal{I}$.

Branching on variables can be seen as the special case in which all considered disjunctions are of the form $(\pi, \pi_0) = (e_j, \lfloor \tilde{x}_j \rfloor)$, e_j being the j-th unit vector. Note that for branching on variables the set of *branching candidates* among which a branching rule chooses is usually the list of fractional variables $\tilde{\mathcal{I}}$. For branching on general disjunctions, the branching candidates consist of a potentially much larger list of disjunctions of form (3). Research on general branching disjunctions has largely been dedicated to determine a short list of promising candidates, see our literature overview in Sec. 2.

Another key component of state-of-the-art MIP solvers is *presolving*. It is applied before the branch-and-bound process and transforms a given MIP instance into a typically smaller instance with a tighter relaxation, which is hopefully easier to solve. These reductions can be based on pure feasibility arguments (keeping the set of feasible solutions unchanged) as well as optimality arguments (excluding also feasible solutions as long as one optimal solution remains).

Important presolving operations are fixings, aggregations, and multi-aggregations of variables. Here, *fixing* means that a variable gets permanently assigned to a constant value, *aggregation* means that a variable is replaced by (a constant value plus a scalar multiple of) another variable, and *multi-aggregation* means that a variable gets replaced by an affine linear combination of several variables. Hence, a *multi-aggregated* variable is a variable that is present in the original formulation, but is represented by an affine linear sum of variables in the presolved problem.

Contribution. The intuitive appeal of branching on general disjunctions is the increased degree of freedom that promises the creation of more balanced subproblems with tighter relaxations. This obvious advantage comes with the main challenge of determining promising candidate disjunctions. We address this difficulty by considering specifically the subset of disjunctions that are defined by the affine combinations stemming from multi-aggregations performed during the presolving stage. These disjunctions are naturally available in state-of-the-art MIP solvers at no cost and branching on them mimics branching on decision variables in the original model formulation.

Note that while the set of all general disjunctions of form (3) is exponentially large even when restricting π to $\{-1, 0, 1\}^n$, the set of multi-aggregated variables provides a list of potential candidates that is linear w.r.t. the size of the original model. Our experiments show that—in combination with standard single-variable disjunctions—this restriction yields not only a managable, but also computationally promising set of candidate disjunctions.

The remainder of the article is organized as follows. In Sec. 2, we give an overview of the literature on branching in MIP, with a particular focus on branching on general disjunctions. Sec. 3 introduces in more detail the concept of multi-aggregation, and Sec. 4 describes the idea of our new branching strategy and details about the implementation in the constraint integer programming framework SCIP [6,7]. In Sec. 5 we presents our computational study and Sec. 6 contains our conclusions and gives an outlook on potential extensions of branching on multi-aggregated variables.

2   Related Work

Various criteria for selecting fractional variables for branching on simple disjunctions have been presented in the literature. Most selection rules focus on the improvement in the dual bound that the branching restrictions produce in the created child nodes since this helps to tighten the global dual bound and prune nodes early. A fundamental strategy of this type is *strong branching* [8], which tentatively restricts the bound of a candidate variable and explicitly computes the resulting dual bound of the potential child node by solving the LP relaxation.

The *full strong branching* rule applies this at every node for each fractional variable. This typically leads to very small branch-and-bound trees, but on the other hand invests considerable effort in analyzing candidates. On average, this usually results in an overall performance deterioration w.r.t. computing time [5]. Nevertheless, the default branching rules in most state-of-the-art MIP solvers use some restricted form of strong branching and combine it with history information to reduce the computational effort for branching in later solving stages. Further strategies based on the same criteria can be found in [4,7,9–12]. Recent research efforts on different criteria for variable-based branching rules include, e.g., [13–18].

Branching on general disjunctions dates back to the 1980s [19], and has been addressed by various researchers in the last 15 years, see, e.g., [20–24]. The main challenge is to find a good class of general disjunctions that can lead to a better and more accurate tightening process of the feasible region, and consequently to a faster convergence of the dual bound to the optimal solution value, ideally without requiring a high computational effort for its generation and evaluation.

Owen and Mehrotra [20] present an algorithm that determines the branching disjunction via a neighborhood search heuristic. They prove that their algorithm is finite, if all variables have finite bounds and the size of the coefficients in the used disjunctions is bounded. As a consequence, they restrict their search to coefficients $\pi_i \in \{-1, 0, 1\}$. Combining this idea with [13], Mahmoud and Chinneck [24] choose a constraint that is active for the current LP optimum and construct a general disjunction with coefficients in $\{-1, 0, 1\}$ that is as perpendicular or as parallel as possible to the chosen active constraint.

Karamanov and Cornuéjols [22] consider disjunctions which correspond to Gomory mixed integer cuts (GMICs) [25]. They filter the GMICs to only keep the ten deepest cuts, and apply a strong-branching-like procedure on the corresponding candidate disjunctions. An extension of [22] is proposed by Cornuéjols et al. [23] who not only consider GMICs on tableau rows, but also on linear combinations of the tableau rows.

On the theoretical side, Mahajan and Ralphs proved that the problem of finding a general disjunction with maximal objective gain is $\mathcal{NP}$-hard [26]. Finally, Local Branching by Fischetti and Lodi [27] is a strategy to interleave variable-based branching with branching on general $\{-1, 0, 1\}$-disjunctions. These disjunctions measure the distance to the incumbent solution.

A typical result when branching on general disjunctions in MIP is that the generated branching trees are smaller on average, but the performance deteriorates w.r.t. running time. One major reason for this computational overhead is that the set of candidate disjunction for branching is much larger, so that a lot of time is spent determining the best one to choose at each node. However, this could in principle be overcome if we had more efficient (implicit) algorithms for evaluating the set of candidates, and it is of course not an issue when such set is still relatively small.

Another, more structural reason is that branching on variables changes a variable bound, which often fixes the variable to the other bound (in particular when branching on binary variables). This *decreases* the size of the LP relaxation for the subproblems by (at least) one column, whereas branching on general disjunctions potentially *increases* the LP's size by one row. This affects the simplex algorithm, which in most cases is the method of choice for solving the LP relaxations during LP-based branch-and-bound. Because the dimension of the basis matrix increases when adding a new row, most simplex implementations will have to recompute its factorization, causing computational overhead. In addition, many performance-relevant components of state-of-the-art MIP solvers such as domain propagation and conflict analysis are currently designed to benefit from branching on variables and become less effective when branching is performed on general disjunctions.

3   Multi-aggregations of Variables

Before the branch-and-bound process is started, state-of-the-art MIP solvers perform a presolving phase during which they analyze the problem and remove redundancies, tighten the formulation, and collect information about the problem structure, see [7, 28–31] for examples. This procedure is exact in the sense that each optimum of the simplified problem can be mapped to an optimal solution of the original problem.

The presolving technique which forms the basis of our newly developed branching rule is the multi-aggregation of variables. It reduces the number of variables by

1. detecting that—in at least one optimal solution—variable x_k equals an affine linear combination of other variables, i.e.,

$$x_k = \sum_{j \in \mathcal{S}_k} \alpha_j^k x_j + \beta^k, \tag{4}$$

   with $\mathcal{S}_k \subseteq \mathcal{N}$, $k \notin \mathcal{S}_k$,
2. replacing every occurrence of x_k in constraints and objective function by the right-hand side in (4), and
3. enforcing the bounds on x_k—if finite—by adding the new constraint

$$\ell_k \leq \sum_{j \in \mathcal{S}_k} \alpha_j^k x_j + \beta^k \leq u_k. \tag{5}$$

Equation (4) may either be explicitly present as one of the problem constraints[1] or implied by a combination of constraints and optimality conditions. An example for the latter is the case when x_k appears in exactly one constraint and its objective function coefficient ensures that this constraint will be fulfilled with equality in an optimal solution. The constraint integer programming framework SCIP, which we use for our computational experiments, has five different presolving operations in which multi-aggregation is performed.

After this step, one of the constraints implying (4) usually becomes void or is modified to enforce (5). If x_k is an integer variable, multi-aggregations are only performed if the integrality is enforced by the multi-aggregation. This holds, e.g., if (4) is an integer combination of integer variables, i.e., $\mathcal{S}_k \subseteq \mathcal{I}$, $\alpha_j^k \in \mathbb{Z}$ for all $j \in \mathcal{S}_k$, and $\beta^k \in \mathbb{Z}$.

In order to avoid a deterioration of performance and potential numerical problems during LP solving, it is crucial to safe-guard against fill-in in the constraint matrix. This can be done a priori by comparing the number of non-zeros that would be removed to the number of non-zeros that would get introduced in the constraint matrix, the latter of which can be bounded from above by the cardinality of $\mathcal{S}$ times the number of occurences of x_k.

To the best of our knowledge, all state-of-the-art MIP solvers use some form of multi-aggregation. For a test set of general MIP instances consisting of the last three MIPLIB [32–34] benchmark sets, the performance of SCIP is deteriorated by 3 % on average when disabling multi-aggregation. Taking into account that multi-aggregations are performed for no more than 15 % of the instances in this test set, this shows that multi-aggregations significantly improve the performance of MIP solvers when applicable.

---

[1] Although in (1) we have formulated MIPs in terms of inequalities, this also includes equality constraints formulated via two inequalities.

In the following, we call a variable *inactive*, if presolving removed it from the problem. This includes variables which are already fixed to some value as well as aggregated and multi-aggregated variables. All other variables are called *active*. During the subsequent solving process, inactive variables are disregarded since their solution value is uniquely defined by the value of the active variables. In the remainder of this article, a MIP of form (1) always refers to the presolved problem containing only active variables. When referencing the original problem, we are using the following notation: the index sets of original and corresponding integer variables are denoted by $\mathcal{N}'$ and $\mathcal{I}'$, respectively. Original variables are written as x'_i and the variable on the left-hand side of a multi-aggregation (4) is an original variable x'_k, while all variables on the right-hand side are active variables x_j.[2]

4   Branching on Multi-aggregated Variables

Simple aggregations of form $x'_k = \alpha^k_j x_j + \beta^k$ performed during presolving do not restrict the choices of variable-based branching rules since branching on the subsequently inactive variable x'_k remains implicitly possible by branching on x_j. In contrast, branching on multi-aggregated variables cannot be realized via branching on active variables. We are not aware of any study that has investigated the effect of multi-aggregation on the performance of branching rules and note that this restriction may indeed have negative performance impact—especially since this effect is currently not considered during presolving.

Our new branching strategy considers the general disjunctions defined by all multi-aggregations (4) for which $k \in \mathcal{I}'$ but $\sum_{j \in \mathcal{S}_k} \alpha^k_j x_j + \beta^k$ evaluates to a fractional value in the current LP solution. In a strong branching fashion, we tentatively test which improvement in the local dual bounds we would obtain by adding one part of the corresponding general disjunction. We compare this to the improvements obtained by simple disjunctions on fractional active variables and choose the best among all branching disjunctions.

The motivation is twofold: first, to compensate for the above drawback, and second, to obtain a set of candidates for general branching disjunctions that is available at no cost in state-of-the-art MIP solvers and computationally manageable. As mentioned earlier, the set of all general disjunctions of form (3) is exponentially large even when restricting π to $\{-1, 0, 1\}^n$, in contrast to that, the number of multi-aggregations is linear w.r.t. the size of the original model.

In an LP-based branch-and-bound algorithm, the multi-aggregated branching rule is called whenever the optimal solution $\tilde{x}$ to the linear relaxation of the current node is fractional. Its procedure is outlined in Algorithm 1.

First, strong branching is performed on all elements in the set of fractional variables $\tilde{\mathcal{I}}$. For each candidate variable x_i, two auxiliary LPs are solved to compute dual bounds $\tilde{z}^-$ and $\tilde{z}^+$ for the potential child nodes. If both are larger

---

[2] Note that nested multi-aggregations can be transferred into this form by (recursively) replacing inactive variables in the right-hand side of a multi-aggregation (4) by the corresponding constant or affine linear combination of variables.

Algorithm 1. Multi-aggregated branching rule

---

    input   :   – a MIP of form (1),
                   – an optimal solution $\tilde{x}$ of the LP relaxation,
                   – an upper bound z^* on the objective value of solutions, and
                   – the index set $\mathcal{A}' \subseteq \mathcal{N}'$ of multi-aggregations of form (4),
                    $x_k' = \sum_{j \in \mathcal{S}^k} \alpha_j^k x_j + \beta^k,\ k \in \mathcal{A}',\ \mathcal{S}^k \subseteq \mathcal{N}$

    output :   – a branching disjunction of form (3) given as $(\tilde{\pi}, \tilde{\pi}_0) \in \mathbb{Z}^n \times \mathbb{Z}$, or
                   – a valid inequality, or
                   – the conclusion that the current node can be pruned

1 **begin**
    | `// 0. initialization`
2 | **for** $k \in \mathcal{A}' \cap \mathcal{I}'$ **do**   `// compute LP values of multi-aggregated vars`
3 | $\lfloor$ $\tilde{x}_k' := \sum_{j \in \mathcal{S}^k} \alpha_j^k x_j + \beta^k$

4 | $\tilde{\mathcal{I}} := \{i \in \mathcal{I} : \tilde{x}_i \notin \mathbb{Z}\}$ `// single-variable candidates`
5 | $\tilde{\mathcal{A}} := \{k \in \mathcal{A}' \cap \mathcal{I}' : \tilde{x}_k' \notin \mathbb{Z}\}$ `// multi-aggregated candidates`
6 | $(\tilde{\pi}, \tilde{\pi}_0) := (0, 0)$ `// incumbent disjunction`
7 | $s_{(\tilde{\pi}, \tilde{\pi}_0)} := -\infty$ `// incumbent score`

    | `// 1. full strong branching on simple disjunctions`
8 | **for** $i \in \tilde{\mathcal{I}}$ **do**
9 | | $\tilde{z}^- \leftarrow \min\{c^T x : Ax \leq b, \ell \leq x \leq u, x_i \leq \lfloor \tilde{x}_i \rfloor\}$
10 | | $\tilde{z}^+ \leftarrow \min\{c^T x : Ax \leq b, \ell \leq x \leq u, x_i \geq \lfloor \tilde{x}_i \rfloor + 1\}$
11 | | **if** $\min\{\tilde{z}^-, \tilde{z}^+\} \geq z^*$ **then return** *current node can be pruned*
12 | | **else if** $\tilde{z}^- \geq z^*$ **then return** *valid inequality* $x_i \geq \lfloor \tilde{x}_i \rfloor + 1$
13 | | **else if** $\tilde{z}^+ \geq z^*$ **then return** *valid inequality* $x_i \leq \lfloor \tilde{x}_i \rfloor$
14 | | **else if** $\text{score}(\tilde{z}^-, \tilde{z}^+) > s_{(\tilde{\pi}, \tilde{\pi}_0)}$ **then**
15 | | | $(\tilde{\pi}, \tilde{\pi}_0) := (e_i, \lfloor \tilde{x}_i \rfloor)$
16 | | $\lfloor$ $s_{(\tilde{\pi}, \tilde{\pi}_0)} := \text{score}(\tilde{z}^-, \tilde{z}^+)$

    | `// 2. full strong branching on multi-aggregated disjunctions`
17 | **for** $k \in \tilde{\mathcal{A}}$ **do**
18 | | $\tilde{z}^- \leftarrow \min\{c^T x : Ax \leq b, \ell \leq x \leq u, \sum_{j \in \mathcal{S}^k} \alpha_j^k x_j \leq \lfloor \tilde{x}_k' \rfloor - \beta^k\}$
19 | | $\tilde{z}^+ \leftarrow \min\{c^T x : Ax \leq b, \ell \leq x \leq u, \sum_{j \in \mathcal{S}^k} \alpha_j^k x_j \geq \lfloor \tilde{x}_k' \rfloor - \beta^k + 1\}$
20 | | **if** $\min\{\tilde{z}^-, \tilde{z}^+\} \geq z^*$ **then return** *current node can be pruned*
21 | | **else if** $\tilde{z}^- \geq z^*$ **then return** $\sum_{j \in \mathcal{S}^k} \alpha_j^k x_j \geq \lfloor \tilde{x}_k' \rfloor - \beta^k + 1$ *valid*
22 | | **else if** $\tilde{z}^+ \geq z^*$ **then return** $\sum_{j \in \mathcal{S}^k} \alpha_j^k x_j \leq \lfloor \tilde{x}_k' \rfloor - \beta^k$ *valid*
23 | | **else if** $\text{score}(\tilde{z}^-, \tilde{z}^+) > s_{(\tilde{\pi}, \tilde{\pi}_0)}$ **then**
24 | | | $(\tilde{\pi}, \tilde{\pi}_0) := (\sum_{j \in \mathcal{S}^k} \alpha_j^k e_j, \lfloor \tilde{x}_k' \rfloor - \beta^k)$
25 | | $\lfloor$ $s_{(\tilde{\pi}, \tilde{\pi}_0)} := \text{score}(\tilde{z}^-, \tilde{z}^+)$

26 | **return** *branching disjunction* $(\tilde{\pi}, \tilde{\pi}_0)$
27 **end**

---

than or equal the given upper bound (usually the objective function value of the incumbent solution), we can stop since no better solution can be found in the current subproblem and the node can be cut off. If only one of the two dual bounds is smaller than the upper bound, the corresponding bound change can directly be applied at the current problem, since the other child node does not contain an improving solution. If both dual bounds are smaller than the upper bound, the score for the candidate variable is computed and the simple disjunction $(e_i, \lfloor \tilde{x}_i \rfloor)$ corresponding to branching on this variable is stored as new best candidate if its score exceeds the best one found so far. The branching score used in SCIP is the product of the objective gains of the two child nodes, more specifically,

$$\text{score}(\tilde{z}^-, \tilde{z}^+) = \max\{\Delta_j^-, \epsilon\} \cdot \max\{\Delta_j^+, \epsilon\} \tag{6}$$

with $\epsilon = 10^{-6}$ and $\Delta_j^- = \tilde{z}^- - c^T \tilde{x}$ and $\Delta_j^+ = \tilde{z}^+ - c^T \tilde{x}$ being the objective gains in the child nodes when branching on x_j.

In the second step of the algorithm, full strong branching is performed on the general disjunctions defined by the multi-aggregated variables of the original problem. To this end, all integer multi-aggregated variables x_k' are taken into account for which the LP solution translates into a fractional solution $\tilde{x}_k'$. Analogously to the first step, two auxiliary LPs are solved with the potential branching disjunction added and the computed dual bounds are compared to the upper bound in order to prune the node or identify valid constraints. The score of the candidate disjunction is evaluated and compared to the best score found so far. If it is higher, the candidate disjunction is updated. Note that possible ties are broken in favor of candidate variables, since those are evaluated first and we are looking for strict improvements.

In the case that a valid bound change or inequality was found, we stop the branching rule, tighten the formulation, and return to the MIP solving process, which will continue by applying domain propagation, reoptimizing the LP, and calling the branching rule again if needed. After the evaluation of all candidate variables and disjunctions, and if no such valid bound or inequality was found, the best disjunction is returned and branching is performed on it.

5   Computational Results

In the following, we present our experiments with branching on multi-aggregated variables. We used the academic constraint integer programming framework SCIP 3.1.0 [6,7] with SoPlex 1.7.0.4 [35] as underlying LP solver and implemented Algorithm 1 as a branching rule plug-in. Our new method builds on the full strong branching scheme and extends it by choosing as the set of candidates to evaluate via strong branching not only candidate variables, but also candidate disjunctions given by multi-aggregations. Therefore, it is consequential to compare our strategy with the basic full strong branching rule of SCIP.

All results were obtained on a cluster of 3.2 GHz Intel Xeon X5672 CPUs with 48 GB main memory, running each job exclusively on one node. To keep the

computation time under control, a time limit of 7200 seconds for each instance
was imposed.

Settings. We compare the methods for two different settings. The first one,
called *pure*, focuses on the main goal of a branching rule, namely proving the
optimality of a solution. To this end, it disables cutting plane separation, primal
heuristics, domain propagation, restarts, and conflict analysis. Additionally, we
provide the optimal objective value as a cutoff bound at the beginning of the
solving process. This is done in order to measure only the impact of branching
without side-effects to and from other solver components. In particular, this
reduces performance variability, cf. [34]. The second setting is called *default* and
runs full strong branching (SB) and multi-aggregated branching (MA) in the
SCIP default environment.

Instances. Our first experiments were performed on a test set of scheduling
[36, 37] instances. More specifically, we were investigating *resource allocation and
scheduling* problems, where jobs are assigned to machines, thereby minimizing
the processing costs which depend on the machine on which a job is performed.
Given sets $\mathcal{J}$ of jobs and $\mathcal{M}$ of machines, the capacity $C \in \mathbb{N}$ of the machines,
and assignment cost $c_{j,m}$, resource allocation and scheduling can be expressed
via the following MIP model [38]:

$$\min \sum_{m \in \mathcal{M}} \sum_{j \in \mathcal{J}} c_{j,m} x_{j,m}$$

$$\text{s.t.} \qquad \sum_{m \in \mathcal{M}} x_{j,m} = 1 \qquad \text{for all } j \in \mathcal{J},$$

$$\sum_{t \in \mathcal{T}_{j,m}} x_{j,m}^t = x_{j,m} \qquad \text{for all } m \in \mathcal{M}, \ j \in \mathcal{J},$$

$$\sum_{j \in \mathcal{J}} \sum_{\bar{t} \in \mathcal{T}_{j,m}^t} c_j x_{j,m}^{\bar{t}} \leq C \qquad \text{for all } m \in \mathcal{M}, \ t \in \mathcal{T},$$

$$x_{j,m}^t \in \{0,1\} \qquad \text{for all } m \in \mathcal{M}, \ j \in \mathcal{J}, \ t \in \mathcal{T}_{j,m},$$

$$x_{j,m} \in \{0,1\} \qquad \text{for all } m \in \mathcal{M}, \ j \in \mathcal{J}.$$

The formulation uses binary variables $x_{j,m}$ and $x_{j,m}^t$, which represent the
decision whether job $j \in \mathcal{J}$ is processed on machine $m \in \mathcal{M}$, and whether the
processing of job $j \in \mathcal{J}$ on machine $m \in \mathcal{M}$ is started at time $t \in \mathcal{T}$, respectively.
We use two subsets of the time periods: $\mathcal{T}_{j,m}$ which contains all time steps in
which job j can start on machine m, and $\mathcal{T}_{j,m}^t$ which further restricts $\mathcal{T}_{j,m}$ to
those starting times causing j to be (still) running in period t. When solving
these instances, the $x_{j,m}$ variables are frequently multi-aggregated, which makes
this problem an interesting test case for our first experiments.

We used a collection of 335 scheduling instances modeled this way in [38]. We
excluded all instances that were solved either during presolving or at the root

node. This left a total of 263 problem instances with the default setting and 276 instances with the pure setting.

In our second experiment, we used a test set of general MIP instances from different sources, including MIPLIB [32–34] and the Cor@l test set [39]. We removed some instances which to the best of our knowledge have never been solved so far and two numerically unstable instances giving slightly different results with both branching rules. Additionally, we restricted the test set to instances in which presolving performed multi-aggregations and removed instances which were solved during presolving or at the root node without branching. This gave us two test sets for the pure and default settings of 76 and 107 instances, respectively.

In the following, we present aggregated results over these test sets. For detailed instance-wise results, we refer to [40].

5.1 Results for Scheduling Instances

Table 1 compares the multi-aggregated branching strategy (MA) against the basic version of full strong branching (SB) available in SCIP with both pure and default settings, as indicated in the first column.

The remainder of the table is split into two parts: The four columns below the "scheduling test set" label display numbers about the performance on the complete scheduling test set. Column "size" shows the number of instances in the test set, "solved" gives the number of instances solved to proven optimality within the time limit of two hours. Column "faster" ("slower") show the number of instances that the MA strategy solved at least 10 % faster (slower) than standard full strong branching.

The right side of the table, labeled "all optimal", shows results for the subset of instances that both variants in the respective setting solved to optimality. Column "size" shows the number of instances in this subset, "nodes" the shifted geometric mean of the B&B nodes and "time (s)" the shifted geometric mean of the running time in seconds. We use shifts of 100 and 10 for the number of nodes and the solving time, respectively. For a discussion of the shifted geometric mean, we refer to [41, Appendix A3].

Let us first look at the results with the pure settings, which focus on the plain branch-and-bound performance. They are promising: 25 more instances (142 vs. 117) can be solved by branching on multi-aggregated variables compared to standard strong branching; this corresponds to an increase of more than 20 %. Furthermore, 100 instances are solved at least 10 % faster with the new method, compared to 13 which slow down by 10 % or more. This corresponds to 70 % of the instances being solved faster with branching on multi-aggregated variables. Looking at the instances that were solved to optimality by both variants, both the number of nodes and the requested time are reduced by a factor of two on average: 58 % less nodes are needed and 49 % less time.

When looking at the results with default settings, the effect is smaller, but still significant: the multi-aggregated branching strategy is able to solve 9 more instances to optimality, with 56 instances being solved faster and 31 slower. On

Table 1. Results for scheduling instances with default and pure settings

setting	scheduling test set			all optimal		
	size	solved	faster	slower	size	nodes	time (s)
SB-pure	276	117			115	472	51.8
MA-pure	276	142	100	13	115	196	26.4
SB-default	263	126			122	349	84.6
MA-default	263	135	56	31	122	221	70.3

instances that both variants solve to optimality, it needs 37 % less nodes and reduces the solving time by 17 %.

One might argue that the multi-aggregation of variables itself could have a negative impact on the performance for the scheduling instances as it restricts standard branching rules from branching on the $x_{j,m}$ variables which can be seen as first-level decisions. However, this is only partly true: When disabling multi-aggregations, the shifted geometric mean of the number of branch-and-bound nodes is indeed decreased by 14 % for the instances solved to optimality both with and without multi-aggregations. On the other hand, the average solving time is increased slightly by 2 %. This shows that the gains obtained by having more branching opportunities with multi-aggregation disabled are compensated by not being able to reduce the problem size so much and having more effort, e.g., in LP solving. Our proposed branching scheme takes the best of both variants, allowing the problem size reductions while still providing the potentially more powerful branching possibilities given by the multi-aggregated variables. This helps to improve both the number of nodes as well as the solving time significantly over the individual best of the two other variants.

Let us note that the positive effect of branching on multi-aggregated variables grows stronger the harder an instance is. This seems reasonable since the additional overhead might not pay off if a standard strong branching is able to solve an instance within a few nodes. When taking into account only instances which needed more than 100 seconds to solve by at least one setting, the reduction in the number of nodes and the solving time goes up to 42 % and 25 %, respectively.

This first computational experiment shows that branching on multi-aggregated variables can significantly improve the performance of SCIP compared to a pure variable-based branching rule: more instances are solved, with less enumeration, in shorter time. Note that in all cases the relative reduction in running time was smaller than the relative reduction in the number of branch-and-bound nodes, which is a typical result for branching strategies that involve general disjunctions (see Sec. 2).

In order to analyze the impact of the new branching rule in more detail, we collected some statistics during the execution of SCIP. On average over the test set, the number of integer multi-aggregations is only 5.7 % of the number of integer variables. Thus, the list of branching candidates is only slightly extended in most cases, which overcomes a typical issue for branching on general disjunctions. Interestingly, despite this relatively small number of multi-aggregations, 39 % of the branching decisions select a multi-aggregated disjunction for branching. Even more, in 85 % of the cases, the first branching on a multi-aggregated disjunction was performed at the root node.

Finally, each time we perform a multi-aggregated branching, we store the ratio of the gain that we would have obtained when branching on the best fractional variable compared to the gain obtained by branching on the current multi-aggregated variable. The gain is computed as the square root of the SCIP branching score value and thus measures the improvement in the score SCIP tries to maximize. On average over all calls where we branched on a multi-aggregated disjunction, the gain would have been reduced to 22 % by branching on the best variable instead.

5.2   Results for General MIP Instances

The results for our collection of general MIP instances are presented in Table 2. The columns and rows show the same statistics as described in Sec. 5.1. We can see that on these instances, multi-aggregated branching is significantly slower and solved one less instance in both settings, compared to standard strong branching. With pure settings, the solving time increases by 25 % while the number of branch-and-bound nodes is decreased by 13 %. Compared to the scheduling instances, multi-aggregated variables are much less effective for branching. That the increased effort in strong branching outweighs the observed node reduction seems plausible. These results confirm our observation from the scheduling instances in the sense that the impact on the number of branch-an-bound nodes was better than the impact on the overall running time. For the scheduling instances, the additional candidates were structurally different and allowed different, higher-level decisions which had an enormous effect on the tree size that even allowed for a running time reduction. For standard MIPs, however, such a large effect is apparently obtained rarely, thus, the performance deteriorates on average. The picture looks even worse for the default settings. Here, the solving time increases by 26 % and the number of nodes now increases by 6 % as well.

Again, we collected statistics to analyze the impact of the multi-aggregated branching scheme. On average over the test set, the amount of integer multi-aggregations is almost twice as high as for the scheduling set, namely 14.4 % of the number of integer variables. However, multi-aggregated variables are selected less often (only for 1.84 % of the branchings) and consequently, also the first branching on a multi-aggregated disjunction was less often performed at the root node (only for 7.4 % of the instances for which multi-aggregated branching was performed). If a multi-aggregated disjunction was selected, selecting the best fractional variable instead would have decreased the gain by 31 % on average, compared to 78 % for the scheduling instances. This shows that multi-aggregated

Table 2. Results for general MIP instances with default and pure settings

setting	MIP test set				all optimal		
	size	solved	faster	slower	size	nodes	time (s)
SB-pure	76	33			32	983	150.9
MA-pure	76	32	0	26	32	852	188.9
SB-default	107	55			49	253	100.4
MA-default	107	57	1	33	49	269	126.3

disjunctions play a smaller role for branching on this test set, but can still be used to improve the quality of branching disjunctions.

Even more surprising is the increase in the number of nodes, which can be explained, however, by the tailoring of many MIP solving algorithms towards variable-based branching. Domain propagation (or *node preprocessing*, see, e.g., [28] for MIP), for example, tries to tighten the local domains of variables by inspecting the constraints and current domains of other variables at the local subproblem. Tightening or fixing variables by branching is naturally beneficial for domain propagation, the impact of adding general disjunctions is rather opaque. Furthermore, techniques like primal heuristics, cutting plane separation, or conflict analysis profit from tightened variable bounds rather than from added general disjunctions. Since all these techniques help to reduce the size of the branch-and-bound tree, branching on general disjunctions with a high branching score can even increase the number of nodes, since as a side effect it makes the named procedure less effective.

We see our results for general MIPs as an important negative result that confirms previous observations by other authors that it is hard to find a branching rule on general disjunctions which is competitive on standard MIP benchmarks. Our results indicate that this holds even when restricting the selection to relatively few additional candidates that are naturally obtained from the problem structure. Finally, adapting procedures like primal heuristics or conflict analysis in such a way that they benefit from added constraints as much as from tightened or fixed variables might be a prerequisite to excel with constraint-based branching schemes in state-of-the-art MIP solvers.

6   Conclusions and Outlook

In this paper, we presented a new branching rule which takes into account a specific type of general disjunctions. These general disjunctions, so-called multi-aggregations, are the affine linear sums of active variables in the presolved problem, which correspond to a decision variable in the original problem. We extended the full strong branching rule of SCIP by taking additionally into account all general disjunctions induced by multi-aggregations. On a set of scheduling instances, this significantly improved the performance of SCIP w.r.t. the tree size as well as the solving time and the number of solved instances.

We tested the same branching rule on standard MIP benchmark sets. The results were much less convincing, but a certain potential for branching on multi-aggregated variables was indicated by the observation that in a "pure" setting, it led to a reduction in the number of branch-and-bound nodes for general MIPs. However, before this potential can be harnessed, we conclude that many advanced solution techniques applied in state-of-the-art MIP solvers—domain propagation, conflict analysis, etc.—must be extended towards a more efficient handling of general disjunctions. An additional performance bias is the slow-down in current simplex implementations when adding and removing constraints. This bottleneck may be alleviated by the recent developments of [42, 43], which improve the underlying linear algebra routines such that the factorization of the basis matrix is preserved when adding new rows. We identify these points as important directions for future research.

Another field for future research would be to find criterions to assess the structure of multi-aggreagations and predict the power of the new scheme for the current instance in order to decide on whether to use it or not. A first basic variant of this would be to heuristically detect scheduling substructures and turn on the branching scheme for the involved multi-aggregations. Many improvements for MIP solving in recent years are based on specific structures, cf. [44, 45]. If this structure is detected, they lead to a significant improvement— as is the case for our scheme for scheduling problems—while the detection is typically fast enough that the performance on other problems is not deteriorated. Therefore, we are convinced that the new strategy can also improve the performance of MIP solvers for general MIP test sets.

The proposed strategy has been studied and implemented for the first time in the constraint integer programming framework SCIP. Since it proved its effectiveness for certain problem classes, it will be available in the next release of SCIP.

Acknowledgements. The authors would like to thank the three anonymous reviewers for helpful comments on the paper. The work for this article has been conducted within the Research Campus Modal funded by the German Federal Ministry of Education and Research (fund number 05M14ZAM).

References

1. Land, A.H., Doig, A.G.: An automatic method of solving discrete programming problems. Econometrica **28**(3), 497–520 (1960)
2. Dakin, R.J.: A tree-search algorithm for mixed integer programming problems. The Computer Journal **8**(3), 250–255 (1965)
3. Mitra, G.: Investigation of some branch and bound strategies for the solution of mixed integer linear programs. Mathematical Programming **4**, 155–170 (1973)
4. Linderoth, J.T., Savelsbergh, M.W.P.: A computational study of search strategies in mixed-integer programming. INFORMS Journal on Computing **11**(2), 173–187 (1999)

5. Achterberg, T., Koch, T., Martin, A.: Branching rules revisited. Operations Research Letters **33**, 42–54 (2005)
6. Achterberg, T., Berthold, T., Koch, T., Wolter, K.: Constraint integer programming: a new approach to integrate CP and MIP. In: Perron, L., Trick, M.A. (eds.) CPAIOR 2008. LNCS, vol. 5015, pp. 6–20. Springer, Heidelberg (2008)
7. Achterberg, T.: SCIP: Solving constraint integer programs. Mathematical Programming Computation **1**(1), 1–41 (2009)
8. Applegate, D.L., Bixby, R.E., Chvátal, V., Cook, W.J.: On the solution of traveling salesman problems. Documenta Mathematica Journal der Deutschen Mathematiker-Vereinigung, Extra Volume ICM **III**, 645–656 (1998)
9. Gauthier, J.M., Ribière, G.: Experiments in mixed-integer linear programming using pseudo-costs. Mathematical Programming **12**(1), 26–47 (1977)
10. Fischetti, M., Monaci, M.: Branching on nonchimerical fractionalities. OR Letters **40**(3), 159–164 (2012)
11. Berthold, T., Salvagnin, D.: Cloud branching. In: Gomes, C., Sellmann, M. (eds.) CPAIOR 2013. LNCS, vol. 7874, pp. 28–43. Springer, Heidelberg (2013)
12. Gamrath, G.: Improving strong branching by domain propagation. EURO Journal on Computational Optimization **2**(3), 99–122 (2014)
13. Patel, J., Chinneck, J.: Active-constraint variable ordering for faster feasibility of mixed integer linear programs. Mathematical Programming **110**, 445–474 (2007)
14. Kılınç Karzan, F., Nemhauser, G.L., Savelsbergh, M.W.: Information-based branching schemes for binary linear mixed integer problems. Mathematical Programming Computation **1**, 249–293 (2009)
15. Achterberg, T., Berthold, T.: Hybrid branching. In: van Hoeve, W.-J., Hooker, J.N. (eds.) CPAIOR 2009. LNCS, vol. 5547, pp. 309–311. Springer, Heidelberg (2009)
16. Fischetti, M., Monaci, M.: Backdoor branching. In: Günlük, O., Woeginger, G.J. (eds.) IPCO 2011. LNCS, vol. 6655, pp. 183–191. Springer, Heidelberg (2011)
17. Gilpin, A., Sandholm, T.: Information-theoretic approaches to branching in search. Discrete Optimization **8**(2), 147–159 (2011)
18. Pryor, J., Chinneck, J.W.: Faster integer-feasibility in mixed-integer linear programs by branching to force change. Computers & OR **38**(8), 1143–1152 (2011)
19. Ryan, D.M., Foster, B.A.: An integer programming approach to scheduling. In: Wren, A. (ed.) Computer Scheduling of Public Transport Urban Passenger Vehicle and Crew Scheduling, pp. 269–280. North Holland, Amsterdam (1981)
20. Owen, J.H., Mehrotra, S.: Experimental results on using general disjunctions in branch-and-bound for general-integer linear programs. Computational Optimization and Applications **20**, 159–170 (2001)
21. Mahajan, A., Ralphs, T.K.: Experiments with branching using general disjunctions. In: Chinneck, J.W., Kristjansson, B., Saltzman, M.J. (eds.) Operations Research and Cyber-Infrastructure. Operations Research/Computer Science Interfaces Series, vol. 47, pp. 101–118. Springer, US (2009)
22. Karamanov, M., Cornuéjols, G.: Branching on general disjunctions. Mathematical Programming **128**, 403–436 (2011)
23. Cornuéjols, G., Liberti, L., Nannicini, G.: Improved strategies for branching on general disjunctions. Mathematical Programming **130**, 225–247 (2011)
24. Mahmoud, H., Chinneck, J.W.: Achieving milp feasibility quickly using general disjunctions. Computers & OR **40**(8), 2094–2102 (2013)
25. Gomory, R.E.: An algorithm for the mixed integer problem. Technical report, RAND Corporation (1960)
26. Mahajan, A., Ralphs, T.K.: On the complexity of selecting disjunctions in integer programming. SIAM Journal on Optimization **20**(5), 2181–2198 (2010)

27. Fischetti, M., Lodi, A.: Local branching. Mathematical Programming **98**(1-3), 23–47 (2003)
28. Savelsbergh, M.W.P.: Preprocessing and probing techniques for mixed integer programming problems. ORSA Journal on Computing **6**, 445–454 (1994)
29. Brearley, A.L., Mitra, G., Williams, H.P.: Analysis of mathematical programming problems prior to applying the simplex algorithm. Mathematical Programming **8**(1), 54–83 (1975)
30. Bixby, R.E., Wagner, D.K.: A note on detecting simple redundancies in linear systems. Operation Research Letters **6**(1), 15–17 (1987)
31. Gamrath, G., Koch, T., Martin, A., Miltenberger, M., Weninger, D.: Progress in presolving for mixed integer programming. Technical Report 13–48, ZIB, Takustr. 7, 14195 Berlin (2013)
32. Bixby, R.E., Ceria, S., McZeal, C.M., Savelsbergh, M.W.P.: An updated mixed integer programming library: MIPLIB 3.0. Optima **58**, 12–15 (1998)
33. Achterberg, T., Koch, T., Martin, A.: MIPLIB 2003. Operations Research Letters **34**(4), 1–12 (2006)
34. Koch, T., Achterberg, T., Andersen, E., Bastert, O., Berthold, T., Bixby, R.E., Danna, E., Gamrath, G., Gleixner, A.M., Heinz, S., Lodi, A., Mittelmann, H., Ralphs, T., Salvagnin, D., Steffy, D.E., Wolter, K.: MIPLIB 2010. Mathematical Programming Computation **3**(2), 103–163 (2011)
35. Wunderling, R.: Paralleler und objektorientierter Simplex-Algorithmus. Ph.D. thesis, Technische Universität Berlin (1996)
36. Baker, K.: Introduction to Sequencing and Scheduling. Wiley (1974)
37. Baker, K.R., Trietsch, D.: Principles of Sequencing and Scheduling. Wiley (2009)
38. Heinz, S., Ku, W.-Y., Beck, J.C.: Recent improvements using constraint integer programming for resource allocation and scheduling. In: Gomes, C., Sellmann, M. (eds.) CPAIOR 2013. LNCS, vol. 7874, pp. 12–27. Springer, Heidelberg (2013)
39. COR@L: MIP Instances (2014). http://coral.ie.lehigh.edu/data-sets/mixed-integer-instances/
40. Gamrath, G., Melchiori, A., Berthold, T., Gleixner, A.M., Salvagnin, D.: Branching on multi-aggregated variables. Technical Report 15–10, ZIB, Takustr. 7, 14195 Berlin (2015)
41. Achterberg, T.: Constraint Integer Programming. Ph.D. thesis, Technische Universität Berlin (2007)
42. Gleixner, A.M.: Factorization and update of a reduced basis matrix for the revised simplex method. ZIB-Report 12–36, Zuse Institute Berlin, October 2012
43. Wunderling, R.: The kernel simplex method. Talk at the 21st International Symposium on Mathematical Programming, Berlin, Germany, August 2012
44. Achterberg, T., Raack, C.: The MCF-separator: detecting and exploiting multicommodity flow structures in MIPs. Mathematical Programming Computation **2**(2), 125–165 (2010)
45. Salvagnin, D.: Detecting and exploiting permutation structures in MIPs. In: Simonis, H. (ed.) CPAIOR 2014. LNCS, vol. 8451, pp. 29–44. Springer, Heidelberg (2014)

Time-Table Disjunctive Reasoning
for the Cumulative Constraint

Steven Gay$^{(\boxtimes)}$, Renaud Hartert, and Pierre Schaus

ICTEAM, UCLouvain, Place Sainte Barbe 2, 1348 Louvain-la-Neuve, Belgium
{steven.gay,renaud.hartert,pierre.schaus}@uclouvain.be

Abstract. Scheduling has been a successful domain of application
for constraint programming since its beginnings. The cumulative
constraint – which enforces the usage of a limited resource by several
tasks – is one of the core components that are surely responsible of this
success. Unfortunately, ensuring bound-consistency for the cumulative
constraint is already NP-Hard. Therefore, several relaxations were proposed to reduce domains in polynomial time such as Time-Tabling, Edge-
Finding, Energetic Reasoning, and Not-First-Not-Last. Recently, Vilim
introduced the Time-Table Edge-Finding reasoning which strengthens
Edge-Finding by considering the time-table of the resource. We pursue
the idea of exploiting the time-table to detect disjunctive pairs of tasks
dynamically during the search. This new type of filtering – which we call
time-table disjunctive reasoning – is not dominated by existing filtering
rules. We propose a simple algorithm that implements this filtering rule
with a $\mathcal{O}(n^2)$ time complexity (where n is the number of tasks) without
relying on complex data structures. Our results on well known benchmarks highlight that using this new algorithm can substantially improve
the solving process for some instances and only adds a marginally low
computation overhead for the other ones.

Keywords: Constraint programming · Scheduling · Cumulative constraint · Time-table · Disjunctive reasoning

1 Introduction

Many real-world scheduling problems involve cumulative resources. A resource
can be seen as an abstraction of any renewable entity – as machinery, electricity, or even manpower – which is used to perform tasks (also called activities).
Although many tasks could be scheduled simultaneously on a same resource, the
total use of a resource cannot exceed a fixed capacity at any moment.

In this paper, we focus on a single cumulative resource with a discrete finite
capacity $C \in \mathbb{N}$ and a set of n tasks $\mathcal{T} = \{1, \ldots, n\}$. Each task i has a starting
time $s_i \in \mathbb{Z}$, a fixed duration $d_i \in \mathbb{N}$, and an ending time $e_i \in \mathbb{Z}$ such that the
equality $s_i + d_i = e_i$ holds. Moreover, each task i consumes a fixed amount of

© Springer International Publishing Switzerland 2015
L. Michel (Ed.): CPAIOR 2015, LNCS 9075, pp. 157–172, 2015.
DOI: 10.1007/978-3-319-18008-3_11

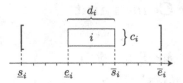

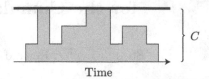

Fig. 1. Task i is characterized by its starting time s_i, its duration d_i, its ending time e_i, and its resource consumption c_i

Fig. 2. Accumulated resource consumption over time. The `cumulative` constraint ensures that the maximum capacity C is not exceeded.

resource $c_i \in \mathbb{N}$ during its processing time. Tasks are non-preemptive, i.e., they cannot be interrupted during their processing time. In the following, we denote by $\underline{s}_i$ and $\overline{s}_i$ the earliest and the latest starting time of task i and by $\underline{e}_i$ and $\overline{e}_i$ the earliest and latest ending time of task i (see Fig. 1). The `cumulative` constraint [1] ensures that the accumulated resource consumption does not exceed the maximum capacity C at any time t (see Fig. 2):

$$\forall t \in \mathbb{Z} : \sum_{i \in T \,:\, s_i \leq t < e_i} c_i \leq C. \tag{1}$$

Unfortunately, ensuring bound consistency for the `cumulative` constraint is already NP-Hard [11]. Therefore, many relaxations were proposed during the last two decades to remove inconsistent starting and ending times in polynomial time. Among them, the Time-Tabling filtering rule has been the subject of much research in the scheduling community [4,9,13]. The idempotent algorithm proposed by Letort in [9] implements Time-Tabling with a $\mathcal{O}(n^2)$ time complexity and has been successfully applied on problems with hundreds of thousands of tasks. The fastest (non-idempotent) known algorithm for Time-Tabling has a time complexity of $\mathcal{O}(n \log n)$ [13]. Despite its scalability, Time-Tabling suffers from limited filtering. On the other extreme, Energetic Reasoning [3,10] achieves a strong filtering at the cost of a prohibitive $\mathcal{O}(n^3)$ time complexity. Between these two extremes, several tradeoffs were proposed to balance strong filtering with low time complexity, e.g., Edge-Finding [6,15], Time-Table Edge-Finding [16], Time-Table Extended-Edge-Finding [13], or Not-First-Not-Last [14]. All the filtering rules listed above are subsumed by the filtering achieved by Energetic Reasoning at the exception of Not-First-Not-Last that is not comparable with Energetic Reasoning.

Surprisingly, Disjunctive Reasoning (DR) [3] has only been partially adapted to the cumulative context. In [2], Baptiste and Le Pape proposed to detect sets of tasks that cannot overlap in time without exceeding the amount of resource available initially. However, this approach is limited as it does not take in account the changes in the amount of resource available over time. This situation is illustrated in Fig. 3 where a task k has been fixed (by search or propagation). It is easy to see that tasks i and j cannot overlap in time due to the amount

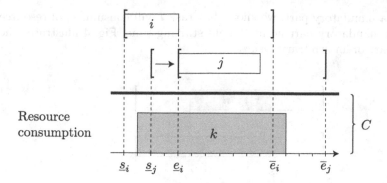

Fig. 3. Tasks i and j cannot overlap in time due to the amount of resource already consumed by task k. As j cannot be scheduled before i, j has to be scheduled after i: $e_i \leq s_j$. This situation is not detected by the approach proposed in [2].

of resource consumed by k. Unfortunately, this situation is not detected by the approach proposed by Baptiste and Le Pape.

In this work, we propose to improve Disjunctive Reasoning by considering changes in the amount of resource available. Similarly to the idea of Vilim in [16], we leverage the time-table – a core concept of Time-Tabling – to detect disjunctive pairs of tasks dynamically. Our new filtering rule – namely *Time-Table Disjunctive Reasoning* – is not subsumed by any known filtering rule. We propose a simple algorithm that implements this filtering rule with a $\mathcal{O}(n^2)$ time complexity without relying on complex data structures. We also propose two ways of improving the filtering of this algorithm. Our results on well known benchmarks from PSPLIB [7] highlight that Time-Tabling Disjunctive Reasoning is a promising filtering rule for state-of-the-art `cumulative` constraints. Indeed, using our algorithm can substantially improve the solving process of some instances and, at worst, only adds a marginally low computation overhead for the other ones.

This paper is structured as follows. Section 2 describes the time-table and the necessary background. Section 3 is dedicated to the Time-Table Disjunctive Reasoning rule and presents the algorithm and two possible extensions. The evaluation of our approach is presented in Section 4. This paper concludes on future works and possible improvements.

2   Mandatory Parts and Time-Table

Even tasks that are not fixed convey some information that can be used by filtering rules. For instance, tasks with a tight execution window must consume some resource during a specific time interval known as *mandatory part*.

Definition 1 (Mandatory part). *Let us consider a task $i \in \mathcal{T}$. The mandatory part of i is the time interval $[\overline{s}_i, \underline{e}_i[$. Task i has a mandatory part only if its latest starting time is smaller than its earliest ending time.*

If task i has a mandatory part, we know that task i will consume c_i of resource during all its mandatory part no matter its starting time. Fig. 4 illustrates the mandatory part of an arbitrary task i.

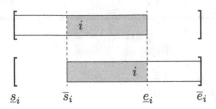

Fig. 4. Task i has a mandatory part $[\bar{s}_i, \underline{e}_i[$ if its latest starting time $\bar{s}_i$ is smaller than its earliest ending time $\underline{e}_i$: $\bar{s}_i < \underline{e}_i$. Task i always consumes the resource during its mandatory part no matter its starting time.

By aggregation, mandatory parts allow to have an optimistic view of the resource consumption over time. This aggregation is known as the *time-table* (also called minimum resource profile).

Definition 2 (Time-Table). *The time-table TT_T is the aggregation of the mandatory part of all the tasks in T. It is formally defined as the following step function:*

$$TT_T = t \in \mathbb{Z} \longrightarrow \sum_{i \in T \mid \bar{s}_i \le t < \underline{e}_i} c_i. \tag{2}$$

The problem is inconsistent if $\exists t \in \mathbb{Z} : TT_T(t) > C$.

The time-table can be computed in $\mathcal{O}(n)$ with a sweep algorithm given the tasks sorted by latest starting time and earliest ending time [4, 9, 16].

3   Time-Table Disjunctive Reasoning

In order to explain Time-Table Disjunctive Reasoning, we will use some additional notations. Let I, J be time intervals. If $I \subseteq J$, we say that J *contains* I. If $I \cap J \ne \emptyset$, we say that I *overlaps* J, or that I and J overlap.

3.1   Disjunctive Reasoning and Minimum Overlapping Intervals

In [3], Baptiste et al. briefly describe Disjunctive Reasoning in the cumulative context as a reasoning on all pairs of tasks $i \ne j$ that enforces bound-consistency on the formula:

$$c_i + c_j \le C \quad \vee \quad e_i \le s_j \quad \vee \quad e_j \le s_i. \tag{3}$$

The filtering rule to update start variables based on this formula is given next.

Proposition 1 (Disjunctive Reasoning). *Let us consider a pair of tasks $i \neq j$ in $\mathcal{T}$, such that $c_i + c_j > C$, $\underline{s}_j < \underline{e}_i$ and $\overline{s}_i < \underline{e}_j$. Then, $\underline{e}_i \leq s_j$ must hold.*

This rule states that if i and j cannot overlap, and if scheduling j at $\underline{s}_j$ would make it overlap i, then $e_i \leq s_j$, so the start of j must be at least $\underline{e}_i$. We say that task i is a *pushing* task while task j is a *pushed* task. A rule for filtering the ending times can be derived by symmetry.

The rule from Prop. 1 does some of the work of time-table filtering: when $c_i + c_j > C$ and i has a mandatory part, the reasoning on pair (i, j) is subsumed by time-tabling [3]. An additional filtering can be achieved by Disjunctive Reasoning when task i does not have a mandatory part. This filtering occurs when placing task j at $\underline{s}_j$ would make it overlap i in every schedule. It is based on the fact that j cannot contain a time interval that i must overlap. When i has no mandatory part, there is a minimum such interval.

Definition 3 (Minimum Overlapping Interval). *The minimum overlapping interval of task i, denoted moi_i, is the smallest time interval that overlaps i no matter the starting time of i. It is defined by the interval $[\underline{e}_i - 1, \overline{s}_i]$. Task i is always executed during at least one time point of moi_i.*

The moi of a task i is illustrated in Fig. 5.

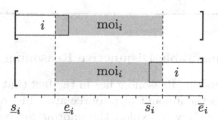

Fig. 5. Minimum overlapping interval of task i. Wherever i is placed, i must overlap moi_i, and moi_i is the smallest such interval.

When a task has a mandatory part, we consider that it has no minimum overlapping interval. Using the concept of minimum overlapping interval, it is possible to rewrite the part of Prop. 1 that is specific to Disjunctive Reasoning.

Proposition 2 (Restricted Disjunctive Reasoning). *Let us consider a pair of tasks $i \neq j$ such that task i has no mandatory part $(\underline{e}_i \leq \overline{s}_i)$ and that $c_i + c_j > C$. If scheduling task j at its earliest starting time makes it completely overlap the minimum overlaping interval of i $(moi_i \subseteq [\underline{s}_j, \underline{e}_j])$, then $\underline{e}_i \leq s_j$ must hold.*

The rule from Prop. 2 is illustrated in Fig. 6. Algorithm 1 directly follows from this rule.

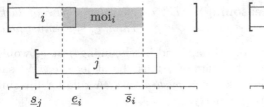 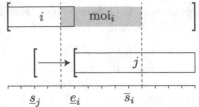

Fig. 6. On the left, tasks i and j cannot fit together for capacity reasons. Setting s_j to $\underline{s}_j$ would make j contain moi_i, and placing i would be impossible. On the right, the inconsistent values of s_j are removed, these are $t \leq \min(\mathrm{moi}_i)$. This filtering is achieved by setting $\underline{s}_j$ to $\underline{e}_i$.

Algorithm 1. $O(n^2)$ algorithm to enforce rule of Prop. 2

Input: a set of tasks $\mathcal{T}$, capacity C
Input: an array s' mapping i to $\underline{s}_i$
Output: array s' mapping i to updated starting time

1 **for** $i \in \mathcal{T}$ such that $\underline{e}_i \leq \overline{s}_i$ **do**
2   **for** $j \in \mathcal{T} - \{i\}$ **do**
3   **if** $c_i + c_j > C \wedge \mathrm{moi}_i \subseteq [\underline{s}_j, \underline{e}_j[$ **then**
4             $s'_j \leftarrow \max(s'_i, \underline{e}_i)$

3.2   Restricted Time-Table Disjunctive Reasoning

One weakness of Disjunctive Reasoning lies in the fact that it does not take into account the changes in the amount of resource available over time (see Fig. 3). In this section, we show how to exploit the information contained in the time-table (see Section 2) to propose an enhanced disjunctive filtering rule called *Time-Table Disjunctive Reasoning*.[1] We first introduce Time-Table Disjunctive Reasoning for the case where tasks i and j have no mandatory part and thus do not contribute to the time-table. This particular case saves us from removing the possible contribution of i or j from the time-table when applying a disjunctive reasoning. This restricting assumption will be relaxed to any pair of tasks later on in section 3.3.

Let us consider a pair of tasks $i \neq j$ with no mandatory parts such that $\mathrm{moi}_i \subseteq [\underline{s}_j, \underline{e}_j[$. Then Prop. 2 only compares $c_i + c_j$ to C. However, tasks in $\mathcal{T} - \{i, j\}$ may not leave C units of resource available during the overlap of i and j. We derive a new rule that leverages the mandatory part of such tasks.

Proposition 3. *Let us consider a pair of tasks $i \neq j \in \mathcal{T}$ such that i and j have no mandatory part and that $c_i + c_j + \min_{t \in \mathrm{moi}_i} \mathrm{TT}_{\mathcal{T}}(t) > C$. If scheduling task j*

---

[1] The idea of leveraging the time-table to strengthen an existing filtering rule has already been applied successfully in [13, 16].

at its earliest starting time makes it completely overlap the minimum overlapping interval of i (moi$_i \subseteq [\underline{s}_j, \underline{e}_j[$), then $\underline{e}_i \leq s_j$ must hold.

Proof. If j contained moi$_i$, then j would increase consumption by c_j during all of moi$_i$, because j does not yet contribute to resource consumption. Then, placing i anywhere would increase consumption by c_i at some point t of moi$_i$, making consumption at t greater than C. Moreover, since moi$_i \subseteq [\underline{s}_j, \underline{e}_j[$, the duration of j is such that scheduling j before $\underline{e}_i$ makes j contain moi$_i$. Hence, these values are inconsistent, and $\underline{e}_i \leq s_j$ must hold.   □

Using this rule, we can only filter values among tasks with no mandatory parts. Next section shows how to apply the same reasoning to every task.

3.3   Time-Table Disjunctive Reasoning

Using the same idea as in [13,16], we strengthen our rule further by splitting every task in two parts, a *free* part and the mandatory part.

Definition 4 (Free part). *Let us consider a task $i \in \mathcal{T}$ such that i has a mandatory part ($\overline{s}_i < \underline{e}_i$). Its free part, denoted i_f, is a separate task with the same earliest starting time and latest ending time as i: $\underline{s}_{i_f} = \underline{s}_i$ and $\overline{e}_{i_f} = \overline{e}_i$. The duration of i_f is equal to the duration of i minus the size of its mandatory part: $d_{i_f} = d_i - (\underline{e}_i - \overline{s}_i)$. If i has no mandatory part, then $i = i_f$.*

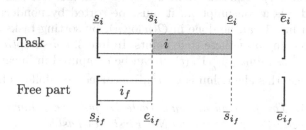

Fig. 7. A task i with a mandatory part and its free part i_f. The free part i_f always has a minimum overlapping interval moi$_{i_f}$.

Free parts have no mandatory part and always have an moi (see Fig. 7). In the remainder, we refer to $\mathcal{T}_f = \{i_f \mid i \in \mathcal{T} \wedge d_{i_f} > 0\}$ as the set of all the free parts of strictly positive duration (i.e. free parts of not assigned tasks).

Using free parts enables us to use any task in the reasoning, without worrying whether or not they contribute to the time-table. Notice that while the update is triggered by computations on free parts of tasks, the actual update should be made on tasks themselves, here s_j.

Proposition 4 (Time-Table Disjunctive Reasoning). *Let us consider a pair of tasks $i_f \neq j_f \in T_f$ such that $c_i + c_j + \min_{t \in \text{moi}_{i_f}} TT_T(t) > C$. If task j_f scheduled at its earliest starting time completely overlaps the minimum overlapping interval of i_f ($\text{moi}_{i_f} \subseteq [\underline{s}_{j_f}, \underline{e}_{j_f}[$), then, $\underline{e}_{i_f} \leq s_j$ must hold.*

Proof. Suppose that the premises are true, and then suppose $s_j \leq \min(\text{moi}_i)$. Since j_f contains moi_{i_f} when left-shifted and $d_j \geq d_{j_f}$, placing j before or at $\min(\text{moi}_{i_f})$ makes it contain moi_{i_f}. Notice that j does not contribute to the time-table during moi_{i_f}, since $\max(\text{moi}_{i_f}) \leq \underline{e}_{j_f} = \min(\underline{e}_j, \bar{s}_j)$.

If i has no mandatory part, it does not contribute to the time-table. This means that $\forall t \in \text{moi}_{i_f}, TT(t) = TT_{T-\{i,j\}}(t)$. Then placing j before or at $\min(\text{moi}_i)$ increases resource consumption on moi_{i_f} by c_j, which prevents $i = i_f$ from being placed on its moi, and is contradictory.

If i has a mandatory part, it contributes to the time-table on $[\min(\text{moi}_{i_f}) + 1, \max(\text{moi}_{i_f}) - 1]$. Placing j before or at $\min(\text{moi}_{i_f})$ increases resource consumption at $\min(\text{moi}_{i_f})$ and at $\max(\text{moi}_{i_f})$ by c_j, This prevents i_f from being left-shifted or right-shifted, which in turn means that i itself cannot overlap these time points. Since it must overlap at least one of these points, this is contradictory.   □

Algorithm 2 is an easy to implement $\mathcal{O}(n^2)$ algorithm combining moi and free parts abstractions. This algorithm enforces the updates of starting times given by Prop. 4. The tasks Pushing are candidate pusher tasks (taking the role of i). The tasks Pushed are candidate pushed tasks (take the role of j). For now, they are both T_f. The time-table is basically an array of pairs (t, c) where t is a time and c is a consumption, it must be sorted by nondecreasing t. Its initialization in line 1 can be done in $\mathcal{O}(n \log n)$, by sorting tasks according to $\bar{s}$ and $\underline{e}$ and sweeping over these time points. In line 3, *consumption*(i, TT) can be implemented[2] as $\min_{t \in \text{moi}_{i_f}} TT(t)$, it can be computed in linear time on the time-table. Hence, this algorithm is $\mathcal{O}(n^2)$. Its correctness follows from Prop. 4.

Proposition 5. *The filtering of Time-Table Disjunctive Filtering is not subsumed by Energetic Reasoning nor by Not-First-Not-Last.*

Proof. Figure 8 shows an example where Time-Table Disjunctive Reasoning can filter some values, but Energetic Reasoning and Not-First-Not-Last cannot. Task i is defined by $(\underline{s}_i, \bar{e}_i, d_i, c_i) = (2, 11, 3, 2)$, task j is defined by $(\underline{s}_j, \bar{e}_j, d_j, c_j) = (1, 20, 9, 1)$. Consumption in the resource could come from task k, defined by $(\underline{s}_k, \bar{e}_k, d_k, c_k) = (2, 11, 9, 1)$. Tasks i and j have no mandatory part, so $i_f = i$ and $j_f = j$. The condition $\text{moi}_i = [4, 8] \subseteq [\underline{s}_j, \underline{e}_j[= [1, 10[$ is satisfied, and the minimum of TT over $\text{moi}_i = [4, 8]$ is 1. It means that the two tasks i and j, consuming respectively 2 and 1 unit of resource, are not allowed to overlap over $[4, 8]$. Hence $\underline{s}_j$ is updated to $\min(\text{moi}_i) + 1 = \underline{e}_i = 5$.   □

---

[2] This primitive is voluntarily let abstract to describe further improvements.

Algorithm 2. Time-Table Disjunctive filtering algorithm

Input: sets of tasks $\mathcal{T}$, Pushing $\subseteq \mathcal{T}_f$ and Pushed $\subseteq \mathcal{T}_f$, capacity C
Input: an array s' initially mapping i to $\underline{s}_i$
Output: array s' with updated starting times

1 TT $\leftarrow initializeTimeTable(\mathcal{T})$
2 **for** $i_f \in$ Pushing **do**
3      $gap \leftarrow C - c_i - \text{consumption}(i, \text{TT})$
4   **for** $j_f \in$ Pushed $-\{i_f\}$ **do**
5   **if** $\text{moi}(i) \subseteq [\underline{s}(j), \underline{e}(j)[$ **then**
6   **if** $c_j > gap$ **then**
7                  $s'_j \leftarrow \max(s'_j, \underline{e}(i))$

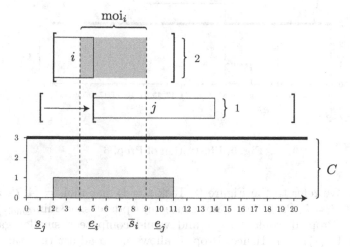

Fig. 8. Due to the time-table, tasks j and i cannot overlap. Since j cannot be scheduled before i, j has to be scheduled after i: $e_i \leq s_j$.

3.4 Improvements

The computation of allowed gap in Prop. 4, reflected at line 3 of Algorithm 2, can be strengthened in some cases, allowing to filter more values.

Pushing Task Does Not Fit Inside Its moi. When the duration of i_f is larger than its moi, we can strengthen the gap allowed by i by taking the minimum of the time-table on extremities of moi_{i_f} instead of taking it on the whole interval:

Proposition 6 (Improvement 1). *Let* $i_f \neq j_f \in \mathcal{T}_f$, *such that* $|\text{moi}_{i_f}| - 1 \leq d_{i_f}$. *Suppose* $\text{moi}_{i_f} \subseteq [\underline{s}_{j_f}, \underline{e}_{j_f}[$ *and*

$$c_{i_f} + c_{j_f} + \min(\text{TT}(\min(\text{moi}_{i_f})), \text{TT}(\max(\text{moi}_{i_f}))) > C$$

then $\underline{e}_{i_f} < s_j$ *must hold.*

Proof. If $s_j \leq \min(moi_{i_f})$, then j contains moi_{i_f}, so it contains the extremities of moi_{i_f}. Thus, j makes the consumption at $\min(moi_{i_f})$ and $\max(moi_{i_f})$ increase by c_j. Because of its duration, i_f must overlap at least one of $\min(moi_{i_f})$ and $\max(moi_{i_f})$. Doing so would overload the resource, so this is contradictory. □

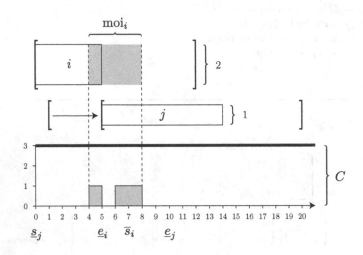

Fig. 9. Illustration of Prop. 6

Example 1. We refer to the Figure 9. Task $i = i_f$ has $moi_i = [4, 7]$, and $3 = |moi_i| - 1 \leq d_i = 5$. Prop. 4 does not cause any update, since $\min_{t \in moi_i} \mathrm{TT}(t) = 0$. However, Prop. 6 would trigger and would compute a smaller gap, since $\min(\mathrm{TT}(4), \mathrm{TT}(7)) = 1$. Hence, Prop. 6 allows us to adjust the starting time of j to 5.

Notice that improving Algorithm 2 using Prop. 6 can be done by changing $\mathrm{consumption}(i, \mathrm{TT})$ at line 3 to compute the minimum only at the extremities of moi_i when $|moi_{i_f}| \leq d_{i_f} + 1$.

Pushing Task Has a Mandatory Part. When task i has a mandatory part, despite i_f's domain, the consumption of i_f will not really be scheduled inside moi_{i_f}. Thus, we can strengthen the gap in the same way as the previous improvement:

Proposition 7 (Improvement 2). *Let* $i_f \neq j_f \in \mathcal{T}_f$, *such that* i *has a mandatory part. Suppose* $moi_{i_f} \subseteq [\underline{s}_{j_f}, \underline{e}_{j_f}[$ *and*

$$c_{i_f} + c_{j_f} + \min(\mathrm{TT}(\min(moi_{i_f})), \mathrm{TT}(\max(moi_{i_f}))) > C$$

then $\underline{e}_{i_f} < s_j$ *must hold.*

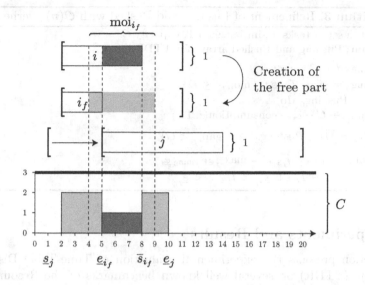

Fig. 10. Illustration of Prop. 7

Proof. The argument is the same as for Prop. 6: i must overlap either $\min(\text{moi}_{i_f})$ or $\max(\text{moi}_{i_f})$, hence computing the gap using only these points is sufficient. □

Example 2. We refer to the Figure 10. Task i has a mandatory part. Since $4 = |\text{moi}_{i_f}| - 1 > d_{i_f} = 2$, we are not in the case of application of Prop. 6. We cannot apply Prop. 4 either, since $\min_{t\in\text{moi}_{i_f}}(\text{TT}(t))$ is 1, allowing i and j to overlap. Nevertheless, since i has a mandatory part, we can apply Prop. 7 and use $\min(\text{TT}(4), \text{TT}(8)) = 2$ to compute the gap, which forbids i and j to overlap in moi_{i_f}. Hence, Prop.7 allows us to adjust the starting time of j to 5.

Once more, improving Algorithm 2 using Prop. 7 can be done by changing consumption(i, TT) at line 3 to compute the minimum only at the extremities of moi_i when i has a mandatory part.

Reducing the Number of Pairs to Consider. Our algorithm has a theoretical $\mathcal{O}(n^2)$ time complexity, but it is possible to reduce it in practice by removing tasks that cannot push, and by removing tasks that cannot be pushed. Refining these sets allows to keep the same filtering, while examining much less than the theoretical $\mathcal{O}(n^2)$ pairs of tasks.

First, pushing tasks must have a tight enough moi and high enough consumption during their moi to push any other task. After refining Pushing, we found useful to refine tasks of Pushed according to their time location: a pushed task must not be schedulable before the minimal earliest start of pushing tasks mois, and must be updatable at least by their maximal earliest end.

This procedure is formalized in Algorithm 3.

Algorithm 3. Refinement of Pushing and Pushed with $\mathcal{O}(n)$ overhead

---

Input: a set of tasks $\mathcal{T}$, initialized TT, capacity C
Output: Pushing and Pushed arrays for TTDR input

1   $D = \max d_{i_f}$, $G = \max c_{i_f}$
2   $\text{Pushing}_0 = \{i_f \in \mathcal{T}_f \text{ s.t. } |\text{moi}_{i_f}| \le D\}$
3   **for** $i_f \in \text{Pushing}_0$ **do**
4     $\text{gap}_{i_f} \leftarrow C - c_i - \text{consumption}(i, \text{TT})$

5   $\text{Pushing} = \{i_f \in \text{Pushing}_0 \text{ s.t. } \text{gap}_{i_f} < G\}$

6   $S = \min_{i_f \in \text{Pushing}} \overline{s}_{i_f}$, $E = \max_{i_f \in \text{Pushing}} \underline{e}_{i_f}$
7   $\text{Pushed} = \{i_f \in \mathcal{T}_f \text{ s.t. } \underline{s}_{j_f} < E \wedge S < \underline{e}_{j_f}\}$

---

4   Experiments and Results

This section presents the experimental evaluation of Time-Table Disjunctive Reasoning (TTDR) on several well known benchmarks of the Resource Constraints Project Scheduling Problem (RCPSP) from PSPLIB [7]. The aim of these experiments is to measure the performance of the cumulative constraint when TTDR is added to a set of filtering rules. To achieve this, we compared the performance of classical filtering rules with and without TTDR. Here is the exhaustive list of the compared algorithms:

- Time-Tabling (TT) is implemented using a fast variant of [9];
- Time-Table Disjunctive Reasoning (TTDR) implemented as described in this paper with all proposed improvements;
- Edge-Finding (EF) is implemented as proposed in [6];
- Energetic Reasoning (ER) is the well known implementation proposed by Baptiste et al. in [3]. We added some improvements proposed by Derrien et Petit to reduce the number of considered intervals [5];
- Not-First-Not-Last (NFNL) is implemented with a $\mathcal{O}(n^3)$ variant of the algorithm proposed by Schutt et al. [14].

All the algorithms were implemented in the open-source OscaR Solver [12]. The priorities chosen for cumulative constraints in the propagation queue are such that TT is executed first, then TTDR, EF, ER and finally NFNL. We used a classic SetTimes search [8], breaking ties by taking a task of minimal duration.

We used a machine with a 4-core, 8 thread Core(TM) i7-2600 CPU @ 3.40GHz processor and 8GB of RAM under GNU/Linux. using Java SE 1.7 JVM.

We report the cumulated distribution $F(\tau)$ of instances solved within computational limit τ in Fig. 11 and Fig. 13. On the left column, τ refers to time, on the right it refers to the size of the search tree; the x-axis is logarithmic in both cases. $F(\tau) = k$ means that k instances where solved under τ ms or nodes. We set a timeout of 90s for every computation.[3] Due to a lack of room, we only

---

[3] Which is why time graphs will not show points further than 2^{17} ms.

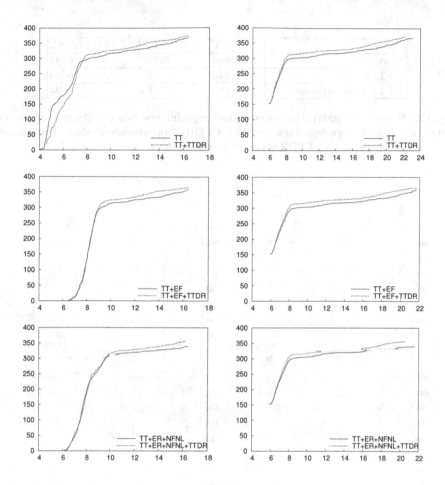

Fig. 11. Plots for all instances of PSPLIB30. y-axis is the cumulative number of solved instances. In the left column, x-axis is $log_2(t)$, with t time in ms. In the right column, x-axis is $log_2(n)$, with n the number of nodes to find optimal and prove optimality.

display results obtained on instances with 30 and 120 tasks. Results obtained on instances with 60 tasks are similar. However, we observed that adding TTDR has a very little effect on instances with 90 tasks in which only two additional instances were closed using TTDR.

In Fig. 12 and Fig. 14, we report results for destructive lower bound experiments, that compute the best lower bound given by propagation alone.

Despite its $\mathcal{O}(n^2)$ theoretical complexity, the algorithm for TTDR is more of a lightweight algorithm. The only computation overhead appears on very small time limits when TTDR is used with TT. However, the additional filtering of TTDR quickly takes over, allowing to solve more instances for a given time limit τ. The PSPLIB instances are well-known to be rather disjunctive than

Stack	TT	TT+EF	TT+ER+NFNL
Score	26364	26712	26765
+TTDR, Score	26543	26815	26845
+TTDR, #Improvements	104	73	65

Fig. 12. Results for destructive lower bound experiments. Score is the sum of proven lower bounds, with the original stack or with TTDR. #Improvements shows the number of instances where adding TTDR gives a strictly higher bound.

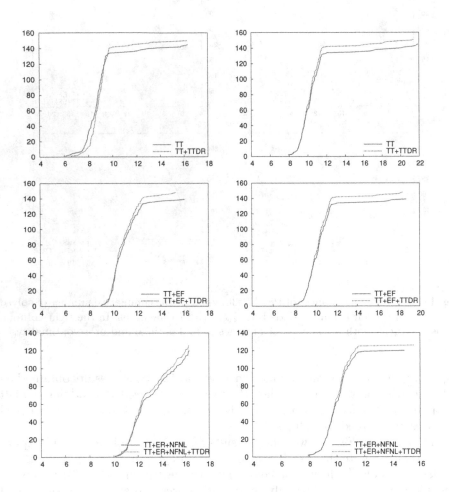

Fig. 13. Plots for all instances of PSPLIB120. y-axis is the cumulative number of solved instances. In the left column, x-axis is $log_2(t)$, with t time in ms. In the right column, x-axis is $log_2(n)$, with n the number of nodes to find optimal and prove optimality.

Stack	TT	TT+EF	TT+ER+NFNL
Score	58365	69074	69509
+TTDR, Score	58575	69117	69536
+TTDR, #Improvements	132	33	22

Fig. 14. Results for destructive lower bound experiments. Score is the sum of proven lower bounds, with the original stack or with TTDR. #Improvements shows the number of instances where adding TTDR gives a strictly higher bound.

cumulative.[4] Adding energy-based reasoning on PSPLIB instances is a risky trade-off. Indeed, energy-based reasoning does not trivialize much the PSPLIB instances, whereas TTDR does improve on the number of instances solved by TT alone. This confirms experimentally that TTDR is complementary to existing energy-based filtering for the cumulative constraint (see Prop. 5)

Finally, we also see that the gain in solved instances does not drop from the 30-task set to the 120-task set. This means that the filtering depends more on the nature of instances than on their size, and that the $\mathcal{O}(n^2)$ algorithm for TTDR scales well.

5   Conclusion

This paper introduces Time-Tabling Disjunctive Reasoning – a new filtering rule that leverages the time-table to detect disjunctive pairs of tasks dynamically. By relying on minimum overlapping intervals, this filtering rules achieve a new type of filtering that is not subsumed by existing filtering rules such as Energetic Reasoning or Not-First-Not-Last. Besides its novelty, Time-Table Disjunctive Reasoning can be implemented with a simple $\mathcal{O}(n^2)$ algorithm that does not rely on complex data-structures. Benefits of using Time-Table Disjunctive Reasoning in combination with other filtering rules were evaluated on well-known benchmarks from PSPLIB. Our results highlight that Time-Table Disjunctive Reasoning is a promising filtering rule to extend state-of-the-art cumulative constraints. Indeed, using our algorithm can substantially improve the solving process of some instances and, at worst, only adds a marginally low overhead for the other ones.

Although the strengthening proposed by Time-Table Disjunctive Reasoning is already a good tradeoff in terms of speed and filtering, it could be improved further. For instance, its practical complexity could be reduced by using sweeping techniques to prevent the examination of non-overlapping pairs of tasks. An even more interesting improvement would be on the filtering side. For instance, one may be able to strengthen the filtering by considering minimum overlapping intervals on more than one task at the same time.

---

[4] A problem is said to be highly disjunctive when many pairs of activities cannot execute in parallel; on the contrary, a problem is highly cumulative if many activities can effectively execute in parallel [2].

Acknowledgments. Steven Gay is financed by project Innoviris 13-R-50 of the Brussels-Capital region. Renaud Hartert is a Research Fellow of the Fonds de la Recherche Scientifiques - FNRS. The authors would like to thank Sascha Van Cauwelaert for sharing his instances and parser, and the reviewers for suggestions of experiments.

References

1. Aggoun, A., Beldiceanu, N.: Extending chip in order to solve complex scheduling and placement problems. Mathematical and Computer Modelling **17**(7), 57–73 (1993)
2. Baptiste, P., Le Pape, C.: Constraint propagation and decomposition techniques for highly disjunctive and highly cumulative project scheduling problems. Constraints **5**(1–2), 119–139 (2000)
3. Baptiste, P., Le Pape, C., Nuijten, W.: Constraint-Based Scheduling: Applying Constraint Programming to Scheduling Problems, vol. 39. Springer (2001)
4. Beldiceanu, N., Carlsson, M.: A new multi-resource *cumulatives* constraint with negative heights. In: Van Hentenryck, P. (ed.) CP 2002. LNCS, vol. 2470, pp. 63–79. Springer, Heidelberg (2002)
5. Derrien, A., Petit, T.: A new characterization of relevant intervals for energetic reasoning. In: O'Sullivan, B. (ed.) CP 2014. LNCS, vol. 8656, pp. 289–297. Springer, Heidelberg (2014)
6. Kameugne, R., Fotso, L.P., Scott, J., Ngo-Kateu, Y.: A quadratic edge-finding filtering algorithm for cumulative resource constraints. Constraints **19**(3), 243–269 (2014)
7. Kolisch, R., Schwindt, C., Sprecher, A.: Benchmark instances for project scheduling problems. In: Project Scheduling, pp. 197–212. Springer (1999)
8. Le Pape, C., Couronné, P., Vergamini, D., Gosselin, V.: Time-Versus-Capacity Compromises in Project Scheduling (1994)
9. Letort, A., Beldiceanu, N., Carlsson, M.: A scalable sweep algorithm for the cumulative constraint. In: Milano, M. (ed.) Principles and Practice of Constraint Programming. LNCS, pp. 439–454. Springer, Heidelberg (2012)
10. Lopez, P., Erschler, J., Esquirol, P.: Ordonnancement de tâches sous contraintes: une approche énergétique. Automatique-productique informatique industrielle **26**(5–6), 453–481 (1992)
11. Nuijten, W.P.M.: Time and resource constrained scheduling: a constraint satisfaction approach. PhD thesis, Technische Universiteit Eindhoven (1994)
12. OscaR Team. OscaR: Scala in OR (2012). https://bitbucket.org/oscarlib/oscar
13. Ouellet, P., Quimper, C.-G.: Time-table extended-edge-finding for the cumulative constraint. In: Schulte, C. (ed.) CP 2013. LNCS, vol. 8124, pp. 562–577. Springer, Heidelberg (2013)
14. Schutt, A., Wolf, A.: A new $O(n^2 \log n)$ not-first/not-last pruning algorithm for cumulative resource constraints. In: Cohen, D. (ed.) CP 2010. LNCS, vol. 6308, pp. 445–459. Springer, Heidelberg (2010)
15. Vilím, P.: Edge finding filtering algorithm for discrete cumulative resources in O(kn log n). In: Gent, I.P. (ed.) CP 2009. LNCS, vol. 5732, pp. 802–816. Springer, Heidelberg (2009)
16. Vilím, P.: Timetable edge finding filtering algorithm for discrete cumulative resources. In: Achterberg, T., Beck, J.C. (eds.) CPAIOR 2011. LNCS, vol. 6697, pp. 230–245. Springer, Heidelberg (2011)

Uncertain Data Dependency Constraints in Matrix Models

C. Gervet[✉] and S. Galichet[✉]

LISTIC, Laboratoire d'Informatique, Systèmes, Traitement de l'Information et de la Connaissance, Université de Savoie, BP 80439, 74944 Annecy-Le-Vieux Cedex, France
{gervetec,sylvie.galichet}@univ-savoie.fr

Abstract. Uncertain data due to imprecise measurements is commonly specified as bounded intervals in a constraint decision or optimization problem. Dependencies do exist among such data, e.g. upper bound on the sum of uncertain production rates per machine, sum of traffic distribution ratios from a router over several links. For tractability reasons existing approaches in constraint programming or robust optimization frameworks assume independence of the data. This assumption is safe, but can lead to large solution spaces, and a loss of problem structure. Thus it cannot be overlooked. In this paper we identify the context of matrix models and show how data dependency constraints over the columns of such matrices can be modeled and handled efficiently in relationship with the decision variables. Matrix models are linear models whereby the matrix cells specify for instance, the duration of production per item, the production rates, or the wage costs, in applications such as production planning, economics, inventory management. Data imprecision applies to the cells of the matrix and the output vector. Our approach contributes the following results: 1) the identification of the context of matrix models with data constraints, 2) an efficient modeling approach of such constraints that suits solvers from multiple paradigms. An illustration of the approach and its benefits are shown on a production planning problem.

Keywords: Data uncertainty · Data constraints · Interval reasoning · Interval linear programs

1 Introduction

Data uncertainty due to imprecise measurements or incomplete knowledge is ubiquitous in many real world applications, such as network design, renewable energy economics, investment and production planning (e.g. [13,18]). Formalisms such as linear programming or constraint programming have been extended and successfully used to tackle certain forms of data uncertainty. Bounded intervals

This research was partly support by the Marie Curie CIG grant, FP7-332683.

L. Michel (Ed.): CPAIOR 2015, LNCS 9075, pp. 173–181, 2015.
DOI: 10.1007/978-3-319-18008-3_12

are commonly used to specify such imprecise parameters, which take the form of coefficients in a given constraint relation. In such problems, uncertain data dependencies do exist, such as an upper bound on the sum of uncertain production rates per machine, the sum of traffic distribution ratios from a router over several links. To our knowledge, existing approaches in Operations Research assume independence of the data when tackling real world problems essentially to maintain computational tractability. This assumption is safe in the sense that no potential solution to the uncertain problem is removed. However, the solution set can be very large even if no solution actually holds once the data dependencies are checked, since the problem structure is lost. Thus accounting for possible data dependencies cannot be overlooked.

In this paper we tackle these issues, by identifying the context of matrix models, where we show how constraints over uncertain data can be handled efficiently. Matrix models are of high practical relevance in many combinatorial optimization problems where the uncertain data corresponds to coefficients of the decision variables. Clearly, the overall problem does not need to be itself a matrix model. With the imprecise data specifying cells of an input matrix, the data constraints correspond to restrictions over the data in each column of the matrix. For instance in a production planning problem, the rows would denote the products to be manufactured and the columns the machines available. A data constraint such as an upper bound on the sum of uncertain production rates per machine, applies to each column of the matrix. In this context, we observe that there is a dynamic relationship between the constraints over uncertain data and the decisions variables that quantify the usage of such data. Uncertain data are not meant to be pruned and instantiated by the decision maker. However, decision variables are, and the solver controls their possible values. This leads us to define a notion of relative consistency of uncertain data constraints, in relationship with the decision variables involved, in order to check and infer consistency of such constraints. For instance, if an uncertain input does not satisfy a dependency constraint, this does not imply that the problem has no solution! It tells us that the associated decision variable should be 0, to reflect the fact that the given machine cannot produce this input.

Our main contribution lies in identifying the context of matrix models, to study the efficient handling of uncertain data constraints. To our knowledge, this is a first efficient handling of uncertain data constraints in combinatorial problems. Our approach contributes the following within this context: 1) identify the role of uncertain data constraints and their impact on the decision variables, 2) propose a new consistency notion of the uncertain data constraints and a model that implements it efficiently. We illustrate the benefits and impacts of our approach on a classical production planning problem with data constraints.

The paper is structured as follows. Section 2 summarizes the related work, while Section 3 gives the intuition of our approach. Section 4 defines our approach, and Section 5 illustrates it. A conclusion is given in Section 6.

2   Related Work

In the past 15 years, the generic CSP formalism has been extended to account for forms of uncertainty: e.g. numerical, mixed, quantified, fuzzy, uncertain CSP and CDF-interval CSPs [6]. The *numerical, uncertain, or CDF-interval* CSPs, extend the classical CSP to approximate and reason with continuous uncertain data represented by intervals; see the real constant type in Numerica [19] or the bounded real type in ECLiPSe [7]. The solution sets produced can be very large. This led to some research to extract the relationship between uncertain data that satisfy dependency constraints and possible solutions by applying regression analysis techniques [11]. The fuzzy and mixed CSP [9] coined the concept of uncontrollable variables, that can take a set of values but their domain is not meant to be pruned during problem solving (unlike decision variables). Some constraints over uncontrollable variables can be expressed and thus some limited form of data dependency modeled, mainly in a discrete environment.

The general QCSP formalism introduces universal quantifiers where the domain of a universally quantified variable (UQV) is not meant to be pruned, and its actual value is unknown a priori. There has been work on QCSP with continuous domains, using one or more UQV and dedicated algorithms [2,5,15]. Discrete QCSP algorithms cannot be used to reason about uncertain data since they apply a preprocessing step enforced by the solver QCSPsolve [10], which essentially determines whether constraints of the form $\forall X, \forall Y, C(X, Y)$, and $\exists Z, \forall Y, C(Z, Y)$, are either always true or false for all values of a UQV. This is a too strong statement, that does not reflect the fact that the data will be refined later on and might satisfy the constraint.

Closer to our approach are the fields of Interval Linear Programming [8,14] and Robust Optimization [3,4], whereby in the former we seek the solution set that encloses all possible solutions whatever the data might be, and in the latter the solution that holds in the larger set of possible data realization. They do offer a sensitivity analysis to study the solution variations as the data changes. However, to our knowledge, uncertain data constraints have been ignored for computational tractability reasons.

3   Intuition

The main novel idea behind our work is based on the study of a problem structure. We identify the context of matrix models where uncertain data correspond to coefficients of the decision variables, and the constraints over these apply to the columns of the input matrix. Such data constraints state restrictions *on the possible usage of the data*, and we show how their satisfaction can be handled efficiently in relationship with the corresponding decision variables.

In this context, the role and handling of uncertain data constraints is to determine "which data can be used, to build a solution to the problem". This is in contrast with standard constraints over decision variables, which role is to determine "what value can a variable take to derive a solution that holds".

We illustrate the context and our new notion of uncertain data constraint satisfaction on a production planning problem inspired from [12].

Example 1. Three types of products are manufactured, P_1, P_2, P_3 on two different machines M_1, M_2. The production rate of each product per machine is imprecise and specified by intervals. Each machine is available 9 hrs per day, and an expected demand per day is specified by experts as intervals. Furthermore we know that the total production rate of each machine cannot exceed 7 pieces per hour. We are looking for the number of hours per machine for each product, to satisfy the expected demand. An instance data model is given below.

Product	Machine M1	Machine M2	Expected demand
P_1	$[2,3]$	$[5,7]$	$[28,32]$
P_2	$[2,3]$	$[1,3]$	$[25,30]$
P_3	$[4,6]$	$[2,3]$	$[31,37]$

The uncertain CSP model is specified as follows:

$$[2,3] * X_{11} + [5,7] * X_{12} = [28,32] \tag{1}$$
$$[2,3] * X_{21} + [1,3] * X_{22} = [25,30] \tag{2}$$
$$[4,6] * X_{31} + [2,3] * X_{32} = [31,37] \tag{3}$$
$$\forall j \in \{1,2\} : X_{1j} + X_{2j} + X_{3j} \leq 9 \tag{4}$$
$$\forall i \in \{1,2,3\}, \forall j \in \{1,2\} : X_{ij} \geq 0 \tag{5}$$

Uncertain data constraints:

$$a_{11} \in [2,3], a_{21} \in [2,3], a_{31} \in [4,6], \quad a_{11} + a_{21} + a_{31} \leq 7 \tag{6}$$
$$a_{12} \in [5,7], a_{22} \in [1,3], a_{32} \in [2,3], \quad a_{12} + a_{22} + a_{32} \leq 7 \tag{7}$$

Consider a state of the uncertain CSP such that $X_{11} = 0$. The production rate of machine M_1 for product P_1 becomes irrelevant since $X_{11} = 0$ means that machine M_1 does not produce P_1 at all in this solution. The maximum production rate of M_1 does not change but now applies to P_2 and P_3. Thus $X_{11} = 0$ infers $a_{11} = 0$. Constraint (6) becomes:

$$a_{21} \in [2,3], a_{31} \in [4,6], \quad a_{21} + a_{31} \leq 7 \tag{8}$$

Assume now that we have a different production rate for P_3 on M_1:

$$a_{11} \in [2,3], a_{21} \in [2,3], a_{31} \in [8,10], \quad a_{11} + a_{21} + a_{31} \leq 7 \tag{9}$$

P_3 cannot be produced by M_1 since $a_{31} \in [8,10] \not\leq 7$, the total production rate of M_1 is too little. This does not imply that the problem is unsatisfiable, but that P_3 cannot be produced by M_1. Thus $a_{31} \not\leq 7$ yields $X_{31} = 0$ and $a_{31} = 0$.   □

4   Our Approach

We now formalize our approach: define the context of matrix models we identified and the handling of uncertain data constraints within it.

4.1   Problem Definition

Definition 1 (interval data). *An interval data, is an uncertain data, specified by an interval $[\underline{a}, \overline{a}]$, where $\underline{a}$ (lower bound) and $\overline{a}$ (upper bound) are positive real numbers, such that $\underline{a} \leq \overline{a}$.*

Definition 2 (Matrix model with column constraints). *A matrix model with uncertain data constraints is a constraint problem or a component of a larger constraint problem that consists of:*

1. *A matrix (A_{ij}) of input data, such that each row i denotes a given product P_i, each column j denotes the source of production and each cell a_{ij} the quantity of product i manufactured by the source j. If the input is bounded, we have an interval input matrix, where each cell is specified by $[\underline{a_{ij}}, \overline{a_{ij}}]$.*
2. *A set of decision variables $X_{ij} \in \mathbb{R}^+$ denoting how many instances of the corresponding input shall be manufactured*
3. *A set of column constraints, such that for each column j: $\Sigma_i [\underline{a_{ij}}, \overline{a_{ij}}] @ c_j$, where $@ \in \{=, \leq\}$, and c_j can be a crisp value or a bounded interval.*

To reason about uncertain matrix models we make use of the robust counterpart transformation of interval linear models into linear ones. We recall it, and define the notion of relative consistency of column constraints.

4.2   Linear Transformation

An Interval Linear Program is a Linear constraint model where the coefficients are bounded real intervals [3, 8]. The handling of such models transforms each interval linear constraint into an equivalent set of atmost 2 standard linear constraints. Equivalence means that both models denote the same solution space. We recall the transformations of an ILP into its equivalent LP counterpart.

Property 1 (Interval linear constraint and equivalence). Let all decision variables $X_{il} \in \mathbb{R}^+$, and all interval coefficients be positive as well. The interval linear constraint $C = \Sigma_i [\underline{a_{il}}, \overline{a_{il}}] * X_{il} @ [\underline{c_l}, \overline{c_l}]$ with $@ \in \{\leq, =\}$, is equivalent to the following set of constraints depending on the nature of $@$. We have:

1. $C = \Sigma_i [\underline{a_{il}}, \overline{a_{il}}] * X_{il} \leq [\underline{c_l}, \overline{c_l}]$ is transformed into: $C = \Sigma_i \underline{a_{il}} * X_{il} \leq \overline{c_l}$
2. $C = \Sigma_i [\underline{a_{il}}, \overline{a_{il}}] * X_{il} = [\underline{c_l}, \overline{c_l}]$ is transformed into:

$$C = \{\Sigma_i \underline{a_{il}} \times X_{il} \leq \overline{c_l} \ \wedge \ \Sigma_i \overline{a_{il}} * X_{il} \geq \underline{c_l}\}$$

Note that case 1 can take a different form depending on the decision maker risk adversity. If he assumes the highest production rate for the smallest demand (pessimistic case), the transformation would be: $C = \Sigma_i \overline{a_{il}} * X_{il} \leq \underline{c_l}$. The solution set of the robust counterpart contains that of the pessimistic model.

Example 2. Consider the following constraint $a_1 * X + a_2 * Y = [120, 150]$ (case 2), with $a_1 \in [0.2, 0.7], a_2 \in [0.1, 0.35], X, Y \in [0, 1000]$. It is rewritten into the system of constraints: $l_1 : 0.7 * X + 0.35 * Y \geq 120 \ \wedge \ l_2 : 0.2 * X + 0.1 * Y \leq 150$.

The transformation procedure also applies to the column constraints, and is denoted `transf`. It evaluates to true or false since there is no variable involved.

4.3   Relative Consistency

We now define in our context, the relative consistency of column constraints with respect to the decision variables. At the unary level this means that if $(X_{ij} = 0)$ then $(a_{ij} = 0)$, if $\neg\ \texttt{transf}(a_{ij}@c_j)$ then $(X_{ij} = 0)$ and if $X_{ij} > 0$ then $\texttt{transf}(a_{ij}@ c_j)$ is true.

Definition 3 (Relative consistency). *A column constraint* $\Sigma_i a_{il}@ c_l$ *over the column* l *of a matrix* $I * J$, *is relative consistent w.r.t. the decision variables* X_{il} *if and only if the following conditions hold (C4. and C5. being recursive):*

C1. $\forall i \in I$ *such that* $X_{il} > 0$, *we have* $\texttt{transf}(\Sigma_i a_{il} @ c_l)$ *is true*
C2. $\forall k \in I$ *such that* $\{\neg\texttt{transf}(\Sigma_{i \neq k} a_{il}@ c_l)$ *and* $\texttt{transf}(\Sigma_i a_{il}@ c_l)\}$ *is true,*
    we have $X_{kl} > 0$
C3. $\forall i \in I$ *such that* X_{il} *we have* $i, \texttt{transf}(a_{il} @ c_l)$ *is true*
C4. $\forall k \in I$, *such that* $\neg\texttt{transf}(a_{kl} @ c_l)$, *we have* $X_{kl} = 0$ *and* $\Sigma_{i \neq k} a_{il}@ c_l$ *is*
    relative consistent
C5. $\forall k \in I$, *such that* $X_{kl} = 0$, *we have* $\Sigma_{i \neq k} a_{il}@ c_l$ *is relative consistent*

Example 3. Consider the Example 1. It illustrates C4. and C5., leading to the recursive call to C3. Let us assume now that the X_{i1} are free, and the column constraint $[2, 3] + [2, 3] + [4, 6] = [7, 9]$. Rewritten into $2 + 2 + 4 \leq 9, 3 + 3 + 6 \geq 7$, we have $X_{31} > 0$, since $3 + 3 \not\geq 7$ and $3 + 3 + 6 \geq 7$. It is relative consistent with $X_{31} > 0$ (C2.).

4.4   Column Constraint Model

Our intent is to model column constraints and infer relative consistency while preserving the computational tractability of the model. We do so by proposing a Mixed Integer Interval model of a column constraint. We show how it allows us to check and infer relative consistency efficiently. This model can be embedded in a larger constraint model. The consistency of the whole constraint system is inferred from the local and relative consistency of each constraint.

Modeling column constraints. Consider the column constraint over column l: $\Sigma_i [\underline{a_{il}}, \overline{a_{il}}] @ c_l$. It needs to be linked with the decision variables X_{il}. Logical implications could be used, but they would not make an active use of consistency and propagation techniques. We propose an alternative MIP model.

To each data we associate a Boolean variable. Each indicates whether: 1) the data must be accounted for to render the column constraint consistent, 2) the data violates the column constraint and needs to be discarded, 3) the decision variable imposes a selection or removal of the data. Thus the column constraint in transformed state is specified as a scalar product of the data and Boolean variables. The link between the decision variables and their corresponding Booleans is specified using a standard mathematical programming technique that introduces a big enough positive constant K, and a small enough constant λ.

Theorem 1 (column constraint model). *Let $X_{il} \in \mathbb{R}^+$ be decision variables of the matrix model for column l. Let $\mathbf{B}_{il}$ be Boolean variables. Let K be a large positive number, and λ a small enough positive number. A column constraint*

$$\Sigma_i[\underline{a_{il}}, \overline{a_{il}}] \; @ \; c_l$$

is relative consistent if the following system of constraints is bounds consistent

$$\mathtt{transf}(\Sigma_i[\underline{a_{il}}, \overline{a_{il}}] \times \mathbf{B}_{il} \; @ \; c_l) \tag{10}$$

$$\forall i, 0 \leq X_{il} \leq K \times \mathbf{B}_{il} \tag{11}$$

$$\forall i, \lambda \times \mathbf{B}_{il} \leq X_{il} \tag{12}$$

The proof is omitted for space reasons.

5 Illustration of the Approach

We illustrate the approach on the production planning problem. The robust model is specified below. Each interval linear core constraint is transformed into a system of two linear constraints, and each column constraint into its robust counterpart.

For the core constraints we have:
$2 * X_{11} + 5 * X_{12} \leq 32, \quad 3 * X_{11} + 7 * X_{12} \geq 28,$
$2 * X_{21} + \quad X_{22} \leq 30, \quad 3 * X_{21} + 3 * X_{22} \geq 25,$
$4 * X_{31} + 2 * X_{32} \leq 37, \quad 6 * X_{31} + 3 * X_{32} \geq 31,$
$\forall j \in \{1, 2\}, X_{1j} + X_{2j} + X_{3j} \leq 9,$
$\forall i \in \{1, 2, 3\}, \forall j \in \{1, 2\} : X_{ij} \geq 0,$
$\forall i, j, X_{ij} \geq 0, B_{ij} \in \{0, 1\},$

And for the column constraints:
$a_{11} \in [2, 3], a_{21} \in [2, 3], a_{31} \in [4, 6], \quad a_{11} + a_{21} + a_{31} \leq 7$ and
$a_{12} \in [5, 7], a_{22} \in [1, 3], a_{32} \in [2, 3], \quad a_{12} + a_{22} + a_{32} \leq 7$ transformed into:

$2 * B_{11} + 2 * B_{21} + 4 * B_{31} \leq 7,$
$5 * B_{12} + \quad B_{22} + 2 * B_{32} \leq 7,$
$\forall i \in \{1, 2, 3\}, j \in \{1, 2\} \quad 0 \leq X_{ij} \leq K * B_{ij},$
$\forall i \in \{1, 2, 3\}, j \in \{1, 2\} \quad \lambda * B_{ij} \leq X_{ij}$

We consider three different models: 1) the robust approach that seeks the largest solution set, 2) the pessimistic approach, and 3) the model without column data constraints. They were implemented using the ECLiPSe ic interval solver [7]. We used the constants K=100 and $\lambda = 1$. The column constraints in the tightest model take the form: $3 * B_{11} + 3 * B_{21} + 6 * B_{31} \leq 7$ and $7 * B_{12} + 3 * B_{22} + 3 * B_{32} \leq 7$.

The solution set results are summarized in the following table with real values rounded up to hundredth for clarity. The tightest model, where the decision maker assumes the highest production rates has no solution.

Variables	With column constraints			Without column constraints
	Robust model		Tightest model	
	Booleans	Solution bounds	Solution bounds	Solution bounds
X_{11}	0	0.0..0.0	−	0.0..7.00
X_{12}	1	4.0..4.5	−	0.99..6.4
X_{21}	1	3.33..3.84	−	0.33..7.34
X_{22}	1	4.49..5.0	−	0.99..8.0
X_{31}	1	5.16..5.67	−	1.66..8.67
X_{32}	0	0.0..0.0	−	0.0..7.0

Results. From the table of results we can clearly see that:

1. Enforcing Bounds Consistency (BC) on the constraint system without the column constraints, is safe since the bounds obtained enclose the ones of the robust model with column constraints. However, they are large, and the impact of accounting for the column constraints, both in the much reduced bounds obtained, and to detect infeasibility is shown.

2. The difference between the column and non column constraint models is also interesting. The solutions show that only X_{11} and X_{32} can possibly take a zero value from enforcing BC on the model without column constraints. Thus all the other decision variables require the usage of the input data resources. Once the column constraints are enforced, the input data a_{11} and a_{32} must be discarded since otherwise the column constraints would fail. This illustrates the benefits of relative consistency over column constraints.

3. The tightest model fails, because we can see from the solution without column constraints that a_{21} and a_{31} must be used since their respective X_{ij} are strictly positive in the solution to the model without column constraints. However from the tight column constraint they can not both be used at full production rate at the same time. The same holds for a_{12} and a_{22}.

All computations were performed in constant time given the size of the problem. This approach can easily scale up, since if we have n uncertain data (thus n related decision variables) in the matrix model, our model generates n Boolean variables and $O(2n + 2) = O(n)$ constraints. This number does not depend on the size or bounds of the uncertain data domain, and the whole problem models a standard CP or MIP problem, making powerful use of existing techniques.

6   Conclusion and Future Work

In this paper we have identified the context of matrix models to account for uncertain data constraints efficiently. Such models are common in many applications ranging from production planning, economics, or inventory management to name a few. In this context, we defined the notion of relative consistency, and a model of uncertain data constraints that implements it effectively. An interesting challenge to our eyes, would be to investigate how the notion of relative consistency can be generalized and applied to certain classes of global constraints in a CP environment, whereby the uncertain data appears as coefficients to the decision variables.

References

1. Benhamou, F.: Inteval constraint logic programming. In: Podelski, A. (ed.) Constraint Programming: Basics and Trends. LNCS, vol. 910, pp. 1–21. Springer, Heidelberg (1995)
2. Benhamou, F., Goualard, F.: Universally quantified interval constraints. In: Dechter, R. (ed.) CP 2000. LNCS, vol. 1894, pp. 67–82. Springer, Heidelberg (2000)
3. Ben-Tal, A., Nemirovski, A.: Robust solutions of uncertain liner programs. Operations Research Letters **25**, 1–13 (1999)
4. Bertsimas, D., Brown, D.: Constructing uncertainty sets for robust linear optimization. Operations Research (2009)
5. Bordeaux, L., Monfroy, E.: Beyond NP: Arc-consistency for quantified constraints. In: Van Hentenryck, P. (ed.) CP 2002. LNCS, vol. 2470, pp. 371–386. Springer, Heidelberg (2002)
6. Brown, K., Miguel, I.: Chapter 21: uncertainty and change. In: Handbook of Constraint Programming. Elsevier (2006)
7. Cheadle, A.M., Harvey, W., Sadler, A.J., Schimpf, J., Shen, K., Wallace, M.G.: ECLiPSe: An Introduction. Tech. Rep. IC-Parc-03-1, Imperial College London, London, UK
8. Chinneck, J.W., Ramadan, K.: Linear programming with interval coefficients. J. Operational Research Society **51**(2), 209–220 (2000)
9. Fargier, H., Lang, J., Schiex, T.: Mixed constraint satisfaction: A framework for decision problems under incomplete knowledge In: Proc. of AAAI-1996 (1996)
10. Gent, I., Nightingale, P., Stergiou, K.: QCSP-solve: A solver for quantified constraint satisfaction problems. In: Proc. of IJCAI (2005)
11. Gervet, C., Galichet, S.: On combining regression analysis and constraint programming. In: Laurent, A., Strauss, O., Bouchon-Meunier, B., Yager, R.R. (eds.) IPMU 2014, Part II. CCIS, vol. 443, pp. 284–293. Springer, Heidelberg (2014)
12. Inuiguchi, M., Kume, Y.: Goal programming problems with interval coefficients and target intervals. European Journl of Oper. Res. **52** (1991)
13. Medina, A., Taft, N., Salamatian, K., Bhattacharyya, S. Diot, C.: Traffic matrix estimation: existing techniques and new directions. In: Proceedings of ACM SIGCOMM02 (2002)
14. Oettli, W.: On the solution set of a linear system with inaccurate coefficients. J. SIAM: Series B, Numerical Analysis **2**(1), 115–118 (1965)
15. Ratschan, S.: Efficient solving of quantified inequality constraints over the real numbers. ACM Trans. Computat. Logic **7**(4), 723–748 (2006)
16. Rossi, F., van Beek, P., Walsh, T.: Handbook of Constraint Programming. Elsevier (2006)
17. Saad, A., Gervet, C., Abdennadher, S.: Constraint reasoning with uncertain data using CDF-intervals. In: Lodi, A., Milano, M., Toth, P. (eds.) CPAIOR 2010. LNCS, vol. 6140, pp. 292–306. Springer, Heidelberg (2010)
18. Tarim, S., Kingsman, B.: The stochastic dynamic production/inventory lot-sizing problem with service-level constraints. International Journal of Production Economics **88**, 105–119 (2004)
19. Van Hentenryck, P., Michel, L., Deville, Y.: Numerica: A Modeling Language for Global Optimization. The MIT Press, Cambridge Mass (1997)
20. Yorke-Smith, N., Gervet, C.: Certainty closure: Reliable constraint reasoning with uncertain data. In: ACM Transactions on Computational Logic **10**(1) (2009)

An Efficient Local Search for Partial Latin Square Extension Problem

Kazuya Haraguchi[✉]

Faculty of Commerce, Otaru University of Commerce, Otaru, Japan
haraguchi@res.otaru-uc.ac.jp

Abstract. A *partial Latin square* (*PLS*) is a partial assignment of n symbols to an $n \times n$ grid such that, in each row and in each column, each symbol appears at most once. The *partial Latin square extension* problem is an NP-hard problem that asks for a largest extension of a given PLS. In this paper we propose an efficient *local search* for this problem. We focus on the local search such that the neighborhood is defined by (p,q)-*swap*, i.e., removing exactly p symbols and then assigning symbols to at most q empty cells. For $p \in \{1,2,3\}$, our neighborhood search algorithm finds an improved solution or concludes that no such solution exists in $O(n^{p+1})$ time. We also propose a novel swap operation, Trellis-swap, which is a generalization of $(1,q)$-swap and $(2,q)$-swap. Our Trellis-neighborhood search algorithm takes $O(n^{3.5})$ time to do the same thing. Using these neighborhood search algorithms, we design a prototype iterated local search algorithm and show its effectiveness in comparison with state-of-the-art optimization solvers such as IBM ILOG CPLEX and LocalSolver.

Keywords: Partial latin square extension problem · Maximum independent set problem · Metaheuristics · Local search

1 Introduction

We address the *partial Latin square extension* (*PLSE*) problem. Let $n \geq 2$ be a natural number. Suppose that we are given an $n \times n$ grid of *cells*. A *partial Latin square* (*PLS*) is a partial assignment of n *symbols* to the grid so that the *Latin square condition* is satisfied. The Latin square condition requires that, in each row and in each column, every symbol should appear at most once. Given a PLS, the PLSE problem asks to fill as many empty cells with symbols as possible so that the Latin square condition remains to be satisfied.

In this paper, we propose an efficient *local search* for the PLSE problem. Let us describe our research background and motivation. The PLSE problem is practically important since it has various applications such as combinatorial design, scheduling, optical routers, and combinatorial puzzles [5,8,18]. The problem is

This work is partially supported by JSPS KAKENHI Grant Number 25870661.

L. Michel (Ed.): CPAIOR 2015, LNCS 9075, pp. 182–198, 2015.
DOI: 10.1007/978-3-319-18008-3_13

NP-hard [7], and was first studied by Kumar et al. [28]. The problem has been studied in the context of constant-ratio approximation algorithms [16,21,23,28]. Currently the best approximation factor is achieved by a local search algorithm [11,13,21]. In that local search the neighborhood is defined by (p, q)-*swap*, where p and q are non-negative integers such that $p < q$. It is the operation of removing exactly p symbols from the current solution and then assigning symbols to at most q empty cells. To the best of the author's knowledge, there is no literature that investigates efficient implementation of local search.

Our local search is based on Andrade et al.'s local search [2] for the *maximum independent set* (*MIS*) problem. The MIS problem is a well-known NP-hard problem as well [14]. We utilize Andrade et al.'s methodology since, as we will see later, the PLSE problem is a special case of the MIS problem.

We improve the efficiency of the local search by utilizing the problem structure peculiar to the PLSE problem. Specifically, for $p \in \{1, 2, 3\}$ and $q = n^2$, our neighborhood search algorithm takes only $O(n^{p+1})$ time to find an improved solution or to conclude that no improved solution exists in the neighborhood, whereas the direct usage of Andrade et al.'s algorithm ($p = 1$ and 2) and Itoyanagi et al.'s algorithm ($p = 3$) [27] requires $O(n^{p+3})$ time to do the same things even for $q = p + 1$. Note that $q = n^2$ is the upper limit of the number of nodes that can be inserted in a swap operation. Our swap operations insert as many nodes to the solution as possible.

We then propose a new type of swap operation that we call *Trellis-swap*. It is a generalization of $(1, n^2)$-swap and $(2, n^2)$-swap, and contains certain cases of $(3, n^2)$-swap. Our Trellis-neighborhood search algorithm takes $O(n^{3.5})$ time to find an improved solution, or to conclude that no improved solution exists in the neighborhood.

We regard our local search as efficient since the time complexities above should be the best possible bounds. For example, when $p = 1$, we may not be able to improve the bound $O(n^2)$ further since in fact the bound is linear with respect to the solution size.

We show how our local search is effective through computational studies. The highlight is that our prototype *iterated local search* (*ILS*) algorithm is likely to deliver a better solution than such state-of-the-art optimization softwares as IP and CP solvers from IBM ILOG CPLEX [26] and a general heuristic solver from LOCALSOLVER [32]. Furthermore, among several ILS variants, the best is one based on Trellis-swap.

The decision problem version of the PLSE problem is known as the *quasigroup completion* (*QC*) problem in AI, CP and SAT communities [3,17,18,38]. The QC problem has been one of the most frequently used benchmark problems in these areas and variant problems are studied intensively, e.g., Sudoku [9,10,29, 31,34,36], mutually orthogonal Latin squares [4,33,39], and spatially balanced Latin squares [19,30,35]. Our local search may be helpful for those who develop exact solvers for the QC problem since the local search itself or metaheuristic algorithms employing it would deliver a good initial solution or a tight lower estimate of the optimal solution size quickly.

The paper is organized as follows. In Sect. 2, preparing terminologies and notations, we see that the PLSE problem is a special case of the MIS problem. We explain the algorithms and the data structure of our local search in Sect. 3 and then present experimental results in Sect. 4. Finally we conclude the paper in Sect. 5.

2　Preliminaries

Let us begin with formulating the PLSE problem. Suppose an $n \times n$ grid of cells. We denote $[n] = \{1, 2, \ldots, n\}$. For any $i, j \in [n]$, we denote the cell in the row i and in the column j by (i, j). We consider a partial assignment of n symbols to the grid. The n symbols to be assigned are n integers in $[n]$. We represent a partial assignment by a set of triples, say $T \subseteq [n]^3$, such that the membership $(v_1, v_2, v_3) \in T$ indicates that the symbol v_3 is assigned to (v_1, v_2). To avoid a duplicate assignment, we assume that, for any two triples $v = (v_1, v_2, v_3)$ and $w = (w_1, w_2, w_3)$ in T ($v \neq w$), $(v_1, v_2) \neq (w_1, w_2)$ holds. Thus $|T| \leq n^2$ holds.

For any two triples $v, w \in [n]^3$, we denote the Hamming distance between v and w by $\delta(v, w)$, i.e., $\delta(v, w) = |\{k \in [3] \mid v_k \neq w_k\}|$. We call a partial assignment $T \subseteq [n]^3$ a *PLS set* if, for any two triples $v, w \in T$ ($v \neq w$), $\delta(v, w)$ is at least two. One easily sees that T is a PLS set iff it satisfies the Latin square condition. We say that two disjoint PLS sets S and S' are *compatible* if, for any $v \in S$ and $v' \in S'$, the distance $\delta(v, v')$ is at least two. Obviously the union of such S and S' is a PLS set. The PLSE problem is then formulated as follows; given a PLS set $L \subseteq [n]^3$, we are asked to construct a PLS set S of the maximum cardinality such that S and L are compatible.

Next, we formulate the MIS problem. An *undirected graph* (or simply a *graph*) $G = (V, E)$ consists of a set V of *nodes* and a set E of unordered pairs of nodes, where each element in E is called an *edge*. When two nodes are joined by an edge, we say that they are *adjacent*, or equivalently, that one node is a *neighbor* of the other. An *independent set* is a subset $V' \subseteq V$ of nodes such that no two nodes in V' are adjacent. Given G, the MIS problem asks for a largest independent set. For any node $v \in V$, we denote the set of its neighbors by $N(v)$. The number $|N(v)|$ of v's neighbors is called the *degree of* v.

Now we are ready to transform any PLSE instance into an MIS instance. Suppose that we are given a PLSE instance in terms of a PLS set $L \subseteq [n]^3$. For any triple $v \in L$, we denote by $N^*(v)$ the set of all triples w's in the entire $[n]^3$ such that $\delta(v, w) = 1$, i.e., $N^*(v) = \{w \in [n]^3 \mid \delta(v, w) = 1\}$. Clearly we have $|N^*(v)| = 3(n - 1)$. The union $\bigcup_{v \in L} N^*(v)$ over L is denoted by $N^*(L)$.

Proposition 1. *A set $S \subseteq [n]^3$ of triples is a feasible solution to the PLSE instance L iff S, as a node set, is a feasible solution to the MIS instance $G_L = (V_L, E_L)$ such that $V_L = [n]^3 \setminus (L \cup N^*(L))$ and $E_L = \{(v, w) \in V_L \times V_L \mid \delta(v, w) = 1\}$.*

We omit the proof due to space limitation. By Proposition 1, we hereafter consider solving the PLSE instance by means of solving the transformed MIS instance $G_L = (V_L, E_L)$. Omitting the suffix L, we write $G = (V, E)$ to represent $G_L = (V_L, E_L)$ for simplicity.

Let us observe the structure of G. We regard each node $v = (v_1, v_2, v_3) \in V$ as a grid point in the 3D integral space. Any grid point is an intersection of three grid lines that are orthogonal to each other. In other words, each node is on exactly three grid lines. A grid line is *in the direction* d if it is parallel to the axis d and perpendicular to the 2D plane that is generated by the other two axes. We denote by $\ell_{v,d}$ the grid line in the direction d that passes v. Two nodes are joined by an edge iff there is a grid line that passes both of them. The nodes on the same grid line form a *clique*. This means that any independent set should contain at most one node among those on a grid line. Since $|N(v)| \leq |N^*(v)| = 3(n-1)$ and $|V| = O(n^3)$, we have $|E| = O(n^4)$.

We introduce notations and terminologies on local search for the MIS problem. We call any independent set in G simply a *solution*. Given a solution $S \subseteq V$, we call any node $x \in S$ a *solution node* and any node $v \notin S$ a *non-solution node*. For a non-solution node v, we call any solution node in $N(v)$ a *solution neighbor of* v. We denote the set of solution neighbors by $N_S(v)$, i.e., $N_S(v) = N(v) \cap S$. Since v has at most one solution neighbor on one grid line and three grid lines pass v, we have $|N_S(v)| \leq 3$. We call the number $|N_S(v)|$ the *tightness of* v and denote it by $\tau_S(v)$. When $\tau_S(v) = t$, we call v t-*tight*. In particular, a 0-tight node is called *free*. When x is a solution neighbor of a t-tight node v, we may say that v is a t-*tight neighbor of* x.

For two integers p, q such that $0 \leq p < q$, the (p, q)-*swap* refers to an operation of removing exactly p solution nodes from S and inserting at most q free nodes into S so that S continues to be a solution. The (p, q)-*neighborhood of* S is a set of all solutions that are obtained by performing a (p, q)-swap on S. We assume $q \leq n^2$ since, for any $q > n^2$, the (p, q)-neighborhood and the (p, n^2)-neighborhood are equivalent. A solution S is called (p, q)-*maximal* if the (p, q)-neighborhood contains no improved solution S' such that $|S'| > |S|$. We call a solution p-*maximal* if it is (p, n^2)-maximal. In particular, we call a 0-maximal solution simply a *maximal* solution. Being p-maximal implies that S is also p'-maximal for any $p' < p$. Equivalently, if S is not p'-maximal, then it is not p-maximal either for any $p > p'$.

3   Local Search

In this section, we present the main component algorithm of the local search. The main component is a *neighborhood search algorithm*. Given a solution S and a neighborhood type being specified, it computes an improved solution in the neighborhood or concludes that no such solution exists. Once a neighborhood search algorithm is established, it is immediate to design a local search algorithm that computes a maximal solution in the sense of the specified neighborhood type; starting with an appropriate initial solution, we repeat moving to an improved solution as long as the neighborhood search algorithm delivers one.

Specifically we present (p, n^2)-neighborhood search algorithms ($p \in \{1, 2, 3\}$) and a Trellis-neighborhood search algorithm. The basic data structure is borrowed from [2], but we improve the efficiency by using the problem structure peculiar to the PLSE problem. The (p, n^2)-neighborhood search algorithms run in $O(n^{p+1})$ time, whereas Trellis-neighborhood search algorithm runs in $O(n^{3.5})$ time. We describe the data structure that is commonly used in all these algorithms in Sect. 3.1 and present the neighborhood search algorithms in Sect. 3.2. Finally in Sect. 3.3, from the viewpoint of approximation algorithms, we mention approximation factors of p-maximal and Trellis-maximal solutions and analyze the time complexities that are needed to compute them.

We claim that our work should be far from trivial. Without using the MIS formulation, one may conceive a $(1, 2)$-neighborhood search algorithm that runs in $O(n^3)$ time, but its improvement is not easy. Concerning previous local search algorithms for the MIS problem, their direct usage would require more computation time. Andrade et al.'s [2] $(1, 2)$-neighborhood search algorithm (resp., $(2, 3)$-neighborhood search algorithm) requires $O(|E|) = O(n^4)$ time (resp., $O(\Delta|E|) = O(n^5)$ time, where Δ denotes the maximum degree in the graph). Itoyanagi et al. [27] extended Andrade et al.'s work to the *maximum weighted independent set problem*. Their $(3, 4)$-neighborhood search algorithm runs in $O(\Delta^2|E|) = O(n^6)$ time.

3.1 Data Structure

We mostly utilize the data structure of Andrade et al.'s [2]. We represent a solution by a permutation of nodes. In the permutation, every solution node is ordered ahead of all the non-solution nodes, and among the non-solution nodes, every free node is ordered ahead of all the non-free nodes. In each of the three sections (i.e., solution nodes, free nodes and non-free nodes), the nodes can be ordered arbitrarily. We also maintain the solution size and the number of free nodes. For each non-solution node $v \notin S$, we store its tightness $\tau_S(v)$ and pointers to its solution neighbors. Since $\tau_S(v) \le 3$, we store at most three pointers for v.

Let us mention the novel settings that we introduce additionally to enhance the efficiency. Regarding each node as a grid point in the 3D integral space, we store the node set V by means of a 3D $n \times n \times n$ array, denoted by C. For each triple $(v_1, v_2, v_3) \in [n]^3$, if $(v_1, v_2, v_3) \in V$, then we let $C[v_1][v_2][v_3]$ have the pointer to the node (v_1, v_2, v_3), and otherwise, we let it have a null pointer. For each solution node $x \in S$, we store the number of its 1-tight neighbors along each of the three grid lines passing x. We denote this number by $\mu_d(x)$ ($d = 1, 2, 3$), where d denotes the direction of the grid line. We emphasize that maintaining $\mu_d(x)$ should play a key role in improving the efficiency of the local search. Clearly the size of the data structure is $O(n^3)$. We can construct it in $O(n^3)$ time, as preprocessing of local search.

Using the data structure, we can execute some significant operations efficiently. For example, we can identify whether S is maximal or not in $O(1)$ time; it suffices to see whether the number of free nodes is zero or not. The neighbor set $N(v)$ of a node $v = (v_1, v_2, v_3)$ can be listed in $O(n)$ time by searching

$C[v'_1][v_2][v_3]$'-s, $C[v_1][v'_2][v_3]$'-s and $C[v_1][v_2][v'_3]$'-s for every $v'_1, v'_2, v'_3 \in [n]$ such that $v'_1 \neq v_1$, $v'_2 \neq v_2$ and $v'_3 \neq v_3$. Furthermore, we can remove a solution node from S or insert a free node into S in $O(n)$ time. We describe how to implement the removal operation. Let us consider removing $x \in S$ from S. For the permutation representation, we exchange the orders between x and the last node in the solution node section. We decrease the number of solution nodes by one and increase the number of free nodes by one; as a result, x falls into the free node section. Its tightness is set to zero since it has no solution neighbor. For each neighbor $v \in N(x)$, we release its pointer to x since x is no longer a solution node, and decrease the tightness $\tau_S(v)$ by one.

- If $\tau_S(v)$ is decreased to zero, then v is now free. To put v in the free node section, we exchange the permutation orders between v and the head node in the non-free node section, and increase the number of free nodes by one.
- If $\tau_S(v)$ is decreased to one, then v is now 1-tight and has a unique solution neighbor, say y. We increase the number $\mu_d(y)$ by one, where d is the direction of the grid line that passes both v and y.

The total time complexity is $O(n)$. The insertion operation can be implemented in an analogous way.

3.2 Neighborhood Search Algorithms

Let us describe the key idea on how to realize efficient neighborhood search algorithms. Suppose that a solution S is given. Removing a subset $R \subseteq S$ from S makes R and certain neighbors free, that is, non-solution nodes such that all solution neighbors are contained in R. If there is an independent set among these free nodes whose size is larger than $|R|$, then there is an improved solution. Of course a larger independent set is preferred. A largest one can be computed efficiently in some cases although the MIS problem is computationally hard in general.

The first case is when $p = |R|$ is a small constant. Below we explain (p, n^2)-neighborhood search algorithms for $p = 1, 2$ and 3. Running in $O(n^{p+1})$ time, the algorithms have the similar structures. Each algorithm searches all R'-s that can lead to an improved solution by means of searching "trigger" nodes, at it were, sweeping the permutation representation. The time complexity is linearly bounded by the number of trigger nodes; For each trigger node, the algorithm collects certain solution nodes around it, which are used as R. This takes $O(1)$ time. On the other hand, the algorithm does not search for the non-solution nodes to be inserted unless it finds the MIS size among the free nodes from $S \setminus R$ larger than $|R|$. Surprisingly we can decide the MIS size in $O(1)$ time. Only when the size is larger than $|R|$, the algorithm searches for the MIS to be inserted, and thereby it obtains an improved solution and terminates. The search for the MIS requires $O(n)$ time, but it does not affect the total time complexity.

Another case such that the MIS problem is solved efficiently is when all solution nodes in R are contained in such a 2D facet that is induced by fixing the value of one dimension in the 3D space. In this case, the MIS problem is solved by means of bipartite maximum matching. This motivates us to invent a novel swap operation, Trellis-swap. In this swap, the size $|R|$ is not a constant but can change from 1 to n. Trellis-neighborhood search algorithm runs in $O(n^{3.5})$ time.

$(1, n^2)$-Swap. Let S be a maximal solution. Its $(1, n^2)$-neighborhood contains an improved solution iff there are a solution node $x \in S$ and non-solution nodes $u, v \notin S$ such that $(S \setminus \{x\}) \cup \{u, v\}$ is a solution. It is clear that u and v should be neighbors of x. They are 1-tight, and their unique solution neighbor is x. The u and v should not be adjacent, which implies that u and v are not on the same grid line. Then, the number of nodes that can be inserted into $S \setminus \{x\}$ is given by $\nu(x) = |\{d \in \{1, 2, 3\} \mid \mu_d(x) > 0\}|$, i.e., the number of grid lines passing x on which a 1-tight node exists.

Theorem 1. *Given a solution S, we can find an improved solution in its $(1, n^2)$-neighborhood or conclude that it is 1-maximal in $O(n^2)$ time.*

Proof. We assume that S is maximal; we can check in $O(1)$ time whether S is maximal or not. If it is not maximal, we have an improved solution by inserting any free node into S. Finding a free node and inserting it into S take $O(n)$ time, and then we have done.

An improved solution exists iff there is $x \in S$ such that $\nu(x) \geq 2$. All solution nodes can be searched by sweeping the first section of the permutation representation. There are at most n^2 solution nodes. For each solution node x, the number $\nu(x)$ can be computed in $O(1)$ time. If x with $\nu(x) \geq 2$ is found, we can determine the 1-tight nodes to be inserted in $O(n)$ time by searching each grid line with $\mu_d(x) > 0$. Removing x from S and inserting the nodes into $S \setminus \{x\}$ take $O(n)$ time. $\square$

$(2, n^2)$-Swap. Let S be a 1-maximal solution. Its $(2, n^2)$-neighborhood contains an improved solution iff there exist $x, y \in S$ and $u, v, w \notin S$ such that $(S \setminus \{x, y\}) \cup \{u, v, w\}$ is a solution. These nodes should satisfy Lemmas 1 to 4 in [2], which are conditions established for the general MIS problem. According to the conditions, at least one node in $\{u, v, w\}$ is 2-tight, and x and y are the solution neighbors of the 2-tight node. Let u be this 2-tight node without loss of generality.

See Fig. 1. Suppose inserting u into $S \setminus \{x, y\}$. The nodes that can be inserted additionally should be on the four solid grid lines, say $\ell_{x,2}$, $\ell_{x,3}$, $\ell_{y,1}$ and $\ell_{y,3}$. Let F be the set of nodes on these grid lines that are free from $(S \cup \{u\}) \setminus \{x, y\}$. Let v be any node in F. For S, v should not have any solution neighbors other than x or y since otherwise it would not be free from $(S \cup \{u\}) \setminus \{x, y\}$. Then we have $N_S(v) \subseteq \{x, y\}$ and thus v is either 1-tight or 2-tight. As can be seen, u_{xy}, the intersection point of $\ell_{x,2}$ and $\ell_{y,1}$, is the only possible 2-tight node. All the other nodes in F are 1-tight. Note that, even though the node u_{xy} exists, it does not necessarily belong to F; it can be 3-tight.

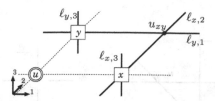

Fig. 1. An illustration of the case such that $(2, n^2)$-swap can occur

Suppose that the node u_{xy} exists and it is 2-tight. In the subgraph induced by F, one can readily see that *every* MIS contains u_{xy} only when no 1-tight node exists on $\ell_{x,2}$ or $\ell_{y,1}$ (i.e., $\mu_2(x) = \mu_1(y) = 0$). Thus, when we count the MIS size in the subgraph, we need to take u_{xy} into account only in this case.

Theorem 2. *Given a solution S, we can find an improved solution in its $(2, n^2)$-neighborhood or conclude that it is 2-maximal in $O(n^3)$ time.*

Proof. Similarly to the proof of Theorem 1, we assume that S is 1-maximal.

Let u be any 2-tight node and x and y be its solution neighbors. The number of 2-tight nodes is at most $|V \setminus S| \leq |V| \leq n^3$. We can recognize its solution neighbors x and y in $O(1)$ time by tracing the pointers from u. The size of an MIS among F can be computed in $O(1)$ time as follows; First we count the number of the four grid lines $\ell_{x,2}$, $\ell_{x,3}$, $\ell_{y,1}$ and $\ell_{y,3}$ such that a 1-tight node exists, where we follow the dimension indices in Fig. 1 without loss of generality. This can be done by checking whether $\mu_d(x) > 0$ (or $\mu_d(y) > 0$) or not. Furthermore, if $\mu_2(x) = \mu_1(y) = 0$, we check whether a 2-tight node u_{xy} exists or not. The check can be done in $O(1)$ time by referring to the 3D array C. If it exists, we increase the MIS size by one. Finally, if the MIS size is no less than two, there exists an improved solution. The non-solution nodes to be inserted other than u are found in $O(n)$ time by searching the four grid lines. $\qquad\square$

$(3, n^2)$-Swap. Given a solution S, consider searching for an improved solution in its $(3, n^2)$-neighborhood. To reduce the search space the following result on the general MIS problem is useful.

Theorem 3. (Itoyanagi et al. [27]) *Suppose that S is a 2-maximal solution and that there are a subset $R \subseteq S$ of solution nodes and a subset $F \subseteq V \setminus S$ of non-solution nodes such that $|R| = 3$, $|F| = 4$ and $(S \setminus R) \cup F$ is a solution. We denote $F = \{u, v, w, t\}$, and without loss of generality, we assume $\tau_S(u) \geq \tau_S(v) \geq \tau_S(w) \geq \tau_S(t)$. Then we are in either of the following two cases:*

(I) *u is 3-tight and $R = N_S(u)$.*

(II) *u and v are 2-tight such that they have exactly one solution node as a common solution neighbor, i.e., $|N_S(u) \cap N_S(v)| = 1$, and $R = N_S(u) \cup N_S(v)$.*

Theorem 4. *Given a solution S, we can find an improved solution in its $(3, n^2)$-neighborhood or conclude that it is 3-maximal in $O(n^4)$ time.*

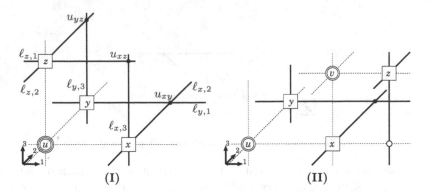

Fig. 2. An illustration of the cases (I) and (II) such that $(3, n^2)$-swap can occur

Proof. We assume that S is 2-maximal, similarly to previous theorems. Below we prove that the search for (I) in Theorem 3 takes $O(n^3)$ time, and that for (II) takes $O(n^4)$ time.

For (I), we search all 3-tight nodes. There are $O(n^3)$ 3-tight nodes. For each 3-tight node u, the solution nodes to be removed are the three solution neighbors of u, which can be decided in $O(1)$ time. Let x, y and z be the three solution neighbors, and F be the set of free nodes from $(S \setminus \{x, y, z\}) \cup \{u\}$. The nodes in F should be on solid grid lines in Fig. 2 (I). There are at most three 2-tight nodes among F, that is, u_{xy}, u_{xz} and u_{yz}. Similarly to the proof of Theorem 2, every MIS among the free nodes contains u_{xy} only when no 1-tight node exists on $\ell_{x,2}$ or $\ell_{y,1}$, i.e., $\mu_2(x) = \mu_1(y) = 0$. The condition on which u_{xz} (or u_{yz}) belongs to every MIS is analogous. Taking these into account, we can decide the MIS size among the free nodes in $O(1)$ time. If it is no less than three, then an improved solution exists; an MIS to be inserted can be decided in $O(n)$ time.

Concerning (II), we search every solution node $x \in S$ and every pair of its 2-tight neighbors that are not on the same grid line. Thus there are $O(n^4)$ pairs in all since there are at most n^2 solution nodes, and for each $x \in S$, there are $O(n^2)$ pairs of 2-tight neighbors. Given x and a pair $\{u, v\}$ of its 2-tight neighbors, the removed solution nodes are three solution nodes in $N_S(u) \cup N_S(v)$, which includes x. They can be found in $O(1)$ time. Let $N_S(u) \cup N_S(v) = \{x, y, z\}$. We illustrate an example of this case in Fig. 2 (ii). The free nodes from $(S \setminus \{x, y, z\}) \cup \{u, v\}$ should be on the solid grid lines in the figure. By means of enumerative argument, we can show that the MIS size among the free nodes can be decided in $O(1)$ time. The argument is rather complicated, and we omit details here. If the MIS size is no less than two, then an improved solution exists; an MIS to be inserted can be decided in $O(n)$ time. $\square$

Trellis-Swap. Now let us introduce Trellis-swap. Note that, in $(2, n^2)$-swap, all removed solution nodes belong to such a 2D facet that is induced by fixing the value of one dimension in the 3D space. We can regard that $(1, n^2)$-swap is also in the case. We invent a more general swap operation for such a case.

Definition 1. *Suppose that a solution S is given. Let $R \subseteq S$ be a maximal subset such that all nodes in R are contained in a 2D facet that is generated by setting $v_d = k$ for some $d \in [3]$ and $k \in [n]$. We denote by F_1 (resp., F_2) the set of all 1-tight (resp., 2-tight) nodes such that the unique solution neighbor is contained in R (resp., both of the solution neighbors are contained in R). We call the subgraph induced by $R \cup F_1 \cup F_2$ a* trellis *(with respect to R).*

One sees that all nodes in $R \cup F_2$ are on the same 2D facet, whereas some nodes in F_1 are also on the facet, but other nodes in F_1 may be out of the facet like a hanging vine. Note that removing R from S makes all nodes in the trellis free. We define the *Trellis-neighborhood of a solution S* as the set of all solutions that can be obtained by removing any R from S and then inserting an independent set among the trellis into $S \setminus R$.

Theorem 5. *Given a solution S, we can find an improved solution in the Trellis-neighborhood or conclude that no such solution exists in $O(n^{3.5})$ time.*

Proof. We explain how we compute an MIS in the trellis for a given R. Let us partition F_1 into $F_1 = F_1' \cup F_1''$ so that F_1' (resp., F_1'') is the subset of nodes on (resp., out of) the considered 2D facet. The node set F_1'' induces a subgraph that consists of cliques, each of which is formed by 1-tight nodes on a grid line perpendicular to the 2D facet. One can easily show that, among MISs in the trellis, there is one such that a solution node is chosen from every clique. Intending such an MIS, we ignore the nodes in F_1'' and their solution neighbors; let $R'' \subseteq R$ be a subset such that $R'' = \bigcup_{u \in F_1''} N_S(u)$. Now that we have chosen nodes from F_1'', we can no longer choose the nodes in R''. Let $R' = R \setminus R''$. The remaining task is to compute an MIS from $R' \cup F_1' \cup F_2$. All the nodes in $R' \cup F_1' \cup F_2$ are on a 2D facet, and we are asked to choose as many nodes as possible so that, from each of $2n$ grid lines on the facet, a node is chosen at most once. We see that the problem is reduced to computing a maximum matching in a certain bipartite graph; a grid line on the 2D facet corresponds to a node in the bipartite graph, and the bipartition of nodes is determined by the directions of grid lines. A node $v \in R' \cup F_1' \cup F_2$ on the 2D facet corresponds to an edge in the bipartite graph such that two nodes are joined if v is on the intersecting point of the corresponding two grid lines.

We have $3n$ 2D facets. For each 2D facet, it takes $O(n^2)$ time to recognize the node sets R, F_1 and F_2 and partitions $R = R' \cup R''$ and $F_1 = F_1' \cup F_1''$, and then to construct the bipartite graph since the bipartite graph has $2n$ nodes and $O(n^2)$ edges. We need $O(n^{2.5})$ time to compute a maximum matching [24].   □

3.3   Approximation Factors and Computation Time

In a maximization problem instance, a *ρ-approximate solution ($\rho \in [0, 1]$)* is a solution whose size is at least the factor ρ of the optimal size. Hajirasouliha et al. [21] analyzed approximation factors of $(p, p+1)$-maximal solutions, using Hurkens and Schrijver's classical result [25] on the general set packing problem. From this and Theorems 1, 2, 4 and 5, and since the solution size is at most n^2, we have the following theorem.

Theorem 6. *Any 1-maximal, 2-maximal, 3-maximal and Trellis-maximal solutions are 1/2-, 5/9-, 3/5- and 5/9-approximate solutions respectively. Furthermore, these can be obtained in $O(n^4)$, $O(n^5)$, $O(n^6)$ and $O(n^{5.5})$ time by extending an arbitrary solution by means of the local search.*

4   Computational Study

In this section we illustrate how our local search computes high-quality solutions efficiently through a computational study. We design a prototype iterated local search (ILS) algorithm and show that it is more likely to deliver better solutions than modern optimization solvers. Furthermore, the best ILS variant is one based on Trellis-swap.

ILS Algorithm. We describe our ILS algorithm in Algorithm 1, which is no better than a conventional one. Let us describe how to construct the initial solution of the next local search by "kicking" S^* (line 7). For this, the algorithm copies S^* to S_0 and "forces" to insert k non-solution nodes into S_0. Specifically, it repeats the followings k times; it chooses a non-solution node $u \notin S_0$, removes the set $N_{S_0}(u)$ of its solution neighbors from the solution (i.e., $S_0 \leftarrow S_0 \backslash N_{S_0}(u)$), and inserts u into the solution (i.e., $S_0 \leftarrow S_0 \cup \{u\}$). If there appear free nodes, one is chosen at random and inserted into S_0 repeatedly until S_0 becomes maximal. The number k is set to $k = \kappa$ with probability $1/2^\kappa$. We choose the k nodes randomly from all the non-solution nodes, except the first one. We choose the first one with great care so that **(i)** trivial cycling is avoided and **(ii)** the diversity of search is attained. For (i), we restrict the candidates to nodes in $N(S_0')$, where $S_0' \subseteq S_0$ is a subset of solution nodes such that $S_0' = \{x \in S_0 \mid \exists d \in [3], \ \mu_d(x) > 0\}$. In other words, S_0' is a subset of solution nodes that have a 1-tight neighbor. Then for (ii), we employ the *soft-tabu* approach [2]; we choose the non-solution node that has been outside the solution for the longest time among $N(S_0')$.[1] We omit the details, but the mechanism for (i) dramatically reduces chances that the next local search returns us to the incumbent solution S^*. In fact, it is substantially effective in enhancing the performance of the algorithm.

We consider four variant ILS algorithms that employ different neighborhoods from each other: 1-ILS, 2-ILS, Tr-ILS and 3-ILS. 1-ILS is the ILS algorithm that iterates 1-LS (i.e., the local search with $(1, n^2)$-neighborhood search algorithm) in the manner of Algorithm 1. The terms 2-LS, 2-ILS, Tr-LS, Tr-ILS, 3-LS and 3-ILS are analogous.

---

[1] When there is no 1-tight node, S_0' becomes an empty set. In this case, we use $N(S_0)$ instead of $N(S_0')$.

Algorithm 1 An ILS algorithm for the PLSE problem

1: $S_0 \leftarrow$ an arbitrary solution and $S^* \leftarrow S_0$ ▷ S^* is the incumbent solution.
2: **while** the computation time does not exceed the given time limit **do**
3:     compute a locally optimal solution S by local search that starts with an initial
       solution S_0
4:     **if** $|S| \geq |S^*|$ **then**
5:         $S^* \leftarrow S$
6:     **end if**
7:     compute the initial solution S_0 of the next local search by "kicking" S^*
8: **end while**
9: output S^*

Experimental Set-up. All the experiments are conducted on a workstation that carries Intel® Core^TM i7-4770 Processor (up to 3.90GHz by means of Turbo Boost Technology) and 8GB main memory. The installed OS is Ubuntu 14.04.1.

Benchmark instances are random PLSs. We generate the instances by utilizing each of the two schemes that are well-known in the literature [6,18]: quasigroup completion (QC) and quasigroup with holes (QWH). Note that a PLS is parametrized by the grid length n and the ratio $r \in [0,1]$ of pre-assigned symbols over the $n \times n$ grid. Starting with an empty assignment, QC repeats assigning a symbol to an empty cell randomly so that the resulting assignment is a PLS, until $\lfloor n^2 r \rfloor$ cells are assigned symbols. On the other hand, QWH generates a PLS by removing $n^2 - \lfloor n^2 r \rfloor$ symbols from an arbitrary Latin square so that $\lfloor n^2 r \rfloor$ symbols remain. Note that a QC instance does not necessarily admit a complete Latin square as an optimal solution, whereas a QWH instance always does. Here we show experimental results only on QC instances due to space limitation. We note that, however, most of the observed tendencies are quite similar between QC and QWH. One can download all the instances and the solution sizes achieved by the ILS algorithms and by the competitors from the author's website (http://puzzle.haraguchi-s.otaru-uc.ac.jp/PLSE/).

Let us mention what kind of instance is "hard" in general. Of course an instance becomes harder when n is larger. Then we set the grid length n to 40, 50 and 60, which are relatively large compared with previous studies (e.g., [18]). For a fixed n, the problem has easy-hard-easy phase transition. Then we regard instances with an intermediate r "hard".

For competitors, we employ two exact solvers and one heuristic solver. For the former, we employ the optimization solver for integer programming model (CPX-IP) and the one for constraint optimization model (CPX-CP) from IBM ILOG CPLEX (ver. 12.6) [26]. It is easy to formulate the PLSE problem by these models (e.g., see [18]). For the latter, we employ LOCALSOLVER (ver. 4.5) [32] (LSSOL), which is a general heuristic solver based on local search. Hopefully our ILS algorithm will outperform LSSOL since ours is specialized to the PLSE problem, whereas LSSOL is developed for general discrete optimization problems. All the parameters are set to default values except that, in CPX-CP,

Table 1. Improved sizes brought by the ILS algorithms and the competitors

n	r	$\lvert L\rvert+\lvert S_0\rvert$	1-ILS	2-ILS	Tr-ILS	3-ILS	CPX-IP	CPX-CP	LSSOL
40	0.3	1597.68	**2.32**	**2.32**	**2.32**	**2.32**	0.00	**2.32**	*0.99
	0.4	1595.28	**4.72**	**4.72**	**4.72**	**4.72**	0.00	**4.72**	*2.72
	0.5	1591.18	**8.82**	*8.80	**8.82**	*8.80	0.00	5.35	5.51
	0.6	1578.89	16.64	*17.01	**17.92**	16.69	5.52	5.85	11.80
	0.7	1550.10	18.35	*18.76	**18.78**	18.33	17.37	8.68	15.01
	0.8	1508.98	*5.05	**5.06**	**5.06**	5.04	**5.06**	3.68	5.00
50	0.3	2496.03	**3.97**	**3.97**	**3.97**	**3.97**	0.00	*3.84	0.32
	0.4	2493.78	**6.22**	**6.22**	**6.22**	*6.20	0.00	4.24	0.87
	0.5	2488.52	11.38	*11.46	**11.48**	11.34	0.00	1.40	4.44
	0.6	2476.21	20.22	*21.38	**21.97**	20.06	0.00	2.66	13.00
	0.7	2442.21	28.03	*28.30	**28.58**	27.68	4.19	8.83	21.24
	0.8	2382.07	*12.18	12.14	12.16	12.08	**12.51**	6.03	11.60
60	0.3	3593.07	**6.93**	**6.93**	**6.93**	*6.85	0.00	5.22	0.13
	0.4	3590.68	**9.32**	**9.32**	**9.32**	*9.20	0.00	1.87	0.49
	0.5	3585.29	*14.65	*14.65	**14.67**	14.06	0.00	0.54	2.21
	0.6	3572.61	24.16	**25.07**	*25.02	22.62	0.00	1.09	12.91
	0.7	3534.62	*37.70	37.54	**39.05**	35.73	0.09	5.83	26.43
	0.8	3456.59	*22.15	**22.16**	*22.15	21.99	21.99	7.55	19.85

`DefaultInferenceLevel` and `AllDiffInferenceLevel` are set to **extended**. We set the time limit of all the solvers (including the ILS algorithms) to 30 seconds.

Results. We show how the ILS algorithms and the competitors improve the initial solution S_0 in Table 1. The S_0 is generated by a constructive algorithm named G5 in [1], which is a "look-ahead" minimum-degree greedy algorithm. We confirmed in our preliminary experiments that G5 is the best among several simple constructive algorithms. For each pair (n, r), a number in the 3rd column is the average of $\lvert L\rvert+\lvert S_0\rvert$ (i.e., the given PLS size $\lvert L\rvert=\lfloor n^2 r\rfloor$ plus the initial solution size) and a number in the 4th to 10th columns is the average of the improved size over 100 instances. A bold number (resp., a number with $*$) indicates the 1st largest (resp., the 2nd largest) improvement among all. An underlined number indicates that an optimal solution is found in all the 100 instances. We can decide a solution S to be an optimal solution if $L \cup S$ is a complete Latin square or an exact solver (i.e., CPX-IP or CPX-CP) reports so.

Obviously the ILS algorithms outperform the competitors in many (n, r)'-s. We claim that Tr-ILS should be the best among the four ILS algorithms. Clearly 3-ILS is inferior to others. The remaining three algorithms seem to be competitive, but Tr-ILS ranks first or second most frequently.

Concerning the competitors, CPX-CP performs well for under-constrained "easy" instances (i.e., $r \leq 0.4$), whereas CPX-IP does well for over-constrained "easy" instances (i.e., $r \geq 0.7$). LSSOL is relatively good for all r'-s and outstanding especially for "hard" instances with $0.5 \leq r \leq 0.7$.

Table 2. Improved sizes in the first LS, 5s, 10s and 30s, and averaged computation times of a single run of LS ($n = 60$); when r is smaller (resp., larger), the number $|V|$ of nodes in the graph becomes larger (resp., smaller), and then LS takes more (resp., less) computation time

r	Algorithm	Improved size				Computation
		1st LS	5s	10s	30s	time of LS (ms)
0.4	1-ILS	0.29	9.24	9.32	9.32	1.44
	2-ILS	1.03	9.28	9.30	9.32	1.84
	Tr-ILS	1.13	9.28	9.32	9.32	2.29
	3-ILS	2.10	6.73	7.91	9.20	42.80
0.5	1-ILS	0.60	14.42	14.58	14.65	0.83
	2-ILS	1.63	14.27	14.47	14.65	1.00
	Tr-ILS	1.65	14.52	14.65	14.67	1.04
	3-ILS	3.08	10.79	12.28	14.06	16.82
0.6	1-ILS	0.75	22.58	23.23	24.16	0.45
	2-ILS	2.69	22.78	23.80	25.07	0.55
	Tr-ILS	2.96	23.38	24.06	25.02	0.45
	3-ILS	4.95	17.86	20.18	22.62	5.91
0.7	1-ILS	1.57	35.33	36.66	37.70	0.24
	2-ILS	4.33	35.16	36.54	37.54	0.28
	Tr-ILS	5.41	35.90	37.52	39.05	0.27
	3-ILS	7.95	29.08	31.91	35.73	2.14

To observe the behavior of the ILS algorithms in detail, we investigate how they improve the solution in the first LS and in the first 5, 10 and 30 seconds in Table 2 ($n = 60$). We also show the averaged computation time of a single run of LS in the rightmost column. Most of the improvements over the 30 seconds are made in earlier periods. It is remarkable that the improvements made by the ILS algorithms in the first 5 seconds are larger than those made by the competitors in 30 seconds in all the shown cases (see Table 1).

The reason why Tr-ILS is the best among the ILS variants must be its efficiency; as can be seen in the rightmost column, Tr-LS is so fast as 2-LS, or even faster in some r'-s, although Trellis-swap is a generalization of $(2, n^2)$-swap. We implemented the Trellis-neighborhood search algorithm so that it runs in $O(\alpha n^{2.5})$ time rather than in $O(n^{3.5})$ time, where α denotes the number of 1-tight nodes. The implementation is expected to be faster since, to the extent of our experiment, the number α is much smaller than n. We will address the detail of this issue in our future papers.

On the other hand, 3-LS is much more time-consuming than the others; the computation time of 3-LS is about 10 to 40 times those of the other three LSs. 3-ILS may not iterate so sufficient a number of 3-LSs that the diversity of search is not attained to a sufficient level. It is true that, in the 1st LS, for all r'-s, 3-LS finds the best solution, followed by Tr-LS, 2-LS and 1-LS; a single LS with a larger neighborhood will find a better solution than one with a smaller neighborhood. By the first 5 seconds pass, however, 3-ILS becomes inferior to the other three ILS algorithms. The solutions that the three ILS algorithms find in 10 seconds are better than those that 3-ILS outputs after 30 seconds.

Then, to enhance the performance of the ILS algorithm, the neighborhood size should not be too large. It is important to run a fast local search with a moderately small neighborhood many times from various initial solutions. We should develop a better mechanism to generate a good initial solution of the local search rather than to investigate a larger neighborhood. This is left for future work.

5   Concluding Remarks

We have designed efficient local search algorithms for the PLSE problem such that the neighborhood is defined by swap operation. The proposed (p, n^2)-neighborhood search algorithm ($p \in \{1, 2, 3\}$) finds an improved solution in the neighborhood or concludes that no such solution exists in $O(n^{p+1})$ time. We also proposed a novel swap operation, Trellis-swap, a generalization of $(1, n^2)$-swap and $(2, n^2)$-swap, whose neighborhood search algorithm takes $O(n^{3.5})$ time.

Our achievement is attributed to observation on the graph structure such that each node is regarded as a 3D integral point, its neighbors are partitioned into $O(1)$ cliques and no two neighbors in different cliques are adjacent. Our idea is never limited to the PLSE problem but can be extended to MIS problems on graphs having the same structure, including some instances of the *maximum strong independent set* problem on hypergraphs [22].

Our ILS algorithm is no better than a prototype and has much room for improvement. Nevertheless it outperforms IBM ILOG CPLEX and LOCAL-SOLVER in most of the tested instances. Of course there are various solvers available, and comparison with them is left for future work. Among these, we consider that SAT based solvers may not be so effective due to our preliminary experiments as follows; we tried to solve the satisfiability problem on QWH instances by SUGAR (ver. 2.2.1) [37], where we used MINISAT (ver. 2.2.1) [12] as the core SAT solver. Note that any QWH instance is satisfiable. SUGAR decides the satisfiability (and thus finds an optimal solution) for about 50% of the instances ($n = 40$ and $r \in \{0.3, 0.4, ..., 0.8\}$) within 30 seconds, while Tr-ILS finds an optimal solution for 78% of the instances within the same time limit.

Alternatively one can conceive another metaheuristic algorithm, utilizing methodologies developed so far [15, 20]. Our local search can be a useful tool for this. For example, in GA, one can enhance the quality of a population by performing our efficient local search on each solution.

Although the local search achieves the best approximation factor for the PLSE problem currently, no one has explored its efficient implementation in the literature. This work resolves this issue to some degree.

References

1. Alidaee, B., Kochenberger, G., Wang, H.: Simple and fast surrogate constraint heuristics for the maximum independent set problem. J. Heuristics **14**, 571–585 (2008)
2. Andrade, D., Resende, M., Werneck, R.: Fast local search for the maximum independent set problem. J. Heuristics **18**, 525–547 (2012). the preliminary version appeared in Proc. 7th WEA (LNCS vol. 5038), pp. 220–234 (2008)
3. Ansótegui, C., Val, A., Dotú, I., Fernández, C., Manyá, F.: Modeling choices in quasigroup completion: SAT vs. CSP. In: Proc. National Conference on Artificial Intelligence, pp. 137–142 (2004)
4. Appa, G., Magos, D., Mourtos, I.: Searching for mutually orthogonal latin squares via integer and constraint programming. European J. Operational Research **173**(2), 519–530 (2006)
5. Barry, R.A., Humblet, P.A.: Latin routers, design and implementation. IEEE/OSA J. Lightwave Technology **11**(5), 891–899 (1993)
6. Barták, R.: On generators of random quasigroup problems. In: Hnich, B., Carlsson, M., Fages, F., Rossi, F. (eds.) CSCLP 2005. LNCS (LNAI), vol. 3978, pp. 164–178. Springer, Heidelberg (2006)
7. Colbourn, C.J.: The complexity of completing partial latin squares. Discrete Applied Mathematics **8**, 25–30 (1984)
8. Colbourn, C.J., Dinitz, J.H.: Handbook of Combinatorial Designs. Chapman & Hall/CRC, 2nd edn. (2006)
9. Crawford, B., Aranda, M., Castro, C., Monfroy, E.: Using constraint programming to solve sudoku puzzles. In: Proc. ICCIT 2008, vol. 2, pp. 926–931 (2008)
10. Crawford, B., Castro, C., Monfroy, E.: Solving sudoku with constraint programming. In: Shi, Y., Wang, S., Peng, Y., Li, J., Zeng, Y. (eds.) MCDM 2009. CCIS, vol. 35, pp. 345–348. Springer, Heidelberg (2009)
11. Cygan, M.: Improved approximation for 3-dimensional matching via bounded pathwidth local search. In: Proc. FOCS 2013, pp. 509–518 (2013)
12. Eén, N., Sörensson, N.: The MiniSat Page, January 20, 2015. http://minisat.se/Main.html
13. Fürer, M., Yu, H.: Approximating the k-set packing problem by local improvements. In: Fouilhoux, P., Gouveia, L.E.N., Mahjoub, A.R., Paschos, V.T. (eds.) ISCO 2014. LNCS, vol. 8596, pp. 408–420. Springer, Heidelberg (2014)
14. Garey, M.R., Johnson, D.S.: Computers and Intractability: A Guide to the Theory of NP-Completeness. W. H. Freeman & Company (1979)
15. Glover, F., Kochenberger, G. (eds.): Handbook of Metaheuristics. Kluwer Academic Publishers (2003)
16. Gomes, C.P., Regis, R.G., Shmoys, D.B.: An improved approximation algorithm for the partial latin square extension problem. Operations Research Letters **32**(5), 479–484 (2004)
17. Gomes, C.P., Selman, B.: Problem structure in the presence of perturbations. In: Proc. AAAI-97, pp. 221–227 (1997)
18. Gomes, C.P., Shmoys, D.B.: Completing quasigroups or latin squares: a structured graph coloring problem. In: Proc. Computational Symposium on Graph Coloring and Generalizations (2002)
19. Gomes, C., Sellmann, M., van Es, C., van Es, H.: The challenge of generating spatially balanced scientific experiment designs. In: Régin, J.-C., Rueher, M. (eds.) CPAIOR 2004. LNCS, vol. 3011, pp. 387–394. Springer, Heidelberg (2004)

20. Gonzalez, T.F. (ed.): Handbook of Approximation Algorithms and Metaheuristics. Chapman & Hall/CRC (2007)
21. Hajirasouliha, I., Jowhari, H., Kumar, R., Sundaram, R.: On completing latin squares. In: Thomas, W., Weil, P. (eds.) STACS 2007. LNCS, vol. 4393, pp. 524–535. Springer, Heidelberg (2007)
22. Halldórsson, M.M., Losievskaja, E.: Independent sets in bounded-degree hypergraphs. Discrete Applied Mathematics 157(8), 1773–1786 (2009)
23. Haraguchi, K., Ono, H.: Approximability of latin square completion-type puzzles. In: Ferro, A., Luccio, F., Widmayer, P. (eds.) FUN 2014. LNCS, vol. 8496, pp. 218–229. Springer, Heidelberg (2014)
24. Hopcroft, J.E., Karp, R.M.: An $n^{5/2}$ algorithm for maximum matchings in bipartite graphs. SIAM J. Computing 2(4), 225–231 (1973)
25. Hurkens, C.A.J., Schrijver, A.: On the size of systems of sets every t of which have an SDR, with an application to the worst-case ratio of heuristics for packing problems. SIAM J. Discrete Mathematics 2(1), 68–72 (1989)
26. IBM ILOG CPLEX, January 20, 2015. http://www-01.ibm.com/software/commerce/optimization/cplex-optimizer/
27. Itoyanagi, J., Hashimoto, H., Yagiura, M.: A local search algorithm with large neighborhoods for the maximum weighted independent set problem. In: Proc. MIC 2011, pp. 191–200 (2011)
28. Kumar, R., Russel, A., Sundaram, R.: Approximating latin square extensions. Algorithmica 24(2), 128–138 (1999)
29. Lambert, T., Monfroy, E., Saubion, F.: A generic framework for local search: application to the sudoku problem. In: Alexandrov, V.N., van Albada, G.D., Sloot, P.M.A., Dongarra, J. (eds.) ICCS 2006. LNCS, vol. 3991, pp. 641–648. Springer, Heidelberg (2006)
30. Le Bras, R., Perrault, A., Gomes, C.P.: Polynomial time construction for spatially balanced latin squares. Tech. rep., Computing and Information Science Technical Reports, Cornell University (2012). http://hdl.handle.net/1813/28697
31. Lewis, R.: Metaheuristics can solve sudoku puzzles. J. Heuristics 13(4), 387–401 (2007)
32. LocalSolver, January 20, 2015. http://www.localsolver.com/
33. Ma, F., Zhang, J.: Finding orthogonal latin squares using finite model searching tools. Science China Information Sciences 56(3), 1–9 (2013)
34. Simonis, H.: Sudoku as a constraint problem, January 20, 2015. http://4c.ucc.ie/hsimonis/sudoku.pdf
35. Smith, C., Gomes, C., Fernandez, C.: Streamlining local search for spatially balanced latin squares. In: Proc. IJCAI 2005, pp. 1539–1541 (2005)
36. Soto, R., Crawford, B., Galleguillos, C., Monfroy, E., Paredes, F.: A hybrid AC3-tabu search algorithm for solving sudoku puzzles. Expert Systems with Applications 40(15), 5817–5821 (2013)
37. Tamura, N.: Sugar: a SAT-based Constraint Solver, January 20, 2015. http://bach.istc.kobe-u.ac.jp/sugar/
38. The International SAT Competitions, January 20, 2015. http://www.satcompetition.org/
39. Vieira Jr., H., Sanchez, S., Kienitz, K.H., Belderrain, M.C.N.: Generating and improving orthogonal designs by using mixed integer programming. European J. Operational Research 215(3), 629–638 (2011)

Enhancing MIP Branching Decisions by Using the Sample Variance of Pseudo Costs

Gregor Hendel[✉]

Konrad Zuse Zentrum für Informationstechnologie, Berlin, Germany
hendel@zib.de

Abstract. The selection of a good branching variable is crucial for small search trees in Mixed Integer Programming. Most modern solvers employ a strategy guided by history information, mainly the variable pseudo-costs, which are used to estimate the objective gain. At the beginning of the search, such information is usually collected via an expensive look-ahead strategy called strong branching until variables are considered reliable.

The reliability notion is thereby mostly based on fixed-number thresholds, which may lead to ineffective branching decisions on problems with highly varying objective gains.

We suggest two new notions of reliability motivated by mathematical statistics that take into account the sample variance of the past observations on each variable individually. The first method prioritizes additional strong branching look-aheads on variables whose pseudo-costs show a large variance by measuring the relative error of a pseudo-cost confidence interval. The second method performs a specialized version of a two-sample Student's t-test for filtering branching candidates with a high probability to be better than the best history candidate.

Both methods were implemented in the MIP-solver SCIP and computational results on standard MIP test sets are presented.

1 Introduction

A *Mixed Integer Program (MIP)* denotes a minimization problem of a linear objective function under linear inequalities and integrality restrictions for a subset of the variables, or to prove that no solution exists. We use the term "mixed" to refer to the occurence of two variable types, continuous and integer variables, in the problem formulation.

Most modern solvers for MIP [1–5] apply a *branch-and-bound* procedure [6,7], which creates a search tree for a MIP P by a successive problem division based on the LP relaxation information at a node. In the most common scheme of variable-based branching, it is crucial to select good branching variables in order to quickly reach terminal nodes and thus keep the required search tree small. A *branching rule* is a scoring mechanism to guide the selection of a branching variable at each inner node of the search tree.

Branching rules [8,9] using variable history information of prior branching decisions have been shown to perform well at later stages of the search, see

© Springer International Publishing Switzerland 2015
L. Michel (Ed.): CPAIOR 2015, LNCS 9075, pp. 199–214, 2015.
DOI: 10.1007/978-3-319-18008-3_14

also [10]. The initial lack of information can be overcome by a computationally expensive *strong branching* initialization [11], which virtually performs a 1-level look-ahead by solving the child node LP relaxations for a subset of the fractional variables and then selects the best candidate.

The current state-of-the-art branching rule for balancing between strong branching and estimation, *reliability branching* [12], uses a fixed number of branching decisions after which the variable information is considered *reliable*. This approach has the disadvantage that it uses the same fixed reliability threshold for all variables. In practice, however, it appears natural that variables that are structurally different inside a MIP model also have different reliability requirements. Another disadvantage of a fixed parameter is that it might not scale well with increasing problem size.

The aim of the present paper is to introduce different notions of reliability by exploiting more statistical information during the branching process. Using the sample variance of past observations, we formulate two criteria for switching between strong branching and estimation that take into account each variable history individually. We perform computational experiments on standard MIP test sets to evaluate the impact of our approach.

The remainder of this article is organized as follows: First, we summarize past and recent related work by other authors from the literature in Section 2. Section 3 introduces the necessary notation and presents the reliability branching rule in more detail. Afterwards, we introduce new notions of reliability in Section 4, and present computational results, which were obtained with an implementation in the Constraint Integer Programming framework SCIP [5] in Section 5. We finish with some concluding remarks in Section 6. The appendix contains an instance-wise summary of our computational experiments.

2   Related Work

Research on branching rules for Mixed Integer Programming has been a focus of interest since the advent of the Branch-and-Bound procedure in the 1960's [6,7]. Note that in this paper, we only consider variable-based branching. This concept is generalizable by incorporating branching on general disjunctions, which was introduced in [13].

Pseudo-costs, which measure the average objective gain for every integer variable, and their use for branching first appeared in [9]. The use of degradation bounds for the pseudo-cost initialization was suggested in [14]. Equipped with more computational power, strong branching was first applied in the context of the Traveling Salesman Problem [11], whereas its first use for general MIP solving is attributed to the commercial MIP solver CPLEX [2]. An important computational study for these techniques, also in the context of node selection, can be found in [10].

In a recent work [15], the strong branching procedure of SCIP could be improved by applying domain propagation techniques at each sub-node in addition to solving the LP relaxation. The authors of [16] observe unnecessary strong branching effort at the presence of *chimerical* variables, i.e. fractional variables with little or no effect on the objective of the LP solutions. They exploit this fact to safely ignore such candidates for the strong branching procedure. Cloud branching [17] has been proposed as a novel approach for better dealing with the frequent degeneracy of LP solutions. It computes several optimal LP solutions and considers variable fractionalities as intervals rather than points. The computational complexity for this approach is comparable to the effort of strong branching because two child relaxations are solved for every fractional variable.

The pseudo-cost branching rule is an effective replacement of the strong branching rule at later stages of the search but lacks information at the beginning. For that reason, combinations of pseudo-cost branching and strong branching have been developed that either use a single strong branching initialization on uninitialized variables, and pseudo-costs for every initialized candidate, or strong branching at the topmost d levels of the tree, and pseudo-cost branching at deeper levels. The state-of-the-art branching scheme, which is applied by most modern MIP solvers albeit the concrete implementation might vary, is *reliability branching* [12], see also Section 3. A threshold number is dynamically adjusted at every node depending on the proportion of Simplex iterations during strong branching and the total number of iterations spent on solving regular nodes, see [18] for further details. Other forms of history information such as inference or cutoff histories have been adopted for general MIP in [19]. A combination of such feasibility-based history information and pseudo-costs into a single score was introduced as *Hybrid reliability branching* [8].

For recent variable branching methods that use other techniques than history information, see, e.g., [20,21]. The branching strategies presented in [22] aim at quickly finding feasible solutions. Therefore, solution densities are approximated by means of normal distributions. Although their approach is quite different from the one presented here, their work has indeed been a motivation to further study links between statistics and optimization. The authors of [23] recently presented a method for restricting the set of branching candidates by calculating so-called *backdoor sets* in advance.

The approach presented here uses variations of past branching information for the decision if strong branching should be continued on a variable or not. It extends the idea of reliability branching by taking into account each variable individually. In the present paper, we further concentrate only on pseudo-costs and do not consider other history information. We do not collect any information prior to the actual search as in [21,23].

3 Reliability Branching with Fixed-Number Thresholds

Let $A \in \mathbb{R}^{m,n}$ be a real matrix, and let $c \in \mathbb{R}^n$ and $b \in \mathbb{R}^m$ be a cost and a right-hand-side vector, respectively. Let further $l, u \in \mathbb{R}^n \cup \{-\infty, \infty\}^n$ denote *bound requirements* for the variables, and let a subset $\mathcal{I} \subseteq \{1, \dots, n\}$ of the variables index set denote *integrality restrictions*.

An optimization problem defined by $c, A, b, l, u, \mathcal{I}$ as

$$c^{\mathrm{opt}} := \inf \left\{ c^t x \; : \; Ax \leq b, \; l \leq x \leq u, \; x \in \mathbb{R}^n, \; x_j \in \mathbb{Z} \text{ for all } j \in \mathcal{I} \right\} \quad \text{(MIP)}$$

is called a *Mixed Integer Program (MIP)*. We denote its *optimal objective value*, which may be infinite, by c^{opt}. Variables indexed by $j \in \mathcal{I}$ are called *integer variables*. If the set of integrality restrictions is empty, we call (MIP) a *Linear Program (LP)*. An LP $\tilde{P}$ is called the *LP relaxation* of a MIP P if it is derived from P by dropping the integrality restrictions. Since the solution space of $\tilde{P}$ is a superset of the solution space of P, its holds that $c_{\tilde{P}}^{\mathrm{opt}} \leq c_{P}^{\mathrm{opt}}$.

The *branch-and-bound* procedure [6,7] creates a search tree for $P =: P^{(0)}$ by a successive problem division called *branching* based on the LP relaxation information at a node. Let $P^{(l)}$ be a feasible (sub-)problem currently processed. We solve the LP relaxation of $P^{(l)}$ and obtain an LP-solution $\tilde{y}$ with objective value $c^t \tilde{y} = \tilde{c}_{P^{(l)}}$. If $\tilde{y}$ violates some of the integrality restrictions $\mathcal{F} \subseteq \mathcal{I}$, branching creates two child problems $P_{-}^{(l)}, P_{+}^{(l)}$ by selecting a *fractional variable* $j \in \mathcal{F}$ and locally restricting the lower and upper bound of j in the child problems to $u_j \leftarrow \lfloor \tilde{y}_j \rfloor$ in $P_{-}^{(l)}$ and $l_j \leftarrow \lceil \tilde{y}_j \rceil$ for $P_{+}^{(l)}$, respectively. Each restriction renders $\tilde{y}$ infeasible. The created problems are then enqueued in a list of open subproblems. The procedure terminates when there is no open subproblem left.

For a fractional variable $j \in \mathcal{F}$ we define its *up-fractionality* and *down-fractionality* as

$$f_j^+ := \lceil \tilde{y}_j \rceil - \tilde{y}_j \quad \text{and} \quad f_j^- := \tilde{y}_j - \lfloor \tilde{y}_j \rfloor,$$

respectively. The decision on which fractional variable to branch is crucial for the success of the branch-and-bound search. A branching rule is characterized by its *score function* $\vartheta : \mathcal{F} \to \mathbb{R}$. It selects a branching variable $j^* \in \mathcal{F}$ that maximizes the score function. All branching rules in this paper calculate branching scores $\vartheta^- (j)$ and $\vartheta^+ (j)$ separately for the two directions. In SCIP, these two score values are then combined to yield a *product score*

$$\vartheta (j) := \max\{\vartheta^+ (j), \epsilon\} \cdot \max\{\vartheta^- (j), \epsilon\} \quad (1)$$

with a small $\epsilon = 10^{-6}$ in order to find a good balance between the two individual scores.

Throughout this paper, we will give definitions and explanations only for the down-branch. The according formula and argumentation for the other direction can be derived analogously.

Let $P_-(j)$ denote the MIP obtained by branching down on $j \in \mathcal{F}_P$. In this paper, we focus on the *gain* in the objective function

$$\vartheta^- (j) = \tilde{c}_{P_-(j)} - \tilde{c}_P \quad (2)$$

in the child node LP relaxation objectives w.r.t. their parent as branching score.

Since this information is unknown by the time a candidate needs to be selected, the strong branching rule determines $\vartheta_{\mathrm{str}}^- (j)$ and $\vartheta_{\mathrm{str}}^+ (j)$ by virtually solving $2 \cdot |\mathcal{F}|$ child node relaxations and evaluating the gains (2). Although strong branching is guaranteed to select the locally best candidate regarding the objective gain, the exhaustive solving of child nodes makes the computational cost of this procedure often too expensive. However, it is well suited as an initialization method for pseudo-costs.

The *pseudo-costs* [9] of a variable are a typical measure to estimate its impact on the children objective gain. Consider a node P with LP solution value $\tilde{c}_P$ and a fractional variable $j \in \mathcal{F}_P$. Let $P_-(j)$ be the down-child of P whose LP relaxation was solved to optimality. The normalization of the objective gain between $P_-(j)$ and P,

$$\varsigma_j^-(P) := \frac{\tilde{c}_{P_-(j)} - \tilde{c}_P}{f_j^-}$$

by the fractionality of j in $\tilde{y}^P$ is called *unit gain*. The pseudo-costs of a variable are the average over all such unit gains,

$$\Psi_j^- := \begin{cases} \dfrac{\gamma_j^-}{\eta_j^-}, & \text{if } \eta_j^- > 0, \\ 0, & \text{else,} \end{cases} \qquad (3)$$

where η_j^- denotes the number of problems Q for which j was selected as branching variable and the child node $Q_-(j)$ has been solved and was feasible, and γ_j^- the sum of obtained unit gains over all these problems. If η_j^- is 0, we call j *uninitialized* in this direction. The *pseudo-cost score function* uses the pseudo-cost information

$$\vartheta_{\text{ps}}^-(j) := \Psi_j^- \cdot f_j^-$$

to estimate the objective gain in the child obtained by branching on j.

We give a definition of reliability branching that is more general than the original definition by Achterberg et al. [12] in that it assumes a subdivision of the fractional variables into reliable and unreliable candidates as input. We will use it later together with our novel notions of reliability.

Definition 1 (Reliability branching). *Let P be a MIP with non-empty set of fractional variables $\mathcal{F}$. Given a subdivision $\mathcal{F} = \mathcal{F}^{rel}\dot{\cup}\mathcal{F}^{url}$ of the fractional variables into* reliable *and* unreliable *branching candidates, we define the* reliability branching score function *of $j \in \mathcal{F}$ as*

$$\vartheta_{rel}^-(j) := \begin{cases} \vartheta_{str}^-(j), & \text{if } j \in \mathcal{F}^{url}, \\ \vartheta_{ps}^-(j), & \text{if } j \in \mathcal{F}^{rel}. \end{cases} \qquad (4)$$

Reliability branching performs strong branching on the set of unreliable candidates $\mathcal{F}^{url}$ to determine their exact gains (2), but uses pseudo-cost estimates for all other branching candidates. It is characterized by its *notion of (un-)reliability*, i.e. a rule how to split the branching candidates into a reliable and an unreliable set.

We refer to the notion of reliability by Achterberg et al. [12], as *fixed-number threshold reliability*:

Definition 2 (Fixed-number threshold reliability). *Given a reliability parameter $\eta > 0$, fixed-number threshold (fnt)-reliability splits the fractionals according to*

$$\mathcal{F}_{fnt}^{url}(\eta) := \{j \in \mathcal{F} : \min\{\eta_j^-, \eta_j^+\} < \eta\}.$$

We call a variable $j \in \mathcal{F} \setminus \mathcal{F}_{fnt}^{url}(\eta)$ (fnt)-reliable.

Using the term "fixed-number", we emphasize that (fnt)-reliability of a variable solely depends on the number of previous branching observations. Achterberg et al. [12] suggested to use 8 as threshold, currently, SCIP uses 5, based on experimentations to yield a good average performance on a variety of MIP instances. In the next section, we introduce novel notions of reliability.

4 Relative-Error and Hypothesis Reliability

The drawback of (fnt)-reliability is that a fixed threshold is supposed to measure the reliability of all variables of the problem equally well. Intuitively, it seems desirable to have a more individual look at the pseudo-cost information of every variable and to continue strong branching on those candidates whose pseudo-costs fail to converge. In the following, we extend the statistical model for pseudo-costs by including the sample variance, which allows for the construction of confidence intervals and testing of hypotheses. There are many textbooks that cover these topics in more detail, see, e.g. [24].

We model the unit gains of a variable $j \in \mathcal{I}$ as independent samples of a normally distributed random variable $C_{j,-} \sim \mathcal{N}(\mu_{j,-}, \sigma_{j,-}^2)$ with unknown *mean* $\mu_{j,-}$ and *variance* $\sigma_{j,-}^2$. It should be noted here that a normal distribution model for unit gains is of limited accuracy because unit gains are always non-negative. The pseudo-costs represent an estimate for $\mu_{j,-}$. By using the *corrected sample variance*, we obtain an estimate for the variance, as well:

Definition 3. *Let $X_1, \ldots, X_n$ be independent, identically distributed samples. The corrected sample variance about the sample mean $\bar{X}$ is given by*

$$s^2 = \frac{1}{n-1} \sum_{i=1}^{n} (X_i - \bar{X})^2 = \frac{1}{n-1} \sum_{i=1}^{n} X_i^2 - \frac{1}{n(n-1)} \left(\sum_{i=1}^{n} X_i \right)^2. \quad (5)$$

The corrected sample variance is an unbiased estimate of the variance of the underlying distribution of the X_i. The right term of Equation (5) allows for constant-time updates of s^2 every time a new sample X is observed.

With increasing n, we can expect $\bar{X}$ to approach the mean of the distribution from the law of large numbers. Under the assumption that the samples

$X_1, \ldots, X_n$ are drawn from a normal distribution with unknown mean μ and variance σ^2, the random variable

$$T := \frac{\bar{X} - \mu}{s/\sqrt{n}}$$

is distributed along a Student's t-distribution with $n - 1$ degrees of freedom. This relation can be used to construct a *confidence interval I*, which contains the true value of μ with a probability of $1 - \alpha$ for any *error rate* $0 < \alpha < 1$:

$$I = \left[\bar{X} - t_{\alpha,n-1} \frac{s}{\sqrt{n}}, \bar{X} + t_{\alpha,n-1} \frac{s}{\sqrt{n}} \right],$$

denoting by $t_{\alpha,n-1} > 0$ the α-percentile of the distribution of T. The distance of the endpoints of I relative to its center $\bar{X} \neq 0$,

$$\varepsilon^{\text{rel}} = t_{\alpha,n-1} \cdot \frac{s}{\sqrt{n}|\bar{X}|}, \tag{6}$$

is called the *relative error* of the estimation.

4.1   Relative-Error Reliability

Applied to pseudo-costs, we determine the relative error for the pseudo-costs associated with each variable. Whenever a new unit gain for variable $j \in \mathcal{F}$ in the down-branching direction at a node P was observed, we increase the counter η_j^- by 1 and update the sum of unit gains γ_j^-. In addition, we keep track of the sum of squared unit gains $(\varsigma_j^-(P))^2$. This enables us to calculate the sample variance $(s_j^-)^2$ whenever $\eta_j^- \geq 2$. At a node Q, we calculate the relative error $\varepsilon_j^{\text{rel},-}$ of the current pseudo-costs Ψ_j^- as

$$\varepsilon_j^{\text{rel},-} := \begin{cases} t_{\alpha,\eta_j^- -1} \cdot \dfrac{s_j^-}{\sqrt{\eta_j^- \Psi_j^-}}, & \text{if } \Psi_j^- > 0 \\ 0, & \text{else.} \end{cases} \tag{7}$$

Recall that pseudo-costs are always non-negative. Hence, we can omit the absolute in the denominator of (6). Furthermore, if the pseudo-costs of j are equal to zero, this also holds for the sample variance $(s_j^-)^2$. We therefore set the relative error to zero in this case.

Definition 4 (Relative-error reliability). *For $\eta > 0$, relative-error (rer)-reliability splits the fractionals according to*

$$\mathcal{F}_{rer}^{url}(\eta) := \{j \in \mathcal{F} : \max\{\varepsilon_j^{\text{rel},+}, \varepsilon_j^{\text{rel},-}\} \geq \eta\}. \tag{8}$$

We call a variable $j \in \mathcal{F} \setminus \mathcal{F}_{rer}^{url}(\eta)$ (rer)-reliable.

The rationale of (rer)-reliability is to continue strong branching on the subset of variables with highly varying objective gains, whereas variables with constant gains are early considered (rer)-reliable. In order to obtain relative errors for the branching directions, we need at least $\eta_j^-, \eta_j^+ \geq 2$ observations in each direction. Note that a variable, which has already been (rer)-reliable, can become (rer)-unreliable again when the relative error rises again above the threshold after new information becomes available. In Section 5, we test an implementation of (rer)-reliability branching. It should be noted that in general there is no containment relation between the variable subsets considered by reliability branching with fixed number thresholds and our approach, i.e. neither is a strict subset of the other.

4.2   Hypothesis Reliability

The disadvantage of (rer)-reliability is that it is likely to spend much strong branching effort on variables with overall low objective gains, but high relative error. In order to overcome this, it is possible to restrict the variables that are selected for strong branching evaluation to only candidates with a probability to be actually better than the best candidate j^{ps} according to pseudo-cost branching. Roughly speaking, we want to ensure that there is little probability that $f_j^- \mu_{j,-} \geq f_{j^{\mathrm{ps}}}^- \mu_{j^{\mathrm{ps}},-}$ for $j^{\mathrm{ps}} \neq j \in \mathcal{F}$.

Therefore, we test against the hypothesis that a fractional $j \in \mathcal{F}$ has an objective gain as least as high as j^{ps}, i.e., $f_j^- \mu_{j,-} \geq f_{j^{\mathrm{ps}}}^- \mu_{j^{\mathrm{ps}},-}$. For two variables $i, j \in \mathcal{F}$ with fractionalities f_i^- and f_j^-, we use the *pooled variance*

$$S_{i,j}^- := \frac{(\eta_i^- - 1)(f_i^-)^2 \left(s_i^-\right)^2 + (\eta_j^- - 1)(f_j^-)^2 \left(s_j^-\right)^2}{\eta_i^- + \eta_j^- - 2}$$

to calculate a *2-sample t-value* for i and j,

$$T_{i,j}^- := \sqrt{\frac{\eta_i^- \eta_j^-}{\eta_i^- + \eta_j^-}} \frac{f_i^- \Psi_i^- - f_j^- \Psi_j^-}{S_{i,j}^-}.$$

Under the hypothesis, $T_{j^{\mathrm{ps}},j}^-$ follows a Student-t distribution with $\eta_{j^{\mathrm{ps}}}^- + \eta_j^- - 2$ degrees of freedom. If, for a given threshold $0 < \alpha < 1$, $T_{j^{\mathrm{ps}},j}^-$ exceeds $t_{\alpha,j^{\mathrm{ps}},j}^- := t_{\alpha,\eta_{j^{\mathrm{ps}}}^- + \eta_i^- -2}$, we can reject the hypothesis with an error probability of at most $\alpha/2$. The division by two is justified because the hypothesis is one-sided. Conversely, if the hypothesis cannot be safely rejected, it is safer to perform strong branching on the two candidates.

The second novel notion of reliability in the present paper rules out fractional variables with little probability to be better than the best pseudo-cost candidate:

Definition 5 (Hypothesis reliability). *Let $j^{ps} \in \mathcal{F}$ be the best pseudo-cost fractional candidate for branching, and let $0 < \alpha < 1$ be a rejection probability.*

The unreliable fractional set for hypothesis reliability *is*

$$\mathcal{F}_{hyp}^{url}(\alpha) := \left\{ j \in \mathcal{F} \,:\, T_{j^{ps},j}^{-} < t_{\alpha,j^{ps},j}^{-} \;and\; T_{j^{ps},j}^{+} < t_{\alpha,j^{ps},j}^{+} \right\}. \tag{9}$$

Variables $j \in \mathcal{F} \setminus \mathcal{F}_{hyp}^{url}(\alpha)$ *are called* (hyp)-reliable.

For practical reasons, we also include variables j with $\min\{\eta_j^-, \eta_j^+\} \leq 1$. It should be noted that the best pseudo-cost candidate j^{ps} is never (hyp)-reliable because $T_{j^{ps},j^{ps}}^{-} = T_{j^{ps},j^{ps}}^{+} = 0$. If no other candidate than j^{ps} is (hyp)-unreliable, this means that no other fractional variable has an estimated objective gain nearly as good as j^{ps}. In this case, we immediately branch on j^{ps} without strong branching. In the experiments in the following section, we tested an error probability of $\alpha = 0.05$, i.e. the error probability for ruling out a better candidate based on the current branching history is $\alpha/2 = 2.5\,\%$.

Note also that the variant of a t-test that we use for (hyp)-reliability is, in theory, only applicable when the two variables can be assumed to have equal variances. In practice, it would be possible to test for equal variances using an F-test and resort to the *Welch*-test if the variances are significantly unequal.

5   Computational Results with SCIP

We implemented the new reliability notions from Section 4 into the existing reliability branching rule of a development version of the Constraint Integer Programming framework SCIP [5] version 3.1.0.2, which we compiled with a gcc compiler version 4.8.2. As underlying LP-solver, we used SOPLEX [25] version 2.0. We used SCIP with default settings except for the following changes: For using a pure objective-based branching score function as in Section 3, tie-breakers such as, e.g., inference scores were deactivated by setting their corresponding weight to 0. Furthermore, we set the known optimal solution values – in case they exist – minus a small threshold 10^{-9} as objective cutoffs, so that only a proof for the optimality/infeasibility of a problem needed to be found. We also disabled all primal heuristics and activated depth-first search node selection as an attempt to minimize performance variability [26,27] due to other factors than the tested branching rules. Finally, the child node selection was changed to use solely pseudo-costs, where SCIP with default settings uses a hybrid approach together with inference scores.

The test bed for our comparison of the different approaches consists of a subset of instances from the three publicly available libraries MIPLIB 3.0 [28], MIPLIB 2003 [29], and MIPLIB 2010 [27], from which we omitted four instances for which an optimal objective value is not known by the time of this writing. Since we are mainly interested in reducing the search tree size, we further dropped all 29 instances that could be solved before or during the processing of the root node. Our final test bed thus contains 135 MIP instances.

The computations were performed on a cluster of 32 computers, each of which runs with a 64bit Intel Xeon X5672 CPUs at 3.20 GHz with 12 MB cache and 48 GB main memory. The operating system was Ubuntu 14.4. Hyperthreading

Table 1. All instances

	98 instances solved by all				135 total	
	t (sec)	%	n	%	t (sec)	%
Settings						
(fnt)-5	81.9	100.0	3402.3	100.0	290.3	100.0
(rer)-0.01	87.0	106.2	2722.3	80.0	303.1	104.4
(rer)-0.05	84.7	103.4	2716.5	79.8	298.1	102.7
(rer)-0.1	85.6	104.5	2902.9	85.3	300.3	103.4
(hyp)	82.7	101.0	2774.0	81.5	289.4	99.7

and Turboboost were disabled. We ran only one job per computer in order to minimize the random noise in the measured running time that might be caused by cache-misses if multiple processes share common resources. Finally, all experiments were run with a time limit of 2h and a 40 GB memory limit.

The newly proposed notions of reliability from Section 4 are represented by four different settings: (hyp) renders candidates (hyp)-unreliable according to the rule (5), whereas (rer)-0.01, (rer)-0.05, and (rer)-0.1 use (rer)-reliability regarding relative errors in pseudo-cost confidence intervals at three different threshold levels 1%, 5%, and 10%. Note that these levels represent different levels of the relative error threshold, whereas the confidence level $1 - \alpha$ is kept fixed at 95% for both (hyp) and (rer)-reliability. We compare them to (fnt)-reliability at a fixed threshold of 5, denoted by (fnt)-5. The latter setting constitutes the default of SCIP except that we disabled the threshold to be dynamically adjusted during the search.

In this section, we only present compressed results of our experiments. For an instance-wise outcome, please refer to Tables 5 and 6 in the Appendix. The first three tables show the aggregated results regarding the solving time t (sec) and the number of explored search tree nodes n for all instances and for only those which could be solved within the time limit by all settings. We consider node results incomparable between settings where the solution status differs and thus only show time results for all instances. We report shifted geometric means with a shift of 10 seconds and 100 nodes, respectively. The column "%" shows the percentage deviation from the result for the reference setting (fnt)-5; values below 100 represent an improvement in this respect.

In Table 1, we compare the results over all instances from the test bed. 98 instances could be solved by all settings within the time limit of 2h, for which the reference run was fastest regarding the solving time, but also required the most branch-and-bound nodes on average. For our novel notion of (rer)-reliability, an evaluation of the different thresholds comes with a surprise: there is no significant difference between the levels 1% and 5% error tolerance. The highest node reduction of 20.2% was obtained with the setting (rer)-0.05, closely followed by (rer)-0.01, which could not reduce the overall solving nodes further in the shifted geometric mean. Regarding the running time, (rer)-0.05 was the fastest to finish the tests among the three (rer)-settings, yet we observe a slight slow

Table 2. Large Trees: $n > 1000$ with at least one setting

	52 instances solved by all				88 total	
	t (sec)	%	n	%	t (sec)	%
Settings						
(fnt)-5	212.6	100.0	49741.5	100.0	897.9	100.0
(rer)-0.01	232.0	109.1	35322.9	71.0	947.4	105.5
(rer)-0.05	221.3	104.1	35104.9	70.6	924.0	102.9
(rer)-0.1	224.9	105.8	38227.1	76.9	932.5	103.9
(hyp)	212.9	100.1	35427.6	71.2	885.9	98.7

down of 3.4 % for the group of instances solved by all settings, and 2.7 % over all 135 instances compared to the reference run. By using (hyp)-reliability, we obtained a node reduction of 18.5 % and an almost performance neutral result for the running time.

The discrepancy between a reduction of the tree size at the cost of more solving time per node is the result of a more aggressive use of strong branching by the novel notions of reliability. The notion of (hyp)-reliability hereby appears to be more effective than relative-error reliability for guiding strong branching effort because it focusses on resolving cases among the top pseudo-cost score branching candidates where the estimation alone may lead to inferior branching decisions.

Table 2 contains only instances for which at least one of the settings needed more than 1000 nodes before termination. With (hyp)-reliability, we could improve the performance of SCIP w.r.t. the reference run by 28.8 % nodes and also obtain a slight time reduction in total, whereas the time on instances in the group containing only solved instances is competitive with the reference run (fnt)-5. Among the (rer)-settings, (rer)-0.05 is fastest regarding the solving time, but is still 4.1 % slower on average than the reference setting. All three settings (rer)-0.05, (rer)-0.01, and (hyp) show a similar decrease in the number of nodes, but at different computational efforts spent per node.

For the sake of completeness, we also present the remaining instances, for which no solver took more than 1000 branch-and-bound nodes before termination, in Table 3. Out of the 47 instances in this group, there is only one, namely stp-3d, that could not be solved by any of the settings. All novel notions of reliability reduce the search tree size, although the effect is less striking than on the instances that required larger search trees. Note that the node reduction obtained with (rer)-0.01 and (rer)-0.05 is now considerably better than the reduction obtained with (hyp)-reliability.

For those 37 instances for which optimality could not be proven within the time limit by at least one of our settings, we computed integrals of the dual gap as a function of time. This measure, which was suggested in [30,31], attempts to compare the convergence of the dual gap towards zero. Table 4 shows the shifted geometric mean integral for all settings using a shift of 1000. All novel notions of reliability decrease the dual integral of the reference run, where the decrease

Table 3. Small trees: All solvers needed $n \leq 1000$ nodes

	46 instances solved by all					47 total	
	t (sec)	%	n	%		t (sec)	%
Settings							
(fnt)-5	23.8	100.0	74.1	100.0		27.8	100.0
(rer)-0.01	24.5	103.0	61.7	83.3		28.6	102.8
(rer)-0.05	24.5	103.0	62.1	83.8		28.6	102.8
(rer)-0.1	24.6	103.4	68.8	92.9		28.7	103.2
(hyp)	24.3	102.4	67.5	91.1		28.5	102.3

Table 4. Shifted geom. mean dual integral for 37 time limit instances

	$\Gamma^*(T)$	%
Settings		
(fnt)-5	73867.8	100.0
(rer)-0.01	66392.1	89.9
(rer)-0.05	73454.5	99.4
(rer)-0.1	72959.7	98.8
(hyp)	63101.8	85.4

is best with (hyp) yielding a reduction of 14.6 %. A similar result is obtained with (rer)-0.01, which outperforms other thresholds for (rer)-reliability in this respect.

6 Conclusions and Future Work

We introduced two novel notions of reliability: the (rer)-reliability based on pseudo-cost confidence intervals, and (hyp)-reliability implementing a variant of Student's t-test. First experimental results with our implementation in SCIP show that these methods are promising for effectively reducing the size of branch-and-bound-trees compared to the current state-of-the-art fixed-number threshold, especially for large trees. Note that node reductions and the resulting reductions of the required memory play an important role, e.g., in the context of solver-parallelization to reduce load-coordination overhead. Together with the presented smaller dual integrals for instances which hit the time limit, we consider the reliability notions useful for proving optimality faster on harder instances.

Our first implementation only considers pseudo-cost information, but can be readily applied to different history information such as, e.g., the inference history of a variable, as well. In the computational study presented, we collected very little history information before using the statistical methods. Combining them with traditional fixed-number threshold reliability might increase the power of the hypothesis and relative error thresholds significantly.

Note that for our computational experiments, we did not allow a dynamic adaption of the fixed-number thresholds depending on the computational expenses on strong branching during the search. Fixed-number thresholds, however, show a superior performance if they are dynamically adjusted during the search, so that the overall strong branching effort is kept reasonably small. For making a more effective use of the suggested approaches, it is necessary to let also the novel approaches dynamically adjust to problems for which strong branching is very expensive.

Acknowledgments. The author would like to thank the three anonymous reviewers for their valuable and constructive comments on the original manuscript, as well as to his colleagues Timo Berthold and Gerald Gamrath for their many valuable comments on earlier versions of this work. The work for this article has been conducted within the Research Campus Modal funded by the German Federal Ministry of Education and Research (fund number 05M14ZAM).

References

1. (COIN-OR branch-and-cut MIP solver). https://projects.coin-or.org/Cbc
2. (IBM ILOG CPLEX Optimizer). http://www-01.ibm.com/software/integration/optimization/cplex-optimizer/
3. (FICO Xpress-Optimizer). http://www.fico.com/en/Products/DMTools/xpress-overview/Pages/Xpress-Optimizer.aspx
4. (GUROBI Optimizer). http://www.gurobi.com/products/gurobi-optimizer/gurobi-overview
5. SCIP. Solving Constraint Integer Programs. (http://scip.zib.de/)
6. Dakin, R.J.: A tree-search algorithm for mixed integer programming problems. The Computer Journal **8**, 250–255 (1965)
7. Land, A.H., Doig, A.G.: An automatic method of solving discrete programming problems. Econometrica **28**, 497–520 (1960)
8. Achterberg, T., Berthold, T.: Hybrid branching. In: van Hoeve, W.-J., Hooker, J.N. (eds.) CPAIOR 2009. LNCS, vol. 5547, pp. 309–311. Springer, Heidelberg (2009)
9. Bénichou, M., Gauthier, J.M., Girodet, P., Hentges, G., Ribière, G., Vincent, O.: Experiments in mixed-integer programming. Mathematical Programming **1**, 76–94 (1971)
10. Linderoth, J.T., Savelsbergh, M.W.P.: A computational study of search strategies for mixed integer programming. INFORMS Journal on Computing **11**, 173–187 (1999)
11. Applegate, D.L., Bixby, R.E., Chvátal, V., Cook, W.J.: Finding cuts in the TSP (A preliminary report). Technical Report 95–05, DIMACS (1995)
12. Achterberg, T., Koch, T., Martin, A.: Branching rules revisited. Operations Research Letters **33**, 42–54 (2004)
13. Ryan, D.M., Foster, B.A.: An integer programming approach to scheduling. In: Wren, A. (ed.) Computer Scheduling of Public Transport Urban Passenger Vehicle and Crew Scheduling, pp. 269–280. North Holland, Amsterdam (1981)

14. Gauthier, J.M., Ribière, G.: Experiments in mixed-integer linear programming using pseudo-costs. Mathematical Programming **12**, 26–47 (1977)
15. Gamrath, G.: Improving strong branching by propagation. In: Gomes, C., Sellmann, M. (eds.) CPAIOR 2013. LNCS, vol. 7874, pp. 347–354. Springer, Heidelberg (2013)
16. Fischetti, M., Monaci, M.: Branching on nonchimerical fractionalities. OR Letters **40**, 159–164 (2012)
17. Berthold, T., Gamrath, G., Salvagnin, D.: Cloud branching. Presentation slides from Mixed Integer Programming Workshop at Ohio State University (2014). https://mip2014.engineering.osu.edu/sites/mip2014.engineering.osu.edu/files/uploads/Berthold_MIP2014_Cloud.pdf
18. Achterberg, T.: Constraint Integer Programming. PhD thesis, Technische Universität Berlin (2007)
19. Achterberg, T.: SCIP: Solving constraint integer programs. Mathematical Programming Computation **1**, 1–41 (2009)
20. Gilpin, A., Sandholm, T.: Information-theoretic approaches to branching in search. Discrete Optimization **8**, 147–159 (2011)
21. Kilinç Karzan, F., Nemhauser, G.L., Savelsbergh, M.W.P.: Information-based branching schemes for binary linear mixed integer problems. Mathematical Programming Computation **1**(4), 249–293 (2009)
22. Pryor, J., Chinneck, J.W.: Faster integer-feasibility in mixed-integer linear programs by branching to force change. Computers & Operations Research **38**, 1143–1152 (2011)
23. Fischetti, M., Monaci, M.: Backdoor branching. In: Günlük, O., Woeginger, G.J. (eds.) IPCO 2011. LNCS, vol. 6655, pp. 183–191. Springer, Heidelberg (2011)
24. Roussas, G.G.: A Course in Mathematical Statistics, Third Edition. Elsevier Science & Technology Books (2014)
25. SoPlex. An open source LP solver implementing the revised simplex algorithm. (http://soplex.zib.de/)
26. Danna, E.: Performance variability in mixed integer programming. Presentation slides from MIP workshop in New York City (2008). http://coral.ie.lehigh.edu/jeff/mip-2008/program.pdf
27. Koch, T., Achterberg, T., Andersen, E., Bastert, O., Berthold, T., Bixby, R.E., Danna, E., Gamrath, G., Gleixner, A.M., Heinz, S., Lodi, A., Mittelmann, H., Ralphs, T., Salvagnin, D., Steffy, D.E., Wolter, K.: MIPLIB 2010. Mathematical Programming Computation **3**, 103–163 (2011)
28. Bixby, R.E., Ceria, S., McZeal, C.M., Savelsbergh, M.W.: An updated mixed integer programming library: MIPLIB 3.0. Optima **58**, 12–15 (1998)
29. Achterberg, T., Koch, T., Martin, A.: MIPLIB 2003. Operations Research Letters **34**, 1–12 (2006)
30. Achterberg, T., Berthold, T., Hendel, G.: Rounding and propagation heuristics for mixed integer programming. In: Klatte, D., Lüthi, H.J., Schmedders, K. (eds.) Operations Research Proceedings 2011, pp. 71–76. Springer, Berlin Heidelberg (2012)
31. Berthold, T.: Measuring the impact of primal heuristics. Operations Research Letters **41**, 611–614 (2013)

Appendix

This appendix contains an instance-wise outcome of our computational experiments described in Section 5. For each of the five settings, we present three columns; the measured dual integral $\Gamma^*(T)$, the number of nodes n, and the solving time in seconds t (sec). Table 5 shows the results for instances which we classified as small tree instances, and Table 6 contains the remaining instances, cf. Tables 3 and 2, respectively.

Table 5. Instance-wise experimental outcome for instances requiring at most 1000 nodes to solve

Settings	(hyp)			(rer)-0.01			(rer)-0.1			(rer)-0.05			(fnt)-5		
	$\Gamma^*(T)$	n	t (sec)	$\Gamma^*(T)$	n	t (sec)	$\Gamma^*(T)$	n	t (sec)	$\Gamma^*(T)$	n	t (sec)	$\Gamma^*(T)$	n	t (sec)
Problem															
30n20b8	12699.9	13	196.4	14587.7	79	221.4	13000.5	18	200.6	15786.8	120	237.6	12756.5	18	196.3
air04	1766.8	8	36.2	1842.3	8	37.6	1842.3	8	37.6	1847.3	8	37.7	1862.4	8	37.9
air05	1275.2	50	25.7	1290.3	52	25.9	1290.4	52	26.0	1275.1	52	25.6	1270.2	62	25.6
app1-2	110453.1	23	1763.5	110730.3	19	1766.7	109918.4	19	1754.8	109878.9	19	1754.2	55296.9	41	875.6
ash608gpia-3col	2010.0	7	20.1	2120.0	9	21.2	2060.0	9	20.6	2080.0	9	20.8	2000.0	7	20.0
blend2	23.4	252	0.6	24.3	117	0.7	28.8	155	0.7	23.4	205	0.6	23.4	240	0.6
dcmulti	5.5	8	1.2	5.4	14	1.0	5.5	14	1.1	5.5	14	1.1	0.3	8	0.8
fast0507	1457.3	598	140.5	4264.0	714	150.5	4570.0	722	149.6	1128.5	646	146.8	4444.4	630	147.8
fiber	7.5	4	1.2	8.0	4	1.2	1.8	4	1.0	7.2	4	1.1	4.1	4	1.0
fixnet6	14.2	18	2.1	19.7	10	2.2	15.3	10	2.2	20.1	10	2.3	11.5	10	1.9
gesa2	5.2	3	0.8	5.1	3	0.8	5.2	3	0.7	5.2	3	0.7	5.0	3	0.4
gesa2-o	5.2	2	1.0	5.2	2	1.1	5.2	2	1.0	5.3	2	1.1	5.1	2	0.9
gesa3	5.2	7	1.3	5.2	7	1.2	5.2	7	1.3	5.1	7	1.0	5.1	7	1.0
gesa3_o	10.1	7	1.3	5.1	7	1.2	5.2	7	1.3	5.1	7	1.2	0.1	7	0.9
khb05250	0.1	4	0.5	0.2	4	0.5	0.2	4	0.5	0.2	4	0.5	0.2	4	0.5
l152lav	31.2	19	1.2	46.6	19	1.6	41.2	19	1.3	41.5	19	1.5	40.9	19	1.1
lseu	21.8	58	0.5	21.8	64	0.5	21.8	64	0.5	21.8	64	0.5	5.9	191	0.2
map18	15193.4	309	291.8	15757.4	275	302.4	15821.5	275	303.6	15777.8	275	302.7	15510.6	285	297.8
map20	12007.8	265	229.9	12518.0	307	239.4	12523.8	307	239.6	12560.5	307	240.2	12353.0	281	236.3
misc03	43.4	77	1.4	37.8	23	1.2	36.3	23	1.1	33.1	23	1.0	24.1	80	0.8
misc06	5.0	4	0.5	5.0	4	0.6	5.0	4	0.6	5.0	4	0.6	5.0	4	0.5
mod008	2.5	7	0.8	0.0	7	0.8	0.0	7	1.0	3.1	7	1.0	2.3	7	0.8
mod010	5.1	2	0.5	5.1	2	0.5	5.1	2	0.5	5.1	2	0.5	5.1	2	0.5
mod011	327.1	671	108.8	329.4	671	109.2	326.5	671	108.6	331.6	853	110.4	333.0	855	116.3
modglob	5.1	21	0.6	5.1	19	0.5	10.1	23	0.6	10.1	25	0.6	0.1	25	0.5
mspp16	131239.4	57	2474.8	88730.3	31	1673.2	95963.6	29	1809.6	98487.9	29	1857.2	98434.8	31	1856.2
mzzv42z	5076.1	132	147.7	5117.1	96	148.4	5101.4	96	147.9	5065.3	96	147.2	5458.9	110	155.5
neos-476283	1971.2	233	73.8	1961.2	105	70.4	2006.2	175	70.1	1976.3	394	74.9	1971.2	110	70.6
neos13	876.7	6	31.2	883.0	8	31.2	860.5	8	30.8	866.8	8	30.7	897.8	8	32.1
ns1208400	33880.0	598	338.8	52850.0	860	528.5	52890.0	860	528.9	52820.0	860	528.2	49760.0	881	497.6
nw04	476.0	8	20.4	480.0	8	20.1	470.0	8	20.0	465.7	8	20.2	496.4	8	20.7
p0201	42.9	11	1.3	37.7	9	1.2	27.5	9	1.0	32.3	9	1.0	32.6	9	1.1
p0282	0.5	3	0.5	0.5	3	0.5	0.5	3	0.5	0.0	3	0.4	0.0	3	0.3
p2756	1.7	3	0.9	1.8	3	1.1	1.8	3	1.1	6.8	3	1.1	5.1	3	0.7
pp08a	32.4	53	1.2	27.4	51	1.1	32.4	51	1.2	27.4	51	1.1	21.2	161	0.7
pp08aCUTS	12.8	53	1.2	23.0	51	1.3	18.1	51	1.3	18.0	51	1.2	1.5	153	0.8
qnet1	16.7	3	2.0	16.7	3	2.0	11.9	3	2.0	17.3	3	2.2	10.2	3	2.0
qnet1_o	4.1	4	1.4	0.0	4	1.3	4.1	4	1.4	0.0	4	1.4	0.0	4	1.3
rail507	760.6	554	145.5	664.3	546	136.9	1207.0	582	144.2	704.4	586	142.6	529.0	644	147.1
rentacar	71.9	4	3.3	77.1	4	3.4	81.7	4	3.8	80.6	4	3.4	80.6	4	3.4
rmatr100-p10	6884.6	723	121.2	7004.4	793	123.3	7028.6	793	123.8	6981.6	791	122.9	6821.8	709	120.1
rmatr100-p5	14864.3	339	245.6	15234.8	367	251.7	15562.9	367	257.1	15224.8	367	251.6	14256.9	349	235.6
set1ch	0.0	3	0.8	0.0	3	0.8	3.0	3	0.8	0.0	3	0.6	0.0	3	0.6
stp3d	145498.8	17	7200.0	145872.3	14	7200.0	145433.0	14	7200.0	145579.5	14	7200.0	145433.0	14	7200.0
tanglegram1	92121.1	31	922.3	77378.6	27	774.7	77468.4	27	775.6	77108.9	27	772.0	99072.9	33	991.9
tanglegram2	783.1	3	7.9	793.1	3	8.0	783.1	3	7.9	793.1	3	8.0	793.1	3	8.0
vpm2	19.1	298	0.9	35.3	50	1.3	24.8	50	1.1	25.0	110	1.1	19.6	272	1.0

Table 6. Instance-wise experimental outcome for instances for which one setting required more 1000 nodes

Settings	(hyp)			(rer)-0.01			(rer)-0.1			(rer)-0.05			(fnt)-5		
	$\Gamma^*(T)$	n	t (sec)	$\Gamma^*(T)$	n	t (sec)	$\Gamma^*(T)$	n	t (sec)	$\Gamma^*(T)$	n	t (sec)	$\Gamma^*(T)$	n	t (sec)
Problem															
a1c1s1	259882.7	1120506	7200.0	259873.5	1364025	7200.0	259879.2	1685774	7200.0	259876.3	2004990	7200.0	259863.8	2070335	7200.0
aflow30a	508.5	1720	12.6	502.7	1908	12.4	507.3	1908	12.4	511.8	1908	12.4	377.5	1246	10.1
aflow40b	20210.8	104858	596.3	29534.9	172263	856.6	31494.3	179651	893.0	9929.0	116421	647.2	16390.8	143682	536.0
arki001	105.7	1120884	7200.0	105.7	592970	7200.0	105.7	754036	7200.0	105.7	1089108	7200.0	105.7	365939	7200.0
atlanta-ip	70465.2	7972	7200.0	70460.7	10011	7200.0	70460.7	10012	7200.0	70469.4	11789	7200.0	70469.7	8529	7200.0
bab5	6707.5	133720	7200.0	6649.7	110708	7200.0	6623.3	105423	7200.0	6644.2	123080	7200.0	6632.3	96735	7200.0
beasleyC3	94243.8	2408887	7200.0	92905.6	2107529	7200.0	92907.7	1377166	7200.0	92907.5	1849676	7200.0	92907.7	2841834	7200.0
bell3a	0.0	20025	3.0	0.0	20027	3.2	1.3	20261	3.1	11.4	21353	3.2	16.3	22611	2.8
bell5	10.0	1025	0.3	10.0	956	0.3	10.0	956	0.3	10.0	956	0.3	10.0	1152	0.3
biella1	34013.5	7109	679.4	19260.3	3629	384.8	19260.3	3629	384.8	15824.9	2771	316.2	33983.5	6883	678.8
bienst2	10925.9	66579	328.4	10702.5	101007	426.9	10718.1	101007	427.7	15947.8	113088	463.4	29985.7	309529	561.7
binkar10_1	596.4	86776	149.4	645.7	107739	163.3	2009.0	106799	161.7	139.4	100593	157.2	175.9	142580	139.1
cap6000	25.0	984	1.6	25.0	738	1.6	25.0	738	2.1	25.0	738	1.7	30.0	1801	1.5
cov1075	238889.6	2029811	4343.6	276493.1	2302521	5027.3	272692.6	2343053	4958.2	239021.6	2057875	4346.0	265179.6	2274377	4821.6
csched010	66451.7	369300	5305.9	84124.4	603007	7200.0	83471.9	595757	7200.0	82408.9	580198	7200.0	84082.4	1035664	7200.0
danoint	28360.6	1587831	6349.2	26431.6	916872	5941.5	32253.7	1563067	7200.0	32305.3	1741600	7200.0	20001.8	1102583	4476.0
dfn-gwin-UUM	1896.5	46931	118.7	1183.9	56297	135.2	1045.1	54593	133.4	1336.6	51457	139.2	1333.5	68085	93.1
ds	273675.0	6949	7200.0	273443.2	8668	7200.0	273433.4	8668	7200.0	273437.3	8554	7200.0	273445.3	7413	7200.0
eil33-2	1769.4	844	34.0	1748.3	1244	38.6	1818.4	1688	40.0	1183.5	1750	36.8	1351.2	340	30.3
eilB101	15408.3	8352	296.7	3224.8	5724	237.5	3264.2	5724	238.4	12732.3	6842	266.3	12660.3	6284	265.1
enlight13	1985.2	35737	21.2	1292.9	19169	14.1	1293.3	19169	14.1	8782.4	212656	97.6	13838.2	522384	158.9
glass4	61870.2	3541756	2966.7	240005.6	6451908	7200.0	70217.8	4150120	3363.6	240005.6	7997381	7200.0	240002.3	12763342	7200.0
gmu-35-40	65.0	21704510	7200.0	65.0	19643305	7200.0	65.0	20060056	7200.0	65.0	20562455	7200.0	60.0	32198403	7200.0
harp2	6822.5	197138	136.7	504.5	243174	161.1	320.6	198128	145.5	6887.6	209303	137.9	233.9	448438	160.7
iis-100-0-cov	19380.1	71291	522.7	26112.1	63349	685.3	30456.6	68587	737.3	25113.1	68335	578.4	24816.6	76155	523.3
iis-bupa-cov	98475.4	176675	2454.5	124920.5	146447	3222.1	65723.4	151437	2874.6	56189.1	153879	2319.6	58125.2	181441	2500.9
iis-pima-cov	14787.3	7531	263.6	13231.9	5821	239.4	14110.9	6339	254.5	13808.1	6059	249.3	13735.8	5869	248.4
macrophage	294853.6	2815744	7200.0	294850.8	1594719	7200.0	294850.8	2450737	7200.0	294854.2	2734477	7200.0	294851.0	2875395	7200.0
markshare1	720000.0	91697134	7200.0	720000.0	101065996	7200.0	720000.0	101801275	7200.0	720000.0	100882365	7200.0	720000.0	102171542	7200.0
markshare2	720000.0	83206634	7200.0	720000.0	88008981	7200.0	720000.0	87343340	7200.0	720000.0	87701931	7200.0	720000.0	88721939	7200.0
mas74	4034.7	2035460	395.3	3886.4	2112420	404.3	3425.2	2476385	355.0	3287.4	2567408	338.0	3306.0	2752472	342.5
mas76	99.8	221799	35.2	143.0	381797	55.0	318.2	395730	54.8	283.6	407343	46.7	244.8	547767	51.4
mcsched	60729.2	101259	1249.8	65736.7	108207	1334.0	56338.3	95019	1160.0	6662.9	60293	826.0	26190.3	46543	603.9
mik-250-1-100-1	1897.8	1072724	434.3	5327.1	3197358	1212.0	2179.0	1076085	438.0	1632.6	1505029	379.8	5995.2	6996847	1372.3
mine-166-5	1724.8	1240	31.1	1196.3	198	22.2	1333.9	328	24.5	1351.9	328	24.8	1250.2	2038	23.1
mine-90-10	2573.4	27903	96.5	8327.5	69247	208.1	10010.1	75221	240.3	8448.2	67032	209.7	20332.3	197329	428.7
misc07	973.8	31204	24.8	1183.4	33068	27.1	1245.4	33905	27.3	1027.9	30592	24.7	1201.8	47624	26.3
mkc	5503.8	1759678	7200.0	5504.2	2780944	7200.0	5503.8	1346673	7200.0	5504.0	1750343	7200.0	5504.2	2175182	7200.0
momentum1	85138.8	224442	7200.0	85144.8	395927	7200.0	85133.0	221818	7200.0	85158.2	264131	7200.0	85140.1	519796	7200.0
momentum2	244940.0	95882	4298.6	146827.4	34806	2723.9	202627.9	218703	7200.0	116352.8	140632	5152.9	226569.4	183471	4366.0
msc98-ip	5196.4	19531	7200.0	5196.4	8595	7200.0	5186.4	88283	7200.0	5196.4	40128	7200.0	5191.4	58884	7200.0
mzzv11	10377.6	1873	332.0	9712.5	1857	319.5	9701.3	1857	318.2	9716.6	1857	318.3	10185.3	2014	324.5
n3div36	54503.3	305537	7200.0	52926.8	350961	7200.0	52929.1	346823	7200.0	54503.6	362117	7200.0	52928.9	513914	7200.0
n3seq24	4800.8	7497	7200.0	4840.6	10605	7200.0	4820.7	9713	7200.0	4825.7	9711	7200.0	4815.7	8601	7200.0
n4-3	40980.4	37963	742.5	47000.5	44277	850.6	47119.4	42763	852.8	47852.6	43169	866.2	9826.5	46719	557.0
neos-1109824	4752.2	16018	88.0	2215.8	5143	41.3	2644.7	7432	49.2	2493.4	6008	46.5	1674.8	17347	84.2
neos-1337307	2851.4	499306	7200.0	2846.1	537959	7200.0	2851.1	501053	7200.0	2856.0	499990	7200.0	2841.1	714364	7200.0
neos-1396125	165458.3	100396	4394.8	107606.3	57401	2582.8	105651.4	59324	2592.5	108259.0	57159	2594.8	37201.4	88307	836.4
neos-686190	2695.3	2190	44.4	2793.4	2400	46.0	2793.4	2400	46.0	2711.4	2263	44.6	2705.3	1865	44.5
neos-916792	124357.4	1936201	7200.0	115050.5	1637517	7200.0	115058.7	1695767	7200.0	117813.7	1044367	7200.0	117042.3	2065470	7200.0
neos18	1316.9	34892	75.9	2794.3	59418	116.6	1699.9	44814	72.4	4982.4	113871	136.1	2054.8	59329	79.2
net12	225620.4	2616	2776.5	289044.4	3642	3556.3	291354.0	3642	3584.8	188654.2	2273	2321.9	231010.4	3139	2842.7
netdiversion	59360.6	1262	7200.4	59214.5	2450	7200.6	59507.7	2440	7200.2	59409.7	2448	7200.2	59752.9	2430	7200.8
newdano	116175.7	1337377	2695.1	202838.0	2102759	4739.1	166152.1	2281511	3972.2	172475.7	2390353	3843.2	166449.7	3194194	3936.2
noswot	7283.7	759174	139.2	7744.2	541939	148.0	5264.0	349424	100.6	9245.9	937041	176.7	16508.7	1607445	315.5
ns1688347	638.9	216	11.5	666.1	227	11.9	765.2	547	13.7	587.2	132	10.6	690.6	1088	12.4
ns1766074	342820.0	741161	3428.2	339300.0	749137	3393.0	342910.0	718315	3429.1	270620.0	774224	2706.2	237360.0	848597	2373.6
ns1830653	16954.7	51131	396.1	30657.6	44597	427.5	30736.6	44597	428.6	31033.3	46837	432.7	9767.1	22244	219.9
nsrand-ipx	11004.5	4470709	7200.0	11118.5	2278695	7200.0	15952.5	3796812	6527.1	11109.1	4328054	7200.0	10543.7	3978440	6502.4
opm2-z7-s2	52472.5	1511	873.2	45429.7	1275	756.0	45772.2	1421	761.7	51655.3	1325	859.6	46319.1	1283	770.8
pg5_34	3338.5	110314	797.7	35711.3	93606	830.2	33931.0	71290	801.6	3277.9	67830	755.8	1535.8	136710	713.1
pigeon-10	72000.0	1903058	7200.0	72000.0	2340172	7200.0	72000.0	2260479	7200.0	72000.0	3267789	7200.0	72000.0	3010254	7200.0
pk1	8503.0	441505	86.6	7850.0	245985	78.5	7230.0	275029	72.3	7341.0	344029	74.6	7282.0	393967	74.0
protfold	169384.3	6724	7200.0	169388.0	3654	7200.0	169388.0	3648	7200.0	169387.9	3643	7200.0	169388.3	6014	7200.0
pw-myciel4	161437.8	504422	2868.5	432078.7	3985751	7200.0	432084.8	3052407	7200.0	432083.7	2963901	7200.0	432085.2	3293759	7200.0
qiu	6514.4	9203	75.4	5908.2	8355	70.0	5790.7	8355	68.6	5799.0	8355	68.7	4489.9	9557	48.7
ran16x16	1579.1	254683	255.4	1497.2	251567	244.4	2064.1	248711	251.3	2212.5	261833	270.0	1222.1	284383	177.9
rd-rplusc-21	719565.9	202482	7200.0	719565.5	365129	7200.0	719565.5	176050	7200.0	719565.9	181324	7200.0	719565.5	563331	7200.0
reblock67	1950.7	38170	79.3	1462.0	27716	67.3	668.6	34065	75.3	1402.4	27858	67.9	789.7	49872	75.6
rmine6	29837.3	379250	643.4	36649.5	264194	787.7	26286.3	275382	588.8	837.0	381988	662.7	28254.0	367462	608.8
rocII-4-11	139430.3	5922	2117.2	326485.0	37845	4944.7	124937.9	20613	3489.0	197450.1	44679	5502.3	219386.9	140524	6949.1
rococoC10-001000	78138.0	94037	7200.0	78138.8	646077	7200.0	78139.8	1134644	7200.0	78140.2	1009680	7200.0	78134.2	2225139	7200.0
roll3000	30539.2	1592527	7200.0	30533.5	1828356	7200.0	30538.8	1883210	7200.0	30528.4	1647039	7200.0	30528.9	2722022	7200.0
rout	527.2	33158	45.0	1253.8	24804	38.4	1232.1	24804	38.0	1438.2	28995	41.8	528.1	19618	22.0
satellites1-25	208591.3	6503	2383.9	215188.8	6168	2459.3	215468.8	6168	2462.5	215582.5	6168	2463.8	149555.0	5749	1709.2
seymour	22803.9	432465	7200.0	22813.7	266060	7200.0	22813.7	298616	7200.0	22808.7	406612	7200.0	22808.8	438631	7200.0
sp97ar	715183.0	2860	7200.0	715182.9	2859	7200.0	715183.0	2859	7200.0	715183.0	2861	7200.0	715182.9	2862	7200.1
sp98ic	643739.2	24910	7200.1	643980.9	25556	7200.1	645279.9	22326	7200.0	645193.7	21939	7200.0	642833.8	33432	7200.0
sp98ir	1717.5	3354	38.1	1550.3	2928	34.7	1560.5	2928	35.0	1591.3	2910	35.9	1661.9	2582	37.1
stein27	51.1	3215	0.8	121.4	921	1.9	115.0	933	1.8	115.0	1449	1.8	38.3	4073	0.6
stein45	646.0	37211	10.2	277.3	41231	10.8	715.7	41669	11.3	614.3	42251	9.7	430.7	49451	6.8
swath	132806.3	955499	7200.0	131192.6	1032573	7200.0	131192.7	1090076	7200.0	131195.7	1120004	7200.0	131547.9	1526206	7200.0
timtab1	10653.6	576355	331.3	18586.8	1024342	535.7	20956.3	1127633	610.5	20065.1	1044816	578.8	10270.7	826333	292.0
timtab2	323569.0	8355916	7200.0	323567.1	8320061	7200.0	323567.7	7890308	7200.0	323567.2	11934864	7200.0	323566.6	13961074	7200.0
tr12-30	661.8	1052649	1724.2	632.6	1371401	2138.7	986.9	1155493	1850.1	554.8	1461519	1985.9	437.4	1072845	1082.6
unitcal_7	3786.0	185919	7200.0	3771.3	186486	7200.0	278109.7	106121	6253.8	148395.5	58245	3666.5	205323.5	112703	4696.1
vpphard	720000.0	5160	7200.0	720000.0	29923	7200.0	720000.0	53957	7200.0	720000.0	41002	7200.0	720000.0	58045	7200.0
zib54-UUE	81108.9	243167	2975.5	90967.0	294299	3264.6	94390.5	315556	3485.8	134064.3	659977	4891.7	66689.8	347875	2336.9

BDD-Guided Clause Generation

Brian Kell[1]([✉]), Ashish Sabharwal[2], and Willem-Jan van Hoeve[3]

[1] Department of Mathematical Sciences, Carnegie Mellon University,
Pittsburgh, PA 15213, USA
bkell@cmu.edu
[2] Allen Institute for Artificial Intelligence, Seattle, WA 98103, USA
AshishS@allenai.org
[3] Tepper School of Business, Carnegie Mellon University, Pittsburgh, PA 15213, USA
vanhoeve@andrew.cmu.edu

Abstract. Nogood learning is a critical component of Boolean satisfiability (SAT) solvers, and increasingly popular in the context of integer programming and constraint programming. We present a generic method to learn valid clauses from exact or approximate binary decision diagrams (BDDs) and resolution in the context of SAT solving. We show that any clause learned from SAT conflict analysis can also be generated using our method, while, in addition, we can generate stronger clauses that cannot be derived from one application of conflict analysis. Importantly, since SAT instances are often too large for an exact BDD representation, we focus on BDD relaxations of polynomial size and show how they can still be used to generated useful clauses. Our experimental results show that when this method is used as a preprocessing step and the generated clauses are appended to the original instance, the size of the search tree for a SAT solver can be significantly reduced.

Introduction

Solvers for Boolean satisfiability (SAT) have become increasingly powerful in recent decades and can now be used to solve large-scale instances involving millions of variables and constraints. Much of the success of modern SAT solvers stems from their ability to quickly learn new constraints from infeasible search states via conflict-directed clause learning (CDCL). Conflict analysis has also been applied in the context of mixed-integer programming (MIP) [1,17] and constraint programming (CP) [15,21,24] as "nogood" learning. In the context of constraint programming, nogood learning techniques have been proposed for specific combinatorial structures that arise from global constraints. For example, Downing et al. [12] study nogoods for global constraints that can be represented as a network flow. However, it remains a challenge to learn effective nogoods for MIP and CP solvers in a more generic context.

In this paper we introduce a generic approach for learning nogoods from decision diagrams, both exact and approximate. Decision diagrams provide a compact representation of the solution space for discrete optimization problems, and have been used to improve constraint propagation [2,11,14] and to derive optimization

© Springer International Publishing Switzerland 2015
L. Michel (Ed.): CPAIOR 2015, LNCS 9075, pp. 215–230, 2015.
DOI: 10.1007/978-3-319-18008-3_15

bounds [6,8]. This work proposes an extension of the use of such decision diagrams to learn nogoods. We specifically focus on clause learning in the context of SAT solving, being perhaps the most general form of nogood learning.

The architecture of today's SAT solvers, combining unit propagation with rapid restarts and CDCL, focuses on techniques with very low overhead and maximizes the number of search nodes that can be processed per second. While this has clearly been beneficial, the unit propagation inference performed by SAT solvers is arguably limited in strength. We therefore investigate a way to generate clauses that are *stronger* than those currently derived from unit propagation and CDCL. We show that these clauses, when added to the original formula, can substantially reduce the search tree size.

Our clause generation scheme is based on a novel application of binary decision diagrams (BDDs) to represent a given propositional formula. In contrast to conventional BDD construction methods that are context-agnostic, we associate a *meaning* with each node of the BDD: the set of clauses that are not yet satisfied. This allows us to apply a top-down compilation scheme [8] in which node equivalence is defined by the set of unsatisfied clauses. This node information provides a sufficient condition for efficiently creating BDDs (that are not necessarily reduced, however).

The key observation in our work is that the BDD node information, for those nodes that do not lead to a satisfying solution, can also be used to generate new clauses. Such clauses can be viewed as "nogoods" that forbid the solver to visit the associated search states. Since a node in a BDD can represent multiple partial assignments, a single nogood generated in this way is as strong as multiple nogoods derived from these separate partial assignments.

We formally characterize the strength of the clauses generated by our method. For example, we show that our clauses can indeed be stronger than one invocation of traditional conflict analysis. We also show the equivalence of our approach to regular and ordered resolution, which are specific restricted forms of resolution proofs.

BDDs that exactly represent a given CNF formula are well known to grow exponentially large in general. This has significantly limited the success of BDD-based techniques for SAT solving. To circumvent this limitation, we explore ways to apply our method to *relaxed* and *restricted* BDDs that represent a superset and subset, respectively, of all solutions instead [7,8]. These approximate BDDs are created by merging non-equivalent nodes so as to respect a given limit on the size of the BDD. We show that the clauses derived from relaxed (or restricted) BDDs are still valid and can be computed efficiently.

We report results of computational experiments performed to evaluate the strength of our generated clauses in practice. We show that, for certain problem classes, our clauses can reduce the search tree size considerably. Interestingly, the solving time is not always reduced accordingly; we attribute this behavior to the length and number of our generated clauses. Nonetheless, the qualitative strength of our clauses demonstrates a great potential for inclusion in SAT solvers, and we propose several suggestions for doing so in the conclusion.

Binary Decision Diagrams

A *binary decision diagram* (BDD) [10,18,19,25] is an edge-labeled acyclic directed multigraph whose nodes are arranged in $n+1$ layers $L_1, \ldots, L_{n+1}$. The layer L_1 consists of a single node, called the *root*. In this paper, every edge in the BDD is directed from a node in layer L_i to a node in layer L_{i+1}. Each node in layers $L_1, \ldots, L_n$ has two outgoing edges, one labeled "true" and the other labeled "false." There are two nodes in layer L_{n+1}, called the *sinks* or *terminals*; one of them, labeled $\top$, is the *true sink*, while the other, labeled $\bot$, is the *false sink*.

A BDD represents a Boolean function f defined on variables $x_1, \ldots, x_n$ as follows. The layers $L_1, \ldots, L_n$ correspond respectively to the variables $x_1, \ldots, x_n$. A path from the root to a sink corresponds to values of these variables; a "true" edge from a node in layer L_i to a node in layer L_{i+1} corresponds to $x_i = 1$, while a "false" edge corresponds to $x_i = 0$. If the path corresponding to the values of $x_1, \ldots, x_n$ ends at the true sink, then $f(x_1, \ldots, x_n) = 1$; otherwise the path ends at the false sink, and $f(x_1, \ldots, x_n) = 0$.

BDDs for SAT

An instance of the Boolean satisfiability (SAT) problem is a propositional formula on variables $x_1, \ldots, x_n$, expressed in conjunctive normal form (CNF), that is, as a conjunction of disjunctions of *literals*, where a literal is a variable x_i or its negation $\bar{x}_i$. Each of these disjunctions is called a *clause*. Because logical conjunction and disjunction are commutative, associative, and idempotent, we may view a SAT instance as a set of clauses, each of which is a set of literals. The objective is to determine whether there exists an assignment of Boolean values to the variables that simultaneously satisfies every clause.

Let I be a SAT instance on the variables $x_1, \ldots, x_n$, and let $\mathcal{S}$ denote the set of satisfying assignments to these variables. Let B be a BDD defined on the variables $x_1, \ldots, x_n$, and let $\mathcal{B}$ denote the set of assignments to these variables represented by B (that is, for which the Boolean function defined by B is true). If $\mathcal{B} = \mathcal{S}$, $\mathcal{B} \supseteq \mathcal{S}$, or $\mathcal{B} \subseteq \mathcal{S}$, then B is said to be an *exact BDD*, a *relaxed BDD*, or a *restricted BDD* for I, respectively [2,7,8,13].

A path in a BDD from the root to a node in the layer L_{i+1} represents a *partial assignment*, i.e., an assignment $y \in \{0,1\}^i$ of values to the variables $x_1, \ldots, x_i$. Let $\mathcal{S}(y)$ denote the set of *satisfying completions* of this partial assignment, that is, $\mathcal{S}(y) = \{ z \in \{0,1\}^{n-i} : (y, z) \text{ is feasible} \}$. If y and y' are partial assignments with $\mathcal{S}(y) = \mathcal{S}(y')$, then we say that y and y' are *equivalent*. Note that in an exact BDD all paths from the root to a fixed node v represent equivalent partial assignments, and conversely if two partial assignments y and y' are equivalent then the paths in an exact BDD that correspond to y and y' can lead to the same node.

In the literature, BDDs are commonly required to be *reduced*, in the sense that any two equivalent partial assignments must be represented by the same node. The BDDs in this paper are not necessarily reduced.

In general, determining whether two partial assignments are equivalent is NP-hard for the SAT problem. However, we can sometimes determine that two partial assignments are equivalent by associating partial assignments with "states" [14, 16]. A *state function* for layer i is a map σ_i from the set $\{0, 1\}^{i-1}$ of partial assignments at layer i into some set S_i of *states*, such that $\sigma_i(y) = \sigma_i(y')$ implies $S(y) = S(y')$. In other words, two partial assignments that lead to the same state have the same set of satisfying completions.

Behle [5] described a top-down algorithm for the construction of threshold BDDs, which are exact representations of solution sets of instances of 0–1 knapsack problems. A general algorithm for a top-down, layer-by-layer construction of a multivalued decision diagram (MDD), which is similar to a BDD except that the labels of the edges may come from any set, was given by Bergman et al. [8]. This algorithm works by maintaining state information for each node, computing the resulting state for each outgoing edge, and reusing nodes (i.e., pointing two edges at the same node) when the resulting states are the same.

To apply this top-down algorithm for the construction of a BDD from a SAT instance, we define $\sigma_i(y)$ for a partial assignment $y = \{y_1, \ldots, y_{i-1}\}$ to be the set of clauses in the instance that are not satisfied by the assignments $x_1 = y_1, \ldots, x_{i-1} = y_{i-1}$. Observe that if two partial assignments at layer i have the same set of unsatisfied clauses, then they have the same set of feasible completions, so this is indeed a state function. The state of the root node is the full set of clauses in the instance, and the state of a child node is formed from the state of its parent by removing all clauses that are satisfied by the variable assignment corresponding to the edge from the parent to the child.

Example 1. Consider a graph coloring problem on a complete graph with three vertices. Vertices 1 and 2 can be colored 0 or 1, while vertex 3 can be colored 0, 1, or 2. All nodes must be colored differently. We introduce variable x_1 for vertex 1, where $\overline{x}_1$ represents color 0 and x_1 represents color 1. Likewise we introduce x_2 for vertex 2. For vertex 3, we introduce three variables x_3, x_4, and x_5 for colors 0, 1, and 2, respectively. Here a positive literal represents that we choose that color, while its negation represents that we do not choose that color (e.g., $\overline{x}_3$ means that vertex 3 is not colored 0). We can formulate this problem as the following SAT instance with 11 clauses:

$$(1)\ x_3 \vee x_4 \vee x_5 \qquad\qquad (7)\ \ \overline{x}_1 \vee \overline{x}_2$$
$$(2)\ \overline{x}_3 \vee \overline{x}_4 \vee \overline{x}_5 \qquad\qquad (8)\ x_1 \vee x_4 \vee x_5$$
$$(3)\ \ \ \overline{x}_3 \vee \overline{x}_4 \qquad\qquad (9)\ \overline{x}_1 \vee x_3 \vee x_5$$
$$(4)\ \ \ \overline{x}_3 \vee \overline{x}_5 \qquad\qquad (10)\ x_2 \vee x_4 \vee x_5$$
$$(5)\ \ \ \overline{x}_4 \vee \overline{x}_5 \qquad\qquad (11)\ \overline{x}_2 \vee x_3 \vee x_5$$
$$(6)\ \ \ x_1 \vee x_2$$

The constructed BDD, using the lexicographic variable ordering, is presented in Figure 1. The state of each node is the set of (indices of) clauses that have not been satisfied by any path from the root to that node. "True" edges are drawn as solid lines, and "false" edges are drawn as dashed lines. Infeasible nodes, that is, nodes from which no path leads to the true sink, are shaded gray.

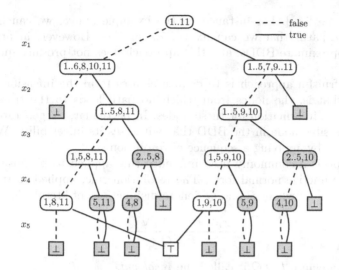

Fig. 1. The exact BDD for the example. The false sink is drawn multiple times for clarity.

Approximate BDDs

In general, exact BDDs can be of exponential size, so the construction of an exact BDD may not be practical. For this reason, it is useful to consider BDDs that represent relaxations or restrictions of the SAT instance. Such BDDs are called *approximate BDDs* because their structure approximates the structure of the exact BDD.

MDDs of limited width were proposed by Andersen et al. [2] to reduce space requirements. In this approach, the MDD is constructed in a top-down, layer-by-layer manner, and whenever a layer of the MDD exceeds some predetermined value W an approximation operation is applied to reduce its size to W before constructing the next layer. One way to perform this approximation is to use a relaxation (or restriction) operation $\oplus$ defined on the states of nodes so that, given nodes v and v', the state given by state$(v) \oplus$ state(v') is a "relaxation" (or "restriction") of both state(v) and state(v'). A subset of nodes in a layer can be merged into a single node by applying this operator to obtain a state for the new node, and this is repeated until the size of the layer is reduced to W [8,14,16].

In our case, the state of a node is a set of unsatisfied clauses, and so the appropriate relaxation operation is the intersection operation (and the appropriate restriction operation is the union operation).

Clause Generation with BDDs

We propose the use of a BDD representation of a SAT instance to generate clauses. One simple way to deduce clauses from a BDD is to project the variable assignments along the satisfying paths in a BDD and to look for variables whose values must be fixed. For instance, in the example above, we can infer from both feasible paths that we can fix $\overline{x}_3$, $\overline{x}_4$, and x_5. However, in practice we must use approximate BDDs, and this approach does not produce much useful information.

A more fruitful approach is to deduce clauses from the infeasible nodes of the BDD (that is, the nodes from which no path leads to the true sink) by using the state information for these nodes. In particular, we generate a clause for each infeasible node in the BDD that witnesses its infeasibility. We do this systematically by applying a sequence of resolution steps.

Resolution is a commonly used inference rule applied to propositional formulas in conjunctive normal form. The resolution rule, applied to two clauses $x_i \vee P$ and $\overline{x}_i \vee Q$, where P and Q denote disjunctions of literals, is

$$\frac{x_i \vee P \qquad \overline{x}_i \vee Q}{P \vee Q}.$$

The resulting clause $P \vee Q$ is called the *resolvent*.

During the top-down construction of a BDD for a SAT instance, infeasibility of a state is detected when an unsatisfied clause contains no variable corresponding to a lower layer of the BDD. When this occurs, we choose one such clause as a witness of the infeasibility of the corresponding node.

After the BDD construction is complete, we perform a single bottom-up pass to identify all infeasible nodes and generate a witness clause for each. For an infeasible node v in layer L_i, we generate a witness clause as follows:

- If one of the child states has a witness clause that does not contain the variable x_i, then choose this clause as the witness clause for v.
- Otherwise, one child has a witness clause containing x_i and the other has a witness clause containing $\overline{x}_i$, so apply the resolution rule to these two clauses with respect to the variable x_i and use the resolvent as the witness clause for v.

At the end, we output the witness clauses for all roots of maximal infeasible subtrees of the BDD.

Example 2. Continuing the graph-coloring example from earlier, consider the infeasible subtree rooted at the node with state $\{2, 3, 4, 5, 8\}$ in layer L_4 in Figure 1. This subtree is redrawn in Figure 2. Setting $x_4 = 0$ satisfies clauses 2, 3, and 5, so the "false" child (i.e., the child along the "false" edge) has state $\{4, 8\}$. However, from this node, setting $x_5 = 0$ means that clause 8 cannot be satisfied, and setting $x_5 = 1$ means that clause 4 cannot be satisfied. Therefore, neither of the children of the node with state $\{4, 8\}$ is feasible, and we have a witness of the infeasibility of each: clause 8, $x_1 \vee x_4 \vee x_5$, for the "false" child, and clause 4, $\overline{x}_3 \vee \overline{x}_5$, for the "true" child.

Likewise, returning to the node with state $\{2, 3, 4, 5, 8\}$, if we set $x_4 = 1$ then clause 3 cannot be satisfied, so clause 3, $\overline{x}_3 \vee \overline{x}_4$, is a witness of the infeasibility of this child.

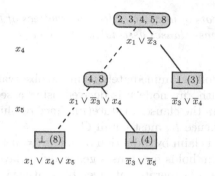

Fig. 2. Witness clauses generated from the infeasible subtree rooted at the node with state $\{2, 3, 4, 5, 8\}$ in layer L_4

Now, in our bottom-up pass, we first determine that the node with state $\{4, 8\}$ is infeasible. Both of its child nodes have witness clauses that contain the variable x_5, so we apply the resolution rule to these two witness clauses with respect to x_5 to obtain the clause $x_1 \vee \overline{x}_3 \vee x_4$, which is a witness of the infeasibility of the node with state $\{4, 8\}$. Likewise, the node with state $\{2, 3, 4, 5, 8\}$ is infeasible, so we apply the resolution rule to these two witness clauses with respect to x_4 to obtain the clause $x_1 \vee x_3$.

Since the node with state $\{2, 3, 4, 5, 8\}$ is the root of a maximal infeasible subtree of the BDD, we produce the clause $x_1 \vee \overline{x}_3$ as output. This is a valid clause for the original SAT instance.

In a similar way, we generate the witness clause $x_2 \vee \overline{x}_3$ for the node with state $\{2, 3, 4, 5, 10\}$ in layer L_4 in the BDD in Figure 1.

Characterization of Generated Clauses

Let us formally define a clause C to be *valid* for a propositional formula F if $F \models C$, i.e., F logically entails C. In other words, $F \wedge C$ has the same set of solutions as F itself. We begin with a few properties of witness clauses generated using the BDD method.

Theorem 1. *Let F be a CNF formula and B be a top-down exact, relaxed, or restricted BDD for F constructed as described above. Then:*

1. *Every witness clause generated from B is valid for F.*
2. *The set of variables in every witness clause generated at layer L_{i+1} is a subset of $\{x_1, x_2, \ldots, x_i\}$.*
3. *If B is an exact or relaxed BDD, the witness clause C generated for a node v of B is falsified by the partial assignment corresponding to every path from the root of B to v. In particular, C does not contain any variable that appears both negatively and positively in paths from the root to v.*
4. *If B is an exact or relaxed BDD, the witness clause C associated with any infeasible node v of B witnesses the infeasibility of v.*

5. *Let U denote the roots of maximal infeasible subtrees of B. If B is exact, then the set G of all witness clauses associated with nodes $v \in U$ is a reformulation of F.*

Proof. The first claim follows immediately from the observation that the witness clause C associated with any node v is derived using a sequence of resolution operations starting from the clauses in state(v). Since resolution is a sound proof system, state(v), and hence F, must entail C.

We prove the second claim by induction on i. For $i = n+1$, the claim trivially holds. Suppose the claim holds for clauses generated at layer L_i, with $i > 1$. By construction, any clause C generated at layer L_{i-1} either is identical to a clause generated at layer L_i, in which case it does not contain the variable x_{i-1}, or else is obtained by resolving two clauses at layer L_i on the variable x_i. In either case, by the induction hypothesis, the variables appearing in C must be a subset of $\{x_1, x_2, \ldots, x_{i-2}\}$.

To prove the third claim, we recall from the definition of the state function that when B is exact or relaxed, the partial assignment y corresponding to any path from the root to v does not satisfy any clause in state$(v) = F_v \subseteq F$. For the sake of contradiction, suppose ℓ is a literal of the witness clause C that is satisfied by y. Since C is derived by applying resolution steps to clauses in F_v, the literal ℓ must appear in at least one clause C' of F_v. Since y satisfies ℓ, it would also satisfy C', a contradiction. Hence, C must be falsified by y. Finally, if C contained a literal ℓ that appears positively and negatively in partial assignments y and y' corresponding to two paths from root to v, then C would clearly be satisfied by at least one of y and y', which, as proved above, cannot happen. Hence, C must not contain any such literal.

For proving the fourth claim, we use the above property that when B is exact or relaxed, the partial assignment y corresponding to any path from the root to v does not satisfy C. Suppose y could be extended to a full assignment (y, z) that satisfies F. Then z must satisfy all clauses in state$(v) = F_v$ as these clauses, by definition of the state function, are not satisfied by y. Since C is derived from F_v by applying a sequence of resolution operations, z must then also satisfy C. However, as observed above, C is a subset of $\{x_1, x_2, \ldots, x_{i-1}\}$, where L_i is the layer containing v, and hence C cannot possibly be satisfied by z. This proves that y cannot be extended to a full assignment satisfying F, and that the generated clause C witnesses this fact as well as the infeasibility of v.

Lastly, when B is exact, we argue that the set G of witness clauses associated with roots of maximal infeasible subtrees of B is logically equivalent to F. If y is a solution to F, then y must satisfy all witness clauses as these clauses are entailed by F. Hence y must also satisfy G. On the other hand, if y is not a solution to F, then let y' be the partial assignment corresponding to the path in B associated with y but truncated at the root v' of a maximal infeasible subtree. By the third property above, y' (and hence y) must falsify the clause C' associated with v', and hence falsify G. It follows that F and G have the same set of solutions and thus G is a reformulation of F. □

In the remainder of this section, we explore how BDD-guided clause generation relates to propagation and inference techniques used in today's SAT solvers. To make this connection precise, we recall the notion of *absorbed clauses* [3,22]. A clause C is said to be absorbed by a CNF formula F if for every literal $\ell \in C$, performing unit propagation[1] on F starting with all literals of C except ℓ set to false either infers ℓ or infers a conflict. The intuition here is that C is absorbed by F if F and $F \wedge C$ have identical entailment power with respect to unit propagation, i.e., whatever one can derive from $F \wedge C$ using unit propagation one can also derive from F itself.

Pipatsrisawat and Darwiche [22] showed that the conflict-directed clause learning (CDCL) mechanism in SAT solvers always produces clauses that are not absorbed by the current theory, that is, by the set of initial clauses of F and those learned thus far during the search. As we show next, this property also holds for clauses generated by the BDD method applied to F, as long as we ignore any states at which unit propagation already identifies a conflict.

More formally, given a BDD B, let us define a *unit-propagated BDD*, denoted B_{up}, as the one obtained by removing from B all nodes v such that unit propagation on state(v) results in a conflict. From a practical standpoint, such nodes can be easily identified in linear time and discarded.

Theorem 2. *Let F be a CNF formula and B be a top-down exact, relaxed, or restricted BDD for F constructed as described above. Let C be the witness clause for the root v of any maximal infeasible subtree in B_{up}. Then C is not absorbed by F.*

Proof. We first show that setting all literals of C to false and performing unit propagation does not result in a conflict. From Theorem 1, the partial assignment y corresponding to the path from the root of B_{up} to y falsifies C. Yet, by design, unit propagation on state(v) does not result in a conflict. Therefore, it must be the case that unit propagation on F starting with the partial assignment y does not result in a conflict. Since y falsifies all literals of C, we infer that falsifying all literals of C and performing unit propagation does not result in a conflict.

To finish the argument that C is not absorbed by F, we next show that there exists a literal ℓ in C such that setting all literals of C except ℓ to false does not allow unit propagation to infer ℓ. Since v is the root of a maximal infeasible subtree in B_{up}, it must be the case that v has a sibling node v' that is *not* identified as being infeasible. Let u be the common parent of v and v'. The witness clause C associated with v must include a literal ℓ corresponding to the branching variable associated with the layer of u. We will use ℓ to demonstrate that C is not absorbed by F.

Suppose, for the sake of contradiction, that setting all literals of C other than ℓ to false and performing unit propagation infers ℓ. Consider the partial

---

[1] *Unit propagation* on a CNF formula is the process of identifying, if there is one, a clause that contains only one literal ℓ, setting ℓ to true, simplifying the formula by removing $\bar{\ell}$ from all clauses and removing all clauses containing ℓ, and repeating. Unit propagation is said to result in a *conflict* if it generates an empty clause.

assignment z corresponding to the path from the root of B_{up} to u. This partial assignment z differs from the partial assignment y identified above in only the literal ℓ. It must then be the case that z falsifies all literals of C except for ℓ, and therefore unit propagation on state(u) starting with z must infer ℓ. In other words, there exists a clause C' in state(u) such that $\ell \in C'$ and z, after unit propagation, falsifies all literals of C' other than ℓ. This, however, implies that C' is also in state(v'), the state corresponding to the sibling v' of v, and further that unit propagation on state(v') must falsify C', resulting in a conflict. This, however, contradicts the fact that v' was not identified as an infeasible node in B_{up}.   □

This establishes that our clause generation approach effectively produces clauses that provide useful information not already captured by unit propagation inference on F.

While a series of potentially exponentially many applications of the CDCL mechanism can eventually let the solver learn any clause entailed by F (including the empty clause in case F is unsatisfiable), we show below that any clause that it can learn with *one* application of conflict analysis starting from a clause set F is a special case of the BDD-generated clauses starting from F. This holds for any clause learning scheme employed by the solver to choose a cut in the underlying conflict graph.[2]

The proof of this claim uses properties of a few different restrictions of general resolution which we briefly recapitulate. A *tree-like resolution* is one where no clause, other than the initial clauses of F, is used in more than one resolution step. A *regular resolution* is one where no variable is resolved upon more than once in any root-to-leaf path. Finally, an *ordered resolution* is one where the order of variables resolved upon is identical across all root-to-leaf paths.

Theorem 3. *For any clause C learned from one application of SAT conflict analysis on F using any clause learning scheme, there exists a variable ordering under which a top-down approximate BDD of width at most $2^{|C|}$ for F generates a clause $C' \subseteq C$.*

Proof. To prove this, we use the resolution-based characterization of CDCL clauses [4], namely, the CDCL derivation of a clause C starting from F and using any clause learning scheme can be viewed as a very simple form of resolution derivation that has a ladder-like structure. More formally, the derivation τ of C is simultaneously a tree-like, regular, linear, and ordered resolution derivation from the clauses in F. This means that each intermediate clause C_{j+1} in τ is obtained by resolving C_j with a clause of F and that the sequence σ of variables resolved upon in τ consists of all distinct variables.

We can use BDDs to derive from F a clause C' that, together with F, absorbs C. To construct such a BDD B, we use as the top-down (partial) variable order first the variables that appear in C (in any order) followed by variables in the reverse order of σ. The first $|C|$ variables result in a BDD of width at

---

[2] The specifics of the SAT conflict analysis terminology are not critical here. The interested reader is referred to relevant surveys such as by Marques-Silva et al. [20].

most $2^{|C|}$. Let v be the node of B in the layer $L_{|C|+1}$ at which all literals of C are falsified. When expanding B from v, the ladder-like structure of τ guarantees that at least one branch on the variables in σ can be labeled directly by a clause of F that is falsified. The corresponding lower part of B starting at v is thus of width 1. For the remaining $2^{|C|} - 1$ nodes of B in the layer $L_{|C|+1}$, we construct an approximate lower portion of the BDD such that the overall width does not increase. This makes the overall width of B be $2^{|C|}$.

While B may have several infeasible nodes, the node v in the layer $L_{|C|+1}$ is guaranteed by the derivation τ to be infeasible. Recall that the path p from the root of B to v falsifies C. Consider the node v' that is the root of the maximal infeasible subtree of B that contains v. Let C' be the BDD-generated clause witnessing the infeasibility of v'. By Theorem 1, C' must be falsified by the path p' from the root of B to v'. Note that p' is a sub-path of p. By construction, C contains all $|C|$ literals mentioned along p, while, by Theorem 1, C' contains a subset of the literals mentioned along p' and hence along p. It follows that $C' \subseteq C$. $\qquad\square$

The above reasoning can be extended to construct an *exact* BDD that generates a subclause of C. However, the width of such a BDD will depend not only on $|C|$ but also on the number of resolution steps involved in conflict analysis during the derivation of C.

Theorem 4. *Clauses generated by applying the BDD method to F correspond to regular and ordered resolution derivations starting from the clauses of F.*

Proof. It is easily seen that the resolution operations performed during clause generation from a BDD respect, by construction, the restrictions of being regular and ordered. Hence, any BDD generated clause C can be derived using regular and ordered resolution starting from F.

On the other hand, let τ be any regular and ordered resolution derivation of C starting from F. An argument similar to the one in the proof of Theorem 3 can be used to show that there exists a natural variable order (namely, first branch on the variables of C, then follow the top-down variable order imposed by τ) under which the top-down BDD B for F contains a node v such that the path from the root of B to v falsifies all literals of C. As before, witness clauses for B may not directly include C as is, but the witness clause C' associated with the root of the maximal infeasible subtree of B containing v would be a subclause of C. $\qquad\square$

We recall again the resolution-based characterization of CDCL clauses, namely, those that can be derived using tree-like, regular, linear, and ordered resolution. This results in linear-size resolution derivations and thus forms a strict subset of all possible derivations that are regular and ordered, but not necessarily tree-like and linear. The above theorem therefore implies the following:

Corollary 1. *There exist BDD-generated clauses that cannot be derived using one application of SAT conflict analysis.*

Implementation and Experimental Results

We implemented the clause generation algorithm described above in C++, as a program called Clausegen. Several implementation decisions needed to be considered.

The variable ordering used in a BDD can have a very significant effect on the size of the BDD (and consequently the quality of an approximate BDD). Unfortunately, determining the optimal variable ordering is very difficult; in general, the problem of determining whether a given variable ordering of a BDD can be improved is NP-complete [9]. For our implementation, we use a simple heuristic to determine the variable ordering: each variable is assigned a score, computed as the quotient between the number of clauses containing the variable and the average arity of those clauses, and the variables are sorted in decreasing order according to this score, so that higher-scoring variables (that is, variables that appear in many mostly short clauses) correspond to layers nearer the top of the BDD.

The construction of a relaxed BDD via merging also requires a rule for determining which nodes to merge in a layer that exceeds the maximum width. Since unsatisfied clauses lead to infeasibility, and our method generates clauses from infeasible subtrees, the following merging rule is used: if a constructed layer exceeds the maximum width W, sort the nodes by the number of unsatisfied clauses in their states, preserve the $W - 1$ nodes with the greatest number of unsatisfied clauses, and merge the other nodes into a single node. (The state of the resulting node is the intersection of the states of the nodes that were merged.) Merging rules similar to this one have been applied before in the context of optimization and scheduling, for example by Cire and van Hoeve [11].

To demonstrate our method, we considered SAT instances produced from randomly generated bipartite graph matching problems, with 15 vertices on each side, in which a random subset of 10 vertices on one side is matched with only 9 vertices on the other side, so that the graph fails to satisfy Hall's condition, thereby making the SAT instance unsatisfiable. We preprocessed the instance with SatELite 1.0 (using the +pre option) and used Minisat 2.2.0 as the SAT solver (with -rnd-freq=0.01). Because Minisat uses a nondeterministic algorithm, it was run 20 times for each test with different random seeds, and the results were averaged. The experiments were run on an Intel Xeon E5345 at 2.33 GHz with 24 GB of RAM running Ubuntu 12.04.5.

For a representative instance of this type, with 225 variables and 748 clauses (80 variables and 405 clauses after preprocessing), Minisat made 864,930 decisions and encountered 714,625 conflicts on average.

Figure 3 shows the results of appending the clauses produced by Clausegen before the instance is given to Minisat. As the maximum BDD width is increased from 10 to 10,000, thus yielding more accurate approximate BDDs, the numbers of decisions and conflicts encountered by Minisat decrease. The clauses generated at BDD width 10,000 produced an improvement in these metrics by over 75% in comparison with the original instance: Minisat averaged 212,158 decisions and 178,101 conflicts.

However, we do not see a corresponding improvement in the running time of Minisat. The stacked area plot in Figure 3 shows the running time of Clausegen and Minisat as the BDD width is increased. On the original instance, Minisat required an average of 7.83 s; this time increased to 17.65 s when the clauses generated at BDD width 10,000 were added. The number of generated clauses increases linearly with the BDD width, from 12 clauses at width 10 to 9745 clauses at width 10,000. The clauses generated at width 10,000 have an average length of 11.8, compared to an average length of 2.1 in the original instance.

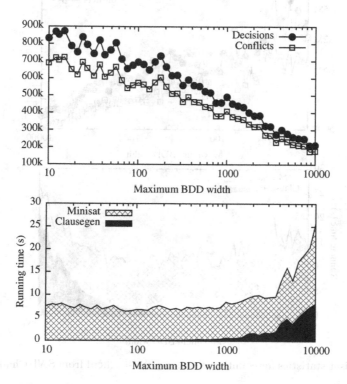

Fig. 3. Minisat statistics for an unsatisfiable bipartite matching instance

Figure 4 shows our results for another instance, counting-clqcolor-unsat-set-b-clqcolor-08-06-07.sat05-1257.reshuffled-07.cnf, from the SAT Challenge 2012 Hard Combinatorial SAT+UNSAT benchmark instances [23]. This instance has 132 variables and 1527 clauses of average length 2.9 (117 variables and 1599 clauses of average length 4.3 after preprocessing with SatELite) and is also unsatisfiable; it represents a graph coloring instance with a hidden clique that is larger than the number of colors available. Minisat averaged 2,072,107 decisions and 1,511,029 conflicts for the original instance, taking 14.18 s on average. When the 3255 clauses of average length 8.7 produced by Clausegen at BDD width 10,000 were added, the average numbers of decisions and conflicts decreased to 713,718

and 515,514, respectively, and the average running time of Minisat decreased to
6.34 s. The minimum total running time of Clausegen and Minisat together was
achieved at a BDD width of 464; Clausegen took 0.57 s to generate 340 clauses
of average length 8.7, and Minisat averaged 1,351,691 decisions and 972,674 conflicts, taking 9.14 s on average to solve the instance, for a total average solving
time of 9.71 s (an improvement of 31.5% over the original instance).

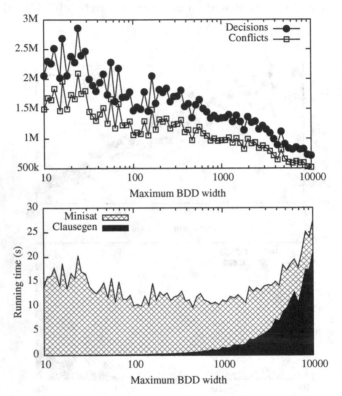

Fig. 4. Minisat statistics for counting-clqcolor-unsat-[. . .].cnf from SAT Challenge 2012

It may be that not all of the generated clauses are necessary. Preliminary
experiments involving the selection of random subsets of the generated clauses
appear to indicate that some subsets are significantly more helpful than others.
A heuristic to select useful subsets of the generated clauses may allow a decrease
in running time to match the decreases in the numbers of decisions and conflicts.

Conclusion

We presented a new algorithm that uses BDDs and resolution to generate valid
clauses from a SAT instance. This algorithm can use approximate BDDs for
instances that are too large for an exact BDD. We compared the strength of

our method to that of SAT conflict analysis and showed that our method can generate strictly stronger clauses than a single application of SAT conflict analysis. Our experimental results show that concatenating these generated clauses to the original instance can significantly reduce the size of the search tree for a SAT solver.

For a practical implementation of our method, we propose the following techniques to improve computational efficiency. First, initial experimentation has shown that not all generated clauses are equally effective. We therefore suggest the development of a heuristic to select and add only a small subset. Second, a large formula may be decomposable into subformulas that are each representable effectively by a BDD. It seems natural to make such a decomposition based on structural properties of the formula (e.g., properties of the constraint graph). Third, when a SAT solver appears to be making very little progress, and the number of remaining free variables is limited, we can interrupt the search and give the rest of the formula to a BDD to generate clauses conditional on the partial assignment represented by the last search state.

References

1. Achterberg, T.: Conflict analysis in mixed integer programming. Discrete Optimization **4**(1), 4–20 (2007)
2. Andersen, H.R., Hadzic, T., Hooker, J.N., Tiedemann, P.: A constraint store based on multivalued decision diagrams. In: Bessière, C. (ed.) CP 2007. LNCS, vol. 4741, pp. 118–132. Springer, Heidelberg (2007)
3. Atserias, A., Fichte, J.K., Thurley, M.: Clause-learning algorithms with many restarts and bounded-width resolution. Journal of Artificial Intelligence Research **40**, 353–373 (2011)
4. Beame, P., Kautz, H., Sabharwal, A.: Understanding and harnessing the potential of clause learning. Journal of Artificial Intelligence Research **22**, 319–351 (2004)
5. Behle, M.: On threshold BDDs and the optimal variable ordering problem. Journal of Combinatorial Optimization **16**, 107–118 (2008)
6. Bergman, B., Cire, A., van Hoeve, W.-J., Hooker, J.: Optimization bounds from binary decision diagrams. INFORMS Journal on Computing **26**(2), 253–268 (2014)
7. Bergman, D., Cire, A.A., van Hoeve, W.-J., Yunes, T.: BDD-based heuristics for binary optimization. Journal of Heuristics **20**(2), 211–234 (2014)
8. Bergman, D., van Hoeve, W.-J., Hooker, J.N.: Manipulating MDD relaxations for combinatorial optimization. In: Achterberg, T., Beck, J.C. (eds.) CPAIOR 2011. LNCS, vol. 6697, pp. 20–35. Springer, Heidelberg (2011)
9. Bollig, B., Wegener, I.: Improving the variable ordering of OBDDs is NP-complete. IEEE Transactions on Computers **45**(9), 993–1002 (1996)
10. Bryant, R.E.: Graph-based algorithms for Boolean function manipulation. IEEE Transactions on Computers **35**(8), 677–691 (1986)
11. Cire, A., van Hoeve, W.-J.: Multivalued decision diagrams for sequencing problems. Operations Research **61**(6), 1411–1428 (2013)
12. Downing, N., Feydy, T., Stuckey, P.: Explaining flow-based propagation. In: Beldiceanu, N., Jussien, N., Pinson, É. (eds.) CPAIOR 2012. LNCS, vol. 7298, pp. 146–162. Springer, Heidelberg (2012)

13. Hadzic, T., Hooker, J., O'Sullivan, B., Tiedemann, P.: Approximate compilation of constraints into multivalued decision diagrams. In: Stuckey, P.J. (ed.) CP 2008. LNCS, vol. 5202, pp. 448–462. Springer, Heidelberg (2008)
14. Hoda, S., van Hoeve, W.-J., Hooker, J.: A systematic approach to MDD-based constraint programming. In: Cohen, D. (ed.) CP 2010. LNCS, vol. 6308, pp. 266–280. Springer, Heidelberg (2010)
15. Katsirelos, G.: Nogood Processing in CSPs. PhD thesis, University of Toronto (2008)
16. Kell, B., van Hoeve, W.-J.: An MDD approach to multidimensional bin packing. In: Gomes, C., Sellmann, M. (eds.) CPAIOR 2013. LNCS, vol. 7874, pp. 128–143. Springer, Heidelberg (2013)
17. Kilinç-Karzan, F., Nemhauser, G., Savelsbergh, M.: Information-based branching schemes for binary linear mixed integer problems. Mathematical Programming Computation 1(4), 249–293 (2009)
18. Knuth, D.E.: The Art of Computer Programming, vol. 4, fascicle 1: Bitwise Tricks & Techniques; Binary Decision Diagrams. Addison-Wesley (2009)
19. Lee, C.Y.: Representation of switching circuits by binary-decision programs. Bell System Technical Journal 38(4), 985–999 (1959)
20. Marques-Silva, J.P., Lynce, I., Malik, S.: CDCL solvers. In: Biere, A., Heule, M., van Maaren, H., Walsh, T., (eds.), Handbook of Satisfiability, chapter 4, pp. 131–154. IOS Press (2009)
21. Ohrimenko, O., Stuckey, P., Codish, M.: Propagation via lazy clause generation. Constraints 14, 357–391 (2009)
22. Pipatsrisawat, K., Darwiche, A.: On the power of clause-learning SAT solvers as resolution engines. Artificial Intelligence 175(2), 512–525 (2011)
23. SAT Challenge 2012. SAT Challenge 2012: Downloads, November 23, 2014. http:// baldur.iti.kit.edu/SAT-Challenge-2012/downloads.html
24. Stuckey, P.: Lazy clause generation: combining the power of SAT and CP (and MIP?) solving. In: Lodi, A., Milano, M., Toth, P. (eds.) CPAIOR 2010. LNCS, vol. 6140, pp. 5–9. Springer, Heidelberg (2010)
25. Wegener, I.: Branching Programs and Binary Decision Diagrams: Theory and Applications. SIAM Monographs on Discrete Mathematics and Applications. SIAM, Philadelphia (2000)

Combining Constraint Propagation and Discrete Ellipsoid-Based Search to Solve the Exact Quadratic Knapsack Problem

Wen-Yang Ku[⊠] and J. Christopher Beck

Department of Mechanical and Industrial Engineering, University of Toronto,
Toronto, Ontario M5S 3G8, Canada
{wku,jcb}@mie.utoronto.ca

Abstract. We propose an extension to the discrete ellipsoid-based search (DEBS) to solve the exact quadratic knapsack problem (EQKP), an important class of optimization problem with a number of practical applications. For the first time, our extension enables DEBS to solve convex quadratically constrained problems with linear constraints. We show that adding linear constraint propagation to DEBS results in an algorithm that is able to outperform both the state-of-the-art MIP solver CPLEX and a semi-definite programming approach by about one order of magnitude on two variations of the EQKP.

1 Introduction

The exact quadratic knapsack problem (EQKP) [1] consists of selecting a subset of elements such that the sum of the distances between the chosen elements is maximized while also satisfying a knapsack constraint. The EQKP is an extension of the well studied maximum diversity problem [2], the quadratic knapsack problem [3], and the exact linear knapsack problem [4], which arise in a wide range of real world applications such as wind farm optimization [5,6] and task allocation [7]. The EQKP consists of one quadratic objective function and two linear constraints, i.e., one knapsack constraint and one cardinality constraint, where all variables are binary. The EQKP was first studied by Létocart [1] who proposed an efficient heuristic method for the problem.

For solving EQKPs exactly, the common generic approach in Operations Research is the use of a commercial MIP solver such as CPLEX or Gurobi. These solvers have been extended to reason about quadratic constraints [8] and they are able to outperform several other exact approaches [9]. The other generic approach is the semi-definite programming (SDP) based branch-and-bound algorithm [10], which is often regarded as the state-of-the-art approach for solving binary quadratic programming problems (BQP). EQKP is an extension of the BQP and it can be solved with the SDP approach. However, the SDP approach has not been evaluated on problems with the EQKP structure.

© Springer International Publishing Switzerland 2015
L. Michel (Ed.): CPAIOR 2015, LNCS 9075, pp. 231–239, 2015.
DOI: 10.1007/978-3-319-18008-3_16

In this paper, we extend discrete ellipsoid-based search (DEBS) method to solve the EQKP. DEBS is a specialized search used in the communications literature (e.g., see [11]) to solve integer least squares problems (ILS) based on the clever enumeration of integer points within the hyper-ellipsoid defining the feasible space. We have previously shown that DEBS can be applied to the BQP efficiently [12]. As EQKPs can be reformulated as ILS problems, we perform this transformation and extend DEBS for solving convex quadratically constrained problems with linear constraints. To our knowledge, this is the first time that DEBS has been applied to problems with linear constraints so our extension enables DEBS to solve a much broader class of problems, i.e., problems with a quadratic objective function and any number of linear constraints. As an initial test-bed, we use the EQKP to show that our extension achieves state-of-the-art for solving the EQKP and a variation, outperforming both CPLEX and the SDP based approach. Interestingly, the three exact approaches agree on problem difficulty: instances that take more time for the DEBS method to solve are also more challenging for the other two approaches.

2   The Exact Quadratic Knapsack Problem

The EQKP problem is defined as follows:

$$\max_{x \in \{0,1\}} \sum_{i=1}^{n-1} \sum_{j=i+1}^{n} h_{ij} x_i x_j, \tag{1}$$

$$\text{s.t.} \quad \sum_{i=1}^{n} x_i = K,$$

$$\sum_{i=1}^{n} a_i x_i \leq B,$$

where $n \in \mathbb{Z}$, $K \in \mathbb{Z}$, $a_i \in \mathbb{R}^+, \forall i$, $B \in \mathbb{R}^+$, $h_{ij} \in \mathbb{R}, \forall i, j$.

For the simplicity of explanation, we reformulate the EQKP in its minimization form with matrix representation as follows:

$$\min_{x \in \{0,1\}} \frac{1}{2} x^T H x + f^T x, \quad \text{s.t.} \ c_1^T x = K, \ c_2^T x \leq B, \tag{2}$$

where $H \in \mathbb{R}^{n \times n}$ is a symmetric semi-definite positive matrix, $f \in \mathbb{R}^n$ is a vector equal to zeros, $c_1 \in \mathbb{R}^n$ is a vector equal to ones, and $c_2 \in \mathbb{R}^n$ is a vector equal to a_is. Note that H can always be made symmetric semi-definite positive to ensure convexity when all variables are binary [9].

To solve the EQKP with the DEBS method, we need to transform the objective function into its equivalent integer least squares (ILS) form through the relationship $H = A^T A$ and $f = -y^T A$. Thus, the equivalent ILS problem can be defined as follows:

$$\min_{x \in \{0,1\}} \|y - Ax\|_2^2, \quad \text{s.t.} \ c_1^T x = K, \ c_2^T x \leq B. \tag{3}$$

3 The DEBS Method

The DEBS method has been developed to address three types of the ILS problems: unconstrained, box-constrained, and ellipsoid-constrained [13–15]. To the best of our knowledge, the DEBS method has not been extended to solve ILS problems with general linear constraints.

The DEBS method consists of two phases: reduction and search. The reduction is a preprocessing step that transforms A to an upper triangular matrix R with properties such that the search is more efficient [13]:

$$\min_{x \in \mathbb{Z}^n} \|y - Ax\|_2^2 \to \min_{z \in \mathbb{Z}^n} \|\bar{y} - Rz\|_2^2. \tag{4}$$

Geometrically, the optimal solution is found by searching discretely inside the ellipsoid defined by the objective function of the reduced ILS problem (right hand side of Equation (4)). Suppose the optimal solution z^* to the ILS problem satisfies the bound $\|\bar{y} - Rz^*\|_2^2 < \beta$, where β is a constant. This expression defines a hyper-ellipsoid with center $R^{-1}\bar{y}$.

Reduction. The reduction phase of DEBS transforms A into an upper triangular matrix R such that the diagonal entries are ordered in non-decreasing order:

$$|r_{11}| \le |r_{22}| \le \ldots \le |r_{kk}| < \ldots \le |r_{n-1,n-1}| \le |r_{nn}|.$$

It has been shown that the above order can greatly affect the efficiency of the DEBS search by reducing the branching factor at the top of the search tree [13].

Transforming A to R can be achieved by finding an orthogonal matrix $Q \in \mathbb{R}^{n \times n}$ and a permutation matrix $P \in \mathbb{Z}^{n \times n}$ such that $Q^T A P = \begin{bmatrix} R \\ 0 \end{bmatrix}$, $Q = [Q_1, Q_2]$. Therefore we have:

$$\|y - Ax\|_2^2 = \left\| Q_1^T y - RP^T x \right\|_2^2 + \left\| Q_2^T y \right\|_2^2.$$

Let $\bar{y} = Q_1^T y$, $z = P^T x$, and $\bar{c}_2 = c_2 P$, the original EQKP (3) is then transformed to the new reduced EQKP:

$$\min_{z \in \{0,1\}} \|\bar{y} - Rz\|_2^2, \quad \text{s.t. } c_1^T x = K, \ \bar{c}_2^T x \le B. \tag{5}$$

Note that applying the permutation matrix P to the original problem changes the order of the variable bounds and the coefficients of the linear constraints. However, since the original lower and upper bounds on the variables are all zeros and ones, and the coefficients of the cardinality constraint are all ones for the EQKP, the new bounds and c_1 remain unchanged after applying the permutation matrix P. For problems with general variable bounds, i.e., $l \le x \le u$, the reordering can be done with $\bar{l} = P^T l$ and $\bar{u} = P^T u$.

After the optimal solution z^* to the new EQKP problem (5) is found, the optimal solution, x^*, to the original EQKP problem (3) can be recovered with the relationship $x^* = Pz^*$.

Search. The search is performed on the reduced problem (5). Among several search strategies in the literature, the Schnorr & Euchner strategy is usually considered the most efficient [16]. The search systematically enumerates all the integer points in the bounded hyper-ellipsoid with specific variable and value ordering heuristics. In addition, the hyper-ellipsoid is contracted during the enumeration process to further constrain the search space. The main ideas of the search are as follows.

Suppose the optimal solution z^* satisfies the bound $\|\bar{y} - Rz\|_2^2 < \beta$, or equivalently $\sum_{k=1}^{n}(\bar{y}_k - \sum_{j=k}^{n} r_{kj}z_j)^2 < \beta$, where β is a constant which can be obtained by substituting any feasible integer solution to equation (5). Let $z_i^n = [z_i, z_{i+1}, \ldots, z_n]^T$. Define the so-far-unknown (apart from c_n) and usually non-integer variables:

$$c_n = \bar{y}_n/r_{nn}, \quad c_k = c_k(z_{k+1}, \ldots, z_n) = (\bar{y}_k - \sum_{j=k+1}^{n} r_{kj}z_j)/r_{kk}, \quad k = n-1, \ldots, 1.$$

Note that c_k is a function of z_{k+1} to z_n, and it is fixed when z_{k+1} to z_n are fixed. The above equation can be rewritten as $\sum_{k=1}^{n} r_{kk}^2(z_k - c_k)^2 < \beta$, which defines the possible values that z_k can take on. This inequality is equivalent to the following n inequalities:

$$\text{level } n: \quad (z_n - c_n)^2 < \frac{1}{r_{nn}^2}\beta,$$

$$\text{level } n-1: \quad (z_{n-1} - c_{n-1})^2 < \frac{1}{r_{n-1,n-1}^2}[\beta - r_{nn}^2(z_n - c_n)^2],$$

$$\vdots$$

$$\text{level } k: \quad (z_k - c_k)^2 < \frac{1}{r_{kk}^2}[\beta - \sum_{i=k+1}^{n} r_{ii}^2(z_i - c_i)^2],$$

$$\vdots$$

$$\text{level } 1: \quad (z_1 - c_1)^2 < \frac{1}{r_{11}^2}[\beta - \sum_{i=2}^{n} r_{ii}^2(z_i - c_i)^2].$$

The search starts at level n, heuristically assigning $z_n = \lfloor c_n \rceil$, the nearest integer to c_n. Given the value of z_n, c_{n-1} can be calculated from the above equation as $c_{n-1} = (\bar{y}_{n-1} - r_{n-1,n}z_n)/r_{n-1,n-1}$. From this value, we can set $z_{n-1} = \lfloor c_{n-1} \rceil$ and search continues. During the search process, z_k is determined at level k, where $z_n, z_{n-1}, \ldots, z_{k+1}$ have already been determined, but $z_{k-1}, z_{k-2}, \ldots, z_1$ are still unassigned. At some level $k - 1$ in the search, it is likely that the inequality cannot be satisfied, requiring the search to backtrack to a previous decision. When we backtrack from level $k - 1$ to level k, we choose z_k to be the next nearest integer to c_k, and so on.

In the Schnorr & Euchner strategy, the initial bound β can be set to ∞. Once the first integer point is found, we can use this integer point to update β, reducing the hyper-ellipsoid thus the search space.

In order to take into account the variable bounds, i.e., $l \leq z \leq u$, the algorithm limits the enumeration of z_k at level k to be within $[l_k, u_k]$.

4   Extending the DEBS Method to Solve the EQKP

From a constraint programming perspective, DEBS does three things. First, the search strategy implies a fixed variable ordering heuristic $(z_n, z_{n-1}, \ldots, z_k, \ldots, z_1)$ that is determined after the reduction phase, where the variables are reordered with the permutation matrix as $z = P^T x$. Second, through the calculation of c_k, DEBS provides a value ordering heuristic: for z_k, integer values closer to c_k are preferred. Third, the inequalities described above induce an interval domain for each z_k. DEBS is essentially the enumeration of these domains under the prescribed variable and value orderings. We use this insight to integrate linear constraints into DEBS in a straightforward way: linear constraints are made arc consistent to further reduce the domains of the z_k variables.

A linear constraint is defined as $\underline{b} \leq a^T z \leq \bar{b}$, where $\underline{b}, \bar{b} \in \mathbb{R}^n \cup \{\pm\infty\}$, and $a \in \mathbb{R}^n$ is the coefficients of the constraint. Given a linear constraint, let

$$\underline{\alpha}_k = \min\{a^T z - a_k z_k \mid l \leq z \leq u\} \quad \text{and} \quad \bar{\alpha}_k = \max\{a^T z - a_k z_k \mid l \leq z \leq u\}$$

be the minimal and maximal values that $a^T z$ can achieve over all variables except z_k with respect to the variable bounds. These values can be computed by substituting the z_ks with their lower or upper bounds, depending on the signs of the a_ks. The propagation rule for each integer variable z_k is then defined as follows:

$$\left\lceil \frac{\underline{b} - \bar{\alpha}_k}{a_k} \right\rceil \leq z_k \leq \left\lfloor \frac{\bar{b} - \underline{\alpha}_k}{a_k} \right\rfloor \quad \text{if } a_k > 0, \tag{6}$$

$$\left\lceil \frac{\bar{b} - \underline{\alpha}_k}{a_k} \right\rceil \leq z_k \leq \left\lfloor \frac{\underline{b} - \bar{\alpha}_k}{a_k} \right\rfloor \quad \text{if } a_k < 0. \tag{7}$$

Although the idea of propagating the linear constraint is straightforward and not novel, the implementation involves maintaining useful information efficiently as the search moves between levels. As a variable is fixed when the search moves down the levels, the values $\underline{\alpha}_k$ and $\bar{\alpha}_k$ actually consist of two parts: the sum of all the variables that are fixed already (z_{k+1} to z_n), and the sum of the maximum or minimum values that the unfixed variables (z_1 to z_{k-1}) can achieve, according to the variable bounds. Below we show how to update $\underline{\alpha}_k$ and $\bar{\alpha}_k$ using the cardinality constraint as an example.

For the cardinality constraint, we have $a_k = 1, \forall k$, and $\underline{b} = \bar{b} = K$. Therefore, using Equation (6), the lower bound L_k and upper bound U_k imposed by the cardinality constraint can be derived as follows:

$$L_k = K - \sum_{i=k+1}^{n} z_i - \sum_{i=1}^{k-1} u_i \quad \text{and} \quad U_k = K - \sum_{i=k+1}^{n} z_i - \sum_{i=1}^{k-1} l_i,$$

where $\sum_{i=1}^{k-1} l_i = 0$ and $\sum_{i=1}^{k-1} u_i = k - 1$, since z_is are binary for the EQKP.

During the search, the sum of the fixed values $S = \sum_{i=k+1}^{n} z_i$ is only changed when moving up or down between two adjacent levels. Therefore, we can use a single variable to keep track this value, efficiently updating S rather than computing from scratch. When the search backtracks from level k to $k+1$, we set $S = S - z_k$ to reverse the assignment of z_k. Similarly, when the search moves down from level k to $k-1$, we set $S = S + z_k$ to take into account the current assignment of z_k.

Equations (6) and (7) are general and they can be applied to all types of linear constraints. In addition, some types of constraints, e.g., the cardinality constraint as shown above, can be specialized from the general linear constraint formulation and be propagated more efficiently. Similarly, each linear constraint keep tracks of its own sum of the current assigned values S. Therefore our extended DEBS can be used to solve problems with any number of linear constraints of any type. The variable domain at each level is therefore the intersection of all the bounds imposed by the linear constraints.

A naive extension of the DEBS method for solving problems with linear constraints can be achieved by using the linear constraints only as "checkers": the constraints simply return true or false when all the variables are instantiated. If all linear constraints agree on an instantiation, it is accepted as an integer solution and the search proceeds correspondingly. However from our preliminary experiments, this naive implementation is, unsurprisingly, orders of magnitude slower than the MIP and SDP approaches. Thus, we do not report the results but note that our simple propagation is critical for results presented below.

Solving Non-Binary Problems. Our extended DEBS method can solve non-binary problems without any modification. However, it requires that the quadratic objective function be inherently convex, as convexifying a non-binary problem is not possible. The reduction and the propagation on the linear constraints are not affected by the variable domains and therefore remain the same.

5   Experimental Results

Experimental Setup. The CPU time limit for each run on each problem instance is 3600 seconds. All experiments were performed on a Intel(R) Xeon(R) CPU E5-1650 v2 3.50GHz machine (in 64 bit mode) with 16GB memory running MAC OS X 10.9.2 with one thread. The DEBS approach is written in C. We use CPLEX Optimization Studio v12.6 and the SDP solver BiqCrunch downloaded from the website [17] for comparison. Both solvers are executed with their default settings. For the SDP solver, there are four specialized versions that deal with problem-specific structures. Three of them are consistent with the EQKP problem and the SDP results presented are the best of the three versions for each individual problem instance, representing the "virtual best" SDP solver. We report the arithmetic mean CPU time "arith", and the shifted geometric mean CPU time "geo" to find and prove optimality for each problem set.[1]

---

[1] The shifted geometric mean time is computed as follows: $\prod(t_i + s)^{1/n} - s$, where t_i is the actual CPU time, n is the number of instances, and s is chosen as 10. Using geometric mean can decrease the influence of the outliers of data [18].

Table 1. A comparison of CPLEX, DEBS and the SDP approach for the EQKP ({0,1} Problems) and its problem variation ({0,1,2} Problems). Bold numbers indicate the best approach for a given problem set. The superscripts indicate the number of instances not solved to optimality within 3600 seconds.

	{0,1} Problems							
n	DEBS		DEBS+Red.		CPLEX		SDP	
	arith	geo	arith	geo	arith	geo	arith	geo
10	**0.00**	0.00	**0.00**	0.00	0.04	0.04	0.14	0.14
20	**0.01**	0.01	**0.01**	0.01	0.18	0.18	4.94	4.35
30	0.72	0.68	**0.69**	0.65	18.82	12.21	336.94	152.53
40	115.64	31.91	**109.63**	30.78	1274.05^2	401.5^2	2374.78^4	1238.91^4
50	**317.83**	49.87	329.31	50.91	1810.34^6	1006.41^6	3237.45^7	3119.11^7

	{0,1,2} Problems					
n	DEBS		DEBS+Red.		CPLEX	
	arith	geo	arith	geo	arith	geo
10	**0.00**	0.00	**0.00**	0.00	0.01	0.01
20	**0.02**	0.02	**0.02**	0.02	0.26	0.25
30	1.05	0.95	**0.99**	0.90	11.49	6.78
40	**276.28**	65.86	308.52	69.78	1203.59^2	281.98^2
50	**2354.50**^{6}	1216.43^6	2406.48^6	1262.99^6	3149.80^8	2700.36^8

Test sets. We use five sets of the benchmark instances with different sizes generated in the same way as Létocart et al. [1], with 10 instances in each set. For additional comparison, we relax the binary domains $x_j \in \{0,1\}$ to $x_j \in \{0,1,2\}$. This means that we have the option of selecting two "copies" of each element when maximizing the quadratic objective function but $h_{jj} = 0$ (see Formulation 1). To the best of our knowledge, this problem has not been studied in the literature.

In order to ensure convexity, we compute the smallest eigenvalue for the H matrix of each problem and let it be λ_{min}. If λ_{min} is negative, i.e., the problem is non-convex, we apply the perturbation vector $u = (-\lambda_{min} + 0.001)e$ such that the original objective function is transformed to: $\min_{x \in \{0,1\}} \frac{1}{2}x^T \bar{H}x + \bar{f}^T x$, where $\bar{H} = H + \text{diag}(u)$ and $\bar{f} = (f - \frac{1}{2}u)^T$. Note that this convex formulation has the same optimal solution as the original one. For the DEBS method, we use Cholesky decomposition on the perturbed matrix $\bar{H}$ to obtain matrix A in the ILS formulation (3), and we obtain y from the relationship $\bar{f} = -y^T A$, which gives us $y = -(\bar{f}A^{-1})^T$. If the problem is originally convex, we obtain A and y with H and f.

Results and Discussion. From Table 1, it is clear that the DEBS method performs the best both for the EQKP and its relaxed problem. For the EQKP,

DEBS is on average one to two orders of magnitude faster than CPLEX and it strictly dominates CPLEX on each problem instance in our experiments. The SDP approach performs the worst among the three approaches. This is surprising given the strong results for solving the BQPs, and the fact that these results are the best per instance results over the three BiqCrunch variations. Analysis of CPLEX and the SDP solving behaviour shows that the reason that both approaches are unable to prove optimality for larger problems is mainly the weakness in the dual bound. This apparently favours the DEBS method as it does not rely on dual bounding. In fact, DEBS completely lacks a dual bounding mechanism, depending on propagation to reduce the search space.

Interestingly, the running times for all three approaches increase significantly when K is increased. The reason is that during the search, when the sum of the already fixed variables at a node is equal to K, we know that the unfixed variables have to be set to zero to satisfy the cardinality constraint. Intuitively, if K is small, such solutions are early in the search tree, as fewer branching decisions are required to reach such nodes. For the same reason, we expect that the running time to be also decreased when K is further increased towards n.

From Table 1, it is observed that the relaxed domain makes the problem much more difficult to solve, but DEBS still dominates CPLEX. The SDP approach cannot be used to solve the relaxed problem without transforming the problem back to binary domain at the cost of introducing additional variables and constraints. Therefore, we expect its performance to be inefficient.

It is interesting to observe that the reduction does not seem to improve the performance for the EQKP as it does on the BQP problems without general linear constraints [12]. Reduction even worsens performance on the relaxed problems. This suggests that it might be of interest to develop new reduction algorithms that take into the account of the linear constraints.

6   Conclusion

We proposed an extension to the discrete ellipsoid-based search (DEBS) method to solve the exact quadratic knapsack problem (EQKP) and a variation. The core of our extension required the modification of the DEBS approach to incorporate linear constraints. We did this by adopting standard linear constraint propagation algorithms. Results showed that our algorithm outperforms both the state-of-the-art MIP and SDP approaches. For future work, we would like to further extend DEBS to solve general mixed integer convex quadratic programming problems and test its ability as a general solution approach. It is also interesting to investigate dual bounding mechanism for the DEBS search algorithm, as it is possible to compute a dual bound at each level to further limit the search interval.

References

1. Létocart, L., Plateau, M.C., Plateau, G.: An efficient hybrid heuristic method for the 0–1 exact k-item quadratic knapsack problem. Pesquisa Operacional **34**(1), 49–72 (2014)
2. Martí, R., Gallego, M., Duarte, A.: A branch and bound algorithm for the maximum diversity problem. European Journal of Operational Research **200**(1), 36–44 (2010)
3. Caprara, A., Pisinger, D., Toth, P.: Exact solution of the quadratic knapsack problem. INFORMS Journal on Computing **11**(2), 125–137 (1999)
4. Caprara, A., Kellerer, H., Pferschy, U., Pisinger, D.: Approximation algorithms for knapsack problems with cardinality constraints. European Journal of Operational Research **123**(2), 333–345 (2000)
5. Turner, S., Romero, D., Zhang, P., Amon, C., Chan, T.: A new mathematical programming approach to optimize wind farm layouts. Renewable Energy **63**, 674–680 (2014)
6. Zhang, P.Y., Romero, D.A., Beck, J.C., Amon, C.H.: Solving wind farm layout optimization with mixed integer programs and constraint programs. EURO Journal on Computational Optimization **2**(3), 195–219 (2014)
7. Lewis, M., Alidaee, B., Kochenberger, G.: Using xqx to model and solve the uncapacitated task allocation problem. Operations Research Letters **33**(2), 176–182 (2005)
8. Bussieck, M.R., Vigerske, S.: MINLP solver software. Wiley Encyclopedia of Operations Research and Management Science. Wiley, Chichester (2010)
9. Billionnet, A., Elloumi, S.: Using a mixed integer quadratic programming solver for the unconstrained quadratic 0–1 problem. Mathematical Programming **109**(1), 55–68 (2007)
10. Krislock, N., Malick, J., Roupin, F.: Improved semidefinite bounding procedure for solving max-cut problems to optimality. Mathematical Programming, 1–26 (2012)
11. Teunissen, P.J., Kleusberg, A., Teunissen, P.: GPS for Geodesy, vol. 2. Springer, Berlin (1998)
12. Ku, W.-Y., Beck, J.C.: Combining discrete ellipsoid-based search and branch-and-cut for binary quadratic programming problems. In: Simonis, H. (ed.) CPAIOR 2014. LNCS, vol. 8451, pp. 334–350. Springer, Heidelberg (2014)
13. Chang, X.W., Han, Q.: Solving box-constrained integer least squares problems. IEEE Transactions on Wireless Communications **7**(1), 277–287 (2008)
14. Chang, X.W., Golub, G.H.: Solving ellipsoid-constrained integer least squares problems. SIAM Journal on Matrix Analysis and Applications **31**(3), 1071–1089 (2009)
15. Ku, W.Y., Beck, J.C.: Combining discrete ellipsoid-based search and branch-and-cut for integer least squares problems. Submitted to IEEE Transactions on Wireless Communications (2014)
16. Schnorr, C.P., Euchner, M.: Lattice basis reduction: Improved practical algorithms and solving subset sum problems. Mathematical programming **66**(1), 181–199 (1994)
17. Krislock, N., Malick, J., Roupin, F.: BiqCrunch online solver (2012) (Retrieved: 11/06/2014). http://lipn.univ-paris13.fr/BiqCrunch/download
18. Achterberg, T.: Constraint Integer Programming. PhD thesis, Technische Universität Berlin (2007)

Large Neighborhood Search for Energy Aware Meeting Scheduling in Smart Buildings

Boon Ping Lim[1]([✉]), Menkes van den Briel[1], Sylvie Thiébaux[1],
Russell Bent[2], and Scott Backhaus[2]

[1] NICTA and The Australian National University, Canberra, Australia
boonping.lim@nicta.com
[2] Los Alamos National Laboratory, Los Alamos, USA

Abstract. One of the main inefficiencies in building management systems is the widespread use of schedule-based control when operating heating, ventilation and air conditioning (HVAC) systems. HVAC systems typically operate on a pre-designed schedule that heats or cools rooms in the building to a set temperature even when rooms are not being used. Occupants, however, influence the thermal behavior of buildings. As a result, using occupancy information for scheduling meetings to occur at specific times and in specific rooms has significant energy savings potential. As shown in Lim et al. [15], combining HVAC control with meeting scheduling can lead to substantial improvements in energy efficiency. We extend this work and develop an approach that scales to larger problems by combining mixed integer programming (MIP) with large neighborhood search (LNS). LNS is used to destroy part of the schedule and MIP is used to repair the schedule so as to minimize energy consumption. This approach is far more effective than solving the complete problem as a MIP problem. Our results show that solutions from the LNS-based approach are up to 36% better than the MIP-based approach when both given 15 minutes.

Keywords: Smart buildings · Scheduling · Large neighborhood search · HVAC control

1 Introduction

In 2010, electricity expenditures in both residential and commercial buildings were over 300 billion dollars in the United States alone. About one third of this consumption can be attributed to space heating, ventilation and air conditioning (HVAC)[1]. Hence, with over 100 billion dollars in annual electricity expenditures even a small percentage improvement in the operation of HVAC systems can lead to significant savings.

While it is possible to reduce energy consumption by retrofitting buildings with more efficient HVAC systems, it is far more cost effective to improve their

---

[1] See Table 1.2.5 at http://buildingsdatabook.eren.doe.gov/

© Springer International Publishing Switzerland 2015
L. Michel (Ed.): CPAIOR 2015, LNCS 9075, pp. 240–254, 2015.
DOI: 10.1007/978-3-319-18008-3_17

control algorithms. A building management system (BMS) predominantly operates on a pre-designed schedule that incorporates a nighttime setback strategy, which relaxes the comfort constraints during the night. Bloomfield and Fisk [2] show that such a strategy leads to significant energy savings. Building management systems, however, could be far more dynamic by adopting strategies that consider occupancy information. Occupants impact the thermal behavior of buildings. For example, a BMS may relax the comfort constraints for those rooms in the building that are not planned on being used.

HVAC control and meeting scheduling have been studied extensively as separate problems and only recent work has started to look at integrating the two problems [15]. This work introduced a mixed integer programming (MIP) model for energy aware meeting scheduling. Given the highly-constrained nature of meeting scheduling with HVAC control, solving the integrated problem as a MIP seems to be a reasonable choice. MIP easily manages the interaction between meeting scheduling and the impact it has on HVAC energy consumption. The problem with the MIP-based approach, however, is that it does not scale. Consequently, in this work we introduce a hybrid solution that combines MIP with large neighborhood search (LNS) so as to scale to problem sizes that, for example, companies and universities may face when scheduling meetings and courses. In order to fine-tune the LNS heuristic we apply automatic parameter tuning, which is a useful step in producing better results on average.

The remainder of this paper is organized as follows. Section 2 reviews related work. Section 3 describes the MIP formulation, which consists of an HVAC model and a scheduling model that are combined into one energy aware meeting scheduling model. Section 4 describes the LNS heuristic and discusses the automatic parameter tuning. Experimental results are presented in Section 5 and we conclude with key observations and opportunities for further work in Section 6.

2   Related Work

Energy aware meeting scheduling has gotten more attention in recent years possibly due to the significant cost saving opportunities. Even simple rules like consolidating meetings in fewer buildings have proven to be very effective. For example, Portland State University consolidated night and weekend classes that were held across 21 buildings into 5 energy efficient buildings. By doing so, they reported a reduction of 18.5% in electricity consumption over the fall season compared to the previous three-year average [22].

Majumdar et al. [16] evaluate the performance of a number of scheduling algorithms that they run through the energy simulation software EnergyPlus [3]. Some criteria considered in their algorithms include minimizing the number of rooms, minimizing time between meetings in the same room, and minimizing room size. The latter performed best when considering only one criterion, but a more involved A* search with successive calls to EnergyPlus performed best overall. Similar criteria are considered in the work by Kwak and others [12,13], but they use MIP to calculate an energy efficient schedule. Pan et al. [21] use the criterion of scheduling meetings back-to-back and observed 30% energy savings when comparing results to existing schedules.

It is important to note that these works minimize energy use without directly modeling the HVAC system. In our work we integrate meeting scheduling with HVAC control, because it is of key importance to determine which rooms need to be heated or cooled and when. For example, sometimes it is more energy efficient to pre-cool a room rather than cool it at the time of a meeting. In order to determine the HVAC operation one needs to decide the HVAC control settings.

Occupancy information, knowing whether or not there are people in a room and ideally also knowing how many people there are in a room, is at the basis of moving away from schedule-based control towards model predictive control (MPC). Schedule-based control is inefficient as it operates HVAC systems according to a fixed schedule that often assumes maximum zone occupancy. Oldewurtel et al. [20] use MPC, which takes into account weather and occupancy forecasts to estimate energy saving opportunities. Goyal et al. [10] also take into account occupancy forecasts and compare the performance of MPC strategies with feedback controllers using occupancy measurements. Their conclusion is that occupancy predictions can lead to additional energy savings over occupancy measurements especially when ventilation standards change. Currently the ASHRAE ventilation standard[1] requires that a certain amount of outside air is mixed with return air from the zones regardless whether a zone is occupied or not.

Lim et al. [15] combine MPC with meeting scheduling and propose a MIP-based solution. This model achieves an energy reduction of over 50% when compared to approaches similar to those presented in [10,13,16]. The drawback of this approach is that it does not scale well, which is why we developed a hybrid solution that combines MIP with LNS. In [15], we presented some of the initial results on applying LNS on this problem. Here, we significantly expand on the details of the LNS algorithm, and present extensive results supporting the quality of the algorithm.

Combining constraint-based methods with neighborhood search methods is not new. For example, LeBras et al. [14] used LNS with MIP in network design for a species conservation problem, Di Gaspero et al. [5] used LNS with CP in balancing bike sharing systems, Rendl et al. [23] applied CP with variable neighborhood search for homecare scheduling, and Danna et al. [4] used LNS in job-shop scheduling. We are, however, unaware of its application in the space of energy aware scheduling in smart buildings.

3   Mixed Integer Programming Model

In this section we present our combined HVAC control and meeting scheduling formulation. In the description of the model, values of constants are given between square brackets. We present key elements of the HVAC model and discuss two updates over Lim et al. [15].

Following Goyal et al. [9,10], we focus on commercial buildings with VAV-based HVAC systems, which serve over 30% of the commercial building floor space in the United States [6]. In particular, we focus on control strategies that can be applied to each VAV box. Specifically, we determine the supply air temperature and the air flow rate, which are manipulated by a heating coil and dampers inside the VAV box. A schematic of a VAV-based HVAC system with two VAV boxes is shown in Figure 1.

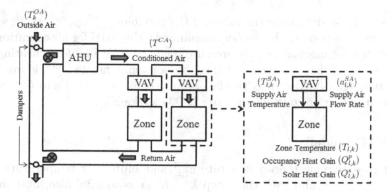

Fig. 1. VAV-based HVAC system

Conditioned air is the air supplied by the air handling unit (AHU) to the VAV boxes. Energy is consumed by the AHU when mixing outdoor air with return air and cooling it to a pre-set conditioned air temperature T^{CA} [12.8 °C]. The AHU consumes less energy when the outdoor air temperature T^{OA} is closer to T^{CA}. Energy is also consumed by the VAV box when regulating the supply air temperature T^{SA} and the supply air flow rate a^{SA} to keep the zone temperature T within comfort bounds. In particular, the supply fan at the AHU tries to maintain a constant air pressure through the supply duct and thus may speed up or slow down depending on air flow rates used by the VAV boxes. Moreover, in order to keep the zone temperature within comfort bounds the conditioned air may need to be reheated in the VAV box.

3.1 Objective Function

We want to schedule meetings in such a way that the HVAC energy consumption is minimized. Thus, the objective function is defined as minimizing the energy use of fan, air-conditioning and heating operations. Let $K = 1, ..., n$ be an ordered set of time steps, where the length of each time step is $t_k - t_{k-1} = \Delta t$. Let L be the set of zones in the building. We assume that each zone corresponds to one room. In general, however, a zone may correspond to a set of rooms. The objective function is defined as

$$Min \sum_{k \in K} \left(p_k^{fan} + p_k^{cond} + \sum_{l \in L} p_{l,k}^{heat} \right) \times \Delta t \qquad (1)$$

where

$$p_k^{fan} = \beta \sum_{l \in L} a_{l,k}^{SA} \quad \forall k \in K \tag{2}$$

$$p_k^{cond} = C^{pa} \left(T_k^{OA} - T^{CA} \right) \sum_{l \in L} a_{l,k}^{SA} \quad \forall k \in K \tag{3}$$

$$p_{l,k}^{heat} = C^{pa} (T_{l,k}^{SA} - T^{CA}) a_{l,k}^{SA} \quad \forall l \in L, k \in K \tag{4}$$

Constraints (2)-(4) determine the values of the variables p_k^{fan}, p_k^{cond}, $p_{l,k}^{heat}$, which respectively correspond to the energy consumed by the AHU for air conditioning, the supply fan for maintaining air pressure, and the VAV box for reheating the air. It is assumed that a forecast of the outdoor temperature T_k^{OA} is available for each time step k. Also, β is the fan power coefficient [0.65] and C^{pa} is the heat capacity of air at constant pressure [1.005 kJ/kg·K].

3.2 HVAC Model

We want to determine the air flow rate $a_{l,k}^{SA}$ and supply air temperature $T_{l,k}^{SA}$ for each zone $l \in L$ and each time step $k \in K$ so as to minimize total energy consumption. In order to keep zone temperature within comfort bounds, we introduce the variables $T_{l,k}$ to represent the indoor temperature in zone l at time step k. We also introduce the variable $z_{l,k}$, which is equal to 1 if zone l is occupied at time step k and equal to 0 otherwise. The meaning of $z_{l,k}$ will become clearer when we describe the scheduling constraints in section 3.3. The variables below are used in the HVAC model to determine how much energy is needed to, for example, cool a zone from say 28 °C (82.4 °F) to 23 °C (73.4 °F).

$$T^{unocc,lb} + \delta^{lb} z_{l,k} \leq T_{l,k} \leq T^{unocc,ub} - \delta^{ub} z_{l,k} \tag{5}$$

$$T^{CA} \leq T_{l,k}^{SA} \leq T^{SA,ub} \quad \forall l \in L, k \in K \tag{6}$$

$$a^{SA,lb} \leq a_{l,k}^{SA} \leq a^{SA,ub} \quad \forall l \in L, k \in K \tag{7}$$

Constraints (5) describe the comfort bounds on zone temperature. When a zone is occupied, the zone temperature must be within a specified comfort interval $(T^{occ,lb}, T^{occ,ub})$ [(21 °C, 23 °C)]. At other times the zone temperature can fluctuate more freely $(T^{unocc,lb}, T^{unocc,ub})$ [(16 °C, 28 °C)]. These bounds can be set to reflect individual building guidelines. We use $T^{occ,lb} = T^{unocc,lb} + \delta^{lb}$ and $T^{occ,ub} = T^{unocc,ub} - \delta^{ub}$, where δ^{lb} and δ^{ub} are appropriately valued constants. Constraints (6) and (7) ensure that the supply air temperature and the air flow rate are bounded by the HVAC operational capacity. The supply air temperature $T_{l,k}^{SA}$ may supply conditioned air at temperature T^{CA}, or heat the conditioned air up to $T^{SA,ub}$ [40 °C]. The air flow rate can fluctuate between $a^{SA,lb}$ [0.108 kg/s] and $a^{SA,ub}$ [5.0 kg/s], where the lower bound is determined by the ASHRAE ventilation standard and the upper bound is reached when the dampers are fully open.

The most involved constraints in the HVAC model come from modeling building thermal dynamics. We use a lumped resistor-capacitor (RC) network, which is commonly used in [7–9] for constructing a model of the transient heat flow through solid surfaces, such as walls and windows. To predict zone temperatures we use the following parameters as inputs: (1) characteristics of the supply air (flow rate and temperature), (2) thermal heat gain due to occupants, (3) thermal heat gain due to solar radiation, and (4) outdoor temperature.

In Lim et al. [15], the temperature in a zone is affected by both internal and external walls. In our experiments, however, we observed that the temperature in a zone is mostly affected by the HVAC, occupants in the zone, and the heat flow through external walls. The heat transfer from neighboring rooms through the internal walls is negligible based on our building settings. Hence, in this model we only consider external walls, that is, those with an outside facing.

The temperature constraints are presented below.

$$
\begin{bmatrix} \dot{T}_{l,k} \\ \dot{T}_{l,k}^1 \\ \cdot \\ \cdot \\ \dot{T}_{l,k}^N \end{bmatrix} =
\begin{bmatrix}
-\frac{1}{C_l}(\sum_{n=1}^{N}\frac{1}{R_l^n} + \frac{1}{R_l^w}) & \frac{1}{C_l R_l^1} & \cdot\ \cdot & \frac{1}{C_l R_l^N} \\
\frac{1}{C_l^1 R_l^1} & -\frac{1}{C_l^1}(\frac{1}{R_l^1+R_1})\ 0\ 0 & & 0 \\
\cdot & \cdot & \cdot & \cdot \\
\cdot & & \cdot\ \cdot & \cdot \\
\frac{1}{C_l^N R_l^N} & 0 & 0\ 0 -\frac{1}{C_l^N}(\frac{1}{R_l^N+R_N}) \end{bmatrix}
\begin{bmatrix} T_{l,k-1} \\ T_{l,k-1}^1 \\ \cdot \\ \cdot \\ T_{l,k-1}^N \end{bmatrix}
$$

$$
+ \begin{bmatrix}
\frac{1}{C_l R_l^w} & 0 & \frac{1}{C^l} \\
\frac{1}{C_l^1 R_l^1} & \frac{1}{C_l^1} & 0 \\
\cdot & \cdot & \cdot \\
\frac{1}{C_l^N R_N^l} & \frac{1}{C_l^N} & 0
\end{bmatrix}
\begin{bmatrix} T_{k-1}^{OA} \\ Q_{l,k-1}^s \\ Q_{l,k-1}^p \end{bmatrix}
+ \begin{bmatrix} \frac{\Delta H_{k-1}^l}{C^l} \\ 0 \\ \cdot \\ 0 \end{bmatrix}
$$

(8)

$$ Q_{l,k}^p = q^p pp_{l,k} \quad \forall l \in L, k \in K \tag{9} $$

$$ \Delta H_{l,k} = C^{pa} a_{l,k}^{SA}(T_{l,k}^{SA} - T_{l,k}) \quad \forall l \in L, k \in K \tag{10} $$

Constraints (8) model the temperature dynamics in zone l when considering a room with N external walls. The expression $\dot{T}_{l,k} = \frac{T_{l,k}-T_{l,k-1}}{\Delta t}$ is the rate of change in zone temperature and $\dot{T}_{l,k}^n$ is the rate of change in temperature of the external wall(s) $n = 1, ..., N$. C_l and C_l^n respectively denote the thermal capacitance of zone l and of the external wall n in zone l. R_l^n and R_l^w respectively represent the thermal resistance of wall n and the windows separating zone l with the outdoors. Constraints (9) determine the internal heat gain due to occupants $Q_{l,k}^p$, which is calculated by multiplying the heat energy q^p [75W] with the number of occupants $pp_{l,k}$. The value of the variable $pp_{l,k}$ is determined in the scheduling model, which is described in the next section. Constraints (10) determine the enthalpy of the heat supplied and extracted by the HVAC.

Note, constraints (4) and (10) contain bilinear terms $a_{l,k}^{SA}T_{l,k}^{SA}$ and $a_{l,k}^{SA}T_{l,k}$, which turns the problem into a mixed integer non-linear programming (MINLP) problem. We use the reformulation method of McCormick [18], which relaxes the original nonlinear equality constraints into a set of linear inequality constraints.

3.3   Scheduling Model

Let M be a set of meetings. Each meeting $m \in M$ has a duration τ_m, a set of allowable start time steps $K_m \subseteq K$, a set of allowable zones $L_m \subseteq L$ and a list of attendees $P_m \subseteq A$, where A is the set of all attendees. The sets K_m and L_m can be used to express a number of constraints, such as, room capacity requirements, room equipment requirements, and time restrictions. Moreover, in order to express meeting conflict constraints due to attendees we let $N \subseteq 2^M$ be the set of meeting sets that have at least one attendee in common, that is $N = \{M_i \subseteq M \mid \forall m, m' \in M_i, P_m \cap P_{m'} \neq \emptyset\}$.

We want to determine $x_{m,l,k}$, which is equal to 1 if meeting m is scheduled to start at time step k in zone l and equal to 0 otherwise. The scheduling model interacts with the HVAC model via the variables $z_{l,k}$ that indicate whether zone l is occupied at time step k or not, and via the variables $pp_{l,k}$ that indicates the number of people in zone l at time step k.

The constraints are as follows.

$$\sum_{l \in L_m, k \in K_m} x_{m,l,k} = 1 \quad \forall m \in M \tag{11}$$

$$\sum_{\substack{m \in M, k' \in K_m: \\ l \in L_m, \ k-\tau_m+1 \leq k' \leq k}} x_{m,l,k'} \leq z_{l,k} \quad \forall l \in L, k \in K \tag{12}$$

$$\sum_{\substack{m \in M, k' \in K_m: \\ l \in L_m, \ k-\tau_m+1 \leq k' \leq k}} x_{m,l,k'} \times |P_m| = pp_{l,k} \quad \forall l \in L, k \in K \tag{13}$$

$$\sum_{\substack{m \in \nu, l \in L_m, k' \in K_m: \\ k-\tau_m+1 \leq k' \leq k}} x_{m,l,k'} \leq 1 \quad \forall k \in K, \nu \in N \tag{14}$$

Constraints (11) require that all meetings must be scheduled exactly once. Constraints (12) state that no more than one meeting can occupy a zone at any time. Observe that the right hand side in this constraint is either zero or one, which limits the number of meetings to at most one. Also, when the left hand side equals one then the zone must be occupied. Constraints (13) determine the number of occupants in a zone $pp_{l,k}$, which is a variable used in the HVAC model and finally constraints (14) ensure that meetings with at least one attendee in common cannot be scheduled in parallel.

One issue with the current model (and the one used in [15]) is that it can have a large number of equivalent solutions when two or more meetings are identical. For example, let meetings 1 and 2 have the same time windows, same number of attendees, and same meeting conflicts. In this case, a solution in which $x_{1,l,k} = 1$ and $x_{2,l,k} = 0$ would be equivalent to one in which $x_{1,l,k} = 0$ and $x_{2,l,k} = 1$. In order to avoid the computational cost of generating both solutions we reduce the number of integer variables by defining meeting types. Meetings that are identical are considered to be of the same meeting type. Since the number of meeting types is smaller than the number of meetings, it allows for a more simplified model that reduces symmetry. Without changing the model in its entirety we simply redefine M to be the set of meeting requests and replace constraint (11) with the one below to state that all meeting types must be scheduled ψ_m times, where ψ_m represents the number of meetings of type m.

$$\sum_{l \in L_m, k \in K_m} x_{m,l,k} = \psi_m \quad \forall m \in M \tag{15}$$

4   Large Neighborhood Search

LNS is a local search metaheuristic, which was originally proposed by Shaw [24]. In LNS, an initial solution is improved iteratively by alternating between a destroy and a repair step. The main idea behind LNS is that a large neighborhood allows the heuristic to easily navigate through the solution space even when the problem is highly-constrained. As opposed to a small neighborhood, which may make escaping a local minimum much harder.

An important decision in the destroy step is determining the amount of destruction. If too little is destroyed the effect of a large neighborhood is lost, but if too much is destroyed then the approach turns into repeated re-optimization. As for the repair step, an important decision is whether the repair should be optimal or not. An optimal repair will typically be slower than a heuristic, but may potentially lead to high quality solutions in a few iterations.

Our LNS approach starts with an initial feasible solution, which is generated using a greedy heuristic. First, this heuristic finds a feasible meeting schedule by minimizing the number of rooms. Second, it determines the HVAC control settings of supply air temperature and supply air flow rate to minimize energy consumption given a fixed schedule. This two-stage approach allows us to come up with an initial solution in reasonable time.

4.1   Destroy and Repair

Our LNS approach considers a neighborhood that contains a subset of the rooms or zones. In particular, we destroy the schedule in two to four randomly selected rooms. This forms a subproblem that can be solved effectively using MIP. When destroying meetings in more than four zones, MIP performance can degrade very quickly and even solving the linear programming relaxation can become quite time consuming. The repair consists of solving an energy aware meeting scheduling problem that is much smaller than the original problem. We do, however, limit MIP runtime to avoid excessive search during a repair step, and to avoid any convergence issues of the MIP problem. Setting a limit on runtime means that we do not necessarily solve the subproblem to optimality, but given that MIP solvers are anytime algorithms, we do improve solution quality in many of the LNS iterations. If we find an improved solution, then the new schedule and control settings are accepted. Otherwise, we maintain the solution that was just destroyed. Given that the LNS starts with a feasible solution and does not accept infeasible solutions, the solution remains feasible throughout the execution of the algorithm.

We should note that we have experimented with a variety of neighborhoods. These include: destroying all meetings in randomly selected time steps, a combination of destroying all meetings in randomly selected rooms and time steps, and simply destroying a set of randomly selected meetings. Our observation was that none of these neighborhoods performed as well as destroying all meetings in a number of randomly selected rooms. Destroying selected rooms means that meetings could be rescheduled at any time during the day. This allows the model to optimize supply air flow rate and supply air temperature over all the time steps. Destroying selected time steps means that meetings may switch rooms, but may need to be scheduled to the same time step due to time window restrictions. This limits the optimization of supply air flow rate and supply air temperature due to the HVAC control constraints on neighboring time steps.

4.2 LNS Parameter Tuning

The parameters that govern the behavior of the LNS heuristic are parameters determining the number (2, 3, or 4) of rooms to destroy and the MIP runtime limit for the repair step. The probabilities on the number of rooms to destroy are defined as a 3-tuple with values ranging between [0,1] and the MIP runtime limit is a parameter with values ranging between 1 and 10 seconds.

While it is possible to reason about certain parameters and their impact on overall performance, there are numerous values that these parameters can take on. Even though we consider only 4 parameters, it is impossible to try all possible configurations because of their continuous domains. Note, even with discretized domains with reasonable level of granularity it remains impractical to try out all configurations. As a result, we use the automated algorithm-configuration method called Sequential Model-based Algorithm Configuration (SMAC) [11] to optimize these parameters.

SMAC can be used to train parameters in order to minimize solution runtime, or to optimize solution quality. In our case, we fix the runtime and minimize energy consumption. We generate problem instances with different degrees of constrainedness and train the parameters to achieve the average best quality for all input scenarios.

Given a list of training instances and corresponding feature vectors, SMAC learns a joint model that predicts the solution quality for combinations of parameter configurations and instance features. These information are useful in selecting promising configurations in large configuration spaces. For each training instance we computed up to 17 features, including: (1) number of constraints, (2) number of variables, (3) number of non-zero coefficients, (4) number of meetings, (5) number of meeting types, (6) scheduling flexibility, (7) average duration of meetings, (8) number of meeting slots per day, (9) total number of meeting slots, (10)-(14) number of rooms in up to 5 building types, and (15)-(17) minimum, maximum, and average difference between outdoor temperature and temperature comfort bounds. These features reflect problem characteristics and are used by SMAC to estimate performance across instances and generate a set of new configurations.

Given a list of promising parameter configurations, SMAC compares them to the current incumbent configuration until a time limit is reached. Each time a promising configuration is compared to the incumbent configuration, SMAC runs several problem instances until it decides that the promising configuration is empirically worse or at least as good as the incumbent configuration. In the latter case the incumbent is updated. In the end, the configuration selected by SMAC is generalized to all problem instances in the training set.

5   Experimental Results

Before describing our main results, we first point out the typical behavior of solving energy aware meeting scheduling as a MIP. Figure 2 shows the performance of the MIP approach on two typical problem instances when given 2 hours of runtime. In general, MIP convergence on larger problems is slow and sometimes MIP fails to converge even after 2 hours. This is exactly why we developed the LNS approach. The typical performance of the LNS approach is also given in Figure 2. In the figure, LNS was given only 15 minutes of runtime but it is capable of returning significantly better results when compared to MIP.

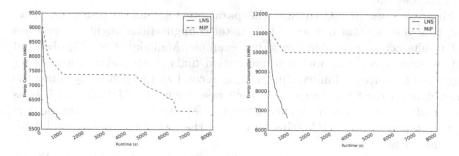

Fig. 2. Typical performance of MIP (2 hours) and LNS (15 minutes) on two benchmark instances

We analyze our LNS approach by considering 8 problem sets. Each problem set contains 10 problem instances that we built by adding energy related information to instances extracted from the PATAT timetabling dataset [19]. The problem sets differ by the number of meetings (M) and the number of rooms (R). Specifically, our problem sets are referred to as 20M-20R, 50M-20R, 100M-20R, 200M-20R, 50M-50R, 100M-50R, 200M-50R, and 500M-50R, where 20M-20R represents the problem set with 20 meetings and 20 rooms. All our experiments were run on a cluster that consists of a 2 × AMD 6-Core Opteron 4334, 3.1GHz with 64GB memory.

Table 1. Total thermal resistance (TR) $\left(\frac{m^2 K}{W}\right)$ and thermal capacitance (TC) $\left(\frac{KJ}{m^2 K}\right)$ of the walls and the window for five types of zones. R and C of each wall can be derived by dividing TR and multiplying TC with area size respectively.

Building Types	External Wall		Internal Wall		Window
	(TR)	(TC)	(TR)	(TC)	(TR)
1	3	120	1.5	120	0.5
2	3	140	1.5	140	0.5
3	3	240	1.5	240	0.5
4	6	120	3	120	0.5
5	6	240	3	240	0.5

In each problem set we must schedule up to 500 meetings whose durations are 1 or 1.5 hours. The meetings must be scheduled over a period of 5 summer days. The available rooms are located in 5 buildings, that differ by their thermal resistance and capacitance as specified in Table 1. We use a 1×4 zone layout where each zone has the same thermal resistance and capacitance as its neighboring zones. Moreover, all rooms have the same geometric area of $6 \times 10 \times 3$ m^3 with a window surface area of 4×2 m^2 and a capacity of 30 people. The solar gain ranges from 50 to 350 W/m^2 during the day. All meetings have between 2 and 30 attendees and we vary the scheduling flexibility for each meeting with an allowable time range of one or two random days (between 09:00-17:00) within the 5 summer days.

First, we used SMAC to tune the parameters for all 80 instances. However, it is possible that one set of parameter configurations might not produce best results across all problem sets. For example, Malitsky et al. [17] observed that instance specific algorithm configuration finds good quality solutions for large sized instances in limited time. Hence, second we independently tuned the parameters for the 8 problem sets. In each case, we used 0.33, 0.33, and 0.34 as the default destroy probabilities and 5 seconds for the default MIP runtime during each repair step. SMAC trains on 60% of the instances and cross-validates with the remaining 40%.

On average, SMAC generated 300 configurations for each problem set. Figure 3 shows the result of the best parameter configurations as determined by SMAC for all 80 instances (AllM-AllR), and for the each of the 8 problem sets. In the end, a MIP runtime of 8.5 seconds, and probabilities of 0.64, 0.28, 0.08 to destroy 2, 3, and 4 rooms respectively were determined to be the best settings by SMAC for all 80 instances. However, the best parameter configurations do vary somewhat for the different problem sets.

Figure 4 shows the performance improvement of MIP and LNS compared to the initial feasible solution, which is referred to as the heuristic solution (HS), over 500 runs. For each run, both the MIP and LNS approach were seeded with HS as an initial solution. Both MIP and LNS were given the same runtime limit of 15 minutes and HS was given 60 seconds. LNS was executed using the parameter configurations that were determined best for each problem set. The results show that LNS significantly improves over MIP when given limited runtime. Overall,

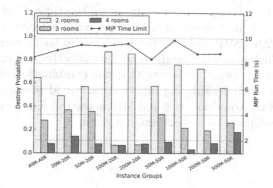

Fig. 3. Parameter configurations as determined by SMAC

the improvement of LNS over MIP is between 14% and 36%. Note that in the largest problem instances, those in 500M-50R, MIP often fails to find even a slight improvement over the given initial solution HS in 15 minutes.

We reran all problem instances in Figure 4 using the AllM-AllR parameter settings. In this case, the performance of LNS decreased by 1 to 2%. Hence, instance specific algorithm configuration did have some impact, but even without it LNS performed significantly better than MIP

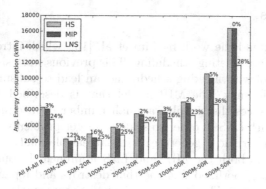

Fig. 4. Performance improvement of MIP and LNS over heuristic solution (HS). MIP and LNS runtime 15 minutes.

Figure 5 provides insight into the optimality gap of the MIP approach and in the variation in solution quality. The MIP optimality gap after 15 minutes runtime is shown on the left. Each box shows the median and the upper and lower quantiles of the optimality gap. The endpoints indicate the maximum and minimum. As observed previously the MIP approach fails to converge on the larger problem instances. Moreover, as can be seen from the figure, MIP's performance substantially degrades as problems become more constrained and exhibit a higher number of meetings to number of rooms ratio.

The variation in solution quality is shown on the right. The median of LNS always falls below the lower quantile of MIP. We note, however, that for problem instances in 20M-20R, MIP and LNS have almost similar performance. For even smaller problem instances MIP tends to be very effective, which is exactly why we have combined the two. We believe that the combination of LNS and MIP is especially good when considering highly constrained problems. LNS is used destroy and repair the solution space and MIP is used to find good local solution, possibly optimal solution, in a short amount of time.

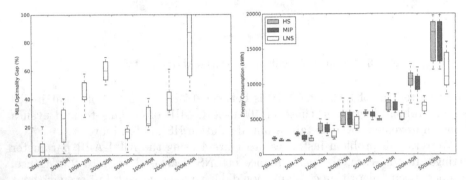

Fig. 5. MIP optimality gap (left) and HS/LNS/MIP solution values (right)

6   Conclusions

In this paper we extend the work by Lim et al. [15], which introduced a MIP model for energy aware meeting scheduling. This previous work shows that combining HVAC control with meeting scheduling can lead to substantial improvements in energy efficiency. The MIP model that is described, however, only solves problem instances that involve a small number of meetings and rooms. We combine MIP with LNS in order to scale to larger problems.

We developed a heuristic to generate an initial feasible solution quickly, which we use to warm start both the MIP and LNS approach. In our experiments, the most effective neighborhood was one that destroys and repairs all meetings scheduled in 2 to 4 rooms. The resulting subproblem was small enough for MIP to solve to (near) optimality and, at the same time, large enough to explore alternate solutions. We studied the performance of MIP and LNS and demonstrated the potential of our LNS approach for effectively tackling large-scale HVAC control and meeting scheduling problems. The LNS achieves 14 to 36% better energy savings than the MIP approach when both given a runtime of 15 minutes. In order to provide an absolute sense of the solution quality, we plan to evaluate our schedules with the EnergyPlus simulator.

We are interested in exploring new algorithmic approaches that allows us to scale even further. We are particularly interested in investigating symmetry breaking in MIP. Symmetry leads to a large number of equivalent solutions, which causes branch-and-bound to be ineffective. While we dealt with symmetries due to meetings with similar characteristics by introducing meeting types, symmetries still exists in our MIP formulation due to rooms with similar characteristics. Introducing room types, however, may not be possible because room temperature at time step t is dependent on time step $t-1$. In future work, we aim to identify branching strategies that can better deal with symmetries in our MIP formulation.

Moreover, we are also interested in investigating an online stochastic approach to our HVAC control and meeting scheduling problem. Such an approach can deal with current requests, future requests, changes and cancelation of requests, but also with the uncertainty around outdoor air temperature and weather conditions.

Acknowledgments. Thanks to Pascal Van Hentenryck for pointing us to optimization problems in the smart buildings space and to Philip Kilby for helpful discussions on LNS. This work is supported by NICTA's Optimisation Research Group as part of the Future Energy Systems project. NICTA is funded by the Australian Government through the Department of Communications and the Australian Research Council through the ICT Centre of Excellence Program.

References

1. ASHRAE: ASHRAE handbook: Fundamentals. American Society of Heating, Refrigerating and Air-Conditioning Engineers (2013)
2. Bloomfield, D., Fisk, D.: The optimisation of intermittent heating. Building and Environment **12**(1), 43–55 (1977)
3. Crawley, D.B., Pedersen, C.O., Lawrie, L.K., Winkelmann, F.C.: Energyplus: Energy simulation program. ASHRAE Journal **42**, 49–56 (2000)
4. Danna, E., Perron, L.: Structured vs. unstructured large neighborhood search: a case study on job-shop scheduling problems with earliness and tardiness costs. In: Proc. International Conference on the Principles and Practice of Constraint Programming (CP), pp. 817–821 (2003)
5. Di Gaspero, L., Rendl, A., Urli, T.: Constraint-based approaches for balancing bike sharing systems. In: Schulte, C. (ed.) CP 2013. LNCS, vol. 8124, pp. 758–773. Springer, Heidelberg (2013)
6. EIA: US Department of Energy, CBECS detailed tables (2003). http://www.eia. gov/consumption/commercial/
7. Gouda, M., Danaher, S., Underwood, C.: Low-order model for the simulation of a building and its heating system. Building Services Engineering Research and Technology **21**(3), 199–208 (2000)
8. Gouda, M., Danaher, S., Underwood, C.: Building thermal model reduction using nonlinear constrained optimization. Building and Environment **37**(12), 1255–1265 (2002)
9. Goyal, S., Barooah, P.: A method for model-reduction of non-linear thermal dynamics of multi-zone buildings. Energy and Buildings **47**, 332–340 (2012)
10. Goyal, S., Ingley, H.A., Barooah, P.: Occupancy-based zone-climate control for energy-efficient buildings: Complexity vs. performance. Applied Energy **106**, 209–221 (2013)

11. Hutter, F., Hoos, H.H., Leyton-Brown, K.: Sequential model-based optimization for general algorithm configuration. In: Coello, C.A.C. (ed.) LION 5. LNCS, vol. 6683, pp. 507–523. Springer, Heidelberg (2011)
12. Kwak, J.Y., Kar, D., Haskell, W., Varakantham, P., Tambe, M.: Building thinc: user incentivization and meeting rescheduling for energy savings. In: Proc. International Conference on Autonomous Agents and Multi-agent Systems (AAMAS), pp. 925–932 (2014)
13. Kwak, J.y., Varakantham, P., Maheswaran, R., Chang, Y.H., Tambe, M., Becerik-Gerber, B., Wood, W.: Tesla: an energy-saving agent that leverages schedule flexibility. In: Proc. International Conference on Autonomous Agents and Multi-agent Systems (AAMAS), pp. 965–972 (2013)
14. LeBras, R., Dilkina, B.N., Xue, Y., Gomes, C.P., McKelvey, K.S., Schwartz, M.K., Montgomery, C.A.: Robust network design for multispecies conservation. In: Proc. AAAI Conference on Artificial Intelligence (AAAI), pp. 1305–1312 (2013)
15. Lim, B.P., van den Briel, M., Thiébaux, S., Backhaus, S., Bent, R.: Hvac-aware occupancy scheduling. In: Proc. AAAI Conference on Artificial Intelligence (AAAI) (2015)
16. Majumdar, A., Albonesi, D.H., Bose, P.: Energy-aware meeting scheduling algorithms for smart buildings. In: Proc. ACM Workshop on Embedded Sensing Systems for Energy-Efficiency in Buildings (BuildSys), pp. 161–168 (2012)
17. Malitsky, Y., Mehta, D., O'Sullivan, B., Simonis, H.: Tuning parameters of large neighborhood search for the machine reassignment problem. In: Gomes, C., Sellmann, M. (eds.) CPAIOR 2013. LNCS, vol. 7874, pp. 176–192. Springer, Heidelberg (2013)
18. McCormick, G.P.: Computability of global solutions to factorable nonconvex programs: Part I- convex underestimating problems. Mathematical Programming 10(1), 147–175 (1976)
19. Melbourne University: Patat dataset (2002). http://goo.gl/XtNwpR
20. Oldewurtel, F., Parisio, A., Jones, C.N., Gyalistras, D., Gwerder, M., Stauch, V., Lehmann, B., Morari, M.: Use of model predictive control and weather forecasts for energy efficient building climate control. Energy and Buildings 45, 15–27 (2012)
21. Pan, D., Yuan, Y., Wang, D., Xu, X., Peng, Y., Peng, X., Wan, P.J.: Thermal inertia: towards an energy conservation room management system. In: Proc. IEEE International Conference on Computer Communications (INFOCOM), pp. 2606–2610 (2012)
22. Portland State University: Efficient class scheduling conserves energy (2012). http://goo.gl/cZwgB
23. Rendl, A., Prandtstetter, M., Hiermann, G., Puchinger, J., Raidl, G.: Hybrid heuristics for multimodal homecare scheduling. In: Beldiceanu, N., Jussien, N., Pinson, É. (eds.) CPAIOR 2012. LNCS, vol. 7298, pp. 339–355. Springer, Heidelberg (2012)
24. Shaw, P.: Using constraint programming and local search methods to solve vehicle routing problems. In: Maher, M.J., Puget, J.-F. (eds.) CP 1998. LNCS, vol. 1520, p. 417. Springer, Heidelberg (1998)

ILP and CP Formulations for the Lazy Bureaucrat Problem

Fabio Furini[1], Ivana Ljubić[2]([✉]), and Markus Sinnl[2]

[1] PSL, Université Paris-Dauphine, CNRS, LAMSADE UMR 7243,
75775 Paris Cedex 16, France
fabio.furini@dauphine.fr
[2] Department of Statistics and Operations Research, University of Vienna,
Vienna, Austria
{ivana.ljubic,markus.sinnl}@univie.ac.at

Abstract. Lazy reformulations of classical combinatorial optimization problems are new and challenging classes of problems. In this paper we focus on the Lazy Bureaucrat Problem (LBP) which is the lazy counterpart of the knapsack problem. Given a set of tasks with a common arrival time and deadline, the goal of a lazy bureaucrat is to schedule a *least* profitable subset of tasks, while having an excuse that no other tasks can be scheduled without exceeding the deadline.

Three ILP formulations and their CP counterparts are studied and implemented. In addition, a dynamic programming algorithm that runs is pseudo-polynomial time and polynomial greedy heuristics are implemented and computationally compared with ILP/CP approaches. For the computational study, a large set of knapsack-type instances with various characteristics is used to examine the applicability and strength of the proposed approaches.

1. Introduction

The lazy bureaucrat problem is a scheduling problem in which a set of jobs has to be scheduled in a most inefficient way. We will consider a special variant of this problem in which all jobs are arriving at the same time and have a common deadline. In this paper, we will simply refer to this problem as the Lazy Bureaucrat Problem (LBP).

The problem is motivated by the following application: a lazy bureaucrat needs to choose a subset of jobs to execute in a single day, in a such a way that no other job fits in his/her working hours and the total profit of selected jobs (e.g., their duration) is minimized (and, hence, for example, the bureaucrat has a good excuse to go home as early as possible). Governments or funding agencies may be interested in applications of the LBP: how to distribute the available budget so that the minimal amount of money is allocated to funding requests, while having a good excuse that no additional funds can be granted without violating the available budget? For further applications and motivations

© Springer International Publishing Switzerland 2015
L. Michel (Ed.): CPAIOR 2015, LNCS 9075, pp. 255–270, 2015.
DOI: 10.1007/978-3-319-18008-3_18

of this problem, see, e.g., [1,5]. The study of the LBP could potentially lead to new interesting insights for knapsack-type problems, see also the conclusion for a discussion on the importance of the LBP.

Definition 1 (Lazy Bureaucrat Problem (with common arrivals and deadlines), LBP). *We are given a set of jobs $I = \{1, \ldots, n\}$, such that each job $i \in I$ is assigned a duration $w_i \in \mathbb{N}$ and a profit $p_i \in \mathbb{N}$. All jobs arrive at the same time, and all have a common deadline $C \in \mathbb{N}$. The goal is to find a least profitable subset of jobs S to be executed so that the schedule cannot be improved by inserting an additional job into it. More precisely, the optimal solution $S^* \subset I$ solves the following problem:*

$$S^* = \arg\min_{S \subset I}\left\{\sum_{i \in S} p_i \mid \sum_{i \in S} w_i \leq C \ and \ \sum_{i \in S} w_i + w_j > C, \ \forall j \notin S\right\}.$$

Note that we assume that the problem instance is non-trivial, so that the optimal solution S is a proper subset of I. The objective considered in this paper is of a very general form and referred to in literature as `weighted-sum`. It generalises the `min-number-of-jobs` objective, which is obtained for $p_i = 1$ and the `time-spent` objective, which is obtained for $p_i = w_i$.

The problem has been introduced in [1] where it was shown that a more general problem variant with individual arrival times and deadlines is NP-hard. For the problem variant with common arrival times and deadlines, [4] show that the problem is weakly NP-hard for the `min-number-of-jobs` objective by reduction from subset-sum. Thus, the problem studied in this paper (with the more general `weighted-sum` objective) is also at least weakly NP-hard. Note that in [1] it is claimed, that the LBP with common arrival times can be solved in pseudo-polynomial time for various objective functions (including `time-spent` and `weighted-sum`), but neither proof nor the corresponding algorithm were explicitly given for this claim. In [5] two greedy heuristics and an FPTAS have been proposed for the `time-spent` objective with common arrival and deadlines. The FPTAS is obtained as an approximate version of an exact enumeration algorithm.

Our Contribution: In this article we provide a first dynamic programming algorithm for the `weighted-sum` LBP (see Section 2) that runs in pseudo-polynomial time. Thus the problem is indeed weakly NP-hard. Besides, we prove properties of optimal solutions that can be used to derive more efficient mathematical models and algorithmic approaches or to derive valid inequalities or stronger constraints in both integer linear programming (ILP) and constraint programming (CP) formulations. Three ILP models including their CP counterparts are presented in Section 3 and greedy heuristics are discussed in Section 4. An extensive computational study is conducted on a large set of diverse benchmark instances, demonstrating the strengths and weaknesses of the proposed approaches.

Notation: Notice that the LBP consists of selecting a subset of jobs, but due to the common arrival times and the common deadline, the order in which the jobs are scheduled remains irrelevant. Therefore, the LBP shares a lot of similarities with the knapsack problem (KP), and can be seen as the problem of packing a set of items in a knapsack in a most inefficient (and therefore, rather counterintuitive) way. Due to this similarity to the knapsack problem, in the remainder of this article we will refer to elements of I as *items*, job durations w_i as *item weights* and p_i as *item profits*. The deadline C will be called the *budget* or *capacity*.

In the following, we will assume that all items are sorted in non-decreasing lexicographic order, first according to their weight, i.e., $w_1 \leq w_2 \leq \cdots \leq w_n$, and then, according to their profits p_i. For each $i \in I$, let $C_i := C - w_i$, so we have $C_1 \geq C_2 \geq \ldots C_n$. We will also denote by $w_{\max} := \max_{i \in I} w_i (= w_n)$ and $w_{\min} := \min_{i \in I} w_i (= w_1)$. Let $W := \sum_{i \in I} w_i$ and $P := \sum_{i \in I} p_i$.

1.1 Solution Properties

In this section we point out some general solution properties that will be used to prove the validity of our models.

Property 1. The capacity used by an arbitrary feasible solution S is bounded from below by $C - w_{\max} + 1$.

Proof. By definition, inserting any item outside of S into the knapsack will exceed its capacity. So, in particular, the capacity of S plus $w_{\max}$ must be $\geq C+1$, i.e., the capacity of S must be at least $C - w_{\max} + 1$. Since $w_{max} \geq w_i, \forall i \in I$, this bound is clearly valid for any item. $\square$

Despite the minimization objective, imposing the upper bound on the size of the knapsack is in general not redundant. Consider the following example in which we are given three items such that $w_1 = 1$, $w_2 = 2$ and $w_3 = 3$, and $p_1 = 10$, $p_2 = p_3 = 1$, and that the knapsack capacity is $C = 4$. Without imposing the knapsack capacity, the optimal solution will be to take the items $\{2, 3\}$ with the total profit of 2, whereas, with the condition of not exceeding the knapsack capacity, the optimal solution will be $\{1, 3\}$ with the total profit of 11. However, in the **time-spent** problem variant, the knapsack constraint is redundant, as shown by the following result:

Property 2. If $w_i = p_i$ for all $i \in I$, capacity of any optimal solution S will not exceed C, even without explicitly imposing this condition.

Definition 2 (Minimal Knapsack Cover). *Given a knapsack problem with capacity C, a cover $\mathcal{C}$ is a set of items, with the property $\sum_{c \in \mathcal{C}} w_c \geq C + 1$, i.e., a subset of items, which exceeds the capacity C. A cover $\mathcal{C}$ is called minimal, iff $\mathcal{C} \setminus \{i\}$ is not a cover for any $i \in \mathcal{C}$. We denote the family of all minimal covers with $\mathcal{K}$.*

Definition 3 (Maximal Knapsack Packing). *Given a knapsack problem with capacity C, a packing $\mathcal{P}$ is a set of items, with the property $\sum_{i \in \mathcal{P}} w_i \leq C$, i.e., a subset of items, which does not exceed the capacity C. A packing $\mathcal{P}$ is called maximal, iff $\mathcal{P} \cup \{i\}$ is not a packing for any $i \notin \mathcal{P}$.*

The following property characterizes the set of feasible LBP solutions.

Property 3. Each feasible LBP solution corresponds to a *maximal feasible packing* of the knapsack with capacity C.

It is worth mentioning that searching for a minimum-profit minimal knapsack cover, and removing an item from it does not lead to an optimal solution of the LBP. Even in the simplest case when $p_i = w_i$, for all $i \in I$, this does not hold: Let the set of items be ordered according to their weights(=profits) $\{31, 32, 33, 40, 45\}$ and let the knapsack capacity be 80. The minimum-profit minimal knapsack cover is $\{40, 45\}$, and after removing one of these items from the cover, the obtained solution is not feasible to LBP (since one can obviously add one of the items from $\{31, 32, 33\}$ without violating the knapsack capacity).

Definition 4 (Critical Weight and Critical Item.). *Assume that items are sorted so that $w_1 \leq w_2 \leq \cdots \leq w_n$. Denote by*

$$i_c = \arg\min\{i \in I \mid \sum_{j \leq i} w_j > C\}$$

the index of a critical item*, i.e., the index of the first item that exceeds the capacity, assuming all $i \leq i_c$ will be taken as well. The* critical weight*, denoted by w_c, is the weight of the critical item, i.e., $w_c = w_{i_c}$.*

Note that the critical item and its weight are uniquely defined, assuming the lexicographic ordering of the items defined above.

Proposition 1. *The weight of the smallest item left out of any feasible LBP solution is bounded above by the critical weight w_c, i.e.:*

$$S \text{ is feasible } \Rightarrow \min_{i \notin S} w_i \leq w_c.$$

Consequently, the size of the knapsack can be bounded from below as:

$$w(S) \geq C - w_c + 1.$$

Proof. Suppose there exists a feasible solution S' with $\min_{i \notin S'} w_i > w_c$. Since the items are ordered by weight, it follows that all items $i \leq i_c$ are in S'. By the definition of the critical item, the sum of the weights of these items exceeds C and thus S' is not feasible. The second part is a refinement of Property 1.   □

2. Dynamic Programming Approach

The latter result indicates a possible iterative approach for solving the LBP. Once, the smallest item left out of the solution is known, the problem reduces to solving the knapsack problem with a lower and upper bound on its capacity (which we will denote LU-KP below). Let S^* be an optimal LBP solution and assume that $i \leq i_c$ is the index of the smallest item not taken into S^* (following the lexicographic ordering of items). Let I_i be the set of the first $i - 1$ items, $I_i = \{1, \ldots, i-1\}$ ($I_1 = \emptyset$), and let $P_i = \sum_{j<i} p_j$ and $W_i = \sum_{j<i} w_j$ be the profits and weights of items from I_i. Then, $S^* = I_i \cup J^*$ where $J^* \subset \{i+1, \ldots, n\}$ solves the following problem:

$$(\text{KP}_i^{\min}) \, J^* = \arg \min_{J \subseteq \{i+1,\ldots,n\}} \{\sum_{j \in J} p_j + P_i \mid C - w_i - W_i + 1 \leq \sum_{j \in J} w_j \leq C - W_i\}$$

The optimal solution of $(\text{KP}_i^{\min})$ can be obtained by solving the complementary maximization problem, which is a LU-KP, as shown in the following result:

Proposition 2. *The optimal LBP solution S^* can be obtained as:*

$$i^* = \arg \min_{i \leq i_c} \max(\text{KP}_i) \text{ and } S^* = I \setminus \arg \max(\text{KP}_{i^*})$$

where $\max(\text{KP}_i)$ is defined as:

$$\max(\text{KP}_i) = \max\{\sum_{j>i} p_j y_j \mid \sum_{j>i} w_j y_j \geq C_i', \sum_{j>i} w_j y_j \leq C'', y_j \in \{0,1\}, \forall j > i\},$$

with $C_i' := W - C - w_i$ and $C'' := W - C + 1$.

Proof. Let J^* be defined as above and let the binary variables y_j be equal to one if $j \notin J^*$ and to zero, otherwise. By rewriting the problem $(\text{KP}_i^{\min})$ stated above as a maximization problem and using y_i variables to indicate items outside of the solution, we end up with the desired reformulation. The value $\text{KP}_{i^*}^{\min}$ is obtained as $P - \text{KP}_{i^*}$. □

Proposition 2 demonstrates that the LBP could be solved in i_c iterations, in each of which a LU-KP is solved to optimality to obtain $\max(\text{KP}_i)$ for an $i \leq i_c$. This can be done in $O(n^2 W)$ time, by applying a standard dynamic programming procedure for the LU-KP that runs in $O(nW)$. We demonstrate below a modification of this algorithm that is based on iteratively solving $\text{KP}_i^{\min}$ by dynamic programming. Similar iterative arguments are used to construct the exact enumeration algorithm underlying the FPTAS for the time-spent variant in [5].

Let $C_i^l := C - w_i - W_i + 1$ and $C_i^u = C - W_i$. For solving $\text{KP}_i^{\min}$ (for a fixed i) we need to apply a "dynamic programming by weights" approach, in which the smallest profit $z(j)$ that can be obtained for each capacity $j \in \{1, \ldots, C_i^u\}$ is computed. The following recursive formulas are used (see, e.g., [3]):

$$\begin{cases} M(0,0) = 0, \quad M(0,j) = -\infty, \quad \forall j \leq C_i^u \\ M(i',j) = \max\{M(i'-1,j), M(i'-1, j-w_{i'}) - p_{i'}\}, \quad i' > i, \forall j \in \{1, \ldots, C_i^u\} \end{cases}$$

The algorithm searches for the filling of the knapsack of capacity j, $j \in \{0, 1, \ldots, C_i^u\}$, that maximizes the sum of negative item profits (which is equivalent to minimizing the sum of the profit of collected items). That way, only the cells containing values $\neq -\infty$ represent feasible solutions, whose total weight is exactly equal to j, i.e., cell $M(i', j)$ contains the maximum possible sum of negative item profits which can be achieved by a solution with weight j using items up to i'. Let $z(j)$ denote the complemented value at the cell $M(|I|, j)$ after applying the DP. The optimal solution of $\text{KP}_i^{\min}$ is obtained as $\min_{C_i^l \leq j \leq C_i^u} z(j)$.

3.   ILP and CP Formulations

The DP presented above relies on solving i_c iterations of the LU-KP, and hence the approach may become prohibitive from the computational perspective if C and/or i_c are very large. In this section we study alternative, *integrative* approaches, in which the LBP is solved as a whole. We propose three ways to formulate the problem, always presenting the valid ILP formulation first, followed by its CP counterpart. We provide some valid inequalities and show how some of them can be lifted to obtain stronger lower bounds. A hybridization of the proposed ILPs and CPs is also discussed in this section as well as a branch-and-cut algorithm based on the ILPs.

3.1   A First Formulation

The first ILP model is obtained by formulating the problem using the binary variables x_i which are set to one if the item is selected, and to zero, otherwise. Each feasible solutions needs to fit into a knapsack, and should exceed its capacity by adding an arbitrary additional item left outside. The formulation uses the Proposition 1, and considers only the items left outside of S whose weight does not exceed the critical weight. The model reads as follows:

$$(\text{ILP}_1) \qquad \min_{x \in \{0,1\}^{|I|}} \sum_{i \in I} p_i x_i$$

$$\sum_{i \in I} w_i x_i \leq C \tag{1}$$

$$\sum_{j \in I, j \neq i} w_j x_j + w_i(1 - x_i) \geq (C + 1)(1 - x_i) \quad \forall i \in I : i \leq i_c \tag{2}$$

Constraint (1) is a knapsack constraint stating that the weight of all selected items cannot exceed the available budget C. Inequalities (2) make sure that adding each additional item i such that $w_i \leq w_c$ will exceed C. We will refer to them as *covering inequalities* associated to items $i \leq i_c$. Notice that the latter inequalities can be rewritten as: $\sum_{j \in I, j \neq i} w_j x_j \geq (C_i + 1)(1 - x_i) \quad \forall i \in I : i \leq i_c$.

Proposition 3. *The model (ILP_1) is a valid formulation for the LBP.*

Strengthening Covering Inequalities. Proposition 3 shows that it is sufficient to consider items not exceeding the critical weight to enforce a feasible LBP solution. It is not difficult to see that covering inequalities associated with remaining items $i > i_c$ are also valid for our problem:

$$\sum_{j \in I, j \neq i} w_j x_j + w_i(1 - x_i) \geq (C + 1)(1 - x_i) \quad \forall i \in I : i > i_c$$

However, all these constraints are dominated by the *global covering constraint*, derived from the global lower bound given in Proposition 1:

$$\sum_{j \in I} w_i x_i \geq C_c + 1, \tag{3}$$

where $C_c = C - w_c$. One can easily construct an example showing that the lower bound of the LP relaxation of the model (ILP_1) is strengthened by adding this constraint: Let the item weights be $w = p = (1, 119, 552, 605, 739, 863)^t$. The optimal LP-solution of the (ILP_1) is $x' = (0.4685, 0.4647, 0.4357, 0.4285, 0.4004, 0.3435)^t$ with the LP-relaxation value equal to 1147.91. However, the optimal LP-solution of the (ILP_1) extended by (3) is $x' = (0.3994, 0.3871, 0.2945, 0.2716, 1.0000, 0.2140)^t$ with the LP-relaxation value equal to 1297. This example also illustrates that the LBP is more difficult than the KP when it comes to the structure of the optimal LP-solution.

Furthermore, coefficients of the covering constraints (2) and (3) can be down-lifted as shown in the following Proposition. Let (for a fixed $i \in I$)

$$\tilde{C} = \begin{cases} C_i + 1, & i \leq i_c \\ C_c + 1, & \text{otherwise} \end{cases}$$

Proposition 4. *For a given $i \in I$, coefficients of the associated covering inequalities* (2) *can be down-lifted to $\sum_{k \in I} \alpha_k x_k \geq \tilde{C}$ where*

$$\alpha_k := \begin{cases} \min\{w_c, \tilde{C}\}, & k = i \\ \min\{w_k, \tilde{C}\}, & \text{otherwise} \end{cases} \quad \forall k \in I$$

Similarly, coefficients of the constraint (3) *can be down-lifted as $\alpha_k = \min\{w_k, \tilde{C}\}$, for all $k \in I$.*

Proof. We only show the first part of the proof. For a given $i \in I$, let us rewrite the covering constraint (2) as:

$$\sum_{j \in I, j \neq i} w_j x_j + \tilde{C} x_i \geq \tilde{C}$$

Lifting down from w_k to $\min\{w_k, \tilde{C}\}$ follows by standard arguments, given that we are dealing with binary decision variables. We can also down-lift the coefficient next to the variable x_i from $\tilde{C}$ to w_c (assuming $w_c < \tilde{C}$), which follows from the following arguments. If $x_i = 0$, the whole constraint makes sure that the capacity of the solution is at least $\tilde{C}$. On the other hand, if $x_i = 1$, by Property (1), we know that the following inequality holds:

$$\sum_{j \in I, j \neq i} w_j x_j + w_i \geq C - w_c + 1.$$

The latter inequality can also be rewritten as $\sum_{j \in I, j \neq i} w_j x_j + w_c \geq \tilde{C}$. Hence, by reordering the coefficients, we end up with $\alpha_i = w_c$.   □

Observe that after lifting, however, the covering inequality (2) associated to i_c and the global covering inequality (3) become identical.

Formulation (CP$_1$). A constraint programming counterpart of the given ILP formulation is derived by following the problem definition. The set of chosen items cannot exceed the given capacity (see constraint (4)), and every item which is left outside, if added to the solution, exceeds C (see constraint (5)). The same set of x of binary variables is used to indicate selected items.

$$(CP_1) \qquad \qquad \min_{x \in \{0,1\}^{|I|}} \sum_{i \in I} p_i x_i$$

$$\text{Pack}_{\leq}(w, x, C) \qquad \qquad (4)$$

$$\text{if } (x_i = 0) \text{ then } \text{Pack}_{\geq}(w, x, C_i + 1) \qquad \forall i \in I \quad (5)$$

In constraint (4), a bin-packing constraint $\text{Pack}_{\leq}(w, x, C)$ of size one (i.e., a knapsack constraint) is used to ensure that the selected items do not exceed the capacity. In our notation, $\text{Pack}_{\leq}(w, x, C)$ means that the set of items whose weights are given by a vector w, need to be packed into a knapsack of capacity C. The constraint is usually implemented by requiring that the items are placed in two bins, one of them having the capacity of C (and corresponding to items i such that $x_i = 1$), and the other of capacity $W - C$, containing the remaining items. Note that efficient constraint propagation techniques for bin-packing type of constraints exist, see, e.g., [8,9]. The remaining set of constraints (5) are a direct translation of inequalities (2) into the CP language. They impose that, for any item i, which is not selected, it must hold that the item cannot be added to the schedule without violating the capacity constraint. The latter is ensured with constraint $\text{Pack}_{\geq}(w, x, C_i + 1)$. Again, by complementing the x variables, $\text{Pack}_{\geq}(w, x, C + 1 - w_i)$ can be modeled as a bin-packing constraint. Note that this constraint can be down-lifted to $\text{Pack}_{\geq}(w \setminus w_i, x \setminus x_i, C_i + 1)$.

Model (CP_1) has $|I| + 1$ constraints. The size of the model can be improved by exploiting Proposition 1, i.e., the size of any feasible solution is bounded below by $C_c + 1$. Since $C_i \leq C_c$ for all $i \geq i_c$, we do not need to impose (5) for these items, but replace it with a single constraint

$$\text{Pack}_{\geq}(w, x, C_c + 1)$$

Hybridization and Branch-and-Cut. One could easily hybridize the (ILP_1) with (CP_1) by replacing the capacity inequalities (2) by the indicator constraints (5).

Concerning the implementation of the (ILP_1), there are two possibilities. The model, involving $i_c + 1$ constraints, can be given as-is (i.e., as a compact model) to a black-box MIP solver. Alternatively, one may start with a simple knapsack constraint (1) and insert the capacity inequalities (2) "on the fly" inside of the branch-and-bound tree, only when these inequalities are violated by the current LP-solution. This procedure, known as branch-and-cut, may be advantageous over the compact model, in particular if the number of items, and respectively the index of the critical item, are very large.

Both observations also hold for the remaining formulations presented below.

3.2   A Second (and Sparser) Formulation

To motivate this new formulation, let us first focus on the capacity inequality for $i = 1$:

$$\sum_{j \in I, j \neq 1} w_j x_j \geq (C_1 + 1)(1 - x_1).$$

Obviously, if $x_1 = 1$, this constraint is not binding, but for $x_1 = 0$, it will dominate the remaining covering constraints (2) of the previous formulation (ILP_1) (since the items are ordered in non-decreasing order according to w_i, which implies $C_1 \geq C_i$, for all $i > 1$).
Similarly, if $x_1 = 1$, and $x_2 = 0$, the constraint $\sum_{j \in I, j \neq 2} w_j x_j \geq (C_2+1)(1-x_2)$ of the previous model will be binding. Hence, for $x_1 = 1$, and $x_2 = 0$, the latter constraint can be replaced by a stronger inequality:

$$\sum_{j \in I, j > 2} w_j x_j + w_1(1 - x_2) \geq (C_2 + 1)(1 - x_2),$$

which also remains valid for $x_2 = 1$ (since it is not binding in this case). One easily observes that we can continue deriving valid inequalities in this fashion, until we reach the critical item. We finally derive the following lifted model for the LBP:

(ILP_2) $$\min_{x \in \{0,1\}^{|I|}} \sum_{i \in I} p_i x_i$$

$$\sum_{i \in I} w_i x_i \leq C$$

$$\sum_{j > i} w_j x_j \geq (C + 1 - \sum_{j \leq i} w_j)(1 - x_i) \quad \forall i \in I, i < i_c \quad (6)$$

The new covering constraints (6) (that we will refer to as *sparser covering constraints*) state that, if the item $i \in I$, $i < i_c$ is the smallest item left outside of the solution, the remaining capacity to be filled by the items j, $j > i$ has to be at least $C + 1 - \sum_{j \leq i} w_i$. We observe that covering constraints (6) for $i \geq i_c$ are redundant, since, by definition, in that case we have $C + 1 - \sum_{j \leq i} w_j < 0$.

Proposition 5. *The model* (ILP_2) *is a valid formulation for the LBP.*

Proof. To show the validity of this model, we need to prove that a) no feasible solutions are cut off by this model, and b) all knapsack solutions that are not maximal are not feasible for this model.

a) Assume that $S \subset I$ is a feasible LBP solution that is cut off by our model. Then, there exists a covering constraint (6) associated to an item $\tilde{i}$ such that S violates that constraint. Obviously, $\tilde{i} \notin S$, and let $\tilde{i} < i_c$ be the item with the smallest index such that constraint (6) is violated. Let x^* be the characteristic vector associated to S. So we have:

$$\sum_{j>\tilde{i}} w_j x_j^* + \sum_{j<\tilde{i}} w_j + w_{\tilde{i}} \leq C$$

This however contradicts the feasibility of S, since the latter inequality says that even if all items from $\{1, \ldots, \tilde{i} - 1\}$ belong to S, S would not be a maximal knapsack solution, since by adding the item $\tilde{i}$ to it, we would still end up with a feasible packing of the knapsack.

b) Let us now assume a subset $S \subset I$ is a feasible solution to our model such that it is a feasible knapsack packing, but not maximal, i.e., it can be extended by at least one more item without violating the capacity constraint. Let x^* be the characteristic vector of S. Let k be the index of the smallest item that does not belong to S. If $k > i_c$, then by definition of i_c, no further items can be added to S, so S is maximal.

   Assume now that there exists an item that can be added to S so that S still remains a feasible packing. If such an item exists, then this certainly holds for the smallest item not in S as well. Since S is a feasible solution, it also satisfies the covering constraint (6) associated to k:

$$\sum_{j>k} w_j x_j^* \geq (C + 1 - \sum_{j \leq k} w_j)(1 - x_k^*)$$

i.e.

$$\sum_{j>k} w_j x_j^* \geq C + 1 - \sum_{j \leq k} w_j$$

Since $x_j^* = 1$ for all $j < k$, the latter inequality can also be rewritten as:

$$\sum_{j>k} w_j x_j^* + \sum_{j<k} w_j x_j^* + w_k \geq C + 1$$

which implies that adding the item k into the solution S would exceed the available knapsack capacity, which is a contradiction.

□

In a similar fashion as above, we can down-lift some of the coefficients associated to the covering constraints (6).

Proposition 6. *For a given $i \in I$, let $\bar{C}_i = C + 1 - \sum_{j \leq i} w_j$. Coefficients of the associated capacity inequality (6) can be down-lifted to $\sum_{k \in I, k > i} \alpha_k x_k + \bar{C}_i x_i \geq \bar{C}_i$ where*

$$\alpha_k := \begin{cases} \min\{w_k, \bar{C}_i\}, & k > i \\ \min\{w_c, \bar{C}_i\}, & k = i \end{cases} \quad \forall k \geq i$$

A potential advantage of (ILP$_2$) over (ILP$_1$) is the sparsity of the covering constraints (6). Since less variables are involved, and even more, the constraint matrix has a triangular structure, we expect that these properties will be effectively exploited by ILP solvers.

Formulation (CP$_2$). A CP counterpart of the (ILP$_2$) can be easily derived using the packing and the indicator constraint. In the formulation (CP$_1$), one has to replace (5) by:

$$\texttt{if } (x_i = 0) \texttt{ then Pack}_\geq((w)_{i+1}^n, (x)_{i+1}^n, C + 1 - \sum_{j \leq i} w_j) \quad \forall i < i_c \quad (7)$$

3.3 A Third (and Extended) Formulation

Recall that for any knapsack solution to be feasible for the LBP it is sufficient that adding the weight of the smallest item left out of the knapsack already violates the capacity C. Our next model encodes this information by extending the previous model with a non-negative continuous variable z modeling the weight of a smallest item that is left out of the solution.

$$(\text{ILP}_3) \quad \min_{x \in \{0,1\}^{|I|}, z \geq 0} \sum_{i \in I} p_i x_i$$

$$\sum_{i \in I} w_i x_i \leq C \quad (8)$$

$$\sum_{i \in I} w_i x_i + z \geq C + 1 \quad (9)$$

$$z \leq w_c - (w_c - w_i)(1 - x_i) \quad \forall i \in I, i \leq i_c \quad (10)$$

This formulation contains only a single additional variable, but significantly simplifies the structure of the constraint matrix. Besides the packing constraint (8) and the covering constraint (9), the remaining matrix has a diagonal structure, plus one column of ones (corresponding to the variable z). Therefore, it is expected that the ILP solvers can be even more efficient when solving (ILP$_3$) than the previous formulations.

Validity of this model can be easily verified. Coupling constraints (10) make sure that $z \geq 0$ is smaller than the weight of the smallest item not included in the solution. Due to the covering constraint (9) and the minimization objective function (with $p_i \geq 0$), it follows that z be exactly the same as the weight of the smallest item left out of the solution. For the same reasons as above (cf. Proposition 1), it is not necessary to impose constraints (10) for $i > i_c$.

Formulation (CP$_3$). CP counterpart (CP$_3$) for the latter ILP model is also obtained with the help of an additional integer variable z.

$$(CP_3) \qquad \min_{x\in\{0,1\}^{|I|},\,z\in\mathbb{N}} \sum_{i\in I} p_i x_i$$

$$\mathrm{Pack}_{\leq}(w,x,C)$$

$$\mathrm{Pack}_{\geq}(w,x,C+1-z) \tag{11}$$

$$z = \min\{w_i \mid i \in I, x_i = 0, i \leq i_c\} \tag{12}$$

The set of constraints (5) (respectively, (7)) is replaced by the combination of constraints (11) and (12). Constraint (12) ensures that z is the weight of the smallest item not selected, and constraint (11) makes sure, that this smallest item can not be added to the selected items without violating the capacity constraint.

4.   Greedy Heuristics

To complete our study, and to verify the difficulty of the studied benchmark instances, we have also implemented a greedy heuristic. Greedy heuristics are the most classical tool for constructing an initial feasible solution to knapsack-related problems. The approach consists of two basic steps: (i) examining the items according to a pre-specified order, and, at each iteration, (ii) adding the current item to the solution iff its weight does not exceed the current residual capacity.

The greedy algorithm for the traditional 0/1 KP sorts the items according to non-increasing p_j/w_j values. For the LBP, since the sum of profits needs to be minimized, different orderings need to be considered. We have tested the following six sorting criteria for creating feasible solutions:

$$1/p_j; \quad 1/w_j; \quad w_j/p_j; \quad 1/(p_j * w_j); \quad 1/(p_j + w_j); \quad p_j/w_j. \tag{13}$$

Note that in the greedy approximation algorithms of [5] for the time-spent variant, the items are ordered non-decreasing according to p_i (which is equivalent to our first variant). Finally, since greedy heuristics are typically executed very fast, we also propose to combine all of them, and return the best obtained solution. This latter approach we call *greedy-comb*.

5.   Computational Study and Conclusion

Benchmark Instances. We used the well-known instance generator of [7] for the classical 0/1 KP problem to randomly generate instances. Such instances are the classical instances used in the literature to test KP algorithms. Nine different classes of instances are obtained, see [6] for more details.

For each instance class and value of $\overline{R} \in \{1000, 10000\}$, we generated 27 LBP instances by considering all combinations of (i) number of items $n \in \{10, 20, 30, 40, 50, 100, 500, 1000, 2000\}$; (ii) capacity $C \in \{\lfloor 0.25W \rfloor, \lfloor 0.50W \rfloor, \lfloor 0.75W \rfloor\}$

(and C increased by 1, if even, for classes 7 and 8); thus obtaining 486 LBP instances. In the next section we present the outcome of our computational study.

Results. All algorithms were coded in C++ and the experiments were performed on a cluster of computers, each consisting of 20 cores (2.3 GHz) and with 64GB RAM available for 20 cores. As ILP and CP solver we used Cplex 12.6 (single thread mode). Default time limit of 600 seconds was imposed to all runs.

We first compare the quality of solutions obtained by the proposed greedy heuristic, with respect to different selection criteria. Table 1 reports percentage gaps to the optimal (or best known solutions) averaged over all instances for all six selection variants presented above, and for the greedy-comb approach. We

Table 1. Average percentage gaps from best known solutions by different heuristic algorithms

Algorithm	Class									
	1	2	3	4	5	6	7	8	9	avg
Greedy heuristics										
greedy$[1/p_j]$	29.25	7.53	11.83	1.55	11.78	2.23	2.24	11 76	2.04	9.01
greedy$[1/w_j]$	00.85	9.08	11.83	1.55	11.78	2.23	2.24	11.76	55.35	19.19
greedy$[w_j/p_j]$	6.71	2.16	1.92	1.55	2.11	2.23	2.24	1.85	2.94	2.63
greedy$[1/(p_j * w_j)]$	56.20	8.35	11.83	1.55	11.78	2.23	2.24	11.76	2.94	12.10
greedy$[1/(p_j + w_j)]$	56.35	8.39	11.83	1.55	11.78	2.23	2.24	11.76	4.20	12.26
greedy$[p_j/w_j]$	71.34	19.67	11.83	22.03	11.78	2.23	2.24	11.76	68.07	24.55
greedy-comb	6.71	1.14	1.03	1.55	1.07	2.23	2.24	0.96	2.82	2.19

observe that among the single selection criteria, the one based on sorting the items according to w_j/p_j performs the best, which is also intuitive, since in this case we prefer items with low profit and high weight. It is worth mentioning that the greedy-comb approach succeeds to improve the quality of solutions obtained by the best greedy approach, for several classes of instances, for which it seems that the other selection criteria perform better. Even for the largest instances, the greedy-comb takes less than one second, which makes it the method of choice for determining high-quality upper bounds.

The quality of lower bounds of the proposed ILP formulations is reported in Table 2. Lifted formulations (ILP_i) are denoted by (ILP_i^l) for $i = 1, 2$. Table 2 shows that lifting significantly reduces the LP-gap for both (ILP_1) and (ILP_2), and that (ILP_3) provides very strong lower bounds, comparable to those obtained after lifting the first two formulations. Looking at different classes of instances, we notice that the first class (uncorrelated weights and profits) is by far the most difficult one, with average LP-gaps of more than 13% for all ILP formulations.

Finally, Table 3 provides a comparison of running times of three basic ILP formulations, (ILP_1^l), (ILP_2^l), two ILP+CP hybrids (in which indicator constraints are used), a branch-and-cut implementation of (ILP_1), constraint programming formulation (CP_3), and the DP algorithm. The first half of the table shows the

Table 2. Average percentage gaps from best known solutions by different linear programming relaxations

Algorithm	Class									
	1	2	3	4	5	6	7	8	9	avg
Linear Relaxation										
ILP_1	20.11	41.52	40.09	45.62	39.90	45.35	45.35	40.17	22.98	37.90
ILP_1^l	13.67	3.19	2.75	2.52	2.68	2.92	2.93	2.84	6.64	4.46
ILP_2	26.20	43.82	41.37	45.90	41.14	46.61	46.62	41.44	43.71	41.87
ILP_2^l	13.75	4.30	3.28	2.46	3.25	2.64	2.64	3.38	6.64	4.71
ILP_3	13.67	3.19	2.79	2.52	2.68	2.95	2.96	2.88	6.64	4.47

Table 3. Comparison between the different exact methods proposed

Items	ILP_1				ILP_2			ILP_3	CP_3	DP
	ILP_1	ILP_1^l	B&C	$ILP_1 + CP$	ILP_2	ILP_2^l	$ILP_2 + CP$			
Avg t[sec.s]										
10	0.0	0.0	0.0	0.0	0.0	0.0	0.0	0.0	0.0	33.3
20	0.1	0.0	0.0	0.1	0.1	0.0	0.1	0.0	6.1	55.6
30	0.3	0.1	0.1	0.5	0.2	0.1	0.3	0.1	432.3	55.6
40	0.6	0.2	0.2	1.3	0.5	0.2	0.7	0.2	555.8	66.7
50	1.1	0.2	0.7	2.7	0.8	0.3	1.3	0.2	557.6	66.7
100	45.6	1.9	30.5	61.2	34.5	26.7	31.6	1.2	600.0	67.2
500	315.9	59.9	341.6	362.0	357.4	169.3	402.9	56.0	600.0	257.3
1000	476.6	60.7	367.5	542.9	531.2	185.9	555.8	34.3	600.0	377.1
2000	552.8	152.3	445.7	567.9	584.1	230.5	578.8	61.5	600.0	555.8
# of TL										
10	0	0	0	0	0	0	0	0	0	3
20	0	0	0	0	0	0	0	0	0	5
30	0	0	0	0	0	0	0	0	36	5
40	0	0	0	0	0	0	0	0	50	6
50	0	0	0	0	0	0	0	0	50	6
100	1	0	1	3	2	1	1	0	54	6
500	22	5	27	24	25	15	30	5	54	22
1000	32	3	31	45	44	16	48	3	54	30
2000	48	3	37	51	52	18	51	5	54	46

average computing times over all instances, sorted according to the number of items. The second half reports the number of unsolved instances per each group (out of 54). Results for (CP_1) and (CP_2) are not reported, as they appear even slower than (CP_3). We notice that the best performing exact approaches are (ILP_1^l) and (ILP_3) which we explain by a fact: very tight lower bounds. Among the whole benchmark set of 486 instances, less then 20 remain unsolved within 10 minutes by running these approaches. On the contrary, for CP and DP algorithms, instances with ≥ 100 already appear very difficult: none of them could be solved to optimality within the time-limit by CP, and less than half of them with ≥ 500 items could not be solved by the DP. The DP did especially struggle to solve instance class 9, where the value of the capacity is high. Surprisingly, running the B&C implementation of (ILP_1^l) is slower than solving the compact model, which may be explained by the strength of general-purpose CPLEX cuts (like knapsack-cover inequalities, MIR or 0-1/2 cuts) that are better exploited if complete information on the structure of the solution is given to the solver, rather then when this information is provided "on the fly".

Conclusions and Future Work. This article considers integrative ILP and CP approaches for solving the LBP problem to optimality. As pointed out at the beginning of this article, one could alternatively solve the problem by resolving a linear number of iterations of a knapsack problem with a lower and upper bound on its capacity. We are currently working on a faster dynamic programming algorithm and alternative formulations that might be competitive against the approaches presented in this paper. Furthermore, we consider the CP formulations presented in this article as a first attempt to model the problem using constraint programming and hope to draw attention of the CP community to this interesting problem for which we believe more efficient approaches could be developed, by hybridizing ILP, CP and DP techniques. In particular, an adaptation of a DP algorithm could be used for propagation (as it was proposed in [9] for the knapsack problem).

"Lazy" reformulations of other standard combinatorial optimization problems (such as, bin packing) have been recently proposed (see, e.g. [2]), but only approximation algorithms have been studied so far. Studying lazy packing or scheduling problems is a particularly interesting field of research for both the IP and the CP community, since it gives rise to a rich set of new solution properties, valid inequalities, and studies of the opposite side of the polytope, as the optimization is driven in the opposite direction when compared to their standard (non-lazy) counterparts. Our computational study showed that the LBP is more difficult than the KP: for the latter, instances with several thousands of items can be easily solved, while for the LBP, a few thousands items make the problem difficult (using ILP formulations). Finally exact methods for the LBP are especially important since, as the lazy counterpart of the KP, the LBP is a basic problem which appears as a subproblem in many lazy reformulations. Effective methods for the LBP can then lead to computational speed-up for a large set of lazy problems.

Acknowledgments. The authors want to thank to Alberto Ceselli, Silvano Martello, Michele Monaci and Ulrich Pferschy for useful discussions. The research of M. Sinnl was supported by the Austrian Research Fund (FWF, Project P26755-N19).

References

1. Arkin, E.M., Bender, M.A., Mitchell, J.S., Skiena, S.S.: The lazy bureaucrat scheduling problem. Information and Computation **184**(1), 129–146 (2003)
2. Boyar, J., Epstein, L., Favrholdt, L.M., Kohrt, J.S., Larsen, K.S., Pedersen, M.M., Wøhlk, S.: The maximum resource bin packing problem. Theoretical Computer Science **362**(1–3), 127–139 (2006)
3. Casazza, M., Ceselli, A.: Mathematical programming algorithms for bin packing problems with item fragmentation. Computers & Operations Research **46**, 1–11 (2014)
4. Gai, L., Zhang, G.: On lazy bureaucrat scheduling with common deadlines. Journal of Combinatorial Optimization **15**(2), 191–199 (2008)

5. Gourvés, L., Monnot, J., Pagourtzis, A.T.: The lazy bureaucrat problem with common arrivals and deadlines: approximation and mechanism design. In: Gasieniec, L., Wolter, F. (eds.) FCT 2013. Lecture Notes in Computer Science, vol. 8070, pp. 171–182. Springer, Heidelberg (2013)
6. Martello, S., Pisinger, D., Toth, P.: Dynamic programming and strong bounds for the 0–1 knapsack problem. Management Science **45**, 414–424 (1999)
7. Pisinger, D.: David Pisinger's optimization codes (2014). http://www.diku.dk/pisinger/codes.html
8. Shaw, P.: A constraint for bin packing. In: Wallace, M. (ed.) CP 2004. LNCS, vol. 3258, pp. 648–662. Springer, Heidelberg (2004)
9. Trick, M.A.: A dynamic programming approach for consistency and propagation for knapsack constraints. Annals of Operations Research **118**(1–4), 73–84 (2003)

The Smart Table Constraint

Jean-Baptiste Mairy[1]([✉]), Yves Deville[1], and Christophe Lecoutre[2]

[1] ICTEAM, Université catholique de Louvain, 1348 Louvain-la-Neuve, Belgium
{jean-baptiste.mairy,yves.deville}@uclouvain.be
[2] CRIL-CNRS UMR 8188, Université d'Artois, F-62307 Lens, France
lecoutre@cril.fr

Abstract. Table Constraints are very useful for modeling combinatorial problems in Constraint Programming (CP). They are a universal mechanism for representing constraints, but unfortunately the size of their tables can grow exponentially with their arities. In this paper, we propose to authorize entries in tables to contain simple arithmetic constraints, replacing classical tuples of values by so-called smart tuples. Smart table constraints can thus be viewed as logical combinations of those simple arithmetic constraints. This new form of tuples allows us to encode compactly many constraints, including a dozen of well-known global constraints. We show that, under a very reasonable assumption about the acyclicity of smart tuples, a Generalized Arc Consistency algorithm of low time complexity can be devised. Our experimental results demonstrate that the smart table constraint is a highly promising general purpose tool for CP.

Table constraints explicitly express the allowed combinations of values as sets of tuples, which are called tables. Table constraints can theoretically encode any kind of constraints and are amongst the most useful ones in Constraint Programming (CP). Indeed, they are often required when modeling combinatorial problems in many application fields. The design of filtering algorithms for such constraints has generated a lot of research effort, see [2,8,14–16,19,21,24,27]. The biggest problem with table constraints are their size. Several approaches have been proposed to reduce this size. Two of them modify the definition of classical tuples: compressed tuples [12,25,30] and short supports applied to table constraints [10]. Compressed tuples allow tuples entries to contain sets. A compressed tuple thus represents all the tuples in the cartesian product of the sets. Short supports applied to table constraints allow variables to be left out of the short tuple. Left-out variables can take any values from their domains. In this paper, we propose to generalize both compressed tuples and short supports in table constraints by authorizing tuples to contain simple arithmetic constraints. We call such tuples *smart tuples*, and tables containing smart tuples *smart tables*. For instance, the following set of tuples $\{(1,2,1), (1,3,1), (2,2,2), (2,3,2), (3,2,3), (3,3,3)\}$ on variables $\{x_1, x_2, x_3\}$ with domains $\{1,2,3\}$ can be represented by a smart table containing only one smart tuple:

x_1	x_2	x_3
$= x_3$	≥ 2	$*$

© Springer International Publishing Switzerland 2015
L. Michel (Ed.): CPAIOR 2015, LNCS 9075, pp. 271–287, 2015.
DOI: 10.1007/978-3-319-18008-3_19

or in an equivalent form by $(x_1 = x_3, x_2 \geq 2)$. A symbol $*$ in the tabular form of a smart tuple means that, if not occurring anywhere else, the corresponding variable is not constrained at all by the tuple (which is not the case here).

As a motivating example, let us consider a car configuration problem. We assume that the cars to configure have 2 colors (one for the body, $colB$, and the other for the roof, $colR$), a model number $modNum$, an option pack $optPack$ and an onboard computer $comp$. A configuration rule might state that, for a particular model number a and some fancy body color set S, an option pack less than a certain pack b implies that the onboard computer cannot be the most powerful one, c, and that the roof color has to be the same as the body color. This configuration constraint can be written as:

$$modNum = a \wedge colB \in S \wedge optPack < b \Rightarrow comp \neq c \wedge colR = colB$$

The encoding of this constraint with a smart table consists of four smart tuples: $(modNum \neq a), (colB \notin S), (optPack \geq b)$ and $(comp \neq c, colR = colB)$, which gives under tabular form:

modNum	colB	colR	optPack	comp
$\neq a$	$*$	$*$	$*$	$*$
$*$	$\notin S$	$*$	$*$	$*$
$*$	$*$	$*$	$\geq b$	$*$
$*$	$*$	$= colB$	$*$	$\neq c$

Encoding this constraint with classical tuples is exponentially larger. Even using compressed tuples or short supports results in a table that is strictly longer. This is because none of these techniques can be used to encode compactly the relation existing between $colB$ and $colR$ (they require, for this case, one distinct tuple for each possible color). Smart table constraints can never be larger than classical table constraints, even using compressed tuples or short support because smart table constraints generalize all of the above. Using reification (decomposition by adding auxiliary variables) of the configuration rule does not guarantee the same level of pruning as the smart table encoding since there is a cycle to handle.

Importantly, smart table constraints can be viewed as a disjunction of conjunctions of basic arithmetic constraints. Indeed, each smart tuple contains a conjunction of basic arithmetic constraints and the table is a disjunction of such tuples, since the variables can satisfy any of the smart tuples. Filtering of logical combinations of constraints has already been studied in the literature [1,3,4,9,11,17,18,28,29]. However, the particular form of our smart tuples leads to a filtering algorithm with a low polynomial time complexity. More precisely, we show how Simple Tabular Reduction (STR) [14,27] can be adapted for smart table constraints to produce an efficient filtering procedure to enforce Generalized Arc Consistency. Smart table constraints can be viewed as a subset of the logic algebra defined in [1]: we impose a particular form on the logical combinations (disjunction of conjunction, conjunctions forming acyclic networks) and we restrict the constraints that can be combined to be simple arithmetic constraints.

The rules for the filtering smart table constraints follow the ones defined in [1]. The reasons for such a subset of the logical algebra are multiple. The choice of disjunction of conjunctions has been made to keep smart tables close to classical tables. The restriction to acyclic conjunctions is a requirement for the filtering rules from [1] to provide the GAC maximal inconsistent sets. Maximal inconsistency sets are mandatory if the filtering rules from [1] are to be used to compute GAC of a smart table constraint. Without this guarantee, a procedure, more complex and more expensive than the filtering rules of [1], would have to be used. See for instance the filtering for general conjunctions defined in [3,11,17]. Despite those restrictions, smart tuples greatly increase the expressive power of classical tuples. The novelty in our approach lies in the introduction of a concrete propagator for such a subset of the logic algebra. The pruning achieved by the disjunction is equivalent to the pruning of constructive disjunction [28,29]. The propagation of a whole table constraint can even be seen as the propagation of a large constructive disjunction. The ability to leave variables out of the constraints in the smart tuple makes their filtering as efficient as the improved constructive disjunction filtering defined in [17]. In [4], the authors propose a filtering for constructive disjunction based on indexicals as well as a stronger filtering, considering disjunctions together with other constraints. In this paper, we do not investigate propagating more than one table constraint at a time. The filtering for disjunctions proposed in this paper is stronger than the one proposed in [9], as we have here the same pruning as the constructive disjunction, which is not the case in [9].

1 Defining Smart Table Constraints

A Constraint Satisfaction Problem (CSP) P is composed of an ordered set of *variables* $X = \{x_1, \ldots, x_n\}$, where each variable x has a domain of possible values denoted by $dom(x)$, and a set of *constraints* $C = \{c_1, \ldots, c_e\}$, where each constraint c corresponds to a relation denoted by $rel(c)$ on a subset of variables of X ; this subset is called the *scope* of c and denoted by $scp(c)$. Each constraint c defines the possible combinations of values satisfying c in $rel(c)$. The *arity* of a constraint c is #$scp(c)$, i.e., the number of variables involved in c. The largest arity is denoted by r, while the size of the largest domain is denoted by d.

A *literal* is a variable value pair (x, a) such that $x \in X$. A literal of a constraint c is a literal (x, a) such that $x \in scp(c)$. A literal (x, a) is *valid* iff $a \in dom(x)$. A *tuple on* an (ordered) subset of variables $Y = \{y_1, \ldots, y_p\} \subseteq X$ is a sequence of literals $((y_1, a_1), \ldots, (y_p, a_p))$, one for each variable $y \in Y$. When there is no ambiguity about Y, we simply write $(a_1, \ldots, a_p)$. A tuple is *valid* iff all its literals are valid. A tuple τ is *allowed* by a constraint c iff $\tau \in rel(c)$. A tuple τ *satisfies* a set of constraints C' iff for every constraint $c' \in C'$, $\tau[scp(c')]$ is allowed by c', where $\tau[Y]$ denotes the restriction of τ on literals referring to variables in Y. The set of solutions of a CSP $P = (X, C)$ is denoted by $sols(P)$; these are the valid tuples on X that satisfy C. A *table constraint* is a constraint whose semantics is defined in extension by listing the set of allowed (or forbidden) tuples. These tuples are *classical*. In this paper, we introduce smart table constraints.

A *smart table constraint sc* is defined semantically from a set of smart tuples, called *smart table* and denoted by *table(sc)*. A *smart tuple* σ is a set of tuple constraints, where a *tuple constraint* can take four possible forms:

1. `<var><op>`a
2. `<var>`$\in S$ or `<var>`$\notin S$
3. `<var><op><var>`
4. `<var><op><var>`$+ b$

where `<var>`is a variable in the scope of the smart table constraint, a and b some constants, S a set of constants and `<op>`an operator in the set $\{<, \leq, =, \neq, \geq, >\}$.

The semantics of smart table constraints is simple and natural: a classical tuple τ is allowed by a smart table constraint sc iff there is at least one smart tuple $\sigma \in table(sc)$ such that τ satisfies σ. Note that when a variable $x \in scp(sc)$ is not involved in any tuple constraint of a smart tuple $\sigma \in table(sc)$ then x can take any value in its domain; such a variable is said to be *unrestricted* on σ and the set of unrestricted variables on σ is denoted by $unres(\sigma)$. Note also that any classical tuple $(a_1, \ldots, a_r)$ on a set of variables $\{x_1, \ldots, x_r\}$ can be re-written as the smart tuple $\{x_1 = a_1, \ldots, x_r = a_r\}$.

As seen in the introduction, smart tuples can help modeling constraints in a compact and natural way, when disjunction is needed. Smart table constraints can also be used to encode some global constraints. The encodings of *Lex*, *Max* and *Element* are smart table constraint versions of the ones proposed in [1]. In the examples below, tuple constraints are written directly inside the tables to ease reading. A tuple constraint of the form x_i `<op>`a (resp. x_i `<op>`x_j+b) is written as `<op>`a (resp. `<op>`$x_j + b$) in the column of the table corresponding to x_i. The following global constraints illustrate the modeling power of the smart table constraint. Their equivalent with classical tuples are exponentially larger. For instance, in the table for element, each smart tuple corresponds to d^m classical tuples. For compressed tuples, if only one variable is the target of all the tuple constraints, each smart tuple can be translated as d compressed tuples. This is the case for all the global constraints presented below except for *Lex*. For this constraint, the smart table is $O(d^m)$ times smaller than the table using compressed tuples. Short supports applied to table constraints can only encode efficiently unrestricted variables, making the encoding of each smart tuple $O(d^m)$ tuples with short supports for *Lex*, *Max* and *AtMost1*. Global constraints are of course not the sole purpose of the smart table constraints but being able to encode efficiently those constraints has many advantages.

$Lex([x_1, \ldots, x_m], [y_1, \ldots, y_m]): \bar{x} > \bar{y}$

x_1	x_2	$\cdots$	x_m	y_1	y_2	$\cdots$	y_m
$> y_1$	$*$	$\cdots$	$*$	$*$	$*$	$\cdots$	$*$
$= y_1$	$> y_2$	$\cdots$	$*$	$*$	$*$	$\cdots$	$*$
$\cdots$	$\cdots$	$\cdots$	$\cdots$	$\cdots$	$\cdots$	$\cdots$	$\cdots$
$= y_1$	$= y_2$	$\cdots$	$> y_m$	$*$	$*$	$\cdots$	$*$

$Max([x_1, x_2, \ldots, x_m], M)$: $max(\bar{\mathbf{x}}) = M$

x_1	x_2	$\ldots$	x_m	M
$*$	$\leq x_1$	$\ldots$	$\leq x_1$	$= x_1$
$\leq x_2$	$*$	$\ldots$	$\leq x_2$	$= x_2$
$\ldots$	$\ldots$	$\ldots$	$\ldots$	$\ldots$
$\leq x_m$	$\leq x_m$	$\ldots$	$*$	$= x_m$

$Element(I, [x_1, x_2, \ldots, x_m], R)$: $\bar{\mathbf{x}}[I] = R$

I	x_1	x_2	$\ldots$	x_m	R
$= 1$	$*$	$*$	$\ldots$	$*$	$= x_1$
$= 2$	$*$	$*$	$\ldots$	$*$	$= x_2$
$\ldots$	$\ldots$	$\ldots$	$\ldots$	$\ldots$	$\ldots$
$= m$	$*$	$*$	$\ldots$	$*$	$= x_m$

$AtMost1([x_1, \ldots, x_m], Y)$: $\#\{1 \leq i \leq m | x_i = Y\} \leq 1$

x_1	x_2	$\ldots$	x_m	Y
$*$	$\neq Y$	$\ldots$	$\neq Y$	$*$
$\neq Y$	$*$	$\ldots$	$\neq Y$	$*$
$\ldots$	$\ldots$	$\ldots$	$\ldots$	$\ldots$
$\neq Y$	$\neq Y$	$\ldots$	$*$	$*$

$NotAllEqual(x_1, \ldots, x_m)$: $\exists 1 \leq i, j \leq m : x_i \neq x_j$

x_1	x_2	x_3	$\ldots$	x_m
$*$	$\neq x_1$	$*$	$\ldots$	$*$
$*$	$*$	$\neq x_1$	$\ldots$	$*$
$\ldots$	$\ldots$	$\ldots$	$\ldots$	$\ldots$
$*$	$*$	$*$	$\ldots$	$\neq x_1$

$Diffn([x_1, \ldots, x_m], [i_1, \ldots, i_m], [y_1, \ldots, y_m], [j_1, \ldots, j_m])$:
no overlap between orthotopes defined in $\mathbb{R}^m$ from points $\bar{\mathbf{x}}$ and $\bar{\mathbf{y}}$ with lengths along axes of $\bar{\mathbf{i}}$ and $\bar{\mathbf{j}}$ respectively.

x_1	x_2	$\ldots$	x_m	y_1	y_2	$\ldots$	y_m
$*$	$*$	$\ldots$	$*$	$\geq x_1 + i_1$	$*$	$\ldots$	$*$
$\geq y_1 + j_1$	$*$	$\ldots$	$*$	$*$	$*$	$\ldots$	$*$
$*$	$*$	$\ldots$	$*$	$*$	$\geq x_2 + i_2$	$\ldots$	$*$
$*$	$\geq y_2 + j_2$	$\ldots$	$*$	$*$	$*$	$\ldots$	$*$
$\ldots$	$\ldots$	$\ldots$	$\ldots$	$\ldots$	$\ldots$	$\ldots$	$\ldots$
$*$	$*$	$\ldots$	$*$	$*$	$*$	$\ldots$	$\geq x_m + i_m$
$*$	$*$	$\ldots$	$\geq y_m + j_m$	$*$	$*$	$\ldots$	$*$

2   Filtering Smart Table Constraints

This section presents a filtering algorithm to establish GAC on smart table constraints. GAC is a property that relies on the concept of support. A *support* of a constraint c is a tuple on $scp(c)$ which is both valid and allowed by c. A support on c for a literal (x, a) of c is a support of c containing (x, a).

Definition 1. *A constraint c is Generalized Arc Consistent (GAC) iff all the literals of the constraint have a support on c. A CSP is GAC iff all its constraints are GAC.*

In general, identifying the set of supports of a constraint allows us to enforce GAC. Actually, for any smart table constraint sc, each smart tuple σ corresponds to a small CSP $P_\sigma = (X_\sigma, C_\sigma)$, with $X_\sigma = \text{scp}(sc)$ and $C_\sigma = \sigma$. The classical tuples that are supports of sc from σ are exactly the solutions in $sols(P_\sigma)$. Hence, the full set of supports of sc is equal to $\bigcup_{\sigma \in table(sc)} sols(P_\sigma)$. This is similar to the way set of supports are computed for constructive disjunction.

   Our objective is to efficiently identify and remove valid literals of sc without any support. It may seem costly to compute $sols(P_\sigma)$ for every smart tuple σ. Obtaining the set of supports for an arbitrary logical combination of constraints is NP-hard [1]. However, we impose that the constraint graph of any CSP P_σ that is associated with a smart tuple σ, is acyclic and P_σ is a conjunction. This restriction allows an efficient processing of the smart tuples when used for filtering.

Property 1. Let σ be a smart tuple of a smart table constraint, P'_σ, the GAC closure of P_σ, is globally consistent, i.e., each literal of P'_σ appears in at least one solution of P_σ.

   This property is derived from [20] and the acyclic nature of the constraint graphs defined by smart tuples. This means that the set of literals appearing in $sols(P_\sigma)$ can be obtained by simply applying GAC on P_σ.

   Obtaining the GAC closure on each of the P_σ and taking their union at the end to have the set of supported literals corresponds exactly to an application of the filtering rules defined in [1], when seeing the smart table constraint as a logical combination of basic arithmetic constraints. The acyclic nature of the conjunctions in the smart tuples guarantees that the set of supported literals computed by this procedure is the set of GAC literals for the logical combination of arithmetic constraints by Theorem 3 in [1]. Hence, this procedure is correct and computes the GAC literals for the smart table constraint. Moreover, the complexity of filtering P_σ can also benefit from the form of the smart tuples, as expressed below.

Property 2. The GAC closure of an acyclic binary CSP can be obtained in $O(e \cdot F)$, where filtering an individual constraint if $O(F)$.

   The procedure for obtaining the GAC closure of an acyclic binary CSP $P = (X, C)$ is the following. The CSP forms a forest (possibly, with only one tree), and each tree of the forest can be filtered independently since no variable is shared between trees. For each tree T, revising constraints in turn from the deepest ones to the shallowest ones, and then the other way around, achieves GAC on T. Each constraint in C is thus revised two times (no fixed point needed). Revising a constraint c consists in removing the literals that have no support on c. We call this procedure GAC_tree. GAC_tree can be viewed as an application of the rules defined in [1] for conjunction. The acyclicity of the networks guarantees that the inconsistency sets computed are maximal [1] and hence that GAC_tree is correct. GAC_tree, as well as properties 1 and 2, are not original to this work, but they justify the filtering procedure of the smart table constraints.

Applying GAC_tree to a smart tuple σ of a constraint sc requires decomposing σ according to its connected components; the result of this decomposition will be denoted by forest(σ). More precisely, for each subset $cc \subseteq \sigma$ that represents a connected component, there is an associated tree T in forest(σ) that defines an independent sub-CSP (X_T, C_T) with $X_T = vars(cc)$ and $C_T = cc$. We shall refer to such sub-CSPs with tree shape as *treeCSPs*. An additional *void* tree T defining a trivial sub-CSP (X_T, C_T) with $X_T = unres(\sigma)$ and $C_T = \emptyset$ is introduced if $unres(\sigma) \neq \emptyset$. This guarantees that $sols(P_\sigma) = \Pi_{T \in trees(\sigma)} sols(T)$, which results from the independence of the trees w.r.t each other.

The filtering algorithm proposed for smart table constraints, called smartSTR, works with the decompositions into treeCSPs instead of working directly with the smart tuples. It is inspired from STR (Simple Tabular Reduction) [14,27]. STR works by scanning constraint tables, going through each tuple sequentially. The validity of each row is checked. When a row is not valid, it is removed from the table. Otherwise, all the literals of the row are marked as having a support. After scanning the whole table, all the literals for which no support has been found are removed. The difference between STR and smartSTR is the way validity checks and the collection of supported literals is performed. A smart tuple σ is valid iff P_σ admits at least one solution. A smart tuple σ is thus valid iff each treeCSP in forest(σ) admits at least one solution. The literals supported by σ are the literals in $sols(P_\sigma)$ (obtained with GAC_tree), computed as the union of the supported literal sets of each individual treeCSP in forest(σ).

Algorithm 1 presents the pseudo-code of smartSTR. In all the algorithms presented in this paper, pre is the precondition and post is the postcondition. SmartSTR uses a data structure sl that contains all the literals without any found support (sl stands for support-less). Line 3 initializes sl with all valid literals (no support has been found yet). Then the algorithm loops over all the smart tuples of the constraint (line 4). The test at line 5 checks the validity of the current smart tuple by testing the validity of all its treeCSPs. If the smart tuple σ is valid, each of its independent treeCSP removes from sl the literals they support (loop at lines 6-7). The loop at line 8 empties the sets sl of all unrestricted variables on σ, as there is no restriction on those variables (actually, this corresponds to dealing with the void tree that is not in practice included in forest(σ)). If the smart tuple is invalid, it is removed from the table (line 9); the table of the constraint is represented using a sparse set, as in STR1 and STR2. After going through all the smart tuples of the constraint, smartSTR removes the literals that are still left without a support (loop at line 10).

```
1     smartSTR(SmartTableConstraint sc):
2         // post: the constraint sc is GAC
3         forall x ∈ scp(sc): sl(sc)[x] ← dom(x)
4         forall σ ∈ table(sc):
5             if (∧_{T∈forest(σ)}T.isValid()):
6                 forall T ∈ forest(σ):
7                     T.collect(sl(sc))
8                 forall x ∈ unres(σ): sl(sc)[x] ← ∅
9             else: remove σ from table(sc)
10        forall x ∈ scp(sc):
11            dom(x) ← dom(x) \ sl(sc)[x]
12
```

Algorithm 1: smartSTR

As seen in Algorithm 1, each treeCSP is responsible to check its validity and to remove from sl the literals it supports. This is done through isValid and collect methods. Those methods correspond to GAC_tree. Their pseudo codes are given below because it is a non standard GAC procedure, efficient and adapted to smart tuples. This also eases the complexity analysis. Their specifications can be found in Interface 1, called TreeCSP. Note that a treeCSP involves a set of variables $vars$ and belongs to the forest of a smart tuple σ.

```
1     interface TreeCSP
2         fields: Variable[] vars
3         isValid()
4             // post: returns true iff the treeCSP is valid
5         collect(Set{Value}[] sl)
6             // pre: the smart tuple σ, such that the treeCSP
7             //      is in forest(σ), is valid
8             // post: ∀x ∈ vars, ∀a ∈ dom(x), (x, a) has a
9             //       support in the treeCSP ⇒ a ∉ sl[x]
10
```

Interface 1: Interface for treeCSPs

From now on, the treeCSPs that are composed of only one constraint will be called *branches*. In the code presented below, we have specific classes for *unary* branches (containing a tuple constraint of the form <var><op>a, <var>∈ S or <var>∉ S), and *binary* branches (containing a tuple constraint of the form <var><op><var>, or <var><op><var>+ b). There is one unary and binary branch class for each value of <op>. We also introduce one class for *simple* trees (trees of height 1 consisting of multiple branches all sharing the same root variable) and another one for *general* trees (trees of height > 1).

Algorithm 2 presents the classes introduced for unary branches with operations =, and <. The pseudo-code for the other operators are very similar. The additional method filterX, not contained in Interface 1, is responsible to filter the (pseudo) domain Dx given as argument. It is used by simple and general trees, where GAC has to be enforced on several branches. Dx is used to avoid filtering directly dom(x), because the effective filtering can only be done when all smart tuples have been processed.

```
1   class UnaryBranchEq:TreeCSP /* x = a */
2       fields: Variable x, Value a
3       isValid(): return a ∈ dom(x)
4       collect(sl): sl[x] ← sl[x] \ {a}
5       filterX(Dx): Dx ← Dx ∩ {a}
6
```

```
1   class UnaryBranchLt:TreeCSP /* x < a */
2       fields: Variable x, Value a
3       isValid(): return min(dom(x)) < a
4       collect(sl):
5           sl[x] ← sl[x] \ {b ∈ dom(x) : b < a}
6       filterX(Dx): Dx ← Dx \ {b ∈ Dx : b ≥ a}
7
```

Algorithm 2: Classes for unary branches = and <

Algorithm 3 presents the classes introduced for binary branches with operations = and <, respectively. Again, the pseudocodes for the other operators are very similar. In those pseudocodes, $S \oplus b$, where S is a set and b a value, represents the addition of the constant to all the values in the set. They all implement the method filterX, as unary branches do, but with two parameters (Dx and Dy). Dx is the copy of the domain of x to filter and Dy is a domain for y to use to filter Dx. The second parameter is needed during the execution of GAC_tree to use an already filtered copy of the domain of y to filter the copy of dom(x). Again, the filtering of the real domains of the variables can only occur after all the smart tuples have been processed. Those classes also implement a filterY method which is the counterpart of filterX for y. They implement a method collectY, used by simple trees to collect values, but only for the second involved variable y with respect to a (pseudo) domain Dx, given as a parameter, for the first involved variable x. It is called after Dx, which is initially a copy of dom(x), has been filtered through the entire simple or general tree.

```
1    class BinaryBranchEq:TreeCSP /* x = y + b */
2        fields: Variable x, y
3               Value b
4        isValid(): return dom(x) ∩ dom(y) ⊕ b ≠ ∅
5        collect(sl):
6            I ← dom(x) ∩ dom(y) ⊕ b
7            sl[x] ← sl[x] \ I
8            sl[y] ← sl[y] \ I
9        collectY(sl, Dx):
10           I ← Dx ∩ dom(y) ⊕ b
11           sl[y] ← sl[y] \ I
12       filterX(Dx,Dy): Dx ← Dx ∩ Dy ⊕ b
13       filterY(Dx,Dy): Dy ← Dx ∩ Dy ⊕ b
14
```

```
1    class BinaryBranchLt:TreeCSP /* x < y + b */
2        fields: Variable x, y
3               Value: b
4        isValid(): return min(dom(x)) < max(dom(y)) + b
5        collect(sl):
6            sl[x] ← sl[x] \ {a ∈ dom(x) : a < max(dom(y)) + b}
7            sl[y] ← sl[y] \ {c ∈ dom(y) : c > min(dom(x)) − b}
8        collectY(sl, Dx):
9            sl[y] ← sl[y] \ {c ∈ dom(y) : c > min(Dx) − b}
10       filterX(Dx,Dy):
11           Dx ← Dx \ {a ∈ Dx : a ≥ max(Dy) + b}
12       filterY(Dx,Dy):
13           Dy ← Dy \ {c ∈ Dy : c ≤ min(Dx) − b}
14
```

Algorithm 3: Classe for binary branches = and <

Algorithm 4 gives the pseudo-code for simple trees, where all involved branches share the same root variable x (see the assertion at line 4). Since we can change the order of the variables in binary branches ($x_1 < x_2 \rightarrow x_2 > x_1$, etc.), this is not a requirement on the form of the smart tuples. This is enforced at the creation of the smart tuples trees. The validity test at line 5 starts by making a copy Dx of $\mathrm{dom}(x)$. Then, Dx is filtered through all branches (loop starting at line 7). The unary branches are treated at lines 8-9 and the binary ones, at lines 10-11. For the binary branches, `filterX` is called with the full domain of y as argument for the copy of y's domain. If Dx does not become empty, that means that the simple tree has at least one solution. The method `collect` at line 13 uses Dx (which has already been filtered by `isValid`). Since all values in Dx have a support in the simple tree, they are removed from sl[x] (line 14). The loop at line 15 goes through every binary branch (i.e., with a scope containing 2 variables) to collect the supported values for the second involved variables (y) from the filtered domain Dx. The supported values for variables y are directly removed from sl instead of copying their domains and filtering

```
1     class SimpleTree:TreeCSP
2         fields: Variable x, TreeCSP[] branches,
3             Domain Dx
4         assert ∀T ∈ branches : T.x = x
5         isValid():
6             Dx ← dom(x)
7             forall T ∈ branches:
8                 if #T.vars = 1
9                     T.filterX(Dx)
10                else
11                    T.filterX(Dx, dom(y))
12            return Dx ≠ ∅
13        collect(sl):
14            sl[x] ← sl[x] \ Dx
15            forall T ∈ branches : #T.vars = 2
16                T.collectY(sl, Dx)
17
```

Algorithm 4: Class for simple trees

them. Note that methods isValid and collect are adaptations of the two pass filtering GAC_tree. During the first pass, only the domain of x (actually, Dx) is filtered. Indeed, as it may change at each new processed branch, filtering the domains of variables y (actually, updating sl) is useless at that time. The validity test is not concerned by the second pass because if x still has values in its domain after the first pass, the simple tree is guaranteed to have at least one solution.

The class for general trees is given in Algorithm 5. This algorithm uses several fields. The array allVars contains all the variables appearing in the tree. The 2 dimensional array branches contains all the branches for each level of the tree, from 1 (branches containing the root variable) to treeHeight. The array domCopy contains the copies of the domains of the variables of the tree that are used during the procedure GAC_tree. For this algorithm, we will suppose that, for all the binary branches, the variable x is always the closest to the root. This is again enforced during the creation of the smart tuples trees. The assertion at line 5 thus checks that all the variables x have a corresponding variable as y at the level below (closer to the root). The isValid method (line 7) realizes the first pass of GAC_tree (using copies of the domains), filtering the domains of the different x variables from the leafs to the root. If the variable at the root (branches[1][1].x) of the tree still has values, it returns true. Its collect method (line 17) then achieves the second pass by filtering the (copies of the) domains of the y variables of the branches. It also removes supported values from sl. At this point, it is important to note that the code presented for unary branches, binary branches and simple trees already covers all the examples given in this paper.

```
1          class GeneralTree:TreeCSP
2              fields: Variable[] allVars, TreeCSP[][] branches,
3                      Value treeHeight, Domain[] domCopy
4              assert ∀1 < l ≤ treeHeight, ∀b ∈ branches[l],
5                      ∃b₂ ∈ branches[l − 1] : b.x = b₂.y
6              isValid():
7                  forall x ∈ allVars:
8                      domCopy[x] ← dom(x)
9                  forall l ∈ treeHeight..1:
10                     forall T ∈ branches[l]:
11                         if #T.vars = 1
12                             T.filterX(domCopy[T.x])
13                         else
14                             T.filterX(domCopy[T.x], domCopy[T.y])
15                 return domCopy[branches[1][1].x] ≠ ∅
16             collect(sl):
17                 forall l ∈ 1..treeHeight:
18                     forall T ∈ branches[l]:
19                         sl[T.x] ← sl[T.x] \ domCopy[T.x]
20                         if #T.vars = 2:
21                             T.filterY(domCopy[T.x], domCopy[T.y])
22                 forall T ∈ branches[treeHeight] : #T.vars = 2
23                     sl[T.y] ← sl[T.y] \ domCopy[T.y]
24
```

Algorithm 5: Class for general trees

We now study the complexity of our approach. The complexity of filtering a smart tuple depends on the complexity of filtering each of its treeCSPs, as they are independent. For a smart tuple σ (on variables with maximal domain size d), the time complexities for the different operators are:

for unary branches

<op>	isValid	collect	filterX
$=$	$O(1)$	$O(1)$	$O(1)$
$\neq$	$O(1)$	$O(1)$	$O(1)$
$> \geq < \leq$	$O(1)$	$O(d)$	$O(d)$
$\in \notin$	$O(d)$	$O(d)$	$O(d)$

for binary branches

<op>	isValid	collect/ collectY	filterX/ filterY
$=$	$O(d)$	$O(d)$	$O(d)$
$\neq$	$O(1)$	$O(1)$	$O(1)$
$> \geq < \leq$	$O(1)$	$O(d)$	$O(d)$

Each tuple constraint is either its own tree or belongs to a larger tree. If the branch is its own tree, the time complexities of isValid and collect are $O(d)$ for any operator. If the branch is included in a simple or general tree, then GAC_tree guarantees that the collectY, filterX and filterY methods are called a constant number of times. The time complexity imputable to the branch is thus $O(d)$ for validity testing and value collection. This makes the treatment of one smart tuple with k tuple constraints $O(k \cdot d + r)$, where r

is the arity of the table constraint. The last term comes from the treatment of unrestricted variables. The initialization of sl at the beginning of smartSTR and the actual filtering of the domains at the end are $O(r \cdot d)$. The total time complexity of one call to smartSTR for a smart table constraint of arity r with t smart tuples is thus $O(r \cdot d + t \cdot k \cdot d + t \cdot r)$. For a classical table constraint of arity r with t' tuples, we have that STR2 has a time complexity of $O(r \cdot d + t' \cdot r)$. In all the examples given, we have that $k \leq r$ (less tuple constraints than variables). We also have that the number of smart tuples is at least $d + 1$ times less than the number of classical tuples. In those conditions, the complexity of filtering the smart table is less than the complexity of using STR2 on the table without smart tuples. Indeed, we have $t \cdot k \cdot d + t \cdot r \leq t' \cdot r$.

3   Experimental Results

Optimization present in STR2 can also be included in smartSTR. The obtained algorithm is then called smartSTR2. Comparing SmartSTR2 with all specialized algorithms developed over the years for the global constraints mentioned earlier is clearly beyond the scope of this paper. However, we shall show the interest of SmartSTR2 on a few case studies. Comparing a propagator F with SmartSTR2 on a global constraint means that, in the same CSP, all the instances of the global constraint are either propagated with F or their encoding in smart table constraint is propagated with SmartSTR2. We have conducted an experimentation (with the solver AbsCon) on a laptop computer, equipped with Intel(R) Core(TM) i7-2820QM CPU @ 2.30GHz, under Linux. Results are given in seconds, or corresponds to number of visited nodes per second. We have checked that all tested approaches were traversing the exact same search trees (most of the time using dom/ddeg as variable ordering heuristic for this purpose).

In natural language processing, one task is to determine whether a given sentence is well-formed (i.e., to what extent, it respects a grammar). A constraint model (R. Coletta, personal communication) has been recently developed for this problem, denoted here by TAL. It involves the Element constraint (with R as a variable as described earlier in the paper). Instances for this optimization problem are defined by entering an input sentence. In this model, Element constraints represent about 8% of the constraints. We compare SmartSTR2 with GACElt that corresponds to the GAC propagator based on watched literals [7]. In this context, the two algorithms are very close in term of performance as shown by Table 1.

A *BIBD* is a standard combinatorial problem. We consider here the model introduced in [22] and the series of instances tested in [5]. There is a lexicographic constraint between any two adjacent rows or columns. We compare SmartSTR2 with GACLex that corresponds to the filtering procedure described in [13] and is a variant of [6]. Table 2 shows the results we have obtained with both algorithms. Interestingly, one can observe that replacing the specialized propagator GACLex with the general-purpose SmartSTR2 has a very limited cost, although SmartSTR2 is generic. Similar results are obtained with the social golfer problem.

Table 1. CPU time to solve *TAL* instances

sentence	GACElt	SmartSTR2
phrase1	3.6	3.7
phrase2	17.6	17.9
phrase3	54.4	54.2
phrase4	46.8	46.8
phrase5	82.4	82.6

Table 2. CPU time to solve *BIBD* instances

v-b-r-k-λ	GACLex	SmartSTR2
6-50-25-3-10	1.3	1.6
6-60-30-3-12	1.5	2.1
6-70-35-3-10	2.2	2.8
10-90-27-3-6	5.8	7.3
9-108-36-3-9	11.4	14.2
15-70-14-3-2	7.4	7.9
12-88-22-3-4	7.0	8.3
9-120-40-3-10	17.9	25.1
10-120-36-3-8	10.6	14.0
13-104-24-3-4	99.1	108.6

The *RectanglePacking* problem [26] consists of packing all squares from size 1×1 to $n \times n$ into a rectangle of size $w \times h$. We adopt the model and search parameters given in [10, 23]. Table 3 reports the nodes searched per second by the algorithms. This measurement has been chosen because some instances trigger timeouts for some algorithms. The node count per second gives, in this context, more information than the runtimes/timeouts. It shows that SmartSTR2 is very efficient on this problem. It clearly outperforms ShortSTR2, and seems to be at least as efficient as the other methods proposed in [23] (not implanted in our system) when we compare their results with ours. Note that GAC-valid (sometimes called GAC-schema) is another general approach, given here as a baseline.

For our last experiment, we consider Case Study 4 in [10], where a problem, denoted by *AllDistinctVectors* here, involves the VectorDiff constraint. An instance *p-a-d* of this problem has exactly p vectors (arrays of variables), each vector of length a and each variable with a domain whose size is equal to d: any pair of vectors must be distinct. In [10], it has been shown that ShortSTR2 is an interesting competitor to HaggisGAC. When we consider Boolean variables only

Table 3. Nodes searched per second for *RectanglePacking* instances

n-w-h	GAC-valid	ShortSTR2	SmartSTR2
18-31-69	1,821	2,784	57,249
19-47-53	2,003	3,166	57,221
20-34-85	1,324	1,579	45,600
21-38-88	849	1,295	40,600
22-39-88	981	1,035	41,162
23-64-68	983	1,292	40,495
24-56-88	446	790	32,758
25-43-129	661	347	30,544
26-70-89	544	703	31,374
27-47-148	326	175	26,786

Table 4. CPU time to enforce GAC on *AllDistinctVectors* instances

p-a-d	ShortSTR2	SmartSTR2
40-100-2	0.07	0.07
40-100-8	1.55	0.18
40-100-16	6.49	0.18
40-100-24	14.7	0.19
40-100-32	28.1	0.20
40-100-40	44.5	0.21

(i.e., $d = 2$), SmartSTR2 is slightly slower than ShortSTR2 (because tables are small). However, when we increase d, Table 4 shows that, just when applying GAC stand-alone, SmartSTR2 is clearly superior to ShortSTR2. This can be explained by the size of the constraint tables. For example, for *40-100-40*, tables contain 156,000 and 100 tuples in ShortSTR2 and SmartSTR2, respectively.

4   Conclusion

Smart tuples generalize (classical) tuples in tables of constraints, as well as short and compressed tuples. They allow a compact and natural representation of many constraints, including important global constraints. Smart table constraints can be seen as a subset of the logical algebra defined in [1]. Restricting smart table constraints to this subset allows an efficient filtering of the constraints. The contribution of this paper is to introduce the smart table constraint and propose a practical GAC filtering algorithm for it. Its practical interest is also demonstrated. We do believe that there exist many optimisations and extensions to this work that still deserve to be explored.

Acknowledgments. The first author is supported as a Research Assistant by the Belgian FNRS. This research is also partially supported by the FRFC project 2.4504.10 of the Belgian FNRS, and by the UCLouvain Action de Recherche Concertée ICTM22C1. The third author benefits from the financial support of both CNRS and OSEO within the ISI project 'Pajero'.

References

1. Bacchus, F., Walsh, T.: Propagating logical combinations of constraints. In: IJCAI, pp. 35–40 (2005)
2. Bessiere, C., Régin, J.-C.: Arc consistency for general constraint networks: preliminary results. In: Proceedings of IJCAI 1997, pp. 398–404 (1997)
3. Bessiere, C., Régin, J.-C.: Local consistency on conjunctions of constraints. In: Proceedings of ECAI 1998 Workshop on Non-binary constraints, pp. 53–59 (1998)
4. Carlson, B., Carlsson, M.: Compiling and executing disjunctions of finite domain constraints. In: Proceedings of ICLP 1995, pp. 117–131 (1995)
5. Frisch, A.M., Hnich, B., Kiziltan, Z., Miguel, I., Walsh, T.: Global constraints for lexicographic orderings. In: Van Hentenryck, P. (ed.) CP 2002. LNCS, vol. 2470, pp. 93–108. Springer, Heidelberg (2002)
6. Frisch, A., Hnich, B., Kiziltan, Z., Miguel, I., Walsh, T.: Propagation algorithms for lexicographic ordering constraints. Artificial Intelligence 170(10), 803–834 (2006)
7. Gent, I.P., Jefferson, C., Miguel, I.: Watched literals for constraint propagation in minion. In: Benhamou, F. (ed.) CP 2006. LNCS, vol. 4204, pp. 182–197. Springer, Heidelberg (2006)
8. Gent, I.P., Jefferson, C., Miguel, I., Nightingale, P.: Data structures for generalised arc consistency for extensional constraints. In: Proceedings of AAAI 2007, pp. 191–197 (2007)
9. Jefferson, C., Moore, N.C.A., Nightingale, P., Petrie, K.E.: Implementing logical connectives in constraint programming. Artificial Intelligence 174(16), 1407–1429 (2010)
10. Jefferson, C., Nightingale, P.: Extending simple tabular reduction with short supports. In: Proceedings of IJCAI 2013, pp. 573–579 (2013)
11. Katsirelos, G., Bacchus, F.: GAC on conjunctions of constraints. In: Walsh, T. (ed.) CP 2001. LNCS, vol. 2239, pp. 610–614. Springer, Heidelberg (2001)
12. Katsirelos, G., Walsh, T.: A compression algorithm for large arity extensional constraints. In: Bessière, C. (ed.) CP 2007. LNCS, vol. 4741, pp. 379–393. Springer, Heidelberg (2007)
13. Lecoutre, C.: Constraint Networks: Techniques and Algorithms. ISTE/Wiley (2009)
14. Lecoutre, C.: STR2: optimized simple tabular reduction for table constraints. Constraints 16(4), 341–371 (2011)
15. Lecoutre, C., Likitvivatanavong, C., Yap, R.H.C.: A path-optimal GAC algorithm for table constraints. In: Proceedings of ECAI 2012, pp. 510–515 (2012)
16. Lecoutre, C., Szymanek, R.: Generalized arc consistency for positive table constraints. In: Benhamou, F. (ed.) CP 2006. LNCS, vol. 4204, pp. 284–298. Springer, Heidelberg (2006)

17. Lhomme, O.: Arc-consistency filtering algorithms for logical combinations of constraints. In: Régin, J.-C., Rueher, M. (eds.) CPAIOR 2004. LNCS, vol. 3011, pp. 209–224. Springer, Heidelberg (2004)
18. Lhomme, O.: Practical reformulations with table constraints. In: Proceedings of ECAI 2012, pp. 911–912 (2012)
19. Lhomme, O., Régin, J.-C.: A fast arc consistency algorithm for n-ary constraints. In: Proceedings of AAAI 2005, pp. 405–410 (2005)
20. Mackworth, A.K., Freuder, E.C.: The complexity of some polynomial network consistency algorithms for constraint satisfaction problems. Artificial intelligence 25(1), 65–74 (1985)
21. Mairy, J.-B., Van Hentenryck, P., Deville, Y.: Optimal and efficient filtering algorithms for table constraints. Constraints 19(1), 77–120 (2014)
22. Meseguer, P., Torras, C.: Solving strategies for highly symmetric CSPs. In: Proceedings of IJCAI 1999, pp. 400–405 (1999)
23. Nightingale, P., Gent, I.P., Jefferson, C.A., Miguel, I.J.: Short and long supports for constraint propagation. Journal of Artificial Intelligence Research 46, 1–45 (2013)
24. Perez, G., Régin, J.-C.: Improving GAC-4 for table and MDD constraints. In: O'Sullivan, B. (ed.) CP 2014. LNCS, vol. 8656, pp. 606–621. Springer, Heidelberg (2014)
25. Régin, J.-C.: Improving the expressiveness of table constraints. In: Proceedings of CP 2011 Workshop on Constraint Modelling and Reformulation (2011)
26. Simonis, H., O'Sullivan, B.: Search strategies for rectangle packing. In: Stuckey, P.J. (ed.) CP 2008. LNCS, vol. 5202, pp. 52–66. Springer, Heidelberg (2008)
27. Ullmann, J.R.: Partition search for non-binary constraint satisfaction. Information Sciences 177(18), 3639–3678 (2007)
28. Van Hentenryck, P., Saraswat, V., Deville, Y.: Design, implementation, and evaluation of the constraint language cc(fd). The Journal of Logic Programming 37(1–3), 139–164 (1998)
29. Würtz, J., Müller, T.: Constructive disjunction revisited. In: Görz, G., Hölldobler, S. (eds.) KI 1996. LNCS, vol. 1137, pp. 377–386. Springer, Heidelberg (1996)
30. Xia, W., Yap, R.H.C.: Optimizing STR algorithms with tuple compression. In: Schulte, C. (ed.) CP 2013. LNCS, vol. 8124, pp. 724–732. Springer, Heidelberg (2013)

Constraint-Based Sequence Mining
Using Constraint Programming

Benjamin Negrevergne[⊠] and Tias Guns

DTAI Research group, KU Leuven, 3000 Leuven, Belgium
{benjamin.negrevergne,tias.guns}@cs.kuleuven.be

Abstract. The goal of constraint-based sequence mining is to find sequences of symbols that are included in a large number of input sequences and that satisfy some constraints specified by the user. Many constraints have been proposed in the literature, but a general framework is still missing. We investigate the use of constraint programming as general framework for this task.

We first identify four categories of constraints that are applicable to sequence mining. We then propose two constraint programming formulations. The first formulation introduces a new global constraint called *exists-embedding*. This formulation is the most efficient but does not support one type of constraint. To support such constraints, we develop a second formulation that is more general but incurs more overhead. Both formulations can use the projected database technique used in specialised algorithms.

Experiments demonstrate the flexibility towards constraint-based settings and compare the approach to existing methods.

Keywords: Sequential pattern mining · Sequence mining · Episode mining · Constrained pattern mining · Constraint programming · Declarative programming

1 Introduction

In AI in general and in data mining in particular, there is an increasing interest in developing general methods for data analysis. In order to be useful, such methods should be easy to extend with domain-specific knowledge.

In pattern mining, the frequent sequence mining problem has already been studied in depth, but usually with a focus on efficiency and less on generality and extensibility. An important step in the development of more general approaches was the cSpade algorithm [20] which supports a variety constraints. It supports many constraints such as constraints on the length of the pattern, on the maximum gap in embeddings or on the discriminative power of the patterns between datasets. Many other constraints have been integrated into specific mining algorithms (e.g. [6,14,17,18]). However, none of these are truly generic in that adding

© Springer International Publishing Switzerland 2015
L. Michel (Ed.): CPAIOR 2015, LNCS 9075, pp. 288–305, 2015.
DOI: 10.1007/978-3-319-18008-3_20

extra constraints usually amounts to changing the data-structures used in the core of the algorithm.

For *itemset* mining, the simplest form of pattern mining, it has been shown that constraint programming (CP) can be used as a generic framework for constraint-based mining [5] and beyond [11,15]. Recent works have also investigated the usage of CP-based approaches for mining sequences with explicit wildcards [3,7,8]. A wildcard represents the presence of exactly one arbitrary symbol in that position in the sequence.

The main difference between mining itemsets, sequences with wildcards and standard sequences lies in the complexity of testing whether a pattern is included in another itemset/sequence, e.g. from the database. For itemsets, this is simply testing the subset inclusion relation which is easy to encode in CP. For sequences with wildcards and general sequences, one has to check whether an *embedding* exists (matching of the individual symbols). But in case only few embeddings are possible, as in sequences with explicit wildcards, this can be done with a disjunctive constraint over all possible embeddings [8]. In general sequence (the setting we address in this paper), a pattern of size m can be embedded into a sequence of size n in $O(n^m)$ different ways, hence prohibiting a direct encoding or enumeration.

The contributions of this paper are as follows:

- We present four categories of user-constraints, this categorization will be useful to compare the generality of the two proposed models.
- We introduce an *exists-embedding* global constraint for sequences, and show the relation to projected databases and *projected frequency* used in the sequence mining literature to speedup the mining process [6,21].
- We propose a more general formulation using a decomposition of the *exists-embedding* constraint. Searching whether an embedding exists for each transaction is not easily expressed in CP and requires a modified search procedure.
- We investigating the effect of adding constraints, and compare our method with state-of-the-art sequence mining algorithms.

The rest of the paper is organized as follows: Section 2 formally introduces the sequence mining problem and the constraint categories. Section 3 explains the basics of encoding sequence mining in CP. Section 4 and 5 present the model with the global constraint and the decomposition respectively. Section 6 presents the experiments. After an overview of related work (Section 7), we discuss the proposed approach and results in Section 8.

2   Sequence Mining

Sequence mining [1] can be seen as a variation of the well-known itemset mining problem proposed in [2]. In itemset mining, one is given a set of *transactions*, where each transaction is a set of items, and the goal is to find patterns (i.e. sets of items) that are included in a large number of transactions. In sequence mining, the problem is similar except that both transactions and patterns are ordered, (i.e. they are sequences instead of sets) and symbols can be repeated. For example, $\langle b, a, c, b \rangle$ and $\langle a, c, c, b, b \rangle$ are two sequences, and the sequence $\langle a, b \rangle$ is one possible pattern included in both.

This problem is known in the literature under multiple names, such as *embedded subsequence mining*, *sequential pattern mining*, *flexible motif mining*, or *serial episode mining* depending on the application.

2.1 Frequent Sequence Mining: Problem Statement

A key concept of any pattern mining setting is the pattern inclusion relation. In sequence mining, a pattern is included in a transaction if there exists an embedding of that sequence in the transaction; where an embedding is a mapping of every symbol in the pattern to the same symbol in the transaction such that the order is respected.

Definition 1 (Embedding in a sequence). *Let $S = \langle s_1, \ldots, s_m \rangle$ and $S' = \langle s'_1, \ldots, s'_n \rangle$ be two sequences of size m and n respectively with $m \leq n$. The tuple of integers $e = (e_1, \ldots, e_m)$ is an **embedding** of S in S' (denoted $S \sqsubseteq_e S'$) if and only if:*

$$S \sqsubseteq_e S' \leftrightarrow e_1 < \ldots < e_m \text{ and } \forall i \in 1, \ldots, m : s_i = s'_{e_i} \tag{1}$$

For example, let $S = \langle a, b \rangle$ be a pattern, then $(2, 4)$ is an embedding of S in $\langle b, a, c, b \rangle$ and $(1, 4), (1, 5)$ are both embeddings of S in $\langle a, c, c, b, b \rangle$. An alternative setting considers sequences of *itemsets* instead of sequences of individual symbols. In this case, the definition is $S \sqsubseteq_e S' \leftrightarrow e_1 < \ldots < e_n$ and $\forall i \in 1, \ldots, n : s_i \subseteq s'_{e_i}$. We do not consider this setting further in this paper, though it is an obvious extension.

We can now define the sequence inclusion relation as follows:

Definition 2 (Inclusion relation for sequences). *Given two sequences S and S', S is included in S' (denoted $S \sqsubseteq S'$) if there exists an embedding e of S in S':*

$$S \sqsubseteq S' \leftrightarrow \exists e \text{ s.t. } S \sqsubseteq_e S'. \tag{2}$$

To continue on the example above, $S = \langle a, b \rangle$ is included in both $\langle b, a, c, b \rangle$ and $\langle a, c, c, b, b \rangle$ but not in $\langle c, b, a, a \rangle$.

Definition 3 (Sequential dataset). *Given an alphabet of symbols Σ, a sequential dataset D is a multiset of sequences defined over symbols in Σ.*

Each sequence in D is called a *transaction* using the terminology from itemset mining. The number of transactions in D is denoted $|D|$ and the sum of the lengths of every transaction in D is denoted $||D||$ ($||D|| = \sum_{i=1}^{|D|} |T_i|$). Furthermore, we use *dataset* as a shorthand for *sequential dataset* when it is clear from context.

Given a dataset $D = \{T_i, \ldots, T_n\}$, one can compute the **cover** of a sequence S as the set of all transactions T_i that contain S:

$$cover(S, D) = \{T_i \in D : S \sqsubseteq T_i\} \tag{3}$$

We can now define frequent sequence mining, where the goal is to find all patterns that are frequent in the database; namely, the size of their cover is sufficiently large.

Definition 4 (Frequent sequence mining). *Given:*

1. *an alphabet Σ*
2. *a sequential dataset $D = \{T_1, \ldots, T_n\}$ defined over Σ*
3. *a minimum frequency threshold θ,*

enumerate all sequences S such that $|cover(S, D)| \geq \theta$.

In large datasets, the number of frequent sequences is often too large to be analyzed by a human. Extra constraints can be added to extract fewer, but more relevant or interesting patterns. Many such constraints have been studied in the past.

2.2 Constraints

Constraints typically capture background knowledge and are provided by the user. We identify four categories of constraints for sequence mining: 1) constraints over the pattern, 2) constraints over the cover set, 3) constraints over the inclusion relation and 4) preferences over the solution set.

Constraints on the Pattern. These put restrictions on the structure of the pattern. Typical examples include size constraints or regular expression constraints.

Size constraints: A size constraint is simply $|S| \gtrless \alpha$ where $\gtrless \in \{=, \neq, >, \geq, <, \leq\}$ and α is a user-supplied threshold. It is used to discard small patterns.

Item constraints: One can constrain a symbol t to surely be in the pattern: $\exists s \in S : s = t$; or that it can not appear in the pattern: $\forall s \in S : s \neq t$, or more complex logical expressions over the symbols in the pattern.

Regular expression constraints: Let R be a regular expression over the vocabulary V and L_R be the language of sequences recognised by R, then for any sequence pattern S over V, the *match-regular* constraint requires that $S \in L_R$ [6].

Constraints on the Cover Set. The *minimum frequency* constraint $|cover(S, D)| \geq \theta$ is the most common example of a constraint over the cover set. Alternatively, one can impose the *maximum frequency* constraint: $|cover(S, D)| \leq \beta$.

Discriminating constraints: In case of multiple datasets, discriminating constraints require that patterns effectively distinguish the datasets from each other. Given two datasets D_1 and D_2, one can require that the ratio between the size of the cover of both is above a threshold: $\frac{|cover(S, D_1)|}{|cover(S, D_2)|} \geq \alpha$. Other examples include more statistical measures such as information gain and entropy [13].

Constraints over the Inclusion Relation. The inclusion relation in definition 2 states that $S \sqsubseteq S' \leftrightarrow \exists e$ s.t. $S \sqsubseteq_e S'$. Hence, an embedding of a pattern can match symbols that are far apart in the transaction. For example, the sequence $\langle a, c \rangle$ is embedded in the transaction $\langle a, b, b, b, \ldots, b, c \rangle$ independently of the distance between a and c in the transaction. This is undesirable when mining datasets with long transactions. The *max-gap* and *max-span* constraints [20] impose a restriction on the embedding, and hence on the inclusion relation. The *max-gap constraint* is satisfied on a transaction T_i if an embedding e maps every two consecutive symbols in S to symbols in T_i that are close to each-other: $max\text{-}gap_i(e) \leftrightarrow \forall j \in 2..|T_i|, (e_j - e_{j-1} - 1) \leq \gamma$. For example, the sequence $\langle abc \rangle$ is embedded in the transaction $\langle adddbc \rangle$ with a maximum gap of 3 whereas $\langle ac \rangle$ is not. The *max-span constraint* requires that the distance between the first and last position of the embedding of all transactions T_i is below a threshold γ: $max\text{-}span_i(e) \leftrightarrow e_{|T_i|} - e_1 + 1 \leq \gamma$.

Preferences over the Solution Set. A pairwise preference over the solution set expresses that a pattern A is preferred over a pattern B. In [11] it was shown that condensed representations like closed, maximal and free patterns can be expressed as pairwise preference relations. Skypatterns [15] and multi-objective optimisation can also be seen as preference over patterns. As an example, let Δ be the set of all patterns; then, the set of all closed patterns is $\{S \in \Delta | \nexists S'$ s.t. $S \sqsubseteq S'$ and $cover(S, D) = cover(S', D)\}$.

3   Sequence Mining in Constraint Programming

In constraint programming, problems are expressed as a constraint satisfaction problem (CSP), or a constraint optimisation problem (COP). A CSP $X = (V, D, C)$ consists of a set of variables V, a finite domain D that defines for each variable $v \in V$ the possible values that it can take, and a set of constraints C over the variables in V. A solution to a CSP is an assignment of each variable to a value from its domain such that all constraints are satisfied. A COP additionally consists of an optimisation criterion $f(V)$ that expresses the quality of the solution.

There is no restriction on what a constraint C can represent. Examples include logical constraints like $\mathbf{X} \wedge \mathbf{Y}$ or $\mathbf{X} \rightarrow \mathbf{Y}$ and mathematical constraints such as $\mathbf{Z} = \mathbf{X} + \mathbf{Y}$ etc. Each constraint has a corresponding *propagator* that ensures the constraint is satisfied during the search. Many *global constraints* have been proposed, such as *alldifferent*, which have a custom propagator that is often more efficient then if one would *decompose* that constraint in terms of simple logical or mathematical constraints. A final important concept used in this paper is that of *reified constraints*. A reified constraint is of the form $\mathbf{B} \leftrightarrow C'$ where $\mathbf{B}$ is a Boolean variable which will be assigned to the truth value of constraint C'. Reified constraints have their own propagator too.

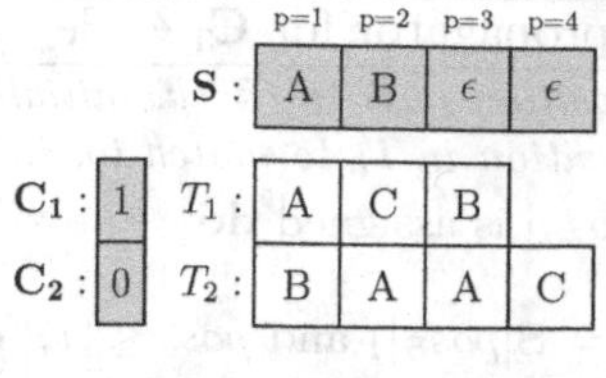

p=1 p=2 p=3 p=4

S : | A | B | ϵ | ϵ |

C_1 : 1 T_1 : | A | C | B |
C_2 : 0 T_2 : | B | A | A | C |

Fig. 1. Example assignment; blue boxes represent variables, white boxes represent data

Variables and domains for modeling sequence mining. Modeling a problem as a CSP requires the definition of a set of variables with a finite domain, and a set of constraints. One solution to the CSP will correspond to one pattern, that is, one frequent sequence.

We model the problem using an array **S** of integer variables representing the characters of the sequence and an array **C** of Boolean variables representing which transactions include the pattern. This is illustrated in Fig. 1:

1. T_1 and T_2 represent two transactions given as input. We denote the number of transactions by n;
2. The array of variables **S** represents the sequence pattern. Each variable S_j represents the character in the jth position of the sequence. The size of **S** is determined by the length of the longest transactions (in the example this is 4). We want to allow patterns that have fewer than $max_i(|T_i|)$ characters, hence we use ϵ to represent an unused position in **S**. The domain of each variable S_j is thus $\Sigma \cup \{\epsilon\}$;
3. Boolean variables C_i represent whether the pattern is included in transaction T_i, that is, whether $S \sqsubseteq T_i$. In the example, this is the case for T_1 but not for T_2.

What remains to be defined is the constraints. The key part here is how to model the inclusion relation; that is, the constraint that verifies whether a pattern is included in the transaction. Conceptually, this is the following reified constraint: $C_i \leftrightarrow \exists e$ s.t. $S \sqsubseteq_e T_i$. As mentioned in the introduction, the number of possible embeddings is exponential in the size of the pattern. Hence, one can not model this as a disjunctive constraint over all possible embeddings (as is done for sequences with explicit wildcards [8]).

We propose two approaches to cope with this problem: one with a global constraint that verifies the inclusion relation directly on the data, and one in which the inclusion relation is decomposed and the embedding is exposed through variables.

4   Sequence Mining with a Global *Exists-Embedding* Constraint

The model consists of three parts: encoding of the pattern, of the minimum frequency constraint and finally of the inclusion relation using a global constraint.

Algorithm 1. Incremental propagator for $C_i \leftrightarrow \exists e$ s.t. $S \sqsubseteq_e T_i$:

---

internal state, pos_S: current position in S to check, initially 1
internal state, pos_e: current position in T_i to match to, initially 1

1: **while** $pos_S \leq |T_i|$ and $S[pos_S]$ is assigned **do**   ▷ note that $|T_i| \leq |S|$
2:     **if** $S[pos_S] \neq \epsilon$ **then**
3:         **while** not $(T_i[pos_e] = S[pos_S])$ and $pos_e \leq |T_i|$ **do**   ▷ find match
4:             $pos_e \leftarrow pos_e + 1$
5:         **end while**
6:         **if** $pos_e \leq |T_i|$ **then**   ▷ match found, on to next one
7:             $pos_S \leftarrow pos_S + 1; pos_e \leftarrow pos_e + 1$
8:         **else**
9:             propagate $C_i = False$ and return
10:     **else**   ▷ previous ones matched and rest is ϵ
11:         propagate $C_i = True$ and return
12: **end while**
13: **if** $pos_S > |S|$ **then**   ▷ previous ones matched and reached end of sequence
14:     propagate $C_i = True$ and return
15: **if** $pos_S > |T_i|$ and $|T_i| < |S|$ **then**
16:     let $R \leftarrow S[|T_i| + 1]$
17:     **if** R is assigned and $R = \epsilon$ **then** ▷ S should not be longer than this transaction
18:         propagate $C_i = True$ and return
19:     **if** ϵ is not in the domain of R **then**
20:         propagate $C_i = False$ and return
21: **if** C_i is assigned and $C_i = True$ **then**
22:     propagate by removing from $S[pos_S]$ all symbols not in $\langle T_i[pos_e]..X_i[|T_i|]\rangle$
        except ϵ

Variable-length pattern: The array S has length k; patterns with $l < k$ symbols are represented with l symbols from Σ and $(k - l)$ times an ϵ value. To avoid enumerating the same pattern with ϵ values in different positions, ϵ values can only appear at the end:

$$\forall j \in 1..(k-1) : S_j = \epsilon \rightarrow S_{j+1} = \epsilon \tag{4}$$

Minimum frequency: At least θ transactions should include the pattern. This inclusion is indicated by the array of Boolean variables C:

$$\sum_{i=1}^{n} C_i \geq \theta \tag{5}$$

Global exists-embedding constraint: The goal is to encode the relation: $C_i \leftrightarrow \exists e$ s.t. $S \sqsubseteq_e T_i$. The propagator algorithm for this constraint is given in Algorithm 1. It is an incremental propagator that should be run when one of the S variables is assigned. Line 1 will loop over the variables in S until reaching an unassigned one at position pos_S. In the sequence mining literature, the sequence $\langle S_1..S_{pos} \rangle$ is called the *prefix*. For each assigned S_j variable, a matching element in the transaction is sought, starting from the position pos_e after the element that matched the previous $\mathbf{S_{j-1}}$ assigned variable. If no such match is found then an embedding can not be found and $\mathbf{C_i}$ is set to false.

Line 11 is called when an $\mathbf{S_j}$ variable is assigned to ϵ. This line can only be reached if all previous values of $\mathbf{S}$ are assigned and were matched in T_i, hence the propagator can set $\mathbf{C_i}$ to true and quit. Similarly for line 14 when the end of the sequence is reached, and lines 15-20 in case the transaction is smaller than the sequence. Lines 21-22 propagate the remaining possible symbols from T_i to the first unassigned $\mathbf{S}$ variable in case $\mathbf{C_i} = True$.

The propagator algorithm has complexity $O(|T_i|)$: the loop on line 1 is run up to $|T_i|$ times and on line 3 at most $|T_i|$ times in total, as pos_e is monotonically increasing.

4.1 Improved Pruning with *Projected Frequency*

Compared to specialised sequence mining algorithms, pos_S in Algorithm 1 points to the first position in $\mathbf{S}$ after the current *prefix*. Dually, pos_e points to the position after the first match of the prefix in the transaction. If one would project the prefix away, only the symbols in the transaction from pos_e on would remain; this is known as *prefix projection* [6]. Given prefix $\langle a, c \rangle$ and transaction $\langle b, a, a, e, c, b, c, b, b \rangle$ the projected transaction is $\langle b, c, b, b \rangle$

The concept of a prefix-projected database can be used to recompute the frequency of all symbols in the projected database. If a symbol is present but not frequent in the projected database, one can avoid searching over it. This is known to speed up specialised mining algorithms considerably [6,17].

To achieve this in the above model, we need to adapt the global propagator so that it exports the symbols that still appear after pos_e. We introduce an auxiliary integer variable $\mathbf{X_i}$ for every transaction T_i, whose domain represents these symbols (the set of symbols is monotonically decreasing). To avoid searching over infrequent symbols, we define a custom search routine (brancher) over the $\mathbf{S}$ variables. It first computes the local frequencies of all symbols based on the domains of the $\mathbf{X_i}$ variables; symbols that are locally infrequent will not not be branched over. See Appendix in extended version [12] for more details.

4.2 Constraints

This formulation supports a variety of constraints, namely on the pattern (type 1), on the cover set (type 2) and over the solution set (type 4). For example, the type 1 constraint *min-size*, constrains the size of the pattern to be larger than a user-defined threshold α. This constraint can be formalised as follows.

$$\sum_{j=1}^{k} [\mathbf{S_j} \neq \epsilon] \geq \alpha \tag{6}$$

Minimum frequency in Equation (5) is an example of a constraint of type 2, over the cover set. Another example is the *discriminative* constraint mentioned in Section 2.2: given two datasets D_1 and D_2, one can require that the ratio between the cover in the two datasets is larger than a user defined threshold α: $\frac{|cover(S,D_1)|}{|cover(S,D_2)|} \geq \alpha$. Let $D = D_1 \cup D_2$ and let $t_1 = \{i|T_i \in D_1\}$ and $t_2 = \{i|T_i \in D_2\}$ then we can extract the discriminant patterns from D by applying the following constraint.

$$\frac{\sum_{i \in t_1} \mathbf{C_i}}{\sum_{i \in t_2} \mathbf{C_i}} \geq \alpha \tag{7}$$

Such a constraint can also be used as an optimisation criterion in a CP framework.

Type 4 constraints a.k.a. preference relations have been proposed in [11] to formalise well-known pattern mining settings such as *maximal* or *closed* patterns. Such preference relations can be enforced dynamically during search for any CP formulation [11]. The preference relation for closed is $S' \succ S \iff S \sqsubset S' \wedge cover(S, D) = cover(S', D)$ and one can reuse the global reified *exists-embedding* constraint for this.

Finally, type 3 constraints over the inclusion relation are not possible in this model. Indeed, a new global constraint would have to be created for every possible (combination of) type 3 constraints. For example for *max-gap*, one would have to modify Algorithm 1 to check whether the gap is smaller than the threshold, and if not, to search for an alternative embedding instead (thereby changing the complexity of the algorithm).

5   Decomposition with Explicit Embedding Variables

In the previous model, we used a global constraint to assign the $\mathbf{C_i}$ variables to their appropriate value, that is: $\mathbf{C_i} \leftrightarrow \exists e$ s.t. $\mathbf{S} \sqsubseteq_e T_i$. The global constraint efficiently tests the existence of one embedding, but does not expose the value of this embedding, thus it is impossible to express constraints over embeddings such as the *max-gap* constraint.

To address this limitation, we extend the previous model with a set of *embedding* variables $\mathbf{E_{i1}}, \ldots, \mathbf{E_{i|T_i|}}$ that will represent an embedding $e = (e_1, \ldots, e_{|T_i|})$ of sequence $\mathbf{S}$ in transaction T_i. In case there is no possible match for a character $\mathbf{S_i}$ in T_i, the corresponding $\mathbf{E_{ij}}$ variable will be assigned a *no-match* value.

5.1   Variables and Constraints

Embedding Variables. For each transaction T_i of length $|T_i|$, we introduce integer variables $\mathbf{E_{i1}}, \ldots, \mathbf{E_{i|T_i|}}$. Each variable $\mathbf{E_{ij}}$ is an index in T_i, and an assignment to $\mathbf{E_{ij}}$ maps the variable $\mathbf{S_j}$ to a position in T_i; see Figure 2, the value of the index is materialized by the red arrows. The domain of $\mathbf{E_{ij}}$ is initialized to all possible positions of T_i, namely $1, \ldots, |T_i|$ plus a *no-match* entry which we represent by the value $|T_i| + 1$.

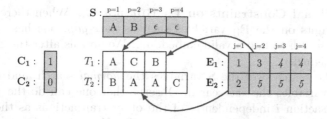

Fig. 2. Example assignment; blue boxes represent variables, white boxes represent data. The cursive values in $\mathbf{E_1}$ and $\mathbf{E_2}$ represent the *no-match* value for that transaction.

The *position-match* constraint. This constraint ensures that the variables $\mathbf{E_i}$ either represent an embedding e such that $\mathbf{S} \sqsubseteq_e T_i$ or otherwise at least one $\mathbf{E_{ij}}$ has the *no-match* value. Hence, each variable $\mathbf{E_{ij}}$ is assigned the value x only if the character in $\mathbf{S_i}$ is equal to the character at position x in T_i. In addition, the constraint also ensures that the values between two consecutive variables $\mathbf{E_{ij}}, \mathbf{E_{i(j+1)}}$ are increasing so that the order of the characters in the sequence is preserved in the transaction. If there exist no possible match satisfying these constraints, the *no-match* value is assigned.

$$\forall i \in 1,\dots,n, \forall j \in 1,\dots,m: \quad (\mathbf{S_j} = T_i[\mathbf{E_{ij}}]) \vee (\mathbf{E_{ij}} = n+1) \tag{8}$$

$$\forall i \in 1,\dots,n, \forall j \in 2,\dots,m: \quad (\mathbf{E_{i(j-1)}} < \mathbf{E_{ij}}) \vee (\mathbf{E_{ij}} = n+1) \tag{9}$$

Here $\mathbf{S_j} = T_i[\mathbf{E_{ij}}]$ means that the symbol of $\mathbf{S_j}$ equals the symbol at index $\mathbf{E_{ij}}$ in transaction T_i. See Appendix in [12] for an effective reformulation of these constraints.

***Is-embedding* constraint.** Finally, this constraint ensures that a variable $\mathbf{C_i}$ is *true* if the embedding variables $\mathbf{E_{i1}}, \dots, \mathbf{E_{i|T_i|}}$ together form a valid embedding of sequence $\mathbf{S}$ in transaction T_i. More precisely: if each character $\mathbf{S_j} \neq \epsilon$ is mapped to a position in the transaction that is different from the *no-match* value.

$$\forall i \in 1,\dots,n: \quad \mathbf{C_i} \leftrightarrow \forall j \in 1,\dots,|T_i|: (\mathbf{S_j} \neq \epsilon) \rightarrow (\mathbf{E_{ij}} \neq |T_i|+1) \tag{10}$$

Note that depending on how the $\mathbf{E_{ij}}$ variables will be searched over, the above constraints are or are not equivalent to enforcing $\mathbf{C_i} \leftrightarrow \exists e$ s.t. $\mathbf{S} \sqsubseteq_e T_i$. This is explained in the following section.

5.2 Search Strategies for Checking the Existence of Embeddings

CP's standard enumerative search would search for all satisfying assignments to the $\mathbf{S_j}, \mathbf{C_i}$ and $\mathbf{E_{ij}}$ variables. As for each sequence of size m, the number of embeddings in a transaction of size n can be $O(n^m)$, such a search would not perform well. Instead, we only need to search whether *one* embedding exists for each transaction.

With Additional Constraints on $\mathbf{E_{ij}}$ But not $\mathbf{C_i}$. When there are additional constraints on the $\mathbf{E_{ij}}$ variables such as *max-gap*, one has to perform backtracking search to find a valid embedding. We do this after the $\mathbf{S}$ variables have been assigned.

We call the search over the $\mathbf{S}$ variables the *normal* search, and the search over the $\mathbf{E_{ij}}$ variables the *sub* search. Observe that one can do the *sub* search for each transaction i independently of the other transactions as the different $\mathbf{E_i}$ have no influence on each other, only on $\mathbf{C_i}$. Hence, one does not need to backtrack across different *sub* searchers.

The goal of a *sub* search for transaction i is to find a valid embedding for that transaction. Hence, that *sub* search should search for an assignment to the $\mathbf{E_{ij}}$ variables with $\mathbf{C_i}$ set to *true* first. If a valid assignment is found, an embedding for T_i exists and the *sub* search can stop. If no assignment is found, $\mathbf{C_i}$ is set to false and the *sub* search can stop too. See Appendix in [12] for more details on the sub search implementation.

With Arbitrary Constraints. The constraint formulation in Equation (10) is not equivalent to $\mathbf{C_i} \leftrightarrow \exists e$ s.t. $\mathbf{S} \sqsubseteq_e T_i$. For example, lets say some arbitrary constraint propagates $\mathbf{C_i}$ to *false*. For the latter constraint, this would mean that it will enforce that $\mathbf{S}$ is such that there does not exists an embedding of it in T_i. In contrast, the constraint in Equation (10) will propagate some $\mathbf{E_{ij}}$ to the *no-match* value, even if there exists a valid match for the respective $\mathbf{S_j}$ in T_i!

To avoid an $\mathbf{E_{ij}}$ being set to the *no-match* value because of an assignment to $\mathbf{C_i}$, we can replace Equation (10) by the half-reified $\forall i : \mathbf{C_i} \rightarrow (\forall j \, (\mathbf{S_j} \neq \epsilon) \rightarrow (\mathbf{E_{ij}} \neq |T_i| + 1)$) during *normal* search.

The *sub* search then has to search for a valid embedding, even if $\mathbf{C_i}$ is set to *false* by some other constraint. One can do this in the *sub* search of a specific transaction i by replacing the respective half-reified constraint by the constraint $\mathbf{C_i'} \leftrightarrow (\forall j \, (\mathbf{S_j} \neq \epsilon) \rightarrow (\mathbf{E_{ij}} \neq |T_i| + 1)$) over a new variable $\mathbf{C_i'}$ that is local to this *sub* search. The *sub* search can then proceed as described above, by setting $\mathbf{C_i'}$ to *true* and searching for a valid assignment to $\mathbf{E_i}$. Consistency between $\mathbf{C_i'}$ and the original $\mathbf{C_i}$ must only be checked after the *sub* search for transaction i is finished. This guarantees that for any solution found, if $\mathbf{C_i}$ is *false* and so is $\mathbf{C_i'}$ then indeed, there exists no embedding of $\mathbf{S}$ in T_i.

5.3   Projected Frequency

Each $\mathbf{E_{ij}}$ variable represents the positions in T_i that $\mathbf{S_j}$ can still take. This is more general than the projected transaction, as it also applies when the previous symbol in the sequence $\mathbf{S_{j-1}}$ is not assigned yet. Thus, we can also use the $\mathbf{E_{ij}}$ variables to require that every symbol of $\mathbf{S_j}$ must be frequent in the (generalised) projected database. This is achieved as follows.

$$\forall j \in 1 \ldots n, \forall x \in \Sigma, \mathbf{S_j} = x \rightarrow |\{i : \mathbf{C_i} \wedge T_i[\mathbf{E_{ij}}] = x\}| \geq \theta \qquad (11)$$

See Appendix in [12] for a more effective reformulation.

5.4 Constraints

All constraints from Section 4.2 are supported in this model too. Additionally, constraints over the inclusion relations are also supported; for example, *max-gap* and *max-span*. Recall from Section 2.2 that for an embedding $e = (e_1, \ldots, e_k)$, we have $max\text{-}gap_i(e) \Leftrightarrow \forall j \in 2 \ldots |T_i|, (e_j - e_{j-1} - 1) \leq \gamma$. One can constrain all the embeddings to satisfy the *max-gap* constraint as follows (note how x is smaller than the *no-match* value $|T_i| + 1$):

$$\forall i \in 1 \ldots n, \forall j \in 2 \ldots |T_i|, x \in 1 \ldots |T_i| : \quad \mathbf{E}_{ij} = x \rightarrow x - \mathbf{E}_{i(j-1)} \leq \gamma + 1 \quad (12)$$

Max-span was formalized as $max\text{-}span_i(e) \Leftrightarrow e_{|T_i|} - e_1 + 1 \leq \gamma$ and can be formulated as a constraint as follows:

$$\forall i \in 1 \ldots n, \forall j \in 2 \ldots |T_i|, x \in 1 \ldots |T_i| : \quad \mathbf{E}_{ij} = x \rightarrow x - \mathbf{E}_{i1} \leq \gamma - 1 \quad (13)$$

In practice, we implemented a simple *difference-except-no-match* constraint that achieves the same without having to post a constraint for each x separately.

6 Experiments

The goal of these experiments is to answer the four following questions: **Q1:** What is the overhead of exposing the embedding variables in the *decomposed* model? **Q2:** What is the impact of using projected frequency in our models? **Q3:** What is the impact of adding constraints on runtime and on number of results? **Q4:** How does our approach compares to existing methods?

Algorithm and execution environment: All the models described in this paper have been implemented in the Gecode solver[1]. We compare our *global* and *decomposed* models (Section 4 and Section 5) to the state-of-the-art algorithms cSpade [20] and PrefixSpan [6]. We use the author's cSpade implementation[2] and a publicly available PrefixSpan implementation by Y. Tabei[3]. We also compare our models to the CP-based approach proposed by [10]. No implementation of this is available so we reimplemented it in Gecode. Gecode does not support non-deterministic automata so we use a more compact DFA encoding that requires only $O(n * |\Sigma|)$ transitions, by constructing it back-to-front. We call this approach *regular-dfa*. Unlike the non-deterministic version, this does not allow the addition of constraints of type 3 such as *max-gap*.

All algorithms were run on a Linux PC with 16 GB of memory. Algorithm runs taking more than 1 hour or more than 75% of the RAM were terminated. The implementation and the datasets used for the experiments are available online[4].

---

[1] http://www.gecode.org
[2] http://www.cs.rpi.edu/~zaki/www-new/pmwiki.php/Software/
[3] https://code.google.com/p/prefixspan/
[4] https://dtai.cs.kuleuven.be/CP4IM/cpsm

Table 1. Dataset characteristics. Respectively: dataset name, number of distinct symbols, number of transactions, total number of symbols in the dataset, maximum transaction length, average transaction length, and density calculated by $\frac{||D||}{|\Sigma| \times |D|}$.

dataset	$|\Sigma|$	$|D|$	$||D||$	$\max_{T \in D} |T|$	avg $|T|$	density
Unix user	265	484	10935	1256	22.59	0.085
JMLR	3847	788	75646	231	96.00	0.025
iPRG	21	7573	98163	13	12.96	0.617
FIFA	20450	2990	741092	100	36.239	0.012

Datasets: The datasets used are from real data and have been chosen to represent a variety of application domains. In **Unix user**[5], each transaction is a series of shell commands executed by a user during one session. We report results on User 3; results are similar for the other users. **JMLR** is a natural language processing dataset; each transaction is an abstract of a paper from the *Journal of Machine Learning Research*. **iPRG** is a proteomics dataset from the application described in [4]; each transaction is a sequence of peptides that is known to cleave in presence of a Trypsin enzyme. **FIFA** is click stream dataset[6] from logs of the website of the FIFA world cup in 98; each transaction is a sequence of webpages visited by a user during a single session. Detailed characteristics of the datasets are given in Table 1. Remark that the characteristic of these datasets are very diverse due to their different origins.

In our experiments, we vary the minimum frequency threshold (*minsup*). Lower values for *minsup* result in larger solution sets, thus in larger execution times.

Experiments: First we compare the *global* and the *decomposed* models. The execution times for these models are shown on Fig. 3, both without and with projected frequency (indicated by *-p.f.*). We first look at the impact of exposing the embedding variables in the *decomposed* model (**Q1**). Perhaps unsurprisingly, the *global* model is up to one order of magnitude faster than the *decomposed* model, which has $O(n * k)$ extra variables. This is the overhead required to allow one to add constraints over the inclusion relation. We also study the impact of the projected frequency on both models (**Q2**). In the *global* model this is done as part of the search, while in the *decomposed* model this is achieved with an elaborate constraint formulation. For *global-p.f.* we always observe a speedup in Fig. 3. Not so for *decomposed-p.f.* for the two largest (in terms of $||D||$) datasets.

We now evaluate the impact of user constraints on the number of results and on the execution time (**Q3**). Fig. 4 shows the number of patterns and the execution times for various combinations of constraints. We can see that adding constraints enables users to control the explosion of the number of patterns, and that the execution times decrease accordingly. The constraint propagation allows early pruning of invalid solutions which effectively compensates the computation

---

[5] https://archive.ics.uci.edu/ml/datasets/
[6] http://www.philippe-fournier-viger.com/spmf/

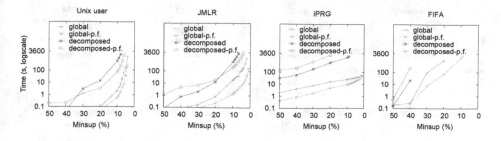

Fig. 3. Global model vs. decomposed model: Execution times (Timeout 1 hour.)

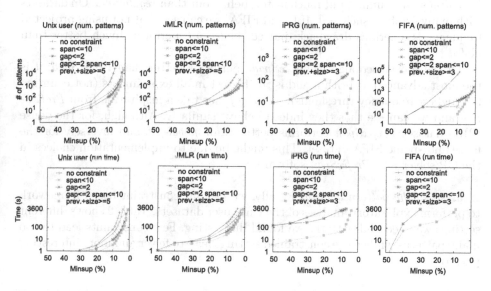

Fig. 4. Number of patterns (top) and execution times (bottom) for the decomposed model with various combinations of constraints

time of checking the constraints. For example, on the Unix user dataset, it is not feasible to mine for patterns at 5% minimum frequency without constraints, let alone do something with the millions of patterns found. On the other hand, by adding constraints one can look for interesting patterns at low frequency without being overwhelmed by the number of results (see also later).

The last experiment compares our models to existing algorithms. Fig. 5 shows the execution times for our *global* model compared with *regular-dfa*, PrefixSpan and cSpade (**Q4**). First, we can observe that *regular-dfa* is always slowest. On iPRG it performs reasonably well, but the number of transitions in the DFAs does not permit it to perform well on datasets with a large alphabet or large transactions, such as Unix user, JMLR or FIFA. Furthermore, it can not make use of projected frequencies.

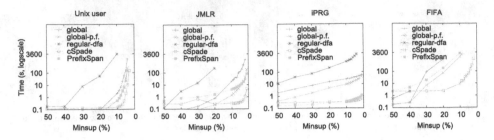

Fig. 5. Global model vs. other approaches. Execution times. (Timeout 1 hour.)

global shows similar, but much faster, behaviour than *regular-dfa*. On datasets with many symbols such as JMLR and FIFA, we can see that not using projected frequency is a serious drawback; indeed, *global-p.f.* performs much better than *global* there.

Of the specialised algorithms, *cSpade* performs better than *PrefixSpan*; it is the most advanced algorithm and is the fastest in all experiments (not counting the highest frequency thresholds). *global-p.f.* has taken inspiration from *PrefixSpan* and we can see that they indeed behave similarly. Although, for the dense iPRG dataset *PrefixSpan* performs better than *global-p.f.* and inversely for the large and sparse FIFA dataset. This might be due to implementation choices in the CP solver and *PrefixSpan* software.

Analysis of the pattern quality. Finally, we use our constraint-based framework to perform exploratory analysis of the Unix user datasets. Table 2 shows different settings we tried and patterns we found interesting. Few constraints lead to too many patterns while more constrained settings lead to fewer and more interesting patterns.

7   Related Work

The idea of mining patterns in sequences dates from earlier work by Agrawal et al. [1] shortly after their well-known work on frequent itemset mining [2]. The

Table 2. Patterns with various settings (User 2): F_1: $minfreq = 5\%$, F_2: $F_1 \wedge$ $min\text{-}size = 3$, F_3: $F_2 \wedge max\text{-}gap = 2 \wedge max\text{-}span = 5$, D_1: $minfreq = 5\% \wedge$ $discriminant = 8$ (w.r.t. all other users), D_2: $minfreq = 0.4\% \wedge discriminant =$ $8 \wedge member(\text{quota})$

setting	# of patterns	interesting pattern	comment
F_1	627	–	Too many patterns
F_2	512	–	Long sequences of cd and ls
F_3	36	⟨latex, bibtex, latex⟩	User2 is using Latex to write a paper
D_1	7	⟨emacs⟩	User2 uses *Emacs*, his/her collaborators use *vi*
D_2	9	⟨quota, rm, ls, quota⟩	User is out of disc quota

problem introduced in [1] consisted of finding frequent sequences of *itemsets*; that is: sequences of sets included in a database of sequences of sets. Mining sequences of individual symbols was introduced later by [9]; the two problems are closely related and one can adapt one to the other [17]. Sequence mining was driven by the application of market basket analysis for customer data spread over multiple days. Other applications include bio-medical ones where a large number of DNA and protein sequence datasets are available (e.g. [19]), or natural language processing where sentences can be represented as sequences of words (e.g. [16]).

Several specialised algorithm have addressed the problem of constrained sequence mining. The cSpade algorithm [20] for example is an extension of the Spade sequence mining algorithm [21] that supports constraints of type 1, 2 and 3. PrefixSpan [6] mentions regular expression constraints too. The LCM-seq algorithm [14] also supports a range of constraints, but does not consider all embeddings during search. Other sequence mining algorithms have often focussed on constraints of type 4, and on closed sequence mining in particular. CloSpan [18] and Bide [17] are both extentions of PrefixSpan to mine *closed* frequent sequences. We could do the same in our CP approach by adding constraints after each solution found, following [8,11].

Different flavors of sequence mining have been studied in the context of a generic framework, and constraint programming in particular. They all study constraints of type 1, 2 and 4. In [3] the setting of sequence patterns with explicit wildcards in a single sequence is studied: such a pattern has a linear number of embeddings. As only a single sequence is considered, frequency is defined as the number of embeddings in that sequence, leading to a similar encoding to itemsets. This is extended in [7] to sequences of itemsets (with explicit wildcards over a single sequence). [8] also studies patterns with explicit wildcards, but in a database of sequences. Finally, [10] considers standard sequences in a database, just like this paper; they also support constraints of type 3. The main difference is in the use of a costly encoding of the inclusion relation using non-deterministic automata and the inherent inability to use projected frequency.

8   Conclusion and Discussion

We have investigated a generic framework for sequence mining, based on constraint programming. The difficulty, compared to itemsets and sequences with explicit wildcards, is that the number of embeddings can be huge, while knowing that one embedding exists is sufficient.

We proposed two models for the sequence mining problem: one in which the exists-embedding relation is captured in a global constraint. The benefit is that the complexity of dealing with the existential check is hidden in the constraint. The downside is that modifying the inclusion relation requires modifying the global constraint; it is hence not generic towards such constraints. We were able to use the same *projected frequency* technique as well-studied algorithms such as PrefixSpan [6], by altering the global exists-embedding constraint and using a specialised search strategy. Doing this does amount to implementing specific propagators and search strategies into a CP solver, making the problem formulation not applicable to other solvers out-of-the-box. On the other hand, it allows for significant efficiency gains.

The second model exposes the actual embedding through variables, allowing for more constraints and making it as generic as can be. However, it has extra overhead and requires a custom two-phased search strategy.

Our observations are not just limited to sequence mining. Other pattern mining tasks such as tree or graph mining also have multiple (and many) embeddings, hence they will also face the same issues with a reified exists relation. Whether a general framework exists for all such pattern mining problems is an open question.

Acknowledgments. The authors would like to thank Siegfried Nijssen, Anton Dries and Rémi Coletta for discussions on the topic, and the reviewers for their valuable comments. This work was supported by the European Commission under project FP7-284715 "Inductive Constraint Programming" and a Postdoc grant by the Research Foundation – Flanders.

References

1. Agrawal, R., Srikant, R.: Mining sequential patterns. In: Proceedings of the Eleventh International Conference on Data Engineering, pp. 3–14. IEEE (1995)
2. Agrawal, R., Srikant, R., et al.: Fast algorithms for mining association rules in large database. In: Proc. 20th Int. Conf. Very Large Data Bases, VLDB, vol. 1215, pp. 487–499 (1994)
3. Coquery, E., Jabbour, S., Sais, L., Salhi, Y.: A sat-based approach for discovering frequent, closed and maximal patterns in a sequence. In: European Conference on Artificial Intelligence (ECAI), pp. 258–263 (2012)
4. Fannes, T., Vandermarliere, E., Schietgat, L., Degroeve, S., Martens, L., Ramon, J.: Predicting tryptic cleavage from proteomics data using decision tree ensembles. Journal of Proteome Research **12**(5), 2253–2259 (2013). http://pubs.acs.org/doi/abs/10.1021/pr4001114
5. Guns, T., Nijssen, S., De Raedt, L.: Itemset mining: A constraint programming perspective. Artificial Intelligence **175**(12–13), 1951–1983 (2011)
6. Han, J., Pei, J., Mortazavi-Asl, B., Pinto, H., Chen, Q., Dayal, U., Hsu, M.: Prefixspan: mining sequential patterns efficiently by prefix-projected pattern growth. ICDE 2001, pp. 215–224, April 2001
7. Jabbour, S., Sais, L., Salhi, Y.: Boolean satisfiability for sequence mining. In: 22nd International Conference on Information and Knowledge Management (CIKM 2013), pp. 649–658. ACM Press, San Francisco (2013)
8. Kemmar, A., Ugarte, W., Loudni, S., Charnois, T., Lebbah, Y., Boizumault, P., Cremilleux, B.: Mining relevant sequence patterns with cp-based framework. In: 2013 IEEE 25th International Conference on Tools with Artificial Intelligence (ICTAI). IEEE (2014)
9. Mannila, H., Toivonen, H., Inkeri Verkamo, A.: Discovery of frequent episodes in event sequences. Data Mining and Knowledge Discovery **1**(3), 259–289 (1997)

10. Métivier, J.P., Loudni, S., Charnois, T.: A constraint programming approach for mining sequential patterns in a sequence database. In: ECML/PKDD 2013 Workshop on Languages for Data Mining and Machine Learning (2013)
11. Negrevergne, B., Dries, A., Guns, T., Nijssen, S.: Dominance programming for itemset mining. In: International Conference on Data Mining (ICDM) (2013)
12. Negrevergne, B., Guns, T.: Constraint-based sequence mining using constraint programming. CoRR abs/1501.01178 (2015)
13. Nijssen, S., Guns, T., De Raedt, L.: Correlated itemset mining in ROC space: A constraint programming approach
14. Ohtani, H., Kida, T., Uno, T., Arimura, H., Arimura, H.: Efficient serial episode mining with minimal occurrences. In: ICUIMC, pp. 457–464 (2009)
15. Ugarte Rojas, W., Boizumault, P., Loudni, S., Crémilleux, B., Lepailleur, A.: Mining (soft-) skypatterns using dynamic CSP. In: Simonis, H. (ed.) CPAIOR 2014. LNCS, vol. 8451, pp. 71–87. Springer, Heidelberg (2014)
16. Tatti, N., Vreeken, J.: The long and the short of it: summarising event sequences with serial episodes. In: KDD, pp. 462–470 (2012)
17. Wang, J., Han, J.: Bide: Efficient mining of frequent closed sequences. In: Proceedings of the 20th International Conference on Data Engineering, pp. 79–90. IEEE (2004)
18. Yan, X., Han, J., Afshar, R.: Clospan: Mining closed sequential patterns in large datasets. In: Proceedings of SIAM International Conference on Data Mining, pp. 166–177 (2003)
19. Ye, K., Kosters, W.A., IJzerman, A.P.: An efficient, versatile and scalable pattern growth approach to mine frequent patterns in unaligned protein sequences. Bioinformatics 23(6), 687–693 (2007)
20. Zaki, M.J.: Sequence mining in categorical domains: incorporating constraints. In: Proceedings of the ninth international conference on Information and knowledge management, pp. 422–429. ACM (2000)
21. Zaki, M.J.: Spade: An efficient algorithm for mining frequent sequences. Machine Learning 42(1), 31–60 (2001)

A Comparative Study of MIP and CP Formulations for the B2B Scheduling Optimization Problem

Gilles Pesant[1,2]([✉]), Gregory Rix[1,2], and Louis-Martin Rousseau[1,2]

[1] École Polytechnique de Montréal, Montréal, Canada
[2] Interuniversity Research Centre on Enterprise Networks,
Logistics and Transportation, Montréal, Canada
{gilles.pesant,greg.rix,louis-martin.rousseau}@polymtl.ca

Abstract. The Business-to-Business Meeting Scheduling Problem was recently introduced to this community. It consists of scheduling meetings between given pairs of participants to an event while taking into account participant availability and accommodation capacity. The challenging aspect of this problem is that breaks in a participant's schedule should be avoided as much as possible. In an earlier paper, starting from two generic CP and Pseudo-Boolean formulations, several solving approaches such as CP, ILP, SMT, and lazy clause generation were compared on real-life instances. In this paper we use this challenging problem to study different formulations adapted either for MIP or CP solving, showing that the `cost_regular` global constraint can be quite useful, both in MIP and CP, in capturing the problem structure.

1 Introduction

Business-to-business events consist of scheduling meetings between pairs of participants with similar interests. The business-to-business scheduling optimization problem (B2BSOP) was formally introduced in [2], with application to the *4th Forum of the Scientific and Technological Park of the University of Girona*[1]. Several constraints must be respected including participant availability by time, accommodation capacity, and fairness constraints between participants. The challenging aspect of this problem is that breaks in a participant's schedule should be avoided as much as possible. The objective is therefore to minimize the number of breaks assigned to participants, cumulatively over all participants.

The authors of [2] mentioned that solving this problem allowed the organizers to significantly improve their efficiency, in particular because the models were able to handle side constraints, such as fairness, and extend partly fixed schedules. Adopting a "model-and-solve" approach with two fairly straightforward models but state-of-the-art solvers using different combinatorial optimization

---

[1] http://www.forumparcudg.com

© Springer International Publishing Switzerland 2015
L. Michel (Ed.): CPAIOR 2015, LNCS 9075, pp. 306–321, 2015.
DOI: 10.1007/978-3-319-18008-3_21

methods, they wondered whether more dedicated SAT or MIP models would perform better. Additionally their CP model did not make use of global constraints, which are known to play an important role in the success of CP solvers. These motivated us to look more closely at the B2BSOP in a attempt to capture its underlying structure and to better exploit it in MIP and CP approaches.

In this paper we thus derive and compare empirically several MIP and CP formulations of the problem. Our main observation is that the use of the cost_regular constraint is powerful in capturing the structure of the problem, producing stronger lower bounds than simpler MIP formulations, scaling better to larger instances by finding better feasible solutions, and providing a more effective cost-aware branching heuristic for CP.

The remainder of this paper is organized as follows. Section 2 recalls the definition of the B2BSOP. Section 3 focuses on the CP approach, presenting the models, computational results and comparisons. Section 4 presents, evaluates and compares four MIP models. Section 5 then discusses the overall results with respect to previously published methodologies.

2   Problem Definition, Instance Description, and Experimental Setting

Let P be the set of participants, M the set of meetings between pairs of participants, $M_p \subseteq M$ the set of meetings involving participant p, L the set of locations for meetings, and T the set of time slots. The problem is to assign the meetings in M to time slots in T and locations in L so that no participant is in more than one meeting at a time and at most $|L|$ meetings are held at any time. A participant's schedule can then be seen as a vector of length T in which each element is either a meeting in M_p (each appearing only once), or a free period ("0"). We refer to a sequence of 0s that occurs between two meetings as a *break*.

A few other constraints must hold. The set of meetings M_p for participant p can be refined into meetings M_p^{AM} that can only be held in the morning and meetings M_p^{PM} that can only be held in the afternoon. These two sets typically have meetings in common, as some meetings are unrestricted. T^{AM} and T^{PM} respectively represent the morning and afternoon time slots. We also define $F_p \subset T$ representing the forbidden time slots for p. We can derive $T_m \subseteq T$, the set of allowed time slots for meeting m, which takes into account the forbidden time slots of both participants as well as any morning or afternoon requirements of meeting m.

The above defines a feasibility problem for B2B scheduling that could be modeled and solved simply as a graph colouring problem. However an important aspect of the problem is to minimize the number of breaks assigned to all participants cumulatively; this is added as an objective function. Additionally, a schedule must be observed to be fair from the point of view of any participant: too many breaks in an individual schedule relative to another participant is viewed as unfair; we define b^{max} to be the maximum deviation in the number of breaks between any two participants.

Table 1. Features of the B2BSOP instances from [2]

feature	T2a	T2c	T3a	T3b	T3c	F3a	F3b	F3c	F4a
#meetings	125	125	180	184	180	154	195	154	302
#participants	42	42	47	46	47	70	76	70	78
#locations	21	16	21	21	19	14	14	12	22
#time slots	8	8	10	10	10	21	21	21	22
#AM slots	0	0	0	0	0	13	13	13	12

Table 1 lists the main features of the instances from [2]. There are two sets **T⋆** and **F⋆**, each originating from an actual event with a mix of real and synthetic instances. The first set has fewer participants, fewer time slots, no meeting restricted to morning or afternoon slots, and no forbidden time slots for participants. The second set is richer and more difficult to solve.

The MIP and CP models were executed on Dual core AMD 2.1 GHz processors with 8 GB of RAM, running IBM ILOG Solver 6.7 as the CP solver and Gurobi 6.0 as the MIP solver. As in [2] we used a 2-hour time limit.

3   CP Models

A natural CP model defines variables $\{s_{pt} : p \in P, t \in T\}$ to represent what participant p is scheduled to do at time t, value 0 corresponding to no meeting. An alternative model could have defined one variable per meeting with the set of allowed time slots for its domain, as proposed in [2]. However such a representation does not allow us to express constraints directly on the sequence of meetings for a participant, which is important to evaluate the cost of an individual schedule and ultimately the objective we seek to minimize: the latter authors had to define auxilliary variables similar to ours and channel them to their main variables.

Here is our model for the feasibility subproblem:

$$\mathsf{gcc}(\{s_{p\star}\}, \langle |T| - |M_p|, 1, \ldots, 1\rangle) \qquad p \in P \qquad (1)$$

$$\mathsf{gcc}(\{s_{\star t}\}, \langle\{|P| - 2|L|, \ldots, |P|\}, \{0, 2\}, \ldots, \{0, 2\}\rangle) \qquad t \in T \qquad (2)$$

$$s_{pt} = 0 \qquad p \in P, t \in F_p \qquad (3)$$

$$s_{pt} \in M_p^{\mathrm{AM}} \cup \{0\} \qquad p \in P, t \in T^{\mathrm{AM}} \qquad (4)$$

$$s_{pt} \in M_p^{\mathrm{PM}} \cup \{0\} \qquad p \in P, t \in T^{\mathrm{PM}} \qquad (5)$$

Constraints (1) use a global cardinality constraint [8] on the decision variables of a given participant to ensure that each of his meetings appears once (the first component of the vector of occurrences, corresponding to value 0, indicates the number of time slots without a meeting). Constraints (2) use a global cardinality constraint on the decision variables of a given time slot to express two things: the first component says that the number of participants not having a meeting

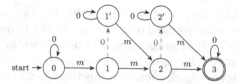

Fig. 1. Automaton $\mathcal{A}_1$ for a participant with three meetings. Arc label "m" stands for any meeting and label "0" for no meeting. Only the red dashed arcs carry a cost, of one unit, to mark the start of a break.

must be at least $|P| - 2|L|$ because we can hold at most $|L|$ meetings and each meeting appears twice (once for each participant); the other components, for each meeting, say that the two participants to a given meeting must attend it in the same time slot and therefore a meeting occurs twice or not at all. Constraints (3) ensure that there is no meeting scheduled for a participant during one of his forbidden time slots.

3.1 Modeling Break Patterns

We now model the optimization component of our problem. We define a variable b_p for each participant p giving the number of breaks in his schedule and seek to minimize the total number of breaks in the schedule. In order to link the b_p variables to the main s_{pt} variables we need to consider the sequence of values taken by the decision variables of a participant: each subsequence of zeros in between scheduled meetings for p corresponds to a break and b_p represents how many such breaks there are in the sequence. For example, patterns 0⋆⋆00⋆00 and ⋆000⋆0⋆0 for eight time slots and three meetings feature respectively one and two breaks.

To express this globally we could enumerate each possible pattern, associate its number of breaks and use a `table` constraint. For given T, M_p, and maximum number of breaks $b^\star$ this makes

$$\sum_{i=0}^{b^\star} \binom{|M_p| - 1}{i} \cdot \binom{|T| - |M_p| + 1}{i + 1}$$

patterns. Even if we restrict ourselves to at most $b^\star = 2$ breaks, the number of patterns is in $\Theta(|M_p|^2(|T| - |M_p|)^3)$ which, when the number of meetings is about half of the number of time slots, simplifies to $\Theta(|M_p|^5)$. Considering that the largest instance has 22 time slots with some participants holding 11 meetings, we could end up generating hundreds of thousands of patterns.

A much more compact way to express this uses an automaton on $2|M_p|$ states that recognizes precisely these patterns. Figure 1 presents such an automaton for a participant with three meetings. Observe however that by concentrating on patterns without distinguishing between meetings we may miss some inferences. For example any assignment from the sequence of domains $\langle \{m_1, m_2, m_3, 0\},$

$\{m_1, m_2, m_3, 0\}, \{m_4, 0\}, \{m_4, 0\}, \{m_1, m_2, m_3, 0\}, \{m_1, m_2, m_3, 0\}\rangle$ correspond- ing to four meetings being scheduled over six time slots will necessarily introduce at least one break but such an automaton will not recognize it. To catch this, a more fine-grain automaton distinguishing between meetings will have $2^{|M_p|+1} - 2$ states essentially representing all subsets of meetings (see Figure 2). Because this automaton has significantly more states we will refrain from using it when the number of meetings is greater than a certain threshold $m^{\max}$. Here is the rest of our model:

$$\min \sum_{p \in P} b_p \qquad \text{s.t.} \tag{6}$$

$$b_p - \min_{p' \in P} b_{p'} \le b^{\max} \qquad p \in P \tag{7}$$

$$b_p = 0 \qquad p \in P : |M_p| \in \{0, 1, |T|\} \tag{8}$$

$$\texttt{cost\_regular}(\langle s_{p\star}\rangle, \mathcal{A}_2, b_p) \qquad p \in P : 1 < |M_p| \le m^{\max} \tag{9}$$

$$\texttt{cost\_regular}(\langle s_{p\star}\rangle, \mathcal{A}_1, b_p) \qquad p \in P : m^{\max} < |M_p| < |T| \tag{10}$$

$$b_p \in \mathbb{N} \qquad p \in P \tag{11}$$

As in [2] we ensure some fairness between individual schedules by requiring that the number of breaks among individual schedules differ by at most $b^{\max} = 2$ (see Constraints (7)). Constraints (8) fix b_p to zero for participants who trivially have no break in their schedule (e.g. they have a single meeting or as many meetings as there are time slots). Note that because in every instance considered there are always such participants, individual schedules will feature at most two breaks. The $\texttt{cost\_regular}$ constraint (10) on automaton $\mathcal{A}_1$ maintains a layered digraph on $2|M_p|(|T|+1)$ vertices and makes variable b_p equal to the sum of the costs of the arcs on the path corresponding to the values taken by the sequence of variables $\langle s_{p\star}\rangle$ [4]. The upper bound on b_p limits the feasible paths in the digraph and possibly removes arcs (i.e. filters values in a domain) that do not belong to any feasible path. Conversely the smallest cost of the possible paths given the current domains of the variables provides a lower bound on b_p. Constraints (9) work similarly but on the larger automaton $\mathcal{A}_2$.

3.2   Branching Heuristics

For the most part, CP branching heuristics are feasibility-driven: an optimization problem is solved as a succession of feasibility problems from which each new improved solution provides a tighter bound on the objective value, formulated as a constraint that is added to the model. A generic heuristic such as max regret [1] does take into account the objective but it is only effective if the cost of a solution can be decomposed at the level of individual variable assignments. For the B2BSOP one needs to consider neighbouring assignments in the sequence or even the whole sequence in order to assess the impact of a particular variable-value assignment on the number of breaks. To make matters worse the objective is a sum, reputed to back propagate poorly.

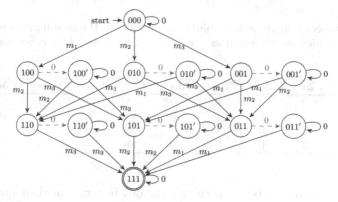

Fig. 2. Automaton $\mathcal{A}_2$ for a participant with three meetings. Arc label "m_i" stands for that particular meeting and label "0" for no meeting. Only the red dashed arcs carry a cost, of one unit, to mark the start of a break.

Table 2. Number of breaks in best solution found within the time limit using different heuristics branching on s_{pt} variables for chronological backtrack search on the same CP model. For reference we add the performance reported in [2] for the CP solver they used (Gecode), albeit on a different model.

heuristic	T2a	T2c	T3a	T3b	T3c	F3a	F3b	F3c	F4a
dom	1	1	27	13	31	64	–	–	81
IBS	0	6	0	0	19	72	–	68	100
maxSD	10	7	–	5	–	–	–	–	–
maxSD*	0	0	3	0	–	1	–	13	23
Gecode [2]	0	–	0	0	–	–	–	–	–

Table 2 reports the performance of a few standard generic branching heuristics on the s_{pt} variables of our CP model. Smallest-domain-first (dom), impact-based search (IBS)[7], and counting-based search (maxSD)[6] perform poorly, as expected, since they are not driven at all by the objective. Some of the smaller instances are solved well by IBS but generally the solutions obtained are quite far from the optimal or best-known solutions and for some instances we get no solution at all within the time limit. This is consistent with the observations of Bofill et al. [2] albeit on a different CP model and running a different CP solver: we report their findings on the last line of the table. We also investigated branching on the b_p variables and trying lower values first, which did help generate somewhat better-quality solutions but not competitive with what we describe below.

Heuristic maxSD, previously shown to be very effective on feasibility problems [6], performs particularly poorly on this optimization problem. It branches on the variable-value pair that has the highest *solution density*, which is the proportion of solutions to an individual constraint that feature this given assignment. Because

Table 3. Number of breaks in best solution found within the time limit using different heuristics branching on s_{pt} variables for limited-discrepancy search on the same CP model

heuristic	T2a	T2c	T3a	T3b	T3c	F3a	F3b	F3c	F4a
dom	0	0	0	0	8	35	73	61	80
IBS	0	0	1	0	15	72	–	76	123
maxSD	0	0	11	0	–	–	–	–	–
maxSD*	0	0	0	0	11	0	23	9	25

it does not discriminate between these solutions in terms of their quality with respect to the objective, such information can be misleading. We experimented with a cost-aware version of this approach: instead of considering solutions of any quality, we compute solution densities among the solutions of lowest cost, thereby integrating optimization into our branching recommendation. For the B2BSOP we retrieve solution densities from the cost_regular constraints, the combinatorial substructure from which the cost (i.e. the number of breaks) associated with individual participants can be computed. Instead of considering all paths in the layered digraph maintained by the constraint, we only consider shortest paths. We branch on the largest cost-aware solution density for scheduling a meeting (values in M). We denote that generic heuristic as maxSD* in Table 2. As anticipated it performs much better than the others, producing higher-quality solutions, though not producing any solution in two instances.

It is well known for backtrack search that regardless of a branching heuristic's quality, chronological backtracking may take a very long time to undo a bad branching decision made early on. There are a few devices commonly used to add robustness to heuristics, e.g. randomized restarts and limited-discrepancy search (LDS) [5]. The latter modifies the order in which the leaves of a search tree are visited according to how often the corresponding path deviates from the search heuristic's recommendation: first the leaf with 0 deviation, then those with $1k$ deviations, followed by those with $2k$ deviations, and so forth, for a given parameter k. This has the effect of more quickly changing decisions close to the root. We add LDS to our branching heuristics (see Table 3). We note a general improvement of solution quality and of robustness, and maxSD* now finds good-quality solutions to every instance.

The $m^{\max}$ threshold, determining when a participant has too many meetings to use the increased-inference but also increased-time-and-space-consumption fine-grain automaton $\mathcal{A}_2$, has an impact on solution quality. Table 4 reports our findings for a few values of that threshold. The first line shows that never using that automaton produces lower-quality solutions for the larger instances. Otherwise a moderate threshold of 5 already makes a difference and the approach does not appear too sensitive to the actual value of the threshold. In our other experiments we used $m^{\max} = 10$.

Table 4. Number of breaks in best solution found within the time limit using different $m^{\max}$ thresholds to select which automaton to use in each cost_regular constraint. The maxSD* branching heuristic is used together with LDS.

$m^{\max}$	T2a	T2c	T3a	T3b	T3c	F3a	F3b	F3c	F4a
0	0	0	0	0	11	7	24	12	27
5	0	0	0	0	11	0	22	10	26
10	0	0	0	0	11	0	23	9	25
12	0	0	0	0	11	0	21	10	25

Table 5. Applying Large Neighbourhood Search on the CP model for the six instances without a proof of optimality after one minute of CP backtrack search

approach	T3a	T3c	F3a	F3b	F3c	F4a
maxSD*	0	11	0	23	9	25
LNS with maxSD* (avg of 10 runs)	0	4.2	2.4	16.8	3.8	17.1
LNS with maxSD* (best of 10 runs)	0	4	0	13	3	13

3.3 Applying Large Neighbourhood Search to the CP Model

Local search is often the method of choice to solve large combinatorial optimization problems. Large Neighbourhood Search (LNS) is a natural way to perform local search on a solution space defined by a CP model [9]: it iteratively freezes part of the current solution and explores the remaining solution space (a potentially large neighbourhood) by applying a (usually incomplete) CP tree search, benefiting from the usual inference and search heuristics.

We evaluated a simple implementation of this idea: we run our exact CP algorithm for one minute in order to get a fair initial solution and then iteratively freeze the schedule of a randomly-selected subset of the participants whose schedule does not contain any break (but we disregard participants who trivially cannot have any break). We explore each neighbourhood with the same exact CP algorithm, stopping at the first improving solution or until 20 seconds have passed. We use the same overall time limit as before.

Table 5 reports the performance of our LNS with respect to our previous CP backtrack search. Since that approach is not deterministic, we give the average and the best objective value out of ten runs. We see a significant improvement of the solutions (except on **F3a** for which an optimal solution was not obtained consistently). In particular an optimal solution to **T3c** (see Section 4) is finally found and we also obtain our best overall solution for **F4a**.

4 MIP Models

We compare four mathematical formulations that exactly solve the B2BSOP. Each formulation contains a common subset of binary variables that fix a meeting to a particular time slot, and a common subset of constraints that enforce a

feasible schedule without consideration of the objective value of break minimization. The difference in the formulations is with respect to the added variables
and constraints that cumulate this objective value, as well as use the objective
function to enforce fairness constraints on the resulting schedule.

The first formulation uses a set of binary variables that determine if a time
slot is the terminating time slot of a break period before a participant begins a
meeting. These variables are linked to the feasibility problem through a series of
linearized logical constraints. The logical constraints to be linearized are those
derived in the pseudoboolean model of [2]. The second formulation defines a network flow problem for each participant, with the flow representing the journey
of the participant between breaks and meetings. The final two formulations are
derived using the `cost_regular` constraint, with two unique deterministic finite
automatons (DFAs) that cumulate the cost of a schedule for each participant.
The first DFA is the previously defined automaton $\mathcal{A}_1$, whereas the second additionally enforces the fairness constraints. However, this final automaton requires
a fixed maximum on the number of breaks in a schedule. As mentioned in Section
3.1, all considered instances contain participants who must have 0 breaks; hence
the maximum number of breaks is equivalent to the maximum deviation $b^{\max}$.
Hence this formulation is applicable in this context and will be used for comparative purposes, but does not solve the B2BSOP as defined in the general sense.
We also note that preliminary experimentation showed that the number of variables associated with the automaton $\mathcal{A}_2$ proved too large to be competitive with
the other formulations.

In order to apply the `cost_regular` constraints to the MIP formulations, we
use the approach of [3]. We first create the layered graph of the `cost_regular`
constraint consistency algorithm. From this layered graph we derive a network
flow formulation, where the source node is the initial state of the DFA, and
the sink node is connected to every final state. The network flow is then easily
translated to a set of linear variables and constraints.

We first give the subproblem present in all considered formulations that
enforces the feasibility of a solution. We let x_{mt} be a binary variable that equals
1 if meeting m is held at time t, and enforce the following constraints.

$$\sum_{t \in T} x_{mt} = 1 \qquad m \in M \tag{12}$$

$$\sum_{m \in M_p} x_{mt} \leq 1 \qquad p \in P, t \in T \tag{13}$$

$$\sum_{m \in M} x_{mt} \leq |L| \qquad t \in T \tag{14}$$

$$x_{mt} = 0 \qquad m \in M, t \in T \setminus T_m \tag{15}$$

Constraints (12) force each meeting to be held. Constraints (13) force each participant to be in at most one meeting at a time. Constraints (14) force the
number of meetings at any time to be at most the number of locations. Constraints (15) force each meeting to be held in an allowed time slot. The following

subsections give the formulation-specific variables and constraints that cumulate the objective function derived from the schedule of each participant.

4.1   Logical

We define the following variables:

$b' \in \mathbb{Z}$ — Variable cumulating the minimum number of breaks assigned to any participant,

$y_{pt} \in \{0,1\}$ — Indicator if participant p has a meeting at time t,

$z_{pt} \in \{0,1\}$ — Equals 1 for time t starting from participant p's first meeting,

$h_{pt} \in \{0,1\}$ — Indicator if time t terminates an idle period for participant p.

The variables h_{pt} contribute a value of 1 to the objective function. The necessary constraints linking these variables to the model appear below.

$$\sum_{t \in T} y_{pt} = |M_p| \qquad p \in P \tag{16}$$

$$x_{mt} \le y_{pt} \qquad p \in P, m \in M_p, t \in T \tag{17}$$

$$y_{pt} < z_{pt} \qquad p \in P, t \in T \tag{18}$$

$$z_{pt} \le z_{p,t+1} \qquad p \in P, t \in T, t \le |T| - 1 \tag{19}$$

$$y_{p(t+1)} - h_{pt} \le y_{pt} + 1 - z_{pt} \qquad p \in P, t \in T \tag{20}$$

$$\sum_{t \in T} h_{pt} \ge b' \qquad p \in P \tag{21}$$

$$\sum_{t \in T} h_{pt} \le b' + b^{\max} \qquad p \in P \tag{22}$$

Constraints (16) through (19) link the y and z variables to the formulation. Constraints (20) then link the h variables to the formulation, forcing a break to be counted after an idle period. Constraints (21) and (22) are fairness constraints that limit the difference in the number of breaks per participant. We then define the problem (P1) as the minimization of the objective function

$$\sum_{p \in P} \sum_{t \in T} h_{pt}$$

subject to constraints (12) through (15) and (16) through (22).

4.2   Network Flow

We define the space-time network $G_p = (N_p, A_p)$ for each participant p, with node set $\{source, sink\} \bigcup (T \times \{0,1\})$. We define v_{tq} to be one such node for time slot t and binary index q. The index $q = 0$ corresponds to the participant on break at time t, and $q = 1$ corresponds to the participant in a meeting. We create an arc from the source to each v_{t1}, from each v_{t1} to the sink, and from each v_{tq} to each $v_{(t+1)q'}$.

Variables y_{pt} are added to the formulation, with the same definition as in problem (P1), and constraints (16) and (17) link these variables to the variables x_{mt}. Additionally, the binary variable f_a^p is defined for all arcs in G_p to represent the source-to-sink flow. This flow is defined by the following constraints.

$$\sum_{(source,v)\in A_p} f^p_{(source,v)} = 1 \quad p \in P \tag{23}$$

$$\sum_{(v,sink)\in A_p} f^p_{(v,sink)} = 1 \quad p \in P \tag{24}$$

$$\sum_{(v,w)\in A_p} f^p_{(v,w)} - \sum_{(w,v)\in A_p} f^p_{(w,v)} = 0 \quad p \in P, v \in T \times \{0,1\} \tag{25}$$

$$f^p_{(source,v_{t1})} + f^p_{(v_{(t-1)0},v_{t1})} + f^p_{(v_{(t-1)1},v_{t1})} = y_{pt} \quad p \in P, t \in T \tag{26}$$

$$\sum_{t\in T, t\leq|T|-1} f^p_{(v_{t1},v_{(t+1)0})} \geq b' \quad p \in P \tag{27}$$

$$\sum_{t\in T, t\leq|T|-1} f^p_{(v_{t1},v_{(t+1)0})} \leq b' + b^{max} \quad p \in P \tag{28}$$

Constraints (23) through (25) define a source-to-sink flow of 1 unit, for each participant. Constraints (26) link this network flow to the variables y_{pt}. Constraints (27) and (28) are fairness constraints that limit the difference in the number of breaks per participant. We then define the problem (P2) as the minimization of the objective function

$$\sum_{p\in P} \sum_{t\in T, t\leq|T|-1} f^p_{(v_{t1},v_{(t+1)0})}$$

subject to constraints (12) through (15), (16), (17), and (23) through (28).

4.3   Cost Regular

We create a MIP cost_regular constraint for each participant $p \in P$ using the previously defined automaton $\mathcal{A}_1$ as illustrated at Figure 1, which links to the variables y_{pt}. This constraint ensures that a covering of the meetings M_p is a word recognized by the DFA, with the objective function measured as the cost of the word. We then define the problem (P3) as the minimization of the objective function defined by the set of these cost_regular constraints, subject to constraints (12) through (15), (16), (17), and the cost_regular constraints defined by the DFA $\mathcal{A}_1$. We enforce the fairness constraints in an analogous manner to the previous two models, bounding the appropriate subsets of the objective function.

4.4   Cost Regular with Integrated Fairness Constraints

We again define a DFA for each participant $p \in P$. Each state corresponds to the triple (n, w, h), where parameter w is 1 if the previous activity was a meeting and 0 otherwise, and parameter h represents the number of breaks taken on the schedule thus far. The DFA presented in Figure 3 is for a participant with three meetings and a maximum of one break. All arcs have zero cost, except again those denoted with a dashed line.

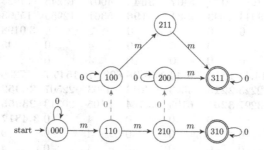

Fig. 3. Automaton $\mathcal{A}_3$ integrating fairness for a participant with three meetings. Arc label "m" stands for any meeting and label "0" for no meeting. Only the red dashed arcs carry a cost, of one unit, to mark the start of a break. Notice that there is no transition labeled "0" out of state "211" because of the limit of one break.

We define the problem (P4) to be the minimization of the derived objective function, subject to constraints (12) through (15), (16), (17), and the MIP cost_regular constraints defined by the DFAs $\mathcal{A}_3$. Note that for the CP approach this augmented automaton is redundant because we already consider fairness in the CP cost_regular constraints by restricting the b_p variables: paths in the digraph will already be of length at most $b^{\max}$.

4.5   Empirical Results

We resolved each of the nine problem instances considered in [2] with each MIP formulation. This was executed for the parameter $b^{\max}$ set to values of both 2 (as used in [2]) and 1. For each resolution, we give the number of variables and constraints present in the formulation, the objective value of the root node linear relaxation, the value of the best found solution by the branch-and-bound, the branch-and-bound lower bound associated with this best solution, and the runtime in seconds. The results for $b^{\max}$ values 2 and 1 appear in Tables 6 and 7, respectively. The formulation that yields either the best objective value or the best runtime is bolded.

All but one formulation are able to solve only the first 6 instances to optimality, with the remaining 3 instances timing out. The exceptions are (P3) and (P4), which additionally proved optimality on instance **F3c** when $b^{\max} = 2$. Under this setting, (P3) also finds the best solution to the instance **F3b** and hence scales

Table 6. MIP Results with $b^{\max} = 2$

Model	Result	T2a	T2c	T3a	T3b	T3c	F3a	F3b	F3c	F4a
P1	$|V|$	**1967**	1967	3164	3175	3164	7575	9231	7575	**11715**
	$|C|$	**3427**	3427	5627	5687	5627	12675	15277	12675	**21160**
	Root	0	0	0	0	0	0	3	0	**2**
	Objective	0	0	0	0	4	0	12	6	**9**
	Bound	0	0	0	0	4	0	5	0	**2**
	Time	**7.62**	15.48	27.04	167.01	303.44	942.61	7200	7200	**7200**
P2	$|V|$	2173	2173	**4007**	**3645**	4007	13245	14923	13245	17961
	$|C|$	3043	3043	**5361**	**5196**	5361	12955	15563	12955	21398
	Root	0	0	**0**	**0**	2	0	3.0198	0	2
	Objective	0	0	**0**	**0**	4	0	45	2	16
	Bound	0	0	**0**	**0**	4	0	5	0	2
	Time	14.43	10.04	**9.44**	156.28	822.97	1545.7	7200	7200	7200
P3	$|V|$	2225	**2225**	4503	4013	4503	22507	**25952**	22507	34408
	$|C|$	3297	**3297**	6105	5764	6105	20313	**23955**	20313	33003
	Root	0	**0**	0	0	4	0	**3.4317**	0	2
	Objective	0	**0**	0	0	4	0	9	1	26
	Bound	0	**0**	0	0	4	0	4	1	2
	Time	22.52	**0.66**	27.24	566.54	37.27	573.12	**7200**	2930.74	7200
P4	$|V|$	3419	3419	7017	6204	**7017**	**94027**	102023	94027	110778
	$|C|$	4328	4328	8126	7567	**8126**	**84051**	91004	84051	97026
	Root	0	0	0	0	**4**	0	3.5125	0	2
	Objective	0	0	0	0	**4**	0	25	1	-
	Bound	0	0	0	0	**4**	0	5	1	2
	Time	9.19	7.89	555.38	281.64	**24.78**	**506.25**	7200	3225.67	7200

the best with problem size while also being competitive in time to optimality on the smaller instances. On the other hand, when $b^{\max} = 1$, formulations (P3) and (P4) are competitive on the 6 smaller instances, proving optimality in the shortest runtime in 2 and 4 instances, respectively. However formulation (P4) is superior to (P3) on all 3 of the more difficult instances, and the best overall formulation under this parameter setting.

While the formulations derived from the `cost_regular` constraints have the greatest number of binary variables, they are tighter formulations that scale well to large problem sizes. Hence they better capture the structure of the B2BSOP. The root node lower bound, derived from a stronger relaxation and the MIP cuts applied by Gurobi, is observably stronger on instances **T3c** and **F3b**. It would be interesting to further observe this on data sets without a trivial optimal lower bound of 0. However, the number of states to be represented in the DFA $\mathcal{A}_3$ increases quickly with the parameter $b^{\max}$; this correlates with an increased number of variables in the MIP formulation. Therefore the simpler DFA appears to be the most appropriate when fairness is less tight.

Table 7. MIP Results with $b^{\max} = 1$

Model	Result	T2a	T2c	T3a	T3b	T3c	F3a	F3b	F3c	F4a
P1	$|V|$	1967	1967	3164	3175	3164	7575	**9231**	**7575**	11715
	$|C|$	3427	3427	5627	5687	5627	12675	**15277**	**12675**	21160
	Root	0	0	0	0	0	0	3	0	2
	Objective	0	0	0	0	4	0	12	2	30
	Bound	0	0	0	0	4	0	3	0	2
	Time	1.22	29.07	401.76	193.24	647.63	833.11	**7200**	**7200**	7200
P2	$|V|$	2173	2173	4007	3645	4007	13245	14923	13245	17961
	$|C|$	3043	3043	5361	5196	5361	12955	15563	12955	21398
	Root	0	0	0	0	2	0	3.0475	0	2
	Objective	0	0	0	0	4	0	-	4	-
	Bound	0	0	0	0	4	0	5	0	2
	Time	10.77	3.4	289.17	243.64	236.33	2313.79	7200	7200	7200
P3	$|V|$	**2225**	**2225**	4503	4013	4503	22507	25952	**22507**	34408
	$|C|$	**3297**	**3297**	6105	5764	6105	20313	23955	**20313**	33003
	Root	**0**	**0**	0	0	4	0	3.5140	0	2
	Objective	**0**	**0**	0	0	4	0	20	2	-
	Bound	**0**	**0**	0	0	4	0	5	1	2
	Time	**1**	**0.93**	469.77	370.42	136.69	1034.5	7200	**7200**	7200
P4	$|V|$	3039	3039	**5973**	**5344**	**5973**	68359	74052	68359	80402
	$|C|$	4030	4030	**7368**	**6930**	**7368**	63235	68735	63235	74547
	Root	0	0	**0**	**0**	4	0	3.6699	0	2
	Objective	0	0	**0**	**0**	4	0	12	2	28
	Bound	0	0	**0**	**0**	4	0	5	1	2
	Time	1.17	1.92	**266.28**	3.15	132.14	552.43	**7200**	**7200**	7200

5 Discussion

Table 8 summarizes the best results obtained with our MIP, CP, and LNS approaches, and recalls those of the two best approaches in [2]: SBDD and clasp are respectively an SMT solver representing the objective function as a binary decision diagram and a conflict-driven answer set solver.

Looking at the different instances, we noticed that although the **T★** set are generally easy, **T3c** is challenging for most of the computational approaches considered. It is quickly solved to optimality by MIP but CP and SBDD only find suboptimal solutions and clasp, none at all. It is interesting that the only difference between instances **T3a** and **T3c** is that the latter has fewer available locations: that type of restriction does not seem to be handled well by those approaches. On the harder **F★** set there is no clear winner but MIP, LNS, and SBDD are generally performing better than the other two.

For CP to be competitive on the harder instances, having a good model did not seem to be sufficient and it required branching heuristics geared toward optimization. This was the case even though an optimization-based global constraint was used in the formulation.

Table 8. An empirical comparison of the computational approaches proposed in this paper as well as some of the previous proposals. For MIP we report the best solution out of the four models presented and for LNS, the best solution out of the ten runs. We give the time in seconds to find the solution value reported, except for SBDD and clasp on non-optimal solutions since these were not provided.

Instance	Approach	Obj.	Time
T2a	MIP	0	7.6
	CP	0	0.5
	LNS	0	0.5
	SBDD	0	2.7
	clasp	0	0.1
T2c	MIP	0	0.7
	CP	0	3.9
	LNS	0	3.9
	SBDD	0	235.4
	clasp	0	977.9
T3a	MIP	0	9.4
	CP	0	98.5
	LNS	0	64.0
	SBDD	0	65.2
	clasp	0	2.1
T3b	MIP	0	156.3
	CP	0	12.0
	LNS	0	12.0
	SBDD	0	24.1
	clasp	0	2.1
T3c	MIP	4	24.8
	CP	11	87.8
	LNS	4	850.6
	SBDD	8	–
	clasp	–	–

Instance	Approach	Obj.	Time
F3a	MIP	0	506.3
	CP	0	1514.9
	LNS	0	5442.9
	SBDD	0	3128.1
	clasp	0	52.4
F3b	MIP	9	5833.0
	CP	23	4795.0
	LNS	13	3189.0
	SBDD	12	–
	clasp	24	–
F3c	MIP	1	2930.7
	CP	9	5070.3
	LNS	3	2507.4
	SBDD	20	–
	clasp	–	–
F4a	MIP	9	7172.0
	CP	25	1296.5
	LNS	13	5996.0
	SBDD	7	–
	clasp	–	–

The main conclusion that can be drawn is that the cost_regular structure is quite useful in building efficient MIP and CP models. Compared to the MIP and CP formulations proposed in [2], models using such structure are more efficient on the instances considered, and often state-of-the-art. For the MIP approach one could ask whether it is possible to tighten the more compact formulations (P1) or (P2) with valid constraints derived from the cost_regular constraints. As an avenue of future research, we believe that embedding some of these constraints into a branch-and-cut algorithm could result in a more robust exact algorithm for the B2BSOP.

Acknowledgments. We wish to thank the reviewers for their constructive criticism. This research was supported in part by the Natural Sciences and Engineering Research Council of Canada (NSERC).

References

1. Balas, E., Saltzman, M.J.: An Algorithm for the Three-Index Assignment Problem. Operations Research **39**(1), 150–161 (1991)
2. Bofill, M., Espasa, J., Garcia, M., Palahí, M., Suy, J., Villaret, M.: Scheduling B2B Meetings. In: O'Sullivan, B. (ed.) CP 2014. LNCS, vol. 8656, pp. 781–796. Springer, Heidelberg (2014)
3. Côté, M.-C., Gendron, B., Rousseau, L.-M.: Modeling the Regular Constraint with Integer Programming. In: Van Hentenryck, P., Wolsey, L.A. (eds.) CPAIOR 2007. LNCS, vol. 4510, pp. 29–43. Springer, Heidelberg (2007)
4. Demassey, S., Pesant, G., Rousseau, L.-M.: A Cost-Regular Based Hybrid Column Generation Approach. Constraints **11**(4), 315–333 (2006)
5. Harvey, W.D., Ginsberg, M.L.: Limited Discrepancy Search. In: Proceedings of the Fourteenth International Joint Conference on Artificial Intelligence, IJCAI 1995, Montréal Québec, Canada, August 20–25, 1995, vol. 2, pp. 607–615. Morgan Kaufmann (1995)
6. Pesant, G., Quimper, C.-G., Zanarini, A.: Counting-Based Search: Branching Heuristics for Constraint Satisfaction Problems. J. Artif. Intell. Res. (JAIR) **43**, 173–210 (2012)
7. Refalo, P.: Impact-Based Search Strategies for Constraint Programming. In: Wallace, M. (ed.) CP 2004. LNCS, vol. 3258, pp. 557–571. Springer, Heidelberg (2004)
8. Régin, J.-C.: Generalized Arc Consistency for Global Cardinality Constraint. In: Proceedings of the Thirteenth National/Eighth Conference on Artificial Intelligence/Innovative Applications of Artificial Intelligence, AAAI-98/IAAI-98, vol. 1, pp. 209–215 (1996)
9. Shaw, P.: Using Constraint Programming and Local Search Methods to Solve Vehicle Routing Problems. In: Maher, M.J., Puget, J.-F. (eds.) CP 1998. LNCS, vol. 1520, pp. 417–431. Springer, Heidelberg (1998)

Constraint-Based Local Search for Golomb Rulers

M.M. Alam Polash[1]([⊠]), M.A. Hakim Newton[1], and Abdul Sattar[1,2]

[1] Institute for Integrated and Intelligent Systems, Griffith University,
Nathan, Australia
mdmasbaul@gmail.com, {mahakim.newton,a.sattar}@griffith.edu.au
[2] Queensland Research Lab, National ICT Australia, Sydney, Australia

Abstract. This paper presents a constraint-based local search algorithm to find an optimal Golomb ruler of a specified order. While the state-of-the-art search algorithms for Golomb rulers hybridise a range of sophisticated techniques, our algorithm relies on simple tabu meta-heuristics and constraint-driven variable selection heuristics. Given a reasonable time limit, our algorithm effectively finds 16-mark optimal rulers with success rate 60 % and 17-mark rulers with 6 % near-optimality.

Keywords: Golomb ruler · Constraints · Local search · Tabu meta-heuristics

1 Introduction

Golomb rulers were described in 1977 by S.W. Golomb [4] although such concept was already conceived in 1953 by W.C. Babcock [3]. A *Golomb ruler* of *order* $m > 0$ and *length* n is a sequence of m integers called *marks* $0 = x_1 < x_2 < \cdots < x_m = n$ such that each $x_j - x_i$ is unique for $1 \leq i < j \leq m$. A Golomb ruler of order m is *optimal* if n is the minimum possible integer. Golomb rulers have a wide variety of applications that include x-ray crystallography [4], radio astronomy [5], information theory [18] and pulse phase modulation [17].

Finding an Optimal Golomb Ruler (OGR) is an extremely difficult task. It takes 36200 CPU hours to find a 19-mark Golomb ruler on a Sun SPARC workstation using an exhaustive parallel search algorithm [8]. OGR is a combinatorial problem whose bounds grow geometrically with respect to the solution size [19]. The major limitation is that each new ruler to be discovered is, by necessity, larger than its predecessor. However, the search space is bounded and, therefore, solvable [13]. Also, for a given order, more than one OGR may exist.

To solve this highly combinatorial problem, a number of approaches have already been developed before. However, the current state-of-the-art results come from a sophisticated hybrid method [7] that combines ideas from greedy randomised adaptive search procedure (GRASP), scatter search (SS), tabu search

© Springer International Publishing Switzerland 2015
L. Michel (Ed.): CPAIOR 2015, LNCS 9075, pp. 322–331, 2015.
DOI: 10.1007/978-3-319-18008-3_22

(TS), clustering techniques and constraint programming. The hybrid algorithm found 16-mark OGRs with success rates 5-10%. Nevertheless, an analysis of the fitness landscape of OGR presented in [7] shows that high irregularities in the neighbourhood structure introduce a drift force towards low-fitness regions of the search space. For higher order rulers, search algorithms thus quickly reach a near-optimal value and then stagnate around it, apparently causing a *cycling problem* and making the search space less accessible. A restarting mechanism is therefore needed at that stage.

In this paper, we present a constraint-based local search approach to find Golomb rulers. Our algorithm takes m and n as input and finds the m integers of the Golomb ruler. For OGRs, we assume x_m to be equal to the optimal n for order m. For near-optimal rulers, we assume x_m to be less than or equal to a given n. Instead of a sophisticated hybridisation of a range of techniques, we rather rely on simple tabu meta-heuristics and constraint-driven variable selection heuristics. Besides the traditional way of enforcing tabu on recently modified variables for a given number of iterations, we use a special type of tabu called *configuration checking* (CC) [6]. The CC strategy for OGR prevents a variable (i.e. a mark) from being selected if it is fully confined by neighbouring variables. The use of CC effectively reduces the number of restarts required during search and thus mitigates the cycling problem of local search for finding an OGR. Experimental results show that within a reasonable time limit, our algorithm effectively achieves significantly high success rate of 60% in finding OGRs of order 16 and about 6% near-optimal rulers of order 17.

The rest of the paper is organised as follows: Section 2 explores related work; Section 3 describes our approach; Section 4 presents the experimental results; and finally, Section 5 draws our conclusions.

2   Related Work

Various techniques have been applied so far to find Golomb rulers. Scientific American algorithm, token passing algorithm, shift algorithm are described and compared in [16]. Geometry tools such as projectile plane construction and affine plane construction are used in a non-systematic method in [10]. A systematic branch and bound algorithm along with the Depth First Search (i.e backtracking algorithm) is proposed in [19]. A genetic algorithm is proposed in [21]. Constraint programming techniques are used in [20]. A combination of constraint programming and sophisticated lower bounds for Golomb rulers are used in [12]. A hybrid of local search and constraint programming is proposed in [15].

Three algorithms, namely a genetic algorithm on its own, then with local search and Baldwinian learning, and with local search and Lamarckian learning are studied in [11]; the best results have the distance between 6.8 and 20.3% from the optimum. A simple hybrid evolutionary algorithm (called GRHEA) is presented in [9] to find an OGR of a specified length. Also, an indirect but effective approach (called GROHEA) is proposed to find near-optimal Golomb rulers. For a given order m, GROHEA starts from an upper bound of n and if a Golomb ruler of length n is found, it then tries to find another one with length $n - 1$. GROHEA systematically finds optimal rulers for up to 11 marks very quickly. It also finds optimal rulers for 12 and 13 marks in less than two minutes, and for 14 marks in about 40 minutes. For 15 and 16 marks, the best solutions of GROHEA are within 4.6% and 5.6% of the optimal rulers.

2.1   A Recent Hybrid Local Search Algorithm

This algorithm [7] combines the greedy randomised adaptive search procedure (GRASP), evolutionary algorithms (EA), scatter search (SS), tabu search (TS), clustering techniques, and constraint programming (CP) to find optimal or near-optimal Golomb rulers. To find OGRs, this algorithm first uses an indirect approach, which incorporates GRASP. However, one major problem of the basic GRASP procedure is that it relies on certain parameter values to select an attribute value from the Ranked Candidate List (RCL). The choice of the parameter value often hinders to find high-quality solutions. GRASP is therefore combined with EA in HEAGRASP so that different parameter values can be used in each application of the ruler construction phase. The plain GRASP and HEAGRASP can find OGRs up to 9 and 10 marks respectively.

The work in [7] then proposes a scatter search (SS) that is basically a memetic algorithm. SS uses an indirect approach in the initialisation and restarting phase (ideas borrowed from HEAGRASP) and a direct approach in the local improvement and recombination phase. More specifically, the TS is used as the local improvement method. SS can find OGRs for up to 15 marks and computes high quality near-optimal solutions for 16 (i.e. 1.1% from the optimum). SS is further enhanced by using a complete search in recombination of individuals and a clustering procedure to achieve higher degree of diversity. As a result, 16-mark OGRs are found with success rates 5-10%.

2.2   A Recent Hybrid Genetic Algorithm

Recently, a hybrid genetic algorithm is presented in [2] to find optimal or near-optimal Golomb rulers. This approach has been able to obtain OGRs for up to 16 marks at the expense of an important execution time. For instance, around 5 hours for 11-mark, 8 hours for 12-mark, and 11 hours for 13-mark ruler. It is also able to find near-optimal rulers for 20 and 23 marks using enormous time. The parallel implementation of this algorithm can be found in [1].

3   Our Approach

Our approach is based on a constraint-based local search algorithm. It is a simple but effective algorithm to find an OGR of a specified order. Given the optimal length n of an m-mark ruler, it searches for a ruler that satisfies the criteria of an optimal one. Note that the first and last marks are fixed to 0 and n respectively. For a near-optimal Golomb ruler, the last mark remains flexible to take a value less than or equal to n', where n' is the optimal length of a $(m + 1)$-mark ruler. To overcome the cycling problem of local search, we use the tabu mechanism and the configuration checking techniques [6]. Nevertheless, below we provide a detailed description of our search algorithm.

3.1 Problem Model

We represent a Golomb ruler of order m and length n by using m variables $x_1, x_2, \cdots, x_m$. Without loss of generality, we fix the value of x_1 at 0 and the value of x_m at n. Initially, the domain of all other marks is defined $x_i \in [1, n-1]$. However, we assume the ordering $x_1 < x_2 < \cdots x_m$. As the search progress, the domain of a mark $x_i (1 < i < m)$ is thus dynamically restricted by the values of its neighbours. Thus, $x_i \in [x_{i-1} + 1, x_{i+1} - 1]$ for each $1 < i < m$.

To define the constraint model, for each $i > j$, we first calculate the difference expression $d_{ij} = x_i - x_j$. The value of d_{ij} is the *distance* between x_i and x_j. We then define an *alldifferent* constraint on the d_{ij}s. To guide the search, the constraint violation metric is calculated as in [9]. Given the current solution R i.e. the values of all x_is, the violation $V_R(d)$ of a distance d is the number of times distance d appears between two marks beyond its allowed occurrences.

$$V_R(d) = max(0, \#\{d_{ij} = d | 1 \leq j < i \leq m\} - 1) \tag{1}$$

The violation $V(R)$ of the current ruler R is simply the sum of $V_R(d)$s:

$$V(R) = \sum_{d=1}^{n} V_R(d) \tag{2}$$

Obviously, a ruler R with $V(R) = 0$ is a solution to the Golomb ruler problem. In each iteration, our algorithm will try to minimise the value of $V(R)$.

To define a variable selection heuristic, we further define the violation metrics for each difference d_{ij} and for each variable $x_i (1 < i < m)$. The violation for each distance d_{ij} will be the violation of its distance value.

$$V_R(d_{ij}) = V_R(d) \tag{3}$$

where $d_{ij} = d$ in R. The violation of a variable is calculated by summing the violations of the distances that depend on that variable.

$$V_R(x_i) = \sum_{k=1}^{i-1} V_R(d_{ik}) + \sum_{k=i+1}^{m} V_R(d_{ki}) \tag{4}$$

During search, we mainly follow max/min style search. At each iteration, a variable $x_i (1 < i < m)$ having the maximum $V_R(x_i)$ is selected first and then a value $v \in [x_{i-1} + 1, x_{i+1} - 1]$ that minimises V_R is selected for x_i.

3.2   Avoiding the Cycling Problem

The cycling problem in local search has been typically tackled by the tabu mechanism. Besides using the tabu mechanism, in this paper, we use another recently emerging strategy called *configuration checking* (CC) [6].

Tabu Mechanism. By maintaining a parameter called tabu tenure, the tabu mechanism prevents the local search to immediately return to a previously visited candidate solution. In our algorithm, we tabu the variable selected in the last iteration and we tabu it for the tabu-tenure period.

Configuration Checking. The CC [6] strategy reduces the cycling problem by checking the circumstance information. The core idea is: A variable's value should not change until at least one of its neighbouring variables has a new value. Since in our algorithm, a variable's value depends on its neighbour as we enforce $x_i \in [x_{i-1} + 1, x_{i+1} - 1]$ during search, CC is relevant here along with the tabu mechanism. However, we use CC in our algorithm in a special case when a variable's range has nothing but its own current value. In this case, the variable is locked for any future changes until any of its neighbouring variables have changed.

3.3   Search Algorithm

Our constraint-based local search algorithm to find Golomb rulers is shown in Algorithm 1. The core of the algorithm is in *Lines* 4–19 where local moves are performed for a number of iterations or until a solution is found. The unlocked variable with the highest number of violations is selected in *Line* 8 and a value for that variable is selected in *Line* 9 such that the number of violations decreases after assigning the value to the variable. The new ruler is generated in *Line* 13 and the tabu is applied on the variable in *Line* 14. Note that the tabu tenure tt is normally within 3 to 5. *Line* 11 and 15 implement the idea of CC. CC locks a variable whenever its domain contains no value except the current one. When a new value is set into a variable, CC unlocks the neighbours of that variable provided they are already locked. *Lines* 16–19 update the best violation metric and plateau size depending on the progress of violation metrics. *Lines* 5–6 restart the search when the current plateau size exceeds a given limit.

Initialisation and Restarting Mechanism. The initial ruler is generated by selecting random values from the initial domain of each mark. Special care is taken so that no two marks have the same value. The marks are then sorted to obtain an ordered ruler. When our search algorithm gets stuck showing no progress, we just use the initialisation procedure to restart the search from scratch. We detect the stagnation situation when the global best violation seen so far does not change for a given number of iterations.

Algorithm 1. Constraint-Based Local Search for Golomb Rulers
1 **Parameters: order** m, **length** n, **tabu tenure** tt
2 Generate an initial solution R using an initialisation procedure
3 plateauSize = 0, iteration = 0, bestViolation = $V(R)$
4 **while** $++iteration \leq maxIteration$ and $bestViolation > 0$ **do**
5   **if** $plateauSize > maxPlateauSize$ **then**
6   Restart from scratch and set plateauSize = 0
7   **else**
8   Select the *unlocked* variable $x_k(1 < k < m)$ with the highest $V_R(x_k)$
9   Select a value $v \in [x_{k-1} + 1, x_{k+1} - 1]$ such that $V(R)$ is minimised
10   **if** v *is* x_k's *current value* **then**
11   lock x_k to stop its future changes //part of CC
12   **else**
13   set $x_k = v$ in the current ruler R
14   apply tabu on x_k for the specified tabu tenure tt
15   unlock x_{k-1} and x_{k+1} if they are locked //part of CC
16   **if** $currentViolation < bestViolation$ **then**
17   bestViolation = currentViolation, plateauSize = 0
18   **else if** $currentViolation == bestViolation$ **then**
19   plateauSize++

---

4 Experiments and Analyses

We implemented our algorithm using C++ and on top of the constraint-based local search system, Kangaroo [14]. The functions and the constraints are defined using invariants in Kangaroo. Invariants are special constructs that are defined by using mathematical operators over the variables. While propagation of violations, simulation of moves, execution and related calculations are performed incrementally by Kangaroo, we mainly focus on the search algorithms.

We ran our experiments on High Performance Computing Cluster Gowonda provided by Griffith University. Each node of the cluster is equipped with Intel Xeon CPU E5-2650 processors @2.60 GHz, FDR 4x infiniBand Interconnect, having system peak performance 18949.2 Gflops. Our search algorithm is run for 25 times with timeout 48 hours for each given order of the Golomb ruler. The tabu tenure is between 3 and 5. For a given order m and its optimal length n, we run our algorithm to find an OGR first. If an OGR is not found, then we consider finding a near-optimal Golomb ruler with an increased ruler length. Our experimental results are shown in Table 1 and Figure 1. Left most two columns in the Table 1(a) show the orders 11–16 and the optimal lengths of the Golomb rulers. Moreover, the columns under "TabuAndCC" show our final results.

4.1 Effectiveness of CC

To investigate the effectiveness of CC, we run a version of our algorithm that does not use the CC strategy. These results are shown in Table 1(a) in the columns under the header "TabuNoCC". Also, the charts in Figure 1 show how the two versions' success rate differ when various timeout limits are assumed. Overall, we observe the TabuAndCC version, our final algorithm, significantly outperforms the TabuNoCC version in obtaining higher order OGRs, in success rates, and in running times. To analyse further, in Table 1(b), we show the number of restarts required by the TabuAndCC and TabuNoCC versions of our algorithm. As we can see, the number of restarts required for the TabuAndCC version is very small compared to the TabuNoCC version. It exhibits that the use of CC effectively reduces the occurrence of stagnation in the search.

Table 1. (a) Performance of our algorithm when compared to GRHEA[9]. Time statistics for GRHEA is collected from the published article while our algorithms are run on our computers. (b) Average numbers of restarts required during search (c) Approximate average numbers of candidate solutions explored/evaluated during search.

Num of Marks	Opt GR Len	TabuAndCC		TabuNoCC		GRHEA[9]	
		Succ Rate	Median Time	Succ Rate	Median Time	Succ Rate	Median Time
11	72	100	1.23 S	100	7.96 S	100	5.86 S
12	85	100	28.09 S	100	1.65 M	99	2.78 M
13	106	100	6.47 M	100	11.39 M	99	15.99 M
14	127	100	2.76 H	96	5.64 H	2	1.07 H
15	151	84	3.26 H	76	22.63 H		
16	177	60	14.83 H				

(a)

Marks	TabuAndCC	TabuNoCC
11	1.2	366.2
12	64.32	18214.12
13	234.08	77233.4

(b)

Marks	TabuAndCC	GRHEA[9]
11	1×10^6	10×10^6
12	4.5×10^7	23×10^7
13	5.1×10^8	11×10^8

(c)

4.2 Optimal Golomb Rulers

We compare our algorithm with other state-of-the-art algorithms for Golomb rulers. As we see in Table 1(a), GRHEA[9] can solve up to order 14 but with success rate for 14 being 2%. HybridGA[2] claims to have solved up to 16 but success rates are not mentioned and run-times are either enormous or not reported. The Hybrid Local Search [7] states to have consistently found OGRs for up to 14 marks and for 16 marks with 5-10% success rate[1]. The OGR problem gets extremely hard from order 16 onward [7]. Notice that our final algorithm (Columns "TabuAndCC") obtains significantly better results with 100%, 84% and 60% success rates for 14, 15 and 16-mark rulers respectively.

We compare the search effort behind the performance of our algorithm and those in [7,9] on the average number of candidate rulers explored and evaluated. Comparison on execution time is not possible because experiments in [7,9] were run on a number of different machines[1]. Note that while the algorithms in [7,9] are variants of memetic algorithms (in other words, local searh is used as a mutation operator in a genetic algorithm), our algorithm is just a constraint-based local search algorithm. So for a fair comparison between these different types of

---

[1] E-mail communication with Antonio J. Fernandez, one of the authors of [7].

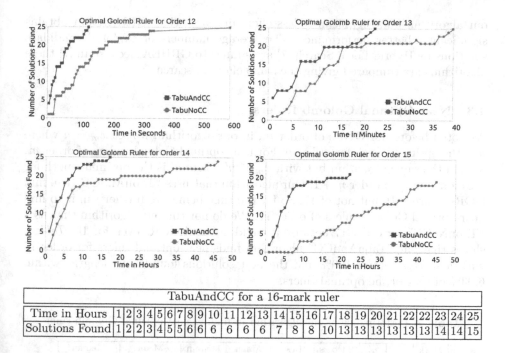

TabuAndCC for a 16-mark ruler																								
Time in Hours	1	2	3	4	5	6	7	8	9	10	11	12	13	14	15	16	17	18	19	20	21	22	22	23	24	25
Solutions Found	1	2	2	3	4	5	5	6	6	6	6	6	6	7	8	8	10	13	13	13	13	13	13	14	14	15

Fig. 1. Number of times the optimal ruler is found when 25 attempts are made within given time limits. The largest timeout was 48 hours for all runs. The times in x-axis are in seconds for order 12, in minutes for 13, and in hours for 14 and 15. The times in the first row of the table is also in hours for the ruler of order 16.

algorithm, particularly when time comparison is not possible, we consider the average numbers of rulers explored and evaluated by each algorithm to be an appropriate criterion; similar notions were used in [7,9].

To obtain the maximum numbers of rulers explored or evaluated by GRHEA [9], we take the maximum number of generations (50), the population size (50), the probability to call the LS procedure for each individual (0.6), the number of iterations in the LS (10000). We get the numbers of LS iterations altogether to be $50 \times 50 \times 0.6 \times 10000 = 15 \times 10^6$. In each LS iteration, GRHEA considers each variable and its value from the range bounded by the two neighbour variables' values. In GRHEA's model, a Golomb ruler with m mark and length n has thus $2(n - m + 1)$ neighbours. So $(n - m + 1) \times 30 \times 10^6$ gives the maximum numbers of rulers explored or evaluated. However, instead of 15×10^6, Table 1(c) uses the average number of local moves made by GRHEA as reported in [9].

For our algorithm, we just take the average numbers of iterations and then consider the fact that in each iteration, unlike in GRHEA, only one variable is selected heuristically and then like GRHEA, the value of the variable is selected only from those bounded by the neighbouring two variables' values. So the number of neighbours on an average is $2(n - m + 1)/(m - 1)$. Table 1(c) shows that our algorithm puts significantly less effort in search than GRHEA but obtains significantly better performance. The average numbers of explored candidate solutions by Hybrid Local Search [7] are similar to GRHEA because in [7] these algorithms are compared giving the same effort in search.

4.3   Near-Optimal Golomb Rulers

As noted before, for OGRs of order m, in our algorithm, we set $x_m = n$ where n is the optimal length for order m. For near-optimal Golomb rulers of order m, we run the same algorithm but with $x_m \leq n'$ where n' is the optimal length for a Golomb ruler of order $m + 1$. Our algorithm that uses "TabuAndCC" can find OGRs of order 16 but not of 17. So for the time being, we further run it to find near-optimal Golomb rulers of order 17. We do not run our algorithm that uses "TabuNoCC" for order 17 because it could not find OGRs even for 16. *Table* 2 shows that our "TabuAndCC" algorithm finds near-optimal rulers for order 17 with a success rate of 100% and the best solutions found have lengths within 6.03% of that of the optimal rulers.

Table 2. Performance of our algorithm in finding near-optimal Golomb ruler

Mark	n	n'	Success Rate	Best Length	Mean Length	Median Length	Mean Time	Median Time
17	199	216	100	211	214	215	6.97 Hours	5.88 Hours

5   Conclusion

We have presented a constraint-based local search algorithm that takes the number of marks and length of an optimal Golomb ruler as input and finds the positions of the marks. Our algorithm is simple, but given a reasonable time limit, it effectively finds 16-mark optimal rulers with 60% success rate and about 6% near-optimal 17-mark rulers.

References

1. Ayari, N., Jemai, A.: Parallel hybrid evolutionary search for Golomb ruler problem. In: International Conference on Metaheuristics and Nature Inspired Computing (META) (2010)
2. Ayari, N., Luong, T., Jemai, A.: A hybrid genetic algorithm for golomb ruler problem. In: IEEE/ACS International Conference on Computer Systems and Applications (AICCSA), pp. 1–4 (2010)
3. Babcock, W.: Intermodulation interface in radio systems. Bell Systems Technical Journal, 63–73 (1953)
4. Bloom, G., Golomb, S.: Application of numbered undirected graphs. In: Proceedings of the IEEE, vol. 65(4), pp. 562–570 (1977)

5. Blum, E., Biraud, F., Ribes, J.: On optimal synthetic linear arrays with applications to radioastronomy. IEEE Transactions on Anntenas and Propagation (1974)
6. Cai, S., Su, K., Sattar, A.: Local search with edge weighting and configuration checking heuristics for minimum vertex cover. Artificial Intelligence **175**(9–10), 1672–1696 (2011)
7. Cotta, C., Dotu, I., Fernandez, A.J., Hentenryck, P.V.: Local search based hybrid algorithm for finding Golomb rulers. Contraints, 263–291 (2007)
8. Dollas, A., Rankin, W., McCracken, D.: A new algorithm for Golomb ruler derivation and proof of the 19-mark ruler. IEEE Transactions on Information Theory, 379–386 (1998)
9. Dotu, I., Hentenryck, P.V.: A simple hybrid evolutionary algorithm for finding Golomb rulers. IEEE Congress on Evolutionary Computation (2005)
10. Drakakis, K.: A review of the available construction methods for Golomb rulers. Advances in Mathematics of Communications **3**(3), 235–250 (2009)
11. Feeny, B.: Determining optimum and near-optimum Golomb rulers using genetic algorithms. Master's thesis, Computer Science, University College Cork, October 2003
12. Galinier, P., Jaumard, B., Morales, R., Pesant, G.: A constraint-based approach to the golomb ruler problem. In: 3rd International Workshop on CPAIOR (2001)
13. Klove, T.: Bounds and construction for difference triangle sets. IEEE Transactions on Information Theory, 879–886 (1989)
14. Newton, M.A.H., Pham, D.N., Sattar, A., Maher, M.: Kangaroo: an efficient constraint-based local search system using lazy propagation. In: Lee, J. (ed.) CP 2011. LNCS, vol. 6876, pp. 645–659. Springer, Heidelberg (2011)
15. Prestwich, S.: Trading completeness for scalability: hybrid search for cliques and rulers. In: 3rd International workshop on CPAIOR, pp. 159–174 (2001)
16. Rankin, W.: Optimal Golomb rulers: An exhaustive parallel search implementation. Master's thesis, Duke University Electrical Engineering Dept., Durham, NC, December 1993
17. Robbins, J., Gagliardi, R., Taylor, H.: Acquisition sequences in PPM communications. IEEE Transactions on Information Theory, 738–744 (1987)
18. Robinson, J., Bernstein, A.: A class of binary recurrent codes with limited error propagation. IEEE Transactions on Information Theory, 106–113 (1967)
19. Shearer, J.: Some new optimum Golomb rulers. IEEE Transactions on Information Theory, 183–184 (1990)
20. Smith, B., Walsh, T.: Modelling the golomb ruler problem. In: Workshop on Non-Binary Constraints (IJCAI), Stockholm (1999)
21. Soliday, S., Homaifar, A., Lebby, G.: Genetic algorithm approach to the search for golomb rulers. In: Eshelman, L.J. (ed.) 6th International Conference on Genetic Algorithm (ICGA 1995), pp. 528–535. Morgan Kaufmann, Pittsburg (1995)

Packing While Traveling: Mixed Integer Programming for a Class of Nonlinear Knapsack Problems

Sergey Polyakovskiy[(✉)] and Frank Neumann

Optimisation and Logistics School of Computer Science,
The University of Adelaide, Adelaide, Australia
{Sergey.Polyakovskiy,Frank.Neumann}@adelaide.edu.au
http://www.cs.adelaide.edu.au/~optlog/

Abstract. Packing and vehicle routing problems play an important role in the area of supply chain management. In this paper, we introduce a non-linear knapsack problem that occurs when packing items along a fixed route and taking into account travel time. We investigate constrained and unconstrained versions of the problem and show that both are $\mathcal{NP}$-hard. In order to solve the problems, we provide a pre-processing scheme as well as exact and approximate mixed integer programming (MIP) solutions. Our experimental results show the effectiveness of the MIP solutions and in particular point out that the approximate MIP approach often leads to near optimal results within far less computation time than the exact approach.

Keywords: Non-linear knapsack problem · NP-hardness · Mixed integer programming · Linearization technique · Approximation technique

1 Introduction

Knapsack problems belong to the core combinatorial optimization problems and have been frequently studied in the literature from the theoretical as well as experimental perspective [8,12]. While the classical knapsack problem asks for the maximizing of a linear pseudo-Boolean function under one linear constraint, different generalizations and variations have been investigated such as the multiple knapsack problem [5] and multi-objective knapsack problems [7].

Furthermore, knapsack problems with nonlinear objective functions have been studied in the literature from different perspectives [4]. Hochbaum [9] considered the problem of maximizing a separable concave objective function subject to a packing constraint and provided an FPTAS. An exact approach for a nonlinear knapsack problem with a nonlinear term penalizing the excessive use of the knapsack capacity has been given in [6].

Nonlinear knapsack problems also play a key-role in various Vehicle Routing Problems (VRP). In recent years, the research on dependence of the fuel consumption on different factors, like a travel velocity, a load's weight and vehicle's

© Springer International Publishing Switzerland 2015
L. Michel (Ed.): CPAIOR 2015, LNCS 9075, pp. 332–346, 2015.
DOI: 10.1007/978-3-319-18008-3_23

technical specifications, in various VRP has gained attention from the operations research community. Mainly, this interest is motivated by a wish to be more accurate with the evaluation of transportation costs, and therefore to stay closer to reality. Indeed, an advanced precision would immediately benefit to transportation efficiency measured by the classic petroleum-based costs and the novel greenhouse gas emission costs. In VRP in general, and in the Green Vehicle Routing Problems (GVRP) that consider energy consumption in particular, given are a depot and a set of customers which are to be served by a set of vehicles collecting (or delivering) required items. While the set of items is fixed, the goal is to find a route for each vehicle such that the total size of assigned items does not exceed the vehicle's capacity and the total traveling cost over all vehicles is minimized. See [11] for an extended overview on VRP and GVRP. Oppositely, we address the situation with one vehicle whose route is fixed but the items can be either collected or skipped. Specifically, this situation represents a class of nonlinear knapsack problems and considers trade-off between the profits of collected items and the traveling cost affected by their total weight. The non-linear packing problem arises in some practical applications. For example, a supplier having a single truck has to decide on goods to purchase going through the constant route in order to maximize profitability of later sales.

Our precise setting is inspired by the recently introduced Traveling Thief Problem (TTP) [3] which combines the classical Traveling Salesperson Problem (TSP) with the 0-1 Knapsack Problem (KP). The TTP involves searching for a permutation of the cities and a packing such that the resulting profit is maximal. The TTP has some relation to the Prize Collecting TSP [2] where a decision is made on whether to visit a given city. In the Prize Collecting TSP, a city-dependent reward is obtained when a city is visited and a city-dependent penalty has to be paid for each non-visited city. In contrast to this, the TTP requires that each given city is visited. Furthermore, each city has a set of available items with weights and profits and a decision has to be made which items to pick. A selected item contributes its profit to the overall profit. However, the weight of an item leads to a higher transportation cost, and therefore has a negative impact on the overall profit.

Our non-linear knapsack problem uses the same cost function as the TTP, but assumes a fixed route. It deals with the problem which items to select when giving a fixed route from an origin to a destination. Therefore, our approach can also be applied to solve the TTP by using the non-linear packing approach as a subroutine to solve the packing part. Our experimental investigations are carried out on the benchmark set for the traveling thief problem [13] where we assume that the route is fixed.

The paper is organized as follows. In Section 2, we introduce the nonlinear knapsack problems and show in Section 3 that they are $\mathcal{NP}$-hard. In Section 3, we provide a pre-processing scheme which allows to identify unprofitable and compulsory items. Sections 5 and 6 introduce our mixed-integer program based approaches to solve the problem exactly and approximately. We report on the results of our experimental investigations in Section 7 and finish with some conclusions.

2   Problem Statement

We consider the following non-linear packing problem inspired by the traveling thief problem [3]. Given is a route $N = (1, 2, \ldots, n+1)$ as a sequence of $n+1$ cities where all cities are unique and distances $d_i > 0$ between pairs of consecutive cities $(i, i+1)$, $1 \leq i \leq n$. There is a vehicle which travels through the cities of N in the order of this sequence starting its trip in the first city and ending it in the city $n+1$ as a destination point. Every city i, $1 \leq i \leq n$, contains a set of distinct items $M_i = \{e_{i1}, \ldots, e_{im_i}\}$ and we denote by $M = \bigcup\limits_{1 \leq i \leq n} M_i$ set of all items available at all cities. Each item $e_{ik} \in M$ has a positive integer profit p_{ik} and a weight w_{ik}. The vehicle may collect a set of items on the route such that the total weight of collected items does not exceed its capacity W. Collecting an item e_{ik} leads to a profit contribution p_{ik}, but increases the transportation cost as the weight w_{ik} slows down the vehicle. The vehicle travels along $(i, i+1)$, $1 \leq i \leq n$, with velocity $v_i \in [v_{\min}, v_{\max}]$ which depends on the weight of the items collected in the first i cities. The goal is to find a subset of M such that the difference between the profit of the selected items and the transportation cost is maximized.

To make the problem precise we give a nonlinear binary integer program formulation. The program consists of one variable x_{ik} for each item $e_{ik} \in M$ where e_{ik} is chosen iff $x_{ik} = 1$. A decision vector $X = (x_{11}, \ldots, x_{nm_n})$ defines the packing plan as a solution. If no item has been selected, the vehicle travels with its maximal velocity v_{max}. Reaching its capacity W, it travels with minimal velocity $v_{min} > 0$. The velocity depends on the weight of the chosen items in a linear way. The travel time $t_i = \frac{d_i}{v_i}$ along $(i, i+1)$ is the ratio of the distance d_i and the current velocity

$$v_i = v_{max} - \nu \sum_{j=1}^{i} \sum_{k=1}^{m_j} w_{jk} x_{jk}$$

which is determined by the weight of the items collected in cities $1, \ldots, i$. Here, $\nu = \frac{v_{max} - v_{min}}{W}$ is a constant value defined by the input parameters. The overall transportation cost is given by the sum of the travel costs along $(i, i+1)$, $1 \leq i \leq n$, multiplied by a given rent rate $R > 0$. In summary, the problem is given by the following nonlinear binary program (NKP^c).

$$\max \sum_{i=1}^{n} \left(\sum_{k=1}^{m_i} p_{ik} x_{ik} - \frac{R d_i}{v_{max} - \nu \sum_{j=1}^{i} \sum_{k=1}^{m_j} w_{jk} x_{jk}} \right) \tag{1}$$

$$\text{s.t.} \sum_{i=1}^{n} \sum_{k=1}^{m_i} w_{ik} x_{ik} \leq W \tag{2}$$

$$x_{ik} \in \{0, 1\}, \ e_{ik} \in M$$

We also consider the unconstrained version NKP^u of NKP^c where we set $W \geq \sum_{e_{ik} \in M} w_{ik}$ such that every selection of items yields a feasible solution. Given a real value B, the decision variant of NKP^c and NKP^u has to answer the question whether the value of (1) is at least B.

3 Complexity of the Problem

In this section, we investigate the complexity of NKP^c and NKP^u. NKP^c is NP-hard as it is a generalization of the classical NP-hard 0-1 knapsack problem [12]. In fact, assigning zero either to the rate R or to every distance value d_i in NKP^c, we obtain KP. Our contribution is the proof that the unconstrained version NKP^u of the problem remains $\mathcal{N}\mathcal{P}$-hard. We show this by reducing the $\mathcal{N}\mathcal{P}$-complete *subset sum problem* (SSP) to the decision variant of NKP^u which asks whether there is a solution with objective value at least B. The input for SSP is given by q positive integers $S = \{s_1, \ldots, s_q\}$ and a positive integer Q. The question is whether there exists a vector $X = (x_1, \ldots, x_q)$, $x_k \in \{0, 1\}$, $1 \leq k \leq q$, such that $\sum_{k=1}^{q} s_k x_k = Q$.

Theorem 1. *NKP^u is $\mathcal{N}\mathcal{P}$-hard.*

Proof. We reduce SSP to the decision variant of NKP^u which asks whether there is a solution of objective value at least B.

We encode the instance of SSP given by the set of integers S and the integer Q as the instance I of NKP^u having two cities. The first city contains q items while the second city is a destination point free of items. We set the distance between two cities $d_1 = 1$, and set $p_{1k} = w_{1k} = s_k$, $1 \leq k \leq q$ and $W = \sum_{k=1}^{q} s_k$. Subsequently, we set $v_{max} = 2$ and $v_{min} = 1$ which implies $v = 1/W$ and define $R^* = W (2 - Q/W)^2$.

Consider the nonlinear function $f_{R^*} : [0, W] \to \mathbb{R}$ defined as

$$f_{R^*}(w) = w - \frac{R^*}{2 - w/W}. \tag{3}$$

f_{R^*} defined on the interval $[0, W]$ is a continuous convex function that reaches its unique maximum in the point $w^* = W \cdot (2 - \sqrt{R^*/W}) = Q$, i.e. $f_{R^*}(w) < f_{R^*}(w^*)$ for $w \in [0, W]$ and $w \neq w^*$. Then $f_{R^*}(Q)$ is the maximum value for f_{R^*} when being restricted to integer input, too. Therefore, we set $B = f_{R^*}(Q)$ and the objective function for NKP^u is given by

$$g_{R^*}(x) = \sum_{k=1}^{q} p_k x_k - \frac{R^*}{2 - \frac{1}{W} \sum_{k=1}^{q} w_k x_k}. \tag{4}$$

There exists an $x \in \{0, 1\}^q$ such that $g_{R^*}(x) \geq B = f_{R^*}(Q)$ iff $\sum_{k=1}^{q} s_k x_k = \sum_{k=1}^{q} w_{1k} x_k = \sum_{k=1}^{q} p_{1k} x_k = Q$. Therefore, the instance of SSP has answer YES iff the optimal solution of the NKP^u instance I has objective value at least $B = f_{R^*}(Q)$. Obviously, the reduction can be carried out in polynomial time which completes the proof. □

4 Pre-processing

We now provide a pre-processing scheme to identify items of a given instance I that can be either directly included or discarded. Removing such items from the optimization process can significantly speed up the algorithms. Our pre-processing will allow to decrease the number of decision variables for mixed integer programming approaches described in Sections 5 and 6. We distinguish between two kinds of items that are identified in the pre-processing: *compulsory* and *unprofitable* items. We call an item *compulsory* if its inclusion in any packing plan increases the value of the objective function, and call an item *unprofitable* if its inclusion in any packing plan does not increase the value of the objective function. Therefore, an optimal solution has to include all compulsory items while all unprofitable items can be discarded.

In order to identify *compulsory* and *unprofitable* items, we consider the total travel cost that a set of items produces.

Definition 1 (Total Travel Cost). *Let $O \subseteq M$ be a subset of items. We define the total travel cost along route N when the items of O are selected as*

$$t_O = R \cdot \sum_{i=1}^{n} \frac{d_i}{v_{max} - \nu \sum_{j=1}^{i} \sum_{e_{jk} \in O_j} w_{jk}},$$

where $O_j = M_j \cap O$, $1 \leq j \leq n$, is the subset of O selected at city j.

We identify compulsory items for the unconstrained case according to the following proposition.

Proposition 1 (Compulsory Item). *Let I be an arbitrary instance of NKP^u. If $p_{ik} > R\left(t_M - t_{M \setminus \{e_{ik}\}}\right)$, then e_{ik} is a compulsory item.*

Proof. We work under the assumption that $p_{ik} > R\left(t_M - t_{M \setminus \{e_{ik}\}}\right)$ holds. In the case of NKP^u, all the existing items can fit into the vehicle at once and all subsets $O \subseteq M$ are feasible. Let $M^* \subseteq M \setminus \{e_{ik}\}$ be an arbitrary subset of items excluding e_{ik}, and consider $t_{M \setminus M^*}$ and $t_{M \setminus M^* \setminus \{e_{ik}\}}$, respectively. Since the velocity depends linearly on the weight of collected items and the travel time $t_i = d_i/v_i$ along $(i, i+1)$ depends inversely proportional on the velocity v_i, we have $\left(t_M - t_{M \setminus \{e_{ik}\}}\right) \geq \left(t_{M \setminus M^*} - t_{M \setminus M^* \setminus \{e_{ik}\}}\right)$. This implies that $p_{ik} > R\left(t_{M \setminus M^*} - t_{M \setminus M^* \setminus \{e_{ik}\}}\right)$ holds for any subset $M \setminus M^*$ of items which completes the proof. □

For the unconstrained variant NKP^u, Proposition 1 is valid to determine whether the item e_{ik} is able to cover by its p_{ik} the largest possible transportation costs it may generate when has been selected in X. Here, the largest possible transportation costs are computed via the worst case scenario when all the possible items are selected along with e_{ik}, and therefore when the vehicle has the maximal possible load and the least velocity.

Based on a given instance, we can identify unprofitable items for the constrained and unconstrained case according to the following proposition.

Proposition 2 (Unprofitable Item, Case 1). *Let I be an arbitrary instance of NKP^c or NKP^u. If $p_{ik} \leq R\left(t_{\{e_{ik}\}} - t_\emptyset\right)$, then e_{ik} is an unprofitable item.*

Proof. We assume that $p_{ik} \leq R\left(t_{\{e_{ik}\}} - t_\emptyset\right)$ holds. Let $M^* \subseteq M \setminus \{e_{ik}\}$ denote an arbitrary subset of items excluding e_{ik} such that $w_{ik} + \sum_{e_{jl} \in M^*} w_{jl} \leq W$ holds. We consider $t_{M^* \cup \{e_{ik}\}}$ and t_{M^*}. Since the velocity depends linearly on the weight of collected items and the travel time $t_i = d_i/v_i$ along $(i, i+1)$ depends inversely proportional on the velocity v_i, the inequality $\left(t_{\{e_{ik}\}} - t_\emptyset\right) \leq \left(t_{M^* \cup \{e_{ik}\}} - t_{M^*}\right)$ holds. Therefore, $p_{ik} \leq R\left(t_{M^* \cup \{e_{ik}\}} - t_{M^*}\right)$ holds for any $M^* \subseteq M \setminus \{e_{ik}\}$ which completes the proof. □

Proposition 2 helps to determine whether the profit p_{ik} of the item e_{ik} is large enough to cover the least transportation costs it incurs when selected in the packing plan X. In this case, the least transportation costs result from accepting the selection of e_{ik} as only selected item in X versus accepting empty X as a solution.

Having all compulsory items included in the unconstrained case according to Proposition 1, we can identify further unprofitable items. This is the case, as the inclusion of compulsory items already increases the travel time and therefore reducing the positive contribution to the overall objective value.

Proposition 3 (Unprofitable Item, Case 2). *Let I be an arbitrary instance of NKP^u and M^c be the set of all compulsory items. If $p_{ik} \leq R\left(t_{M^c \cup \{e_{ik}\}} - t_{M^c}\right)$, then e_{ik} is an unprofitable item.*

Proof. We assume that $p_{ik} \leq R\left(t_{M^c \cup \{e_{ik}\}} - t_{M^c}\right)$ holds. Recall that in the case of NKP^u, all the existing items can fit into the vehicle at once and all subsets $O \subseteq M$ are feasible. Let $M^* \subseteq M \setminus \{M^c \cup \{e_{ik}\}\}$ be an arbitrary subset of M that does not include any item of $M^c \cup \{e_{ik}\}$ and consider $t_{M^c \cup M^*}$ and $t_{M^c \cup M^* \cup \{e_{ik}\}}$. Since the velocity depends linearly on the weight of collected items and the travel time $t_i = d_i/v_i$ along $(i, i+1)$ depends inversely proportional on the velocity v_i, we have $\left(t_{M^c \cup \{e_{ik}\}} - t_{M^c}\right) \leq \left(t_{M^c \cup M^* \cup \{e_{ik}\}} - t_{M^c \cup M^*}\right)$. Hence, we have $p_{ik} \leq R\left(t_{M^c \cup M^* \cup \{e_{ik}\}} - t_{M^c \cup M^*}\right)$ for any $M^* \subseteq M \setminus \{M^c \cup \{e_{ik}\}\}$ which completes the proof. □

Proposition 3 determines for the NKP^u problem whether the profit p_{ik} of the item e_{ik} is large enough to cover the least transportation costs resulted from its selection along with all known compulsory items. Specifically, in Proposition 3 the list transportation costs follow from accepting the selection of e_{ik} along with the set of compulsory items M^c in X versus accepting just the selection of M^c as a solution.

It is important to note that Proposition 2 can reduce NKP^c problem to NKP^u by excluding items such that the sum of the weights of all remaining items does not exceed the weight bound W. In this case, Propositions 1 and 3 can be applied iteratively to the remaining set of items until no compulsory or unprofitable item is found. Before applying our approaches given in Section 5 and 6, we remove all unprofitable and compulsory items using these preprocessing steps.

5   Exact Solution

Both NKP^c and NKP^u contain nonlinear terms in the objective function, and therefore are nonlinear binary programs. They belong to the specific class of fractional binary programming problems for which several efficient reformulation techniques exist to handle nonlinear terms. We follow the approach of [10] and [16] to reformulate NKP^c and NKP^u as a linear mixed 0-1 program.

The denominator of each fractional term in (1) is not equal to zero since $v_{min} > 0$. We introduce the auxiliary real-valued variables y_i, $i = 1, \dots, n$, such that $y_i = 1 / \left(v_{max} - \nu \sum_{j=1}^{i} \sum_{k=1}^{m_j} w_{jk} x_{jk} \right)$. The variables y_i express the travel time per distance unit along $(i, i+1)$. According to [10], we can reformulate NKP^c as a mixed 0-1 quadratic program by replacing (1) with (5) and adding the set of constraints (6) and (7).

$$\max \sum_{i=1}^{n} \left(\sum_{k=1}^{m_i} p_{ik} x_{ik} - R d_i y_i \right) \tag{5}$$

$$\text{s.t. } v_{max} y_i + \nu \sum_{j=1}^{i} \sum_{k=1}^{m_j} w_{jk} x_{jk} y_i = 1, \ i = 1, \dots, n \tag{6}$$

$$y_i \in \mathbb{R}_+, \ i = 1, \dots, n \tag{7}$$

If $z = xy$ is a polynomial mixed 0-1 term where x is binary and y is a real variable, then it can be linearized via the set of linear inequalities: (i) $z \leq Ux$; (ii) $z \geq Lx$; (iii) $z \leq y + L(x-1)$; (iiii) $z \geq y + U(x-1)$ (see [16]). U and L are the upper and lower bounds on y, i.e. $L \leq y \leq U$. We can linearize the $x_{jk} y_i$ term in (6) by introducing a new real variable $z_{jk}^i = x_{jk} y_i$. Furthermore, let p_i^c and w_i^c denote the total profit and the total weight of the compulsory items in city i according to Proposition 1. Variable y_i, $i = 1, \dots, n$, can be bounded from below by $L_i = 1 / \left(v_{max} - \nu \sum_{j=1}^{i} w_j^c \right)$. Similarly, let w_i^{max} be the total weight of the items (including all the compulsory items) in city i. We can bound y_i, $i = 1, \dots, n$, from above by $U_i = 1 / \left(v_{max} - \nu \cdot min \left(\sum_{j=1}^{i} w_j^{max}, W \right) \right)$ and formulate NKP^c as the following linear mixed 0-1 program (NKP^e):

$$\max \sum_{i=1}^{n} \left(p_i^c + \sum_{k=1}^{m_i} p_{ik} x_{ik} - R d_i y_i \right)$$

$$\text{s.t.} \, v_{max} y_i + \nu \left(w_i^c + \sum_{j=1}^{i} \sum_{k=1}^{m_j} w_{jk} z_{jk}^i \right) = 1, \, i = 1, \dots, n$$

$$z_{jk}^i \le U_i x_{jk}, \, i, j = 1, \dots, n, \, j \le i, \, e_{jk} \in M_j$$

$$z_{jk}^i \ge L_i x_{jk}, \, i, j = 1, \dots, n, \, j \le i, \, e_{jk} \in M_j$$

$$z_{jk}^i \ge y_i + U_i (x_{jk} - 1), \, i, j = 1, \dots, n, \, j \le i, \, e_{jk} \in M_j$$

$$z_{jk}^i \le y_i + L_i (x_{jk} - 1), \, i, j = 1, \dots, n, \, j \le i, \, e_{jk} \in M_j$$

$$\sum_{i=1}^{n} \sum_{k=1}^{m_i} w_{ik} x_{ik} \le W \tag{8}$$

$$x_{ik} \in \{0, 1\}, \, e_{ik} \in M$$

$$z_{jk}^i \in \mathbb{R}_+, \, i, j = 1, \dots, n, \, j \le i, \, e_{jk} \in M_j$$

$$y_i \in \mathbb{R}_+, \, i = 1, \dots, n$$

We now introduce a set of inequalities in order to obtain tighter relaxations. The reformulation-linearization technique by [15] uses $3n$ additional inequalities for the capacity constraint (8). Multiplying (8) by y_l, $U_l - y_l$ and $y_l - L_l$, $l = 1, \dots, n$, we obtain the inequalities

$$\sum_{i=1}^{n} \sum_{k=1}^{m_i} w_{ik} z_{ik}^l \le W y_l;$$

$$U_l \sum_{i=1}^{n} \sum_{k=1}^{m_i} w_{ik} x_{ik} - \sum_{i=1}^{n} \sum_{k=1}^{m_i} w_{ik} z_{ik}^l \le U_l W - W y_l;$$

$$\sum_{i=1}^{n} \sum_{k=1}^{m_i} w_{ik} z_{ik}^l - L_l \sum_{i=1}^{n} \sum_{k=1}^{m_i} w_{ik} x_{ik} \le W y_l - L_l W.$$

Another set of inequalities can be derived from the fact that the item e_{il} in the city i should not be selected if in the same city there exists unselected item e_{ik} with $p_{il} < p_{ik}$ and $w_{il} > w_{ik}$. Furthermore, the item e_{jl} in the city j should not be selected if there exists unselected item e_{ik} in the city i, with $j < i$, $p_{jl} - \Delta_l^{ji} < p_{ik}$ and $w_{jl} > w_{ik}$ where

$$\Delta_l^{ji} = R \sum_{a=j}^{i-1} d_a \left(\frac{1}{v_{max} - \nu \left(w_{jl} + \sum_{b=1}^{a} w_b^c \right)} - \frac{1}{v_{max} - \nu \sum_{b=1}^{a} w_b^c} \right)$$

is a lower bound on the transportation cost to deliver e_{jl} from j to i. Similarly, the item e_{ik} in the city i should not be selected if there exists unselected item e_{jl} in the city j, with $j < i$, $p_{jl} - \overline{\Delta}_l^{ji} > p_{ik}$ and $w_{jl} < w_{ik}$ where

$$\overline{\Delta}_l^{ji} = R \sum_{a=j}^{i-1} d_a \left(\frac{1}{v_{max} - \nu \cdot min\left(w_{jl} + \sum_{b=1}^{a} w_b^{max}, W\right)} - \frac{1}{v_{max} - \nu \cdot min\left(\sum_{b=1}^{a} w_b^{max}, W\right)} \right)$$

is an upper bound on the transportation cost to deliver e_{jl} from j to i. This leads to the following inequalities for $i, j = 1, \ldots, n$:

$$x_{il} \leq x_{ik}, \; e_{il}, e_{ik} \in M_i \; : \; l \neq k, \; p_{il} < p_{ik}, \; w_{il} > w_{ik}; \tag{9}$$

$$x_{jl} \leq x_{ik}, \; j < i, \; e_{jl} \in M_j, \; e_{ik} \in M_i, \; : \; p_{jl} - \Delta_l^{ji} < p_{ik}, \; w_{jl} > w_{ik}; \tag{10}$$

$$x_{jl} \geq x_{ik}, \; j < i, \; e_{jl} \in M_j, \; e_{ik} \in M_i, \; : \; p_{jl} - \overline{\Delta}_l^{ji} > p_{ik}, \; w_{jl} < w_{ik}. \tag{11}$$

6   Approximate Solution

In practice, the use of approximations is an efficient way to deal with nonlinear terms. Although the approximate solution is likely to be different from the exact one, it might be close enough and obtainable in a reasonable computational time.

Consider an arbitrary pair $(i, i+1)$, $i = 1, \ldots, n$, and the traveling time $t_i' \in [t_{min}, t_{max}]$ per distance unit along it. Here $t_{max} = 1/v_{min}$ and $t_{min} = 1/v_{max}$ denote the maximum and minimum travel time per unit, respectively. We partition the interval $[t_{min}, t_{max}]$ into τ equal-sized sub-intervals and determine thus a set $T = \{T_1, \ldots, T_\tau\}$ of straight line segments to approximate the curve $t(v)$ as illustrated in Figure 1. Each segment $a \in T$ is characterized by its minimal velocity v_a^{min} and its corresponding maximum traveling time per distance unit t_a^{max}, and by its maximum velocity v_a^{max} and its corresponding minimum traveling time per distance unit t_a^{min}. Specifically, $\left(v_a^{min}, t_a^{max}\right)$ and $\left(v_a^{max}, t_a^{min}\right)$ are the endpoints of $a \in T$ referred to as breakpoints. We approximate t_i' by the linear combination of t_a^{max} and t_a^{min} if $v_i \in \left[v_a^{min}, v_a^{max}\right]$.

Our approximation model uses three types of decision variables in addition to the binary variable x_{ik} for each item $e_{ik} \in M$ from Section 2. Let w_i be a real variable equal to the total weight of selected items when traveling along the $(i, i+1)$. Let p_i be a real variable equal to the difference of the total profit of selected items and their total transportation costs when delivering them to city $i+1$. We set $w_0 = p_0 = 0$. Let $A_i \subseteq T$, $1 \leq i \leq n$, denote a set of possible segments to which velocity v_i of the vehicle may belong, i.e. $A_i = \{a \in T \; : \; \left(v_a^{min} \in \left[v_i^{min}, v_i^{max}\right]\right) \lor \left(v_a^{max} \in \left[v_i^{min}, v_i^{max}\right]\right)\}$, where $v_i^{max} = v_{max} - \nu \sum_{j=1}^{i} w_j^c$ is the maximal possible velocity that the vehicle can move along $(i, i+1)$ when packing in all compulsory items only, and $v_i^{min} = v_{max} - \nu \cdot min\left(\sum_{j=1}^{i} w_j^{max}, W\right)$ the minimum possible velocity along $(i, i+1)$ after having packed in all items available in cities $1, \ldots, i$. Actually, we have $v_i \in \left[v_i^{min}, v_i^{max}\right]$.

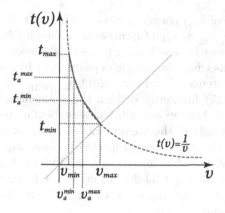

Fig. 1. Piecewise linear approximation of $t(v) = 1/v$

When $v_i \in \left[v_a^{\min}, v_a^{\max}\right]$ for $a \in T$, any point in between endpoints of a is a weighted sum of them. Let B_i denote a set of all breakpoints that the linear segments of A_i have. Then the value of the real variable $y_{ib} \in [0,1]$ is a weight assigned to the breakpoint $b \in B_i$, $b \sim (v_b, t_b)$. NKP^c (and NKP^u) can be approximated by the following linear mixed 0-1 program (NKP$_\tau^a$):

$$\max p_n \tag{12}$$

$$\text{s.t. } p_i = p_{i-1} + p_i^c + \sum_{e_{ik} \in M_i} p_{ik} x_{ik} - R d_i \sum_{b \in B_i} t_b y_{ib}, \ i = 1, \ldots, n \tag{13}$$

$$w_i = w_{i-1} + w_i^c + \sum_{e_{ik} \in M_i} w_{ik} x_{ik}, \ i = 1, \ldots, n \tag{14}$$

$$\nu w_i + \sum_{b \in B_i} v_b y_{ib} = v_{max}, \ i = 1, \ldots, n \tag{15}$$

$$\sum_{b \in B_i} y_{ib} = 1, \ i = 1, \ldots, n \tag{16}$$

$$w_n \leq W \tag{17}$$

$$x_{ik} \in \{0,1\}, \ e_{ik} \in M \tag{18}$$

$$y_{ib} \in [0,1], \ i = 1, \ldots, n, \ b \in B_i \tag{19}$$

$$p_i \in \mathbb{R}, \ i = 1, \ldots, n \tag{20}$$

$$w_i \in \mathbb{R}_{\geq 0}, \ i = 1, \ldots, n \tag{21}$$

$$w_0 = p_0 = 0 \tag{22}$$

Equation (12) defines the objective p_n as the difference of the total profit of selected items and their total transportation costs delivered to city $n+1$. Since the transportation costs are approximated in NKP$_\tau^a$, the actual objective value for NKP^c (and NKP^u) should be computed on values of decision variables of vector X. Equation (13) computes the difference p_i of the total profit of selected items and their total transportation costs when arriving at city $i+1$ by summing up the value of p_{i-1} concerning $(i-1,i)$, the profit of compulsory items p_i^c and the profit $\sum_{e_{ik} \in M_i} p_{ik} x_{ik}$ of items selected in city i, and subtracting the approximated transportation costs along $(i, i+1)$. Equation (14) gives the weight w_i of the selected items when the vehicle departs city i by summing up w_{i-1}, the weight of compulsory items w_i^c and the weight $\sum_{e_{ik} \in M_i} w_{ik} x_{ik}$ of items selected in city i. Equation (15) implicitly defines the segment $a \in A_i$ to which the velocity of the vehicle v_i belongs and sets the weights of its breakpoints. Equation (16) forces the total weight of the breakpoints of B_i be exactly 1. Equation (17) imposes the capacity constraint, and Eq. (18) declares x_{ik} as binary. Equation (19) states y_{ib} as a real variable defined in $[0, 1]$. Finally, Equation (20) declares p_i as a real variable, while Eq. (21) defines w_i as a non-negative real. A solution of NKP_τ^a can be used as a starting solution for NKP^e in the case that all sets of inequalities (9), (10) and (11) are met.

7   Computational Experiments

We now investigate the effectiveness of proposed approaches by experimental studies. On the one hand, we evaluate our MIP models NKP^e and NKP_τ^a in terms of solution quality and running time. On the other hand, we assess the advantage of the pre-processing scheme in terms of quantity of discarded items and auxiliary decision variables. The program code is implemented in JAVA using the CPLEX 12.6 library with default settings. The experiments have been carried out on a computational cluster with 128 Gb RAM and 2.8 GHz 48-cores AMD Opteron processor.

The test instances are adopted from the benchmark set B of [13]. This benchmark set is constructed on TSP instances from TSPLIB (see [14]). In addition, it contains for each city but the first one a set of items. We use the set of items available at each city and obtain the route from the corresponding TSP instance by running the Chained Lin-Kernighan heuristic (see [1]). Given the permutation $\pi = (\pi_1, \pi_2, \ldots, \pi_n)$ of cities computed by the Chained Lin-Kernighan heuristic, where π_1 is free of items, we use $N = (\pi_2, \pi_3, \ldots, \pi_n, \pi_1)$ as the route for our problem. We consider the uncorrelated, uncorrelated with similar weights, and bounded strongly correlated types of items' generation, and set v_{min} and v_{max} to 0.1 and 1 as proposed for B.

The results of our experiments are shown in Tables 1 and 2. First, we investigate three families of small size instances based on the TSP problems eil51, eil76, and eil101 with 51, 76 and 101 cities, respectively. Note that all instances of a family have the same route N. We considered instances with 1, 5, and 10 items per city. The postfixes 1, 6 and 10 in the instances' names indicate the capacity W. Column 2 specifies the total number of items m. A ratio $\alpha = 100\,(m - m')\,/m$ in Column 3 denotes a percentage of items discarded in pre-processing step, where m' is the number of items left after pre-processing. Column 4 identifies by "u" whether NKP^c has been reduced to NKP^u by pre-processing. Columns 5 and 6 report a computational time in seconds and a relative gap reached in percents for NKP^e. The time limit of 1 day has been given

Table 1. Results of Computational Experiments on Small Size Instances

instance	m	α	ver	t^e	gap^e	ρ^{100}	t^{100}	β^{100}	ρ^{1000}	t^{1000}	β^{1000}
					instance family eil51						
uncorr_01	50	42.0	c	1	0.00	1.0000	0	56.9	1.0000	1	55.9
uncorr_06	50	14.0	c	3	0.00	1.0000	0	39.9	1.0000	0	38.7
uncorr_10	50	50.0	u	1	0.00	1.0000	0	11.3	1.0000	0	9.4
uncorr-s-w_01	50	30.0	c	0	0.00	1.0000	0	79.0	1.0000	1	78.0
uncorr-s-w_06	50	24.0	c	3	0.00	1.0000	0	36.5	1.0000	0	35.2
uncorr-s-w_10	50	34.0	u	3	0.00	1.0000	0	13.4	1.0000	0	11.9
b-s-corr_01	50	4.0	c	2	0.00	1.0000	0	91.5	1.0000	2	90.5
b-s-corr_06	50	0.0	c	249	0.00	1.0000	0	54.5	1.0000	1	53.3
b-s-corr_10	50	0.0	c	139	0.00	1.0000	0	26.2	1.0000	0	24.9
uncorr_01	250	39.2	c	1855	0.00	1.0000	0	66.8	1.0000	1	65.7
uncorr_06	250	16.4	c	-	10.66	1.0000	0	39.0	1.0000	0	37.8
uncorr_10	250	54.4	u	268	0.00	1.0000	0	11.2	1.0000	0	9.5
uncorr-s-w_01	250	20.8	c	22	0.00	1.0000	0	89.8	1.0000	1	88.8
uncorr-s-w_06	250	14.0	c	-	25.20	1.0000	0	45.5	1.0000	0	44.2
uncorr-s-w_10	250	19.2	u	73472	0.00	1.0000	0	16.0	1.0000	0	14.6
b-s-corr_01	250	0.0	c	-	0.89	1.0000	0	92.0	1.0000	1	91.1
b-s-corr_06	250	0.0	c	-	53.48	1.0000	0	56.9	1.0000	1	55.7
b-s-corr_10	250	0.0	c	-	60.94	1.0000	0	27.3	1.0000	0	25.9
uncorr_01	500	37.0	c	-	14.82	1.0000	0	69.1	1.0000	1	68.0
uncorr_06	500	15.2	c	-	21.26	1.0000	0	39.6	1.0000	0	38.3
uncorr_10	500	51.4	u	-	1.27	1.0000	0	11.8	1.0000	0	10.1
uncorr-s-w_01	500	20.2	c	-	1.80	1.0000	0	90.8	1.0000	1	89.9
uncorr-s-w_06	500	15.2	c	-	37.83	0.9999	0	45.1	1.0000	0	43.9
uncorr-s-w_10	500	18.6	u	-	4.44	1.0000	0	16.4	1.0000	0	15.0
b-s-corr_01	500	0.0	c	-	5.97	1.0000	0	93.1	1.0000	2	92.1
b-s-corr_06	500	0.0	c	-	49.28	1.0000	0	56.5	1.0000	0	55.4
b-s-corr_10	500	0.0	c	-	71.87	1.0000	0	26.6	1.0000	0	25.2
					instance family eil76						
uncorr_01	75	26.7	c	4	0.00	1.0000	0	77.7	1.0000	1	76.7
uncorr_06	75	14.7	c	50	0.00	1.0000	0	34.3	1.0000	0	33.1
uncorr_10	75	48.0	u	15	0.00	1.0000	0	11.5	1.0000	0	9.6
uncorr-s-w_01	75	26.7	c	1	0.00	1.0000	0	79.2	1.0000	3	78.2
uncorr-s-w_06	75	17.3	c	82	0.00	1.0000	0	41.3	1.0000	1	40.1
uncorr-s-w_10	75	16.0	u	9	0.00	1.0000	0	16.8	1.0000	0	15.4
b-s-corr_01	75	0.0	c	6	0.00	1.0000	0	94.7	1.0000	1	93.8
b-s-corr_06	75	0.0	c	-	8.53	1.0000	0	59.7	1.0000	2	58.5
b-s-corr_10	75	0.0	c	4555	0.00	1.0000	0	25.9	1.0000	0	24.5
uncorr_01	375	38.1	c	-	15.49	1.0000	0	67.2	1.0000	2	66.1
uncorr_06	375	16.0	c	-	18.04	1.0000	0	37.5	1.0000	0	36.2
uncorr_10	375	49.3	u	-	0.57	1.0000	0	12.0	1.0000	0	10.2
uncorr-s-w_01	375	14.9	c	30376	0.00	1.0000	0	90.9	1.0000	5	89.9
uncorr-s-w_06	375	12.3	c	-	48.36	1.0000	0	47.4	1.0000	1	46.2
uncorr-s-w_10	375	14.9	u	-	3.70	1.0000	0	17.3	1.0000	0	15.9
b-s-corr_01	375	0.0	c	-	9.32	1.0000	0	95.4	1.0000	2	94.4
b-s-corr_06	375	0.0	c	-	60.98	1.0000	0	57.4	1.0000	1	56.2
b-s-corr_10	375	0.0	c	-	69.90	1.0000	0	27.8	1.0000	1	26.6
uncorr_01	750	32.5	c	-	19.52	1.0000	0	72.3	1.0000	2	71.2
uncorr_06	750	14.8	c	-	33.14	1.0000	0	39.5	1.0000	0	38.3
uncorr_10	750	43.1	u	-	5.25	1.0000	0	13.1	1.0000	0	11.4
uncorr-s-w_01	750	16.7	c	-	11.31	1.0000	0	89.8	1.0000	2	88.9
uncorr-s-w_06	750	13.5	c	-	60.27	1.0000	0	46.3	1.0000	1	45.1
uncorr-s-w_10	750	14.4	u	-	6.88	1.0000	0	17.2	1.0000	0	15.9
b-s-corr_01	750	0.0	c	-	10.46	1.0000	0	95.0	1.0000	2	94.0
b-s-corr_06	750	0.0	c	-	62.42	1.0000	0	56.1	1.0000	1	54.9
b-s-corr_10	750	0.0	c	-	84.45	1.0000	0	26.2	1.0000	0	24.9
					instance family eil101						
uncorr_01	100	49.0	c	9	0.00	1.0000	0	61.3	1.0000	1	60.2
uncorr_06	100	16.0	c	714	0.00	0.9999	0	40.1	1.0000	2	38.8
uncorr_10	100	57.0	u	21	0.00	1.0000	0	10.2	1.0000	0	8.5
uncorr-s-w_01	100	25.0	c	3	0.00	1.0000	0	91.2	1.0000	1	90.3
uncorr-s-w_06	100	17.0	c	446	0.00	1.0000	0	42.3	1.0000	1	41.0
uncorr-s-w_10	100	15.0	u	68	0.00	1.0000	0	17.4	1.0000	0	16.0
b-s-corr_01	100	0.0	c	532	0.00	1.0000	0	95.4	1.0000	4	94.4
b-s-corr_06	100	0.0	c	-	44.03	1.0000	0	56.8	1.0000	2	55.7
b-s-corr_10	100	0.0	c	-	28.96	0.9999	0	28.5	1.0000	1	27.2
uncorr_01	500	38.8	c	-	13.92	1.0000	0	66.6	1.0000	3	65.5
uncorr_06	500	14.4	c	-	20.49	1.0000	0	39.6	1.0000	1	38.4
uncorr_10	500	51.4	u	-	1.94	1.0000	0	11.5	1.0000	0	9.8
uncorr-s-w_01	500	20.4	c	-	7.00	1.0000	1	89.3	1.0000	14	88.3
uncorr-s-w_06	500	14.2	c	-	40.92	1.0000	0	45.3	1.0000	1	44.1
uncorr-s-w_10	500	16.4	u	-	7.20	1.0000	0	16.4	1.0000	0	15.1
b-s-corr_01	500	0.0	c	-	13.73	1.0000	1	94.4	1.0000	3	93.5
b-s-corr_06	500	0.0	c	-	68.68	1.0000	0	55.3	1.0000	2	54.1
b-s-corr_10	500	0.0	c	-	77.57	1.0000	0	26.3	1.0000	0	25.1
uncorr_01	1000	37.0	c	-	26.74	0.9999	0	67.2	1.0000	3	66.1
uncorr_06	1000	15.1	c	-	30.91	1.0000	0	39.5	1.0000	1	38.3
uncorr_10	1000	50.4	u	-	4.69	1.0000	0	11.8	1.0000	0	10.1
uncorr-s-w_01	1000	19.7	c	-	10.46	0.9999	248	89.3	1.0000	6144	88.3
uncorr-s-w_06	1000	13.7	c	-	57.02	1.0000	0	45.6	1.0000	1	44.4
uncorr-s-w_10	1000	15.9	u	-	13.54	1.0000	0	16.7	1.0000	0	15.3
b-s-corr_01	1000	0.0	c	-	14.41	1.0000	1	93.9	1.0000	7	93.0
b-s-corr_06	1000	0.0	c	-	80.39	1.0000	0	55.8	1.0000	2	54.6
b-s-corr_10	1000	0.0	c	-	97.54	1.0000	0	27.1	1.0000	1	25.8

to NKP^e. Thus, Column 5 either contains a required time or "-" if the time limit is reached. Results for NKP_τ^a with $\tau = 100$ are demonstrated in Columns 7 and 8, while the case of $\tau = 1000$ is shown in Columns 10 and 11. Columns 7 and 10 report ρ^τ as a ratio between the best lower bounds obtained by NKP_τ^a and NKP^e. Within the experiments, NKP_τ^a with $\tau = 100$ produces an initial solution for NKP^e. Columns 8 and 11 contain running times of NKP_τ^a. The time limit of 2 hours has been given to NKP_τ^a. Finally, columns 9 and 12 show a rate β^τ which is a percentage of auxiliary decision variables y_{ib} for $i = 1, \ldots, n$ and $b \in B_i$ used in practice by NKP_τ^a. At most τn variables is required by NKP_τ^a. Thus, β is computed as $\beta = 100 \left(\sum_{i=1}^n |B_i| \right) / (\tau n)$.

The results show that only the instances of small size are solved by NKP^e to optimality within the given time limit. At the same time, the unconstrained cases of the problem turn out to be easier to handle. They either are solved to optimality or have a low relative gap comparing to the constrained cases, even when latter have less number of items m. Generally, the instances with large W are liable to reduction. Because W is large, they have more chances to loose enough items so that the total weight of rest items becomes less or equal to W. However, the pre-processing scheme does not work for bounded strongly correlated type of the instances. No instance of this type is reduced to NKP^u. Moreover, the results show that this type is presumably harder to solve comparing to others as expected in [13]. In fact, the relative gap is significantly larger concerning this type.

NKP_τ^a is particularly fast and its model is solved to optimality for all the small size instances within the given time limit. The ratio ρ^τ is very close to 1 which leads to two observations. Firstly, NKP_τ^a obtains approximately the same result as the optimal solution of NKP^e has but in a shorter time. Secondly, NKP^e cannot find much better solutions even within large given time. Therefore, we can conclude that NKP_τ^a gives an advanced trade-off in terms of computational time and solution's quality comparing to NKP^e. It looks very swift even with instances of hard bounded strongly correlated type. Moreover, NKP_τ^a produces very good approximation even for reasonably small $\tau = 100$. Only one instance of the whole test suite causes a considerable difficulty for NKP_τ^a in terms of a running time. The rate β^τ demonstrates that in practice NKP_τ^a uses a very reduced set of auxiliary decision variables. The medians over all entries of β^{100} and β^{1000} are 45.3 and 44.1, respectively. In general, β^τ is significantly small when W is large, since latter results in a slower growth of diapason $\left[v_i^{\min}, v_i^{\max} \right]$ in NKP_τ^a, for $i = 1, \ldots, n$. In other words, the instances with large W require less number of auxiliary decision variables comparing to the instances where W is smaller.

The goal of our second experiment is to understand how fast NKP_τ^a handles instances of larger size. We use the same settings as for the first experiment, but now give NKP_τ^a the time limit of 6 hours and set $\tau = 100$. We investigate two families of largest size instances of B of [13], namely those based on the TSP problems pla33810 and pla85900 with 33810 and 85900 cities, respectively. Table 2 reports the results. NKP_τ^a needs less than ~ 40 minutes to solve any instance of family `pla33810`. Almost all instances of family `pla85900` can be solved within 2 hours; it takes no longer than ~ 5.5 hours for any of them. Therefore, NKP_T^a proves its ability to master large problems in a reasonable time.

Table 2. Results of Computational Experiments on Large Size Instances

instance	m	α	ver	t^{100}	β^{100}	instance	m	α	ver	t^{100}	β^{100}
instance family pla33810						instance family pla85900					
uncorr_01	33809	29.0	c	522	77.7	uncorr_01	85899	32.4	c	2582	72.8
uncorr_06	33809	12.8	c	337	41.9	uncorr_06	85899	13.5	c	3888	40.6
uncorr_10	33809	35.9	u	32	14.2	uncorr_10	85899	40.8	u	140	12.9
uncorr-s-w_01	33809	19.3	c	425	88.5	uncorr-s-w_01	85899	16.4	c	1707	89.2
uncorr-s-w_06	33809	11.2	c	634	46.8	uncorr-s-w_06	85899	12.3	c	2053	45.7
uncorr-s-w_10	33809	8.7	c	33	17.3	uncorr-s-w_10	85899	13.6	u	152	16.3
b-s-corr_01	33809	0.0	c	419	92.6	b-s-corr_01	85899	0.0	c	4021	92.6
b-s-corr_06	33809	0.0	c	582	55.3	b-s-corr_06	85899	0.0	c	1619	55.4
b-s-corr_10	33809	0.0	c	696	25.6	b-s-corr_10	85899	0.0	c	3550	25.4
uncorr_01	169045	30.6	c	601	75.6	uncorr_01	429495	32.5	c	3506	72.7
uncorr_06	169045	12.8	c	1276	41.8	uncorr_06	429495	13.6	c	6416	40.5
uncorr_10	169045	35.8	u	72	13.9	uncorr_10	429495	40.4	u	538	13.0
uncorr-s-w_01	169045	15.2	c	389	89.5	uncorr-s-w_01	429495	16.3	c	2470	89.2
uncorr-s-w_06	169045	11.7	c	600	46.3	uncorr-s-w_06	429495	12.8	c	7918	46.3
uncorr-s-w_10	169045	9.0	c	774	17.1	uncorr-s-w_10	429495	13.2	u	585	16.5
b-s-corr_01	169045	0.0	c	1526	92.7	b-s-corr_01	429495	0.0	c	3492	92.6
b-s-corr_06	169045	0.0	c	433	55.4	b-s-corr_06	429495	0.0	c	5835	55.2
b-s-corr_10	169045	0.0	c	830	25.4	b-s-corr_10	429495	0.0	c	6834	25.4
uncorr_01	338090	31.6	c	2079	74.5	uncorr_01	858990	33.2	c	7213	71.6
uncorr_06	338090	12.8	c	1272	41.7	uncorr_06	858990	13.6	c	5752	40.4
uncorr_10	338090	35.9	u	1264	13.8	uncorr_10	858990	40.6	u	1895	13.1
uncorr-s-w_01	338090	15.2	c	1266	89.6	uncorr-s-w_01	858990	16.4	c	5036	89.2
uncorr-s-w_06	338090	11.9	c	1225	46.2	uncorr-s-w_06	858990	12.7	c	11793	46.3
uncorr-s-w_10	338090	9.0	c	2509	17.1	uncorr-s-w_10	858990	13.2	u	15593	17.4
b-s-corr_01	338090	0.0	c	851	92.6	b-s-corr_01	858990	0.0	c	6066	92.6
b-s-corr_06	338090	0.0	c	971	55.4	b-s-corr_06	858990	0.0	c	14733	56.2
b-s-corr_10	338090	0.0	c	1300	25.4	b-s-corr_07	858990	0.0	c	19346	26.4

8   Conclusion

We have introduced a new non-linear knapsack problem where items during a travel along a fixed route have to be selected. We have shown that both the constrained and unconstrained version of the problem are $\mathcal{NP}$-hard. Our proposed pre-processing scheme can significantly decrease the size of instances making them easier for computation. The experimental results show that small sized instances can be solved to optimality in a reasonable time by the proposed exact approach. Larger instances can be efficiently handled by our approximate approach producing near-optimal solutions.

As a future work, this problem has several natural generalizations. First, it makes sense to consider the case where the sequence of cities may be changed. This variant asks for the mutual solution of the traveling salesman and knapsack problems. Another interesting situation takes place when cities may be skipped because are of no worth, for example any item stored there imposes low or negative profit. Finally, the possibility to pickup and delivery the items is for certain one another challenging problem.

Acknowledgments. This research was supported under the ARC Discovery Project DP130104395.

References

1. Applegate, D., Cook, W.J., Rohe, A.: Chained lin-kernighan for large traveling salesman problems. INFORMS Journal on Computing 15(1), 82–92 (2003)
2. Balas, E.: The prize collecting traveling salesman problem. Networks 19(6), 621–636 (1989)
3. Bonyadi, M.R., Michalewicz, Z., Barone, L.: The travelling thief problem: the first step in the transition from theoretical problems to realistic problems. In: Proceedings of the IEEE Congress on Evolutionary Computation, CEC 2013, pp. 1037–1044. IEEE, Cancun, June 20–23, 2013
4. Bretthauer, K.M., Shetty, B.: The nonlinear knapsack problem - algorithms and applications. European Journal of Operational Research 138(3), 459–472 (2002)
5. Chekuri, C., Khanna, S.: A polynomial time approximation scheme for the multiple knapsack problem. SIAM J. Comput. 35(3), 713–728 (2005)
6. Elhedhli, S.: Exact solution of a class of nonlinear knapsack problems. Oper. Res. Lett. 33(6), 615–624 (2005)
7. Erlebach, T., Kellerer, H., Pferschy, U.: Approximating multi-objective knapsack problems. In: Dehne, F., Sack, J.-R., Tamassia, R. (eds.) WADS 2001. LNCS, vol. 2125, p. 210. Springer, Heidelberg (2001)
8. Garey, M., Johnson, D.: Computers and Intractability: A Guide to the Theory of NP-Completeness. W. H. Freeman (1979)
9. Hochbaum, D.S.: A nonlinear knapsack problem. Oper. Res. Lett. 17(3), 103–110 (1995)
10. Li, H.L.: A global approach for general 0–1 fractional programming. European Journal of Operational Research 73(3), 590–596 (1994)
11. Lin, C., Choy, K., Ho, G., Chung, S., Lam, H.: Survey of green vehicle routing problem: past and future trends. Expert Systems with Applications 41(4, Part 1), 1118–1138 (2014)
12. Martello, S., Toth, P.: Knapsack Problems: Algorithms and Computer Implementations. John Wiley & Sons (1990)
13. Polyakovskiy, S., Bonyadi, M.R., Wagner, M., Michalewicz, Z., Neumann, F.: A comprehensive benchmark set and heuristics for the traveling thief problem. In: Arnold, D.V. (ed.) GECCO, pp. 477–484. ACM (2014)
14. Reinelt, G.: TSPLIB - A traveling salesman problem library. ORSA Journal on Computing 3(4), 376–384 (1991)
15. Sherali, H., Adams, W.: A Reformulation Linearization Technique for Solving Discrete and Continuous Nonconvex Problems. J Kluwer Academic Publishing, Boston (1999)
16. Tawarmalani, M., Ahmed, S., Sahinidis, N.: Global optimization of 0–1 hyperbolic programs. Journal of Global Optimization 24(4), 385–416 (2002)

MaxSAT-Based Cutting Planes for Learning Graphical Models

Paul Saikko, Brandon Malone, and Matti Järvisalo[✉]

Helsinki Institute for Information Technology, Department of Computer Science,
University of Helsinki, Helsinki, Finland
matti.jarvisalo@cs.helsinki.fi

Abstract. A way of implementing domain-specific cutting planes in branch-and-cut based Mixed-Integer Programming (MIP) solvers is through solving so-called sub-IPs, solutions of which correspond to the actual cuts. We consider the suitability of using Maximum satisfiability solvers instead of MIP for solving sub-IPs. As a case study, we focus on the problem of learning optimal graphical models, namely, Bayesian and chordal Markov network structures.

1 Introduction

A central element contributing to the success of mixed-integer programming (MIP) solvers are algorithms for deriving cutting planes which prune the search space within a branch-and-cut routine. One way of implementing domain-specific cutting planes is through solving so-called sub-IPs, solutions of which correspond to the actual cuts. We consider the suitability of using Maximum satisfiability (MaxSAT) solvers instead of the more typical choice of using MIP solvers for solving sub-IPs. As a case study, we focus on important NP-hard optimization problems of learning probabilistic graphical models, namely, optimal Bayesian [4,6,12,16,18,19] and chordal Markov network structures [5,11]. The GOBNILP system [3,6], which implements a practical MIP-based branch-and-cut approach using specific sub-IPs for deriving domain-specific cutting planes, is a state-of-the-art exact approach for these problem domains. We point out that GOBNILP's sub-IPs can be naturally expressed as MaxSAT, and thereby a MaxSAT solver can be harnessed for solving the sub-IPs instead of relying on a MIP solver such as IBM CPLEX. This results in a hybrid MIP-MaxSAT approach which allows for fine-grained control over the number and structure of the derived cutting planes, as well as enables deriving a set of optimal cutting planes wrt the sub-IP cost function. We present results of a preliminary empirical evaluation of the behavior of such a hybrid approach. The preliminary results suggest that MaxSAT can achieve similar performance as GOBNILP while finding fewer but higher quality cutting planes than the MIP-based sub-IP procedure within GOBNILP. We hope this encourages looking into possibilities of harnessing MaxSAT solvers within other domains in which sub-IPs are used for deriving domain-specific cutting planes within MIP-based approaches.

© Springer International Publishing Switzerland 2015
L. Michel (Ed.): CPAIOR 2015, LNCS 9075, pp. 347–356, 2015.
DOI: 10.1007/978-3-319-18008-3_24

2 Preliminaries

Bayesian Network Structure Learning. Given a set $X = \{X_1, \ldots, X_N\}$ of nodes (representing random variables), an element of $\mathcal{P}_i = 2^{X \setminus \{X_i\}}$ is a *candidate parent set* of X_i. For a given DAG $G = (X, E)$, the parent set of node X_i is $\{X_j \mid (X_j, X_i) \in E\}$, i.e., it consists of the parents of X_i in G. Picking a single $P_i \in \mathcal{P}_i$ for each X_i gives rise to the (not necessarily acyclic) graph in which, for each X_i, there is an edge (X_j, X_i) iff $X_j \in P_i$. In case this graph is acyclic, the choice of P_is corresponds to a Bayesian network structure (DAG) [17]. With these definitions, the Bayesian network structure learning problem (BNSL) [9] is as follows. Given a set $X = \{X_1, \ldots, X_N\}$ of nodes and, for each X_i, a non-negative local score (cost) $s_i(P_i)$ for each $P_i \in \mathcal{P}_i$ as input, the task is to find a DAG G^* such that

$$G^* \in \argmin_{G \in \text{DAGs}(N)} \sum_{i=1}^{N} s_i(P_i), \tag{1}$$

where P_i is the parent set of X_i in G and $\text{DAGs}(N)$ the set of DAGs over X^1.

MaxSAT. Maximum satisfiability (MaxSAT) [2,13,15] is a well-known optimization variant of SAT. For a Boolean variable x, there are two literals, x and $\neg x$. A clause is a disjunction ($\vee$, logical OR) of literals. A truth assignment is a function from Boolean variables to $\{0, 1\}$. A clause C is satisfied by a truth assignment τ ($\tau(C) = 1$) if $\tau(x) = 1$ for a literal x in C, or $\tau(x) = 0$ for a literal $\neg x$ in C. A set F of clauses is satisfiable if there is an assignment τ satisfying all clauses in F ($\tau(F) = 1$), and unsatisfiable ($\tau(F) = 0$ for every assignment τ) otherwise. An instance $F = (F_h, F_s, c)$ of the *weighted partial MaxSAT* problem consists of two sets of clauses, a set F_h of *hard* clauses and a set F_s of *soft* clauses, and a function $c : F_s \rightarrow \mathbb{R}^+$ that associates a non-negative cost with each of the soft clauses. Any truth assignment τ that satisfies F_h is a *solution* to F. The *cost* of a solution τ to F is

$$\text{COST}(F, \tau) = \sum_{\substack{C \in F_s: \\ \tau(C) = 0}} c(C),$$

i.e., the sum of the costs of the soft clauses not satisfied by τ. A solution τ is (globally) *optimal* for F if $\text{COST}(F, \tau) \leq \text{COST}(F, \tau')$ holds for any solution τ' to F. Given an instance F, the weighted partial MaxSAT problem asks to find an optimal solution to F. We will refer to weighted partial MaxSAT instances simply as MaxSAT instances.

3 The GOBNILP Approach to BNSL

In this section, we give an overview of the GOBNILP solver for the BNSL problem. GOBNILP [3,6] is based on formulating BNSL as an integer program (IP),

---

[1] For scoring functions with negative scores (e.g., BD [9]), the problem is instead to *maximize* the score. Flipping the signs gives the equivalent minimization problem considered here.

and implements a branch-and-cut search algorithm using the constrained integer programming framework SCIP [1], enabling the use of, e.g., the state-of-the-art IBM CPLEX IP solver for solving the IP instances encountered during search.

Assume an arbitrary BNSL instance $(X = \{X_1, \ldots, X\}, \{s_i\}_{i=1}^N)$, where $s_i :$ $\mathcal{P}_i \to \mathbb{R}^+$. GOBNILP is based on the following binary IP formulation of BNSL:

$$\text{minimize} \quad \sum_{X_i \in X} \sum_{S \in \mathcal{P}_i} s_i(S) \cdot P_i^S \tag{2}$$

$$\text{subject to} \quad \sum_{S \in \mathcal{P}_i} P_i^S = 1 \qquad \forall i = 1..N \tag{3}$$

$$\sum_{X_i \in C} \sum_{S \cap C = \emptyset} P_i^S \geq 1 \quad \forall C \subset X \tag{4}$$

$$P_i^S \in \{0, 1\} \qquad \forall i = 1..N,\ S \in \mathcal{P}_i \tag{5}$$

In words, binary "parent set" variables P_i^S are indicators for choosing $S \in \mathcal{P}_i$ as the parent set of node X_i (Eq. 5). The BNSL cost function (Eq. 1) is directly represented as Eq. 2 under minimization. The fact that for each X_i exactly one parent set $P_i \in \mathcal{P}_i$ has to be selected is encoded as Eq. 3. Finally, and most importantly, acyclicity of the graph G^* corresponding to the choice of parent sets is ensured by the so-called *cluster constraints* [10] in Eq. 4, stating that for each possible *cluster* C (a subset of nodes), there is at least one variable in C whose parent-set is either outside C or empty.

Instead of directly declaring and solving the integer program consisting of Eqs. 2–5, GOBNILP implements a branch-and-cut approach, a basic outline of which is presented as Algorithm 1. Essentially, the search starts with the linear programming (LP) relaxation consisting of Eqs. 2–3. Cyclic subgraphs are ruled out during search by deriving cutting planes based on a found solution to the LP relaxation consisting of Eqs. 2–3 and the already added cluster constraints. At each search node, an LP relaxation consisting of the current set of constraints (Line 3) is solved. If the solution x to the LP relaxation has worse cost than a best already found solution x^* (initialized to a known upper bound solution), the search backtracks (Line 4). Otherwise, x^* is updated to x, and one or more clusters C for which the cluster constraints are violated under this new x^* are identified, and *cutting planes* are added to the current LP relaxation (Lines 6 and 7) based on C. If no cutting planes are found (i.e., no clusters C are identified) and x is integral, then it is the optimal solution for that branch (Line 8). Failing that, a variable with a non-integral value in x is selected for branching (Lines 10–13).

Two of the main components of the algorithm are solving the LP relaxation and computing cutting planes. GOBNILP uses an off-the-shelf LP solver, such as CPLEX or SoPlex, to solve the LP relaxation. It looks for standard cutting planes, including Gomory, strong Chvátal-Gomory and zero-half cuts. However, the primary strength of GOBNILP is in using a custom routine to find violated cluster constraints, which are added as cutting planes to the current LP relaxation. We will next describe how this is implemented within GOBNILP.

Algorithm 1. Branch-and-cut

1: **procedure** SOLVE(objective function f, constraints c)
2:      **while** True **do**
3:          $x \leftarrow$ SOLVELPRELAXATION(f, c)
4:          **if** $f(x) \geq f(x^*)$ **then return** x^* ▷ $f(x^*) = \infty$ *if* x^* *undefined*
5:          $x^* \leftarrow x$
6:          $c_{new} \leftarrow$ FINDCUTTINGPLANES(x^*)
7:          **if** $c_{new} \neq \emptyset$ **then** $c \leftarrow c \cup c_{new}$
8:          **else if** x^* is integral **then return** x^*
9:          **else** break
10:     $y \leftarrow$ a variable that x^* assigns a non-integral value
11:     $x^*_{y=0} \leftarrow$ Solve$(f, c \cup \{y = 0\})$
12:     $x^*_{y=1} \leftarrow$ Solve$(f, c \cup \{y = 1\})$
13:     **return** arg max$_{x \in \{x^*_{y=0}, x^*_{y=1}\}} f(x)$
14: **end procedure**

GOBNILP implements FINDCUTTINGPLANES (Line 6 of Alg. 1), by solving exactly a nested integer program, referred to as a *sub-IP*. A solution to the sub-IP corresponds to a subset of the nodes (i.e., a cluster C) for which the cluster constraint is violated (Eq. 4). Each identified cluster C gives rise to the cutting plane

$$\sum_{v \in C} \sum_{S: S \cap C = \emptyset} P_v^S \geq 1. \tag{6}$$

We will now detail the sub-IP formulation used within GOBNILP. Intuitively, solutions to the sub-IP represent cyclic subgraphs over the set X of nodes. For the following, let $x^*(P_i^S)$ indicate the value of P_i^S in the current best solution x^* to the outer LP relaxation. Note that, by construction, $\sum_{S \in \mathcal{P}_i} x^*(P_i^S) \leq 1$ holds generally for any solution x^* and node X_i. Furthermore, if for each X_i there is an $S \in \mathcal{P}_i$ such that $x^*(P_i^S) = 1$, then x^* represents a (possibly cyclic) directed graph.

Two types of binary variables are used in the sub-IP: (1) for each X_i, a binary variable C_i indicates whether X_i is in a cluster C found; and (2) for each $x^*(P_i^S) > 0$, where $S \neq \emptyset$, a binary variable J_i^S indicates whether the set S of nodes are the parents of X_i in the cyclic subgraph found, such that at least one of the parents are in C whenever X_i is in C. Using these variables, the sub-IP formulation is the following.

$$\text{maximize} \quad \sum_{X_i \in X} \sum_{S \in \mathcal{P}_i} x^*(P_i^S) \cdot J_i^S - \sum_{X_i \in X} C_i \tag{7}$$

$$\text{subject to} \quad J_i^S \rightarrow C_i \qquad \forall i = 1..N, \ x^*(P_i^S) > 0 \tag{8}$$

$$J_i^S \rightarrow \bigvee_{s \in S} C_s \qquad \forall i = 1..N, \ x^*(P_i^S) > 0 \tag{9}$$

$$\sum C_i \geq 2 \qquad \forall i = 1..N \tag{10}$$

$$C_i, J_i^S \in \{0, 1\} \qquad \forall i = 1..N, \ x^*(P_i^S) > 0 \tag{11}$$

Intuitively, the objective function (Eq. 7) under maximization balances between finding small clusters (the term $-C_i$ contributing a unit penalty) and including nodes from parent-sets with high x^* values. Eq. 8 declares that a node X_i must be in C whenever at least one parent set is chosen for X_i; and Eq. 9 states that at least one node in any chosen parent set must be in C. Finally, Eq. 10 requires that any found cluster must be non-trivial, i.e., contain at least two nodes.

As argued in [6], any feasible solution to the sub-IP has cost greater than -1, and corresponds to a valid cutting plane (following Eq. 6). During search, GOBNILP solves the sub-IPs in a way which generates multiple non-optimal feasible solutions before finding an optimal solution. GOBNILP generates cutting planes according to Eq. 6 for each of the found solutions. Eqs. 8–9 are implemented using SCIP's `logicor` construct.

4   Solving Sub-IPs via MaxSAT

We formulate the GOBNILP sub-IP as MaxSAT using the same set of binary variables, and describe how a MaxSAT solver can be used to provide k best solutions to the sub-IP under different side-constraints over the next solutions w.r.t. the already found clusters. Eq. 8 is represented as the hard clause

$$\neg J_i^S \vee C_i, \tag{12}$$

and Eq. 9 as the hard clauses

$$\neg J_i^S \vee \bigvee_{s \in S} C_s. \tag{13}$$

The non-trivial cluster constraint (Eq. 10, i.e., $\sum C_i \geq 2$) can be equivalently expressed *using the J_i^S variables* as the hard clause

$$\bigvee_{S \in \mathcal{P}_i} J_i^S. \tag{14}$$

This is due to the fact that, for any X_i, if $J_i^S = 1$ for some $S \in \mathcal{P}_i$, Eq. 12 and Eq. 13 together imply that $C_i = 1$ as well as $C_s = 1$ for some $s \in S$ (and by the BNSL problem definition we have that $s \neq i$ for all $s \in S \in \mathcal{P}_i$).

Finally, the sub-IP objective function (Eq. 7) is represented in two parts with soft clauses.

- The first part $\sum_{X_i \in X} \sum_{S \in \mathcal{P}_i} x^*(P_i^S) \cdot J_i^S$ is represented by introducing the soft clause

$$J_i^S \text{ with cost } x^*(P_i^S) \text{ , for each } X_i \text{ and } S \in \mathcal{P}_i. \tag{15}$$

- The second part $-\sum_{X_i \in X} C_i$ is represented by the soft clause

$$\neg C_i \text{ with cost } 1, \text{ for each } X_i. \tag{16}$$

Given a solution τ to the sub-IP, we have that $\tau(C_i) = 1$ if and only if $X_i \in C$, i.e., node X_i is in the cluster corresponding to τ. We will here consider different strategies for ruling out C from the set of candidate clusters when finding $k > 1$ solutions to the sub-IP via MaxSAT.

Ruling out Exactly the Found Cluster. By adding the hard clause

$$\bigvee_{\tau(C_i)=1} \neg C_i \vee \bigvee_{\tau(C_i)=0} C_i,$$

we rule out exactly the cluster C from the remaining solutions to the sub-IP. In other words, C will not correspond to any optimal solution after adding this hard clause.

Ruling out Cluster Supersets and Subsets. Given two clusters, C and C', such that $C \subset C'$, the cutting plane resulting from C can result in stronger pruning of the outer LP search space than the cutting plane resulting from C', since the cutting plane constraint becomes more restrictive. Given a cluster C, adding the hard clause

$$\bigvee_{\tau(C_i)=1} \neg C_i$$

results in ruling out all supersets of C from the set of solutions to the sub-IP; the clause guarantees that all remaining solutions will correspond to clusters which include at least one node which is not in C. Analogously, adding the hard clause

$$\bigvee_{\tau(C_i)=0} C_i$$

results in ruling out all subsets of C, ensuring that all remaining solutions be *orthogonal* to C in the sense that they will involve variables not mentioned in the cutting plane corresponding to C.

Ruling out Overlapping Clusters. Even more orthogonal solutions—in terms of *non-overlapping* clusters, involving non-overlapping subsets of nodes—to the sub-IP, provide cutting planes which together prune different dimensions of the search space of the outer LP relaxation. To guarantee finding a set of non-overlapping clusters via MaxSAT, after each found solution τ corresponding to a cluster C, one can add the hard unit clauses

$$\neg C_i \quad \text{for each } C_i \text{ such that } \tau(C_i) = 1,$$

guaranteeing that none of the nodes in C will occur in any of the remaining solutions.

For integrating the sub-IP search via MaxSAT within GOBNILP, we use our own prototype re-implementation of the MaxHS MaxSAT solver [8] that in preliminary experiments showed good performance compared to other MaxSAT solvers in this domain. The MaxHS algorithm is a hybrid SAT-MIP approach based on iteratively solving a sequence of SAT instances and extracting unsatisfiable cores, and using the IBM CPLEX MIP solver to solve a sequence of minimum hitting set problems over the extracted cores. The search progresses bottom-up by proving increasingly tight lower bounds for the optimal solutions. We have implemented an API for the solver which allows for incrementally querying for k best solutions without having to restart the search from scratch after each found solution. Furthermore, the API enables adding arbitrary hard clauses after each solution, using which we can apply the different set-based strategies for finding multiple best solutions to the sub-IPs. Our implementation also natively supports real-valued weights for the MaxSAT soft clauses. We used Minisat 2.2.0 as the underlying SAT solver.

5   Experiments

For a preliminary empirical evaluation of using MaxSAT to solve sub-IPs within GOBNILP (version 1.4.1), we used a set of 530 Bayesian network structure learning instances from [14] and 285 chordal Markov network learning instances from [11][2] over 17–61 nodes. The experiments were run on a cluster of 2.53-GHz Intel Xeon quad core machines with 32-GB memory and Ubuntu Linux 12.04. A timeout of 2 h and a memory limit of 30 GB were enforced for solving each benchmark instance.

An overview of the results is presented in Fig. 1. The upper two plots give views to the total per-instance running times of the default GOBNILP ("GOB-NILP", using CPLEX to solve the sub-IP encountered during search) and our modified GOBNILP variants ("MaxSAT") which use our MaxSAT solver on the MaxSAT formulations presented in Sect. 4 to solve the encountered sub-IPs. In the plot keys, the numerical parameter (1/5/10) after "MaxSAT" gives the number of best solutions looked for, where "all" ("opt", respectively) refers to finding *all* (respectively, all optimal) solutions regardless of how many exist for the individual sub-IPs. In case of "super+subset" and "overlap", the set of solutions to the sub-IPs are incrementally refined after each solution by ruling out solutions which are either superset or subsets ("super+subset") of the found solution, or overlap with the solution, using the clauses detailed in Sect. 4.

Fig. 1 (top left) gives the number of Bayesian network structure learning instances solved (x-axis) within different timeouts (y-axis). We observe that the best-performing MaxSAT-based variants for solving the sub-IPs show generally very similar performance as the default GOBNILP. Here we emphasize that the default MIP-based sub-IP solving strategy within GOBNILP has been carefully hand-tuned for these kinds of instances. The MaxSAT-based variant which aims at finding orthogonal (non-overlapping, i.e., variable-disjoint) cuts, especially the one which incrementally finds a maximal disjoint set of optimal sub-IP solutions performs very similarly to GOBNILP. Surprisingly, even finding only a single optimal solution to the sub-IP using MaxSAT comes close to the performance

---

[2] Setting the option `gobnilp/noimmoralities` to true in GOBNILP allows for learning chordal Markov networks with GOBNILP [7] without changing the sub-IP model.

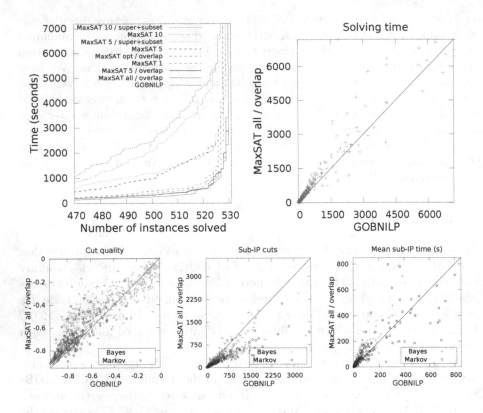

Fig. 1. Top left: number of solved instances using different timeouts on Bayesian networks; right: running time comparison of GOBNILP and MaxSAT finding all cuts under the disjoint refinement on Markov networks. Bottom left: average cut quality; middle: number of cuts; right: average sub-IP solving time.

of GOBNILP. These observations seem to suggest that high-quality cuts (in terms of the sub-IP objective function) are very important in pruning the search space. In contrast, the variants which look for many cuts (5-10) with the less-restrictive refinement strategies (ruling out either all supersets and all subsets, or only the exact solutions found), perform noticeably worse. Here we note that, due to the fact that our MaxSAT solver implementation allows for adding the hard refinement clauses incrementally without having to start the solver from scratch, we observed that the running time cost of finding many solutions is rather negligible, often a fraction of a second. Hence it seems that the refinement strategy plays a key role. Focusing on the best-performing MaxSAT-based variant (finding a maximal disjoint set of optimal solutions), Fig. 1 (top right) gives a per-instance running time comparison with GOBNILP on the Markov network learning instances, again showing performance close to that of GOB-NILP. In fact, MaxSAT results in solving one more instance within the timeout, and the average time spent in solving the sub-IPs is less than that of GOBNILP on several Markov network instances (Fig. 1 top right). As can be seen from Fig. 1 (bottom left), the cuts found with MaxSAT tend to be of better quality on average compared to those found by GOBNILP. Here it is important to note that since GOBNILP and the MaxSAT variants find different cuts, the overall search performed by the different solvers, especially, the sub-IPs encountered during search, differ. While the MaxSAT-based sub-IP routine results in overall performance similar to GOBNILP, this is achieved by adding notably fewer cuts, as shown in Fig. 1 (bottom middle). The price paid for finding better quality cuts, on the other hand, is reflected in the average running time of solving the per-instance sub-IPs, as can be seen from Fig. 1 (bottom right).

A current challenge is to further speed up solving the sub-IPs with MaxSAT, e.g. by by devising domain-specific search heuristics. Similar modifications to GOBNILP's MIP-based sub-IP routine, as well as studying alternative sub-IP objective functions, would also be of interest. It would also be interesting to apply MaxSAT solvers to sub-IPs within MIP-based approaches to other problem domains.

Acknowledgments. The authors thank James Cussens for discussions on GOBNILP and Kustaa Kangas for the Markov network structure learning instances. Work funded by Academy of Finland, grants 251170 Centre of Excellence in Computational Inference Research, 276412, and 284591; and Research Funds of the University of Helsinki.

References

1. Achterberg, T.: Constrained Integer Programming. Ph.D. Thesis, TU Berlin (2007)
2. Ansotegui, C., Bonet, M.L., Levy, J.: SAT-based MaxSAT algorithms. Artificial Intelligence **196**, 77–105 (2013)
3. Bartlett, M., Cussens, J.: Advances in Bayesian network learning using integer programming. In: Proc. UAI, pp. 182–191. AUAI Press (2013)
4. de Campos, C.P., Zeng, Z., Ji, Q.: Structure learning of Bayesian networks using constraints. In: Proc. ICML, pp. 113–120. ACM (2009)
5. Corander, J., Janhunen, T., Rintanen, J., Nyman, H., Pensar, J.: Learning chordal Markov networks by constraint satisfaction. In: Proc. NIPS, pp. 1349–1357. Curran Associates, Inc. (2013)
6. Cussens, J.: Bayesian network learning with cutting planes. In: Proc. UAI, pp. 153–160. AUAI Press (2011)
7. Cussens, J., Bartlett, M.: GOBNILP 1.4.1 user/developer manual (2013)
8. Davies, J., Bacchus, F.: Exploiting the power of MIP solvers in MAXSAT. In: Järvisalo, M., Van Gelder, A. (eds.) SAT 2013. LNCS, vol. 7962, pp. 166–181. Springer, Heidelberg (2013)
9. Heckerman, D., Geiger, D., Chickering, D.M.: Learning Bayesian networks: the combination of knowledge and statistical data. Machine Learning **20**, 197–243 (1995)
10. Jaakkola, T., Sontag, D., Globerson, A., Meila, M.: Learning Bayesian network structure using LP relaxations. In: Proc. AISTATS, pp. 358–365. JMLR.org (2010)
11. Kangas, K., Niinimäki, T., Koivisto, M.: Learning chordal Markov networks by dynamic programming. In: Proc. NIPS, pp. 2357–2365. Curran Associates, Inc. (2014)

12. Koivisto, M., Sood, K.: Exact Bayesian structure discovery in Bayesian networks. Journal of Machine Learning Research, 549–573 (2004)
13. Li, C.M., Manyà, F.: MaxSAT, hard and soft constraints. In: Handbook of Satisfiability, Frontiers in Artificial Intelligence and Applications, vol. 185, chap. 19, pp. 613–631. IOS Press (2009)
14. Malone, B., Kangas, K., Järvisalo, M., Koivisto, M., Myllymäki, P.: Predicting the hardness of learning Bayesian networks. In: Proc. AAAI, pp. 1694–1700. AAAI Press (2014)
15. Morgado, A., Heras, F., Liffiton, M.H., Planes, J., Marques-Silva, J.: Iterative and core-guided MaxSAT solving: a survey and assessment. Constraints **18**(4), 478–534 (2013)
16. Ott, S., Miyano, S.: Finding optimal gene networks using biological constraints. Genome Informatics **14**, 124–133 (2003)
17. Pearl, J.: Probabilistic Reasoning in Intelligent Systems: Networks of Plausible Inference. Morgan Kaufmann Publishers Inc. (1988)
18. Silander, T., Myllymäki, P.: A simple approach for finding the globally optimal Bayesian network structure. In: Proc. UAI, pp. 445–452. AUAI Press (2006)
19. Yuan, C., Malone, B.: Learning optimal Bayesian networks: a shortest path perspective. Journal of Artificial Intelligence Research **48**, 23–65 (2013)

A Multistage Stochastic Programming Approach to the Dynamic and Stochastic VRPTW

Michael Saint-Guillain[1]([⊠]), Yves Deville[1], and Christine Solnon[2]

[1] ICTEAM, Université catholique de Louvain, Louvain-la-Neuve, Belgium
michael.saint@uclouvain.be
[2] Université de Lyon, CNRS, INSA-Lyon, LIRIS, UMR5205, 69621 Lyon, France

Abstract. We consider a dynamic vehicle routing problem with time windows and stochastic customers (DS-VRPTW), such that customers may request for services as vehicles have already started their tours. To solve this problem, the goal is to provide a decision rule for choosing, at each time step, the next action to perform in light of known requests and probabilistic knowledge on requests likelihood. We introduce a new decision rule, called Global Stochastic Assessment (GSA) rule for the DS-VRPTW, and we compare it with existing decision rules, such as MSA. In particular, we show that GSA fully integrates nonanticipativity constraints so that it leads to better decisions in our stochastic context. We describe a new heuristic approach for efficiently approximating our GSA rule. We introduce a new waiting strategy. Experiments on dynamic and stochastic benchmarks, which include instances of different degrees of dynamism, show that not only our approach is competitive with state-of-the-art methods, but also enables to compute meaningful offline solutions to fully dynamic problems where absolutely no a priori customer request is provided.

1 Introduction

Dynamic (or *online*) vehicle routing problems (D-VRPs) arise when information about demands is incomplete, *e.g.*, whenever a customer is able to submit a request during the online execution of a solution. D-VRP instances usually indicate the deterministic requests, *i.e.*, those that are known before the online process if any. Whenever some additional (stochastic) knowledge about unknown requests is available, the problem is said to be *stochastic*. We focus on the *Dynamic* and *Stochastic* VRP with *Time Windows* (DS-VRPTW). These problems arise in many practical situations, as door-to-door or door-to-hospital transportation of elderly or disabled persons. In many countries, authorities try to set up dial-a-ride services, but escalating operating costs and the complexity of satisfying all customer demands become rapidly unmanageable for solution methods based on human choices [10]. However, such complex dynamic problems need reliable and efficient algorithms that should first be assessed on reference problems, such as the DS-VRPTW.

In this paper, we present *a new heuristic method for solving the DS-VRPTW*, based on a Stochastic Programming modeling. By definition, our approach enables

© Springer International Publishing Switzerland 2015
L. Michel (Ed.): CPAIOR 2015, LNCS 9075, pp. 357–374, 2015.
DOI: 10.1007/978-3-319-18008-3_25

a *higher level of anticipation* than heuristic state-of-the-art methods. The resulting new online decision rule, called Global Stochastic Assessment (GSA), comes with a theoretical analysis that clearly defines the nature of the method. We propose a *new waiting strategy* together with a heuristic algorithm that embeds GSA. We compare GSA with the state-of-the-art method MSA from [7], and provide a *comprehensive experimental study* that highlights the contributions of existing and new waiting and relocation strategies.

This paper is organized as follows. Section 2 describes the problem. Section 3 presents the state-of-the-art method we compare to and briefly discuss related works. GSA is then presented in Section 4. Section 5 describes an implementation that embeds GSA, based on heuristic local search. Finally, section 6 resumes the experimental results. A conclusion follows in section 7.

2   Description of the DS-VRPTW

Notations. We note $[l, u]$ the set of all integer values i such that $l \leq i \leq u$. A sequence $< x^i, x^{i+1}, \ldots, x^{i+k} >$ (with $k \geq 0$) is noted $x^{i..i+k}$, and the concatenation of two sequences $x^{i..j}$ and $x^{j+1..k}$ (with $i \leq j < k$) is noted $x^{i..j}.x^{j+1..k}$. Random variables are noted $\boldsymbol{\xi}$ and their realizations are noted ξ. We note $\xi \in \boldsymbol{\xi}$ the fact that ξ is a realization of $\boldsymbol{\xi}$, and $p(\boldsymbol{\xi} = \xi)$ the probability that the random variable $\boldsymbol{\xi}$ is realized to ξ. Finally, we note $\mathbb{E}_{\boldsymbol{\xi}}[f(\xi)]$ the expected value of $f(\boldsymbol{\xi})$ which is defined by $\mathbb{E}_{\boldsymbol{\xi}}[f(\xi)] = \sum_{\xi \in \boldsymbol{\xi}} p(\boldsymbol{\xi} = \xi) \cdot f(\xi)$.

Input Data of a DS-VRPTW. We consider a discrete time horizon $[1, H]$ such that each online event or decision occurs at a discrete time $t \in [1, H]$, whereas each offline event or decision occurs at time $t = 0$. The DS-VRPTW is defined on a complete and directed graph $G = (V, E)$. The set of vertices $V = [0, n]$ is composed of a depot (vertex 0) and n customer regions (vertices 1 to n). To each arc $(i, j) \in E$ is associated a travel time $t_{i,j} \in \mathbb{R}_{\geq 0}$, that is the time needed by a vehicle to travel from i to j, with $t_{i,j} \neq t_{j,i}$ in general. To each customer region $i \in [1, n]$ is associated a load q_i, a service duration $d_i \in [1, H]$ and a time window $[e_i, l_i]$ with $e_i, l_i \in [1, H]$ and $e_i \leq l_i$.

The set of all customer requests is $R \subseteq [1, n] \times [0, H]$. For each request $(i, t) \in R$, t is the time when the request is revealed. When $t = 0$, the request is known before the online execution and it is said to be *deterministic*. When $t > 0$, the request is revealed during the online execution at time t and it is said to be *online* (or dynamic). There may be several requests for a same vertex i which are revealed at different times. During the online execution, we only know a subset of the requests of R (*i.e.*, those which have already been revealed). However, for each time $t \in [1, H]$, we are provided a probability vector P^t such that, for each vertex $i \in [1, n]$, $P^t[i]$ is the probability that a request is revealed for i at time t.

There are k vehicles and all vehicles have the same capacity Q.

Solution of a DS-VRPTW. At the end of the time horizon, a solution is a subset of requests $R_a \subseteq R$ together with k routes (one for each vehicle). Requests in

R_a are said to be *accepted*, whereas requests in $R \setminus R_a$ are said to be *rejected*. The routes must satisfy the constraints of the classical VRPTW restricted to the subset R_a of accepted requests, *i.e.*, each route must start at the depot at a time $t \geq 1$ and end at the depot at a time $t' \leq H$, and for each accepted request $(i, t) \in R_a$, there must be exactly one route that arrives at vertex i at a time $t' \in [e_i, l_i]$ with a current load $l \leq Q - q_i$ and leaves vertex i at a time $t'' \geq t' + d_i$. The goal is to minimize the number of rejected requests.

As not all requests are known at time 0, the solution cannot be computed offline, and it is computed during the online execution. More precisely, at each time $t \in [1, H]$, an action a^t is computed. Each action a^t is composed of two parts: first, for each request $(i, t) \in R$ revealed at time t, the action a^t must either accept the request or reject it; second, for each vehicle, the action a^t must give operational decisions for this vehicle at time t (*i.e.*, service a request, travel towards a vertex, or wait at its current position). Before the online execution (at time 0), some decisions are computed offline. Therefore, we also have to compute a first action a^0.

A solution is a sequence of actions $a^{0..H}$ which covers the whole time horizon. This sequence must satisfy VRPTW constraints, *i.e.*, the actions of $a^{0..H}$ must define k routes such that each request accepted in $a^{0..H}$ is served once by one of these routes within the time window associated with the served vertex and without violating capacity constraints. We define the objective function ω such that $\omega(a^{0..t})$ is $+\infty$ if $a^{0..t}$ does not satisfy VRPTW constraints, and $\omega(a^{0..t})$ is the number of requests rejected in $a^{0..t}$ otherwise. Hence, a solution is a sequence $a^{0..H}$ such that $\omega(a^{0..H})$ is minimal at the end of the horizon.

Stochastic program. There are different notations used for formulating stochastic programs; we mainly use those from [8]. For each time $t \in [1, H]$, we have a vector of random variables $\boldsymbol{\xi}^t$ such that, for each vertex $i \in [1, n]$, $\boldsymbol{\xi}^t[i]$ is realized to 1 if a request for i is revealed at time t, and to 0 otherwise. The probability distribution of $\boldsymbol{\xi}^t$ is defined by P^t, *i.e.*, $p(\boldsymbol{\xi}^t[i] = 1) = P^t[i]$ and $p(\boldsymbol{\xi}^t[i] = 0) = 1 - P^t[i]$. We note $\boldsymbol{\xi}^{1..t}$ the random matrix composed of the random vectors $\boldsymbol{\xi}^1$ to $\boldsymbol{\xi}^t$. A realization $\xi^{1..H} \in \boldsymbol{\xi}^{1..H}$ is called a *scenario*.

At each time $t \in [1, H]$, the action a^t must contain one accept or reject for each request which is revealed in ξ^t. Therefore, we note $A(\xi^t)$ the set of all actions that contain an accept or a reject for each vertex $i \in [1, n]$ such that $\xi^t[i] = 1$. Of course, these actions also contain other decisions related to the k vehicles. We also note $A(\xi^{t1..t2})$ the sequence of sets $< A(\xi^{t1}), \ldots, A(\xi^{t2}) >$ where $t1 \leq t2$.

Hence, at each time t, given the sequence $a^{0..t-1}$ of past actions, the best action a^t is obtained by solving the multistage stochastic problem defined by eq. (1):

$$a^t = \underset{a^t \in A(\xi^t)}{\operatorname{argmin}} \mathbb{E}_{\xi^{t+1}} \left[\underset{a^{t+1} \in A(\xi^{t+1})}{\min} \mathbb{E}_{\xi^{t+2}} \left[\cdots \underset{a^{H-1} \in A(\xi^{H-1})}{\min} \mathbb{E}_{\xi^H} \left[\underset{a^H \in A(\xi^H)}{\min} \omega(a^{0..H}) \right] \cdots \right] \right] \tag{1}$$

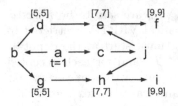

Time	2	3	4	5	6	7	8	9
Scenario $\xi_1^{2..5}$	$\emptyset$	$\emptyset$	$\{d,e,f\}$	$\emptyset$	$\emptyset$	$\emptyset$	$\emptyset$	$\emptyset$
Scenario $\xi_2^{2..5}$	$\emptyset$	$\emptyset$	$\{g,h,i\}$	$\emptyset$	$\emptyset$	$\emptyset$	$\emptyset$	$\emptyset$

At time $t = 1$, there is only 1 vehicle which is on vertex a, and we have to choose between 2 possible actions: *travel to b* or *travel to c*

Fig. 1. A simple example of nonanticipation. The graph is displayed on the left. Time windows are displayed in brackets. For every couple of vertices (i, j), if an arrow $i \to j$ is displayed then $t_{i,j} = 2$; otherwise $t_{i,j} = 20$. To simplify, we consider only 2 equiprobable scenarios (displayed on the right). These scenarios have the same prefix (at times 2 and 3 no request is revealed) but reveal different requests at time 4. When using eq. (1) at time $t = 1$, we choose to travel to c as the expected cost with nonanticipativity constraints is 1: At time 4, only one scenario will remain and if this scenario is ξ_1 (resp. ξ_2), request $(d, 4)$ (resp. $(g, 4)$) will be rejected. When using eq. (2), we choose to travel to b as the expected cost without nonanticipativity constraints is 0 (for each possible scenario, there exists a sequence of actions which serves all requests: travel to d, e, and f for ξ_1 and travel to g, h, and i for ξ_2). However, if we travel to b, at time 3 we will have to choose between traveling to d or g and at this time the expected cost of both actions will be 1.5: If we travel to d (resp. g), the cost with scenario ξ_1 is 0 (resp. 3) and the cost with scenario ξ_2 is 3 (resp. 0). In this example, the nonanticipativity contraints of multistage problem (1) thus leads to a better action than the two-stage relaxation (2).

Note that this multistage stochastic problem is different from the two-stage stochastic problem defined by eq. (2):

$$a^t = \underset{a^t \in A(\xi^t)}{\arg\min} \, \mathbb{E}_{\boldsymbol{\xi}^{t+1..H}} \Big[\underset{a^{t+1..H} \in A(\xi^{t+1..H})}{\min} \omega(a^{0..H}) \Big] \qquad (2)$$

Indeed, eq. (1) enforces *nonanticipativity constraints* so that, at each time $t' > t$, we consider the action $a^{t'}$ which minimizes the expectation with respect to $\boldsymbol{\xi}^{t'}$ only, without considering the possible realizations of $\boldsymbol{\xi}^{t'+1..H}$. Eq. (2) does not enforce these constraints and considers the best sequence $a^{t+1..H}$ for each realization $\xi^{t+1..H} \in \boldsymbol{\xi}^{t+1..H}$. Therefore, eq. (1) may lead to a larger expectation of ω than eq. (2), as it is more constrained. However, the expectation computed in eq. (1) leads to better decisions in our context where some requests are not revealed at time t. This is illustrated in Fig. 1.

3   Related Work

The first D-VRP is proposed in [29], which introduces a single vehicle Dynamic Dial-a-Ride Problem (D-DARP) in which customer requests appear dynamically. Then, [20] introduced the concept of immediate requests that must be serviced as soon as possible, implying a replanning of the current vehicle route. Complete reviews on D-VRP may be found in [18,21]. In this section, we more

---

Algorithm 1. The ChooseRequest-ε Expectation Algorithm

---

1 **for** $a^t \in A(\xi^t)$ **do** $f(a^t) \leftarrow 0$;
2 Generate a set S of α scenarios using Monte Carlo sampling
3 **for** *each scenario* $s \in S$ *and each action* $a^t \in A(\xi^t)$ **do**
4 $\quad\lfloor\ f(a^t) \leftarrow f(a^t) +$cost of (approximate) solution to scenario s starting with a^t
5 **return** $\arg\min_{a^t \in A(\xi^t)} f(a^t)$

---

particularly focus on stochastic D-VRP. [18] classifies approaches for stochastic D-VRP in two categories, either based on *stochastic modeling* or on *sampling*. Stochastic modeling approaches formally capture the stochastic nature of the problem, so that solutions are computed in the light of an overall stochastic context. Such holistic approaches usually require strong assumptions and efficient computation of complex expected values. Sampling approaches try to capture stochastic knowledge by sampling scenarios, so that they tend to be more focused on local stochastic evidences. Their local decisions however allow sample-based methods to scale up to larger problem instances, even under challenging timing constraints. One usually needs to find a good compromise between having a high number of scenarios, providing a better representation of the real distributions, and a more restricted number of these leading to less computational effort.

[7] studies the DS-VRPTW and introduces the Multiple Scenario Approach (MSA). A key element of MSA is an adaptive memory that stores a pool of solutions. Each solution is computed by considering a particular scenario which is optimized for a few seconds. The pool is continuously populated and filtered such that all solutions are consistent with the current system state. Another important element of MSA is the *ranking function* used to make operational decisions involving idle vehicles. The authors designed 3 algorithms for that purpose:

- *Expectation* [3,4] samples a set of scenarios and selects the next request to be serviced by considering its average cost on the sampled set of scenarios. Algorithm 1 [27] depicts how it chooses the next action a^t to perform. It requires an optimization for each action $a^t \in A(\xi^t)$ and each scenario $s \in S$ (lines 3-4), which is computationally very expensive, even with a heuristic approach.
- *Regret* [3,6] approximates the expectation algorithm by recognizing that, given a solution sol_s^* to a particular scenario s, it is possible to compute a good approximation of the local loss inquired by performing another action than the next planned one in sol_s^*.
- *Consensus* [4,7] selects the request that appears the most frequently as the next serviced request in the solution pool.

Quite similar to the consensus algorithm is the Dynamic Sample Scenario Hedging Heuristic introduced by [14] for the stochastic VRP. Also, [15] designed a Tabu Search heuristic for the DS-VRPTW and introduced a vehicle-waiting strategy computed on a future request probability threshold in the near region.

Finally, [5] extends MSA with waiting and relocation strategies so that the vehicles are now able to relocate to promising but unrequested yet vertices. As the performances of MSA has been demonstrated in several studies [5,12,19,22], it is still considered as a state-of-the-art method for dealing with DS-VRPTW.

Other studies of particular interest for our paper are [13], on the dynamic and stochastic pickup and delivery problem, and [22], on the DS-DARP. Both consider local search based algorithms. Instead of a solution pool, they exploit one single solution that minimizes the expected cost over a set of scenarios. However, in order to limit computational effort, only near future requests are sampled within each scenario. Although the approach of [22] is similar to the one of [13], the set of scenarios considered is reduced to one scenario. Although these later papers show some similarities with the approach we propose, they do not provide any mathematical motivation and analysis of their methods.

4   The global Stochastic Assessment decision rule

The two-stage stochastic problem defined by eq. (2) may be solved by a sampling solving method such as MSA, which solves a deterministic VRPTW for each possible scenario (i.e., realization of the random variables) and selects the action a^t which minimizes the sum of all minimum objective function values weighted by scenario probabilities. However, we have shown in Section 2 that eq. (2) does not enforce nonanticipativity constraints because the different deterministic VRPTW are solved independently. To enforce nonanticipativity constraints while enabling sampling methods, we push these constraints in the computation of the optimal solutions for all different scenarios: Instead of computing these different optimal solutions independently, we propose to compute them all together so that we can ensure that whenever two scenarios share a same prefix of realizations, the corresponding actions are enforced to be equal.

At each time $t \in [0, H]$, let r be the number of different possible realizations of $\xi^{t+1..H}$, and let us note $\xi_1^{t+1..H}, \ldots, \xi_r^{t+1..H}$ these realizations. Given the sequence $a^{0..t-1}$ of past actions, we choose action a^t by using eq. (3)

$$a^t = \arg\min_{a^t \in A(\xi^t)} \mathcal{Q}(a^{0..t}, \{\xi_1^{t+1..H}, \ldots, \xi_r^{t+1..H}\}) \tag{3}$$

which is called the deterministic equivalent form of eq. (1).
$\mathcal{Q}(a^{0..t}, \{\xi_1^{t+1..H}, \ldots, \xi_r^{t+1..H}\})$ solves the deterministic optimization problem

$$\min_{a_1^{t+1..H} \in A(\xi_1^{t+1..H}), \ldots, a_r^{t+1..H} \in A(\xi_r^{t+1..H})} \sum_{i=1}^{r} p(\boldsymbol{\xi}^{t+1..H} = \xi_i^{t+1..H}) \cdot \omega(a^{0..t}.a_i^{t+1..H}) \tag{4}$$

$$s.t. \ (\xi_i^{t+1..t'} = \xi_j^{t+1..t'}) \Rightarrow (a_i^{t+1..t'} = a_j^{t+1..t'}), \ \forall t' \in [t+1, H], \ \forall i, j \in [1, r] \tag{5}$$

The nonanticipativity constraints (5) state that, when 2 realizations $\xi_i^{t+1..H}$ and $\xi_j^{t+1..H}$ share a same prefix from $t+1$ to t', the corresponding actions must be equal [23].

Solving eq. (3) is computationally intractable for two reasons. First, since the number r of possible realizations of $\xi^{t+1..H}$ is exponential in the number of vertices and in the remaining horizon size $H - t$, considering every possible scenario is intractable in practice. We therefore consider a smaller set of α scenarios $S = \{s_1, ..., s_\alpha\}$ such that each scenario $s_i \in S$ is a realization of $\xi^{t+1..H}$, i.e., $\forall i \in [1, \alpha], s_i \in \xi^{t+1..H}$. This set S is obtained by Monte Carlo sampling [2]. All elements of S share the same probability, i.e., $p(\xi^{t+1..H} = s_1) = ... = p(\xi^{t+1..H} = s_\alpha)$.

Second, solving eq. (3) basically involves solving to optimality problem Q for each possible action $a^t \in A(\xi^t)$. Each problem Q involves solving a VRPTW for each possible scenario of S, while ensuring nonanticipativity constraints between the different solutions. As the VRPTW problem is an $\mathcal{NP}$-hard problem, we propose to compute an upper bound $\overline{Q}$ of Q based on a given sequence $a_R^{t+1..H}$ of future route actions. Because we impose the sequence $a_R^{t+1..H}$, the set of possible actions at time t is limited to those directly compatible with it, denoted $\tilde{A}(\xi^t, a_R^{t+1..H}) \subseteq A(\xi^t)$. That limitation enforces $\omega(a^{0..H}) < +\infty$. This finally leads to the GSA decision rule:

$$(GSA) \qquad a^t = \underset{a^t \in \tilde{A}(\xi^t, a_R^{t+1..H})}{\arg\min} \overline{Q}(a^{0..t}, a_R^{t+1..H}, S) \qquad (6)$$

which, provided realization ξ^t, sampled scenarios S and future route actions $u_R^{t+1..H}$, selects the action a^t that minimizes the expected approximate cost over scenarios S. Notice that almost all the anticipative efficiency of the GSA decision rule relies on the sequence $a_R^{t+1..H}$, which directly affects the quality of the upper bound $\overline{Q}$.

Sequence $a_R^{t+1..H}$ of future route actions. This sequence is used to compute an upper bound of Q. For each time $t' \in [t + 1, H]$, the route action $a_R^{t'}$ only contains operational decisions related to vehicle routing (i.e., for each vehicle, travel towards a vertex, or wait at its current position) and does not contain decisions related to requests (i.e., request acceptance or rejection). The more flexible $a_R^{t'}$ with respect to S, the better the bound $\overline{Q}$. We describe in Section 5 how a flexible sequence is computed through local search.

Computation of an upper bound $\overline{Q}$ of Q. Algorithm 2 depicts the computation of an upper bound $\overline{Q}$ of Q given a sequence $a_R^{t+1..H}$ of route actions consistent with past actions $a^{0..t}$. For each scenario s_i of S, Algorithm 2 builds a sequence $b^{0..H}$ for s_i, which starts with $a^{0..t}$, and whose end $b^{t+1..H}$ is computed from $a_R^{t+1..H}$ in a greedy way. At each time $t' \in [t + 1..H]$, each request revealed at time t' in scenario s_i is accepted if it is possible to modify $b^{t'..H}$ so that one vehicle can service it; it is rejected otherwise. One can consider $b^{t'..H}$ as being a set of vehicle routes, each defined by a sequence of planned vertices. Each planned vertex comes with specific decisions: a waiting time and whether a service is performed. In this context, *trytoServe* performs a deterministic linear time modification of $b^{t'..H}$ such that (j, t') corresponds to the insertion of the vertex j in one of the routes defined by $b^{t'..H}$, at the best position with respect to VRPTW constraints and travel times, without modifying the order of the remaining vertices. At the end, Algorithm 2 returns the average number of rejected requests for all scenarios. Note that, when modifying a sequence of actions so that a request can be accepted (line 6), actions $b^{t'..H}$ can be modified, but $b^{0..t'-1}$ are not modified. This ensures that $\overline{\mathcal{Q}}$ preserves the nonanticipativity constraints. Indeed, the fact that two identical scenarios prefixes could be assigned two different subsequences of actions implies that either $trytoServe((j,t'), b^{t'..H})$ is able to modify an action $b^{t<t'}$ or is a nondeterministic function. In both cases, there is a contradiction. Finally, notice that contrary to other local search methods based on Monte Carlo simulation as in [13, 22], GSA considers the whole timing horizon when evaluating a first-stage solution against a scenario.

Comparison to MSA. GSA has two major differences with MSA. Given a set of scenarios, GSA maintains only one solution, namely the sequence $a_R^{t+1..H}$, that best suits to a pool of scenarios whilst MSA computes a set of solutions, each specialized to one scenario from the pool. Furthermore, by preserving nonanticipativity GSA approximates the multistage problem of equations (1,3). In contrary, MSA relaxes these constraints and therefore approximates the two-stage problem (2) [27].

In particular, given a pool of scenarios obtained by Monte Carlo sampling, MSA Expectation Algorithm 1 reformulates eq. (2) as a *sample average approximation* (SAA, [1,28]) problem. The SAA tackles each scenario as a separate deterministic problem. For a specific scenario $\xi^{t+1..H}$, it considers the recourse cost of a solution starting with actions $a^{0..t}$. Because the scenarios are not linked by nonanticipativity constraints, two scenarios i and j that share the same prefix $\xi^{t+1..t'}$ can actually be assigned two solutions performing completely different actions $a_i^{0..t'}$ and $a_j^{0..t'}$, for some $t' > t$. The evaluation of action a^t over the set of scenarios is therefore too optimistic, leading to a suboptimal choice. By definition, the *Regret* algorithm approximates the Expectation algorithm. The *Regret*

Algorithm 2. The $\overline{\mathcal{Q}}(a^{0..t}, a_R^{t+1..H}, S)$ approximation function

1   Precondition: $a_R^{t+1..H}$ is a sequence of route actions consistent with $a^{0..t}$
2   **for** *each scenario* $s_i \in S$ **do**
3      $nbRejected[i] \leftarrow 0$; $b^{0..t} \leftarrow a^{0..t}$; $b^{t+1..H} \leftarrow a_R^{t+1..H}$
4   **for** $t' \in [t+1..H]$ **do**
5   **for** *each request* (j,t') *revealed at time* t' *for a vertex* j *in scenario* s_i **do**
6              $c^{t'..H} \leftarrow trytoServe((j,t'), b^{t'..H})$
7   **if** $b^{t+1..t'-1} \cdot c^{t'..H}$ *is feasible* **then** $b^{t'..H} \leftarrow c^{t'..H}$
8   **else** add the decision *reject(j,t')* to $b^{t'}$ and increment $nbRejected[i]$;

9   **return** $\frac{1}{|S|} \cdot \sum_{s_i \in S} nbRejected[i]$

algorithm then also approximates a two-stage problem. The *Consensus* algorithm selects the most suggested action among plans of the pool. By selecting the most frequent action in the pool, *Consensus* somehow encourages nonanticipation. However, the nonanticipativity constraints are not enforced as each scenario is solved separately. *Consensus* also approximates a two-stage problem.

5   Solving the Dynamic and Stochastic VRPTW

GSA alone does not permit to solve a DS-VRPTW instance. In this section, we now show how the decision rule, as defined in eq. 6, can be embedded in an online algorithm that solves the DS-VRPTW. Finally, we present the different waiting and relocation strategies we exploit, including a new waiting strategy.

5.1   Embedding GSA

In order to solve the DS-VRPTW, we design Algorithm 3, which embeds the GSA decision rule.

Main Algorithm. It is parameterized by: α which determines the size of the pool S of scenarios; β which determines the frequency for re-initializing S; and δ_{ins} which limits the time spent for trying to insert a request in a sequence.

It runs in *real time*. It is started before the beginning of the time horizon, in order to compute an initial pool S of α scenarios and an initial solution $a_R^{1..H}$ with respect to offline requests (revealed at time 0). It runs during the whole time horizon, and loops on lines 3 to 11. It is stopped when reaching the end of the time horizon. The *real time* is discretized in H time units, and the variable t represents the current time unit: It is incremented when real time exceeds the end of the tth time unit. In order to be correct, Algorithm 3 requires the real computation time of lines 4 to 11 to be smaller than the real time spent in one time unit. This is achieved by choosing suitable values for parameters α and δ_{ins}.

Lines 4 and 5 describe what happens whenever the algorithm enters a new time unit: Function handleRequests (described below) chooses the next action a^t and updates $a_R^{t+1..H}$; Finally, S is updated such that it stays coherent with respect to realization ξ^t. Each scenario $\xi^{t..H} \in S$ is composed of a sequence of sampled requests. To each customer region i is associated an upper bound $\bar{r}_i = \min(l_0 - t_{i,0} - d_i, l_i - t_{0,i})$ on the time unit at which a request can be revealed in that region, like in [7]. That constraint prevents tricky or inserviceable requests to be sampled. At time t, a sampled request (i, t) which doesn't appear in ξ^t is either removed if $t \geq \bar{r}_i$ or randomly delayed in $\xi^{t+1..H} \in S$ otherwise.

The algorithm spends the rest of the time unit to iterate over lines 7 to 10, in order to improve the sequence of future route actions $a_R^{t+1..H}$. We consider a hill climbing strategy: The current solution $a_R^{t+1..H}$ is shaked to obtain a new candidate solution $b_R^{t+1..H}$, and if this solution leads to a better upper bound $\overline{Q}$ of Q, then it becomes the new current solution. Shaking is performed by the

---

Algorithm 3. LS-based GSA

---

1 Initialize S with α scenarios and compute initial solution $a_R^{1..H}$ w.r.t. known
   requests

2 $t \leftarrow 1$;

3 **while** *real time has not reached the end of the time horizon* **do**

   /* Beginning of the time unit   */

4     $(a^t, a_R^{t+1..H}) \leftarrow$ handleRequests$(a^{0..t-1}, a_R^{t..H}, \xi^t)$

5   execute action a^t and update the pool S of scenarios w.r.t. to ξ^t

   /* Remaining of the time unit   */

6   **while** *real time has not reached the end of time unit t* **do**

7        $b_R^{t+1..H} \leftarrow$ shakeSolution$(a_R^{t+1..H})$

8   **if** $\overline{\mathcal{Q}}(a^{0..t}, b_R^{t+1..H}, S) < \overline{\mathcal{Q}}(a^{0..t}, a_R^{t+1..H}, S)$ **then** $a_R^{t+1..H} \leftarrow b_R^{t+1..H}$;

9   **if** *the number of iterations since the last re-initialization of S is equal to*
         β **then**

10   Re-initialize the pool S of scenarios w.r.t. $\boldsymbol{\xi}^{t+1..H}$

11    $t \leftarrow t + 1$ /* Skip to next time unit   */

12 **Function** handleRequests$(a^{0..t-1}, a_R^{t..H}, \xi^t)$

13    $b^{0..t-1} \leftarrow a^{0..t-1}$; $b^{t..H} \leftarrow a_R^{t..H}$

14   **for** *each request revealed for a vertex j in realization ξ^t* **do**

15   **if** *we find, in less than δ_{ins}, how to modify $b^{t..H}$ s.t. request (j,t) is
         served* **then**

16   modify $b^{t..H}$ to accept request (j,t)

17   **else**

18   modify $b^{t..H}$ to reject request (j,t)

19   **return** $(b^t, b^{t+1..H})$

---

shakeSolution function. This function considers different neighborhoods, corresponding to the following move operators: relocate, swap, inverted 2-opt, and cross-exchange (see [16,26] for complete descriptions). As explained in Section 5.2, depending on the chosen waiting and relocation strategy, additional move operators are exploited. At each call to the shakeSolution function, the considered move operator is changed, such that the operators are equally selected one after another in the list. Every β iterations, the pool S of scenarios is re-sampled (lines 9-10). This re-sampling introduces diversification as the upper bound computed by $\overline{\mathcal{Q}}$ changes. We therefore do not need any other meta-heuristic such as Simulated Annealing.

Function handleRequest is called at the beginning of a new time unit t, to compute action a^t in light of online requests (if any). It implements the GSA decision rule defined in eq. (6). The function considers each request revealed at time t for a vertex j, in a sequential way. For each request, it tries to insert it into the sequence $a_R^{t..H}$ (*i.e.*, modify the routes so that a vehicle visits j during its time window). As in shakeSolution, local search operations are performed during that computation. The time spent to find a feasible solution including the new request is limited to δ_{ins}. If such a feasible solution is found, then the request is accepted, otherwise it is rejected. If there are several online requests for the same discretized time t, we process these requests in their real-time order of arrival, and we assume that all requests are revealed at different real times.

5.2 Waiting and Relocation strategies

As defined in section 2, a vehicle that just visited a vertex usually has the choice between traveling right away to the next planned vertex or first waiting for some time at its current position. Unlike in the static (and deterministic) case, in the dynamic (and stochastic) VRPTW these choices may have a significant impact on the solution quality.

Waiting and relocation strategies have attracted a great interest on dynamic and stochastic VRP's. In this section, we present and describe how waiting and relocation strategies are integrated to our framework, including a new waiting strategy called *relocation-only*.

Relocation strategies. Studies in [8,9] already showed that for a dynamic VRP with no stochastic information, it is optimal to relocate the vehicle(s) either to the center (in case of single-vehicle) or to strategical points (multiple-vehicle case) of the service region. The idea evolved and has been successfully adapted to routing problems with customer stochastic information, in reoptimization approaches as well as sampling approaches.

Relocation strategies explore solutions obtained when allowing a vehicle to move towards a customer vertex even if there is no request received for that vertex at the current time slice. Doing so, one recognizes the fact that, in the context of dynamic and stochastic vehicle routing, a higher level of anticipation can be obtained by considering to reposition the vehicle after having serviced a request to a more stochastically fruitful location. Such a relocation strategy has already been applied to the DS-VRPTW in [5].

Waiting strategies. In a dynamic context, the planning of a vehicle usually contains more time than needed for traveling and servicing requests. When it finishes to service a request, a vehicle has the choice between waiting for some time at its location or leaving for the next planned vertex. A good strategy for deciding where and how long to wait can potentially help at anticipating future requests and hence increase the dynamic performances. We consider three existing waiting strategies and introduce a new one:

- *Drive-First (DF)*: The basic strategy aims at leaving each serviced request as soon as possible, and possibly wait at the next vertex before servicing it if the vehicle arrives before its time window.
- *Wait-First (WF)*: Another classical waiting strategy consists in delaying as much as possible the service time of every planned requests, without violating

their time windows. After having serviced a request, the vehicle hence waits as long as possible before moving to the next planned request.

- *Custom-Wait (CW)*: A more tailored waiting strategy aims at controlling the waiting time at each vertex, which becomes part of the online decisions.
- *Relocation-Only waiting (RO)*: In order to take maximum benefit of relocation strategy while avoiding the computational overhead due to additional decision variables involved in custom waiting, we introduce a new waiting strategy. It basically applies *drive-first* scheduling to every request and then applies *wait-first* waiting only to those requests that follow a relocation one. By doing so, a vehicle will try to arrive as soon as possible at a planned *relocation request*, and wait there as long as possible. In contrary, it will spend as less time as possible at non-relocation request vertices. Note that if it is not coupled to a relocation strategy, RO reduces to DF. Furthermore, RO also reduces to the dynamic waiting strategy described in [17] if we define the service zones as being delimited by relocation requests. However, our strategy differs by the fact that service zones in our approach are computed in light of stochastic information instead of geometrical considerations.

Depending on the waiting strategy we apply and whether we use relocation or not, additional LS move operators are exploited. Specifically, among the waiting strategies, only *custom-wait* requires additional move operators aiming at either increasing or decreasing the waiting time at a random planned vertex. *Relocation* also requires two additional move operators that modify a given solution by either inserting or removing a relocation action at a random vertex.

6   Experimentations

We now describe our experimentations and compare our results with those of the state of the art MSA algorithm of [7].

6.1   Algorithms

Different versions of Algorithm 3 have been experimentally assessed, depending on which waiting strategy is implemented and whether in addition we use the relocation strategy or not.

Surprisingly, the *wait-first* waiting strategy, as well as its version including *relocation*, produced very bad results in comparison to other strategies, rejecting more than twice more online requests in average. Because of its computational overhead, the *custom-wait* strategy also produced bad results, even with relocation. For conciseness we therefore do not report these strategies in the result plots.

The 3 different versions of Algorithm 3 we thus consider are the following: GSA*df*, which stands for GSA with *drive-first* waiting strategy, GSA*dfr* which stands for GSA with *drive-first* and *relocation* strategies, and finally GSA*ro* with means GSA using *relocation-only* strategy. Recall that, by definition, the *relocation-only* strategy involves relocation. In addition to those 3 algorithms, as a baseline we consider the GLS*df* algorithm, which stands for *greedy local search* with *drive-first* waiting. This algorithm is similar to the dynamic LS described in [22], to which we coupled a Simulated Annealing metaheuristic. In this algorithm, stochastic information about future request is not taken into account and a neighboring solution is solely evaluated by its total travel cost.

Finally, GSA and GLS are compared to two MSA algorithms, namely MSA*d* and MSA*c* depending on whether the *travel distance* or the *consensus function* are used as ranking functions.

6.2 Benchmarks

The selected benchmarks are borrowed from [7] which considers a set of benchmarks initially designed for the static and deterministic VRPTW in [25], each of these containing 100 customers. In our stochastic and dynamic context, each customer becomes a request region, where dynamic requests can occur during the online execution.

The original problems from [7] are divided into 4 classes of 15 instances. Each class is characterized by its degree of dynamism (DOD, the ratio of the number of dynamic requests revealed at time $t > 0$ over the number of a priori request known at time $t = 0$) and whether the dynamic requests are known early or lately along the online execution. The time horizon $H = 480$ is divided into 3 time slices. A request is said to be early if it is revealed during the first time slice $t \in [1, 160]$. A late request is revealed during the second time slice $t \in [161, 320]$. There is no request revealed during the third time slice $t \in [321, 480]$, but the vehicles can use it to perform customer operations.

In Class 1 there are many initial requests, many early requests and very few late requests. Class 2 instances have many initial requests, very few early requests and some late requests. Class 3 is a mix of classes 1 and 2. In Class 4, there are few initial requests, few early requests and many late requests. Finally, classes 1, 2 and 3 have an average DOD of 44%, whilst Class 4 has an average DOD of 57%.

In [5], a fifth class is proposed with a higher DOD of 81% in average. Unfortunately, we were not able to get those Class 5 instances. We complete these classes by providing a sixth class of instance, with DOD of 100%. Each instance hence contains no initial request, an early request with probability 0.3 and a late request with probability 0.7.

Figure 2 summarizes the different instance classes.

6.3 Results

Computations are performed on a cluster composed of 32 64-bits AMD Opteron (tm) Processor 6284 SE cores, with CPU frequencies ranging from 1400 to 2600 MHz. Executables were developed with C++ and compiled on a Linux Red Hat environment with GCC 4.4.7. Average results over 10 runs are reported. In [7],

	DOD	$t=0$	$t \in [1,160]$	$t \in [161,320]$	$t \in [321,480]$
Class 1,2,3	44%	$P^0[i] = 0.5$	$P^{[1,160]}[i] = 0.25$	$P^{[161,320]}[i] = 0.25$	$P^{[321,480]}[i] = 0$
Class 4	57%	$P^0[i] = 0.2$	$P^{[1,160]}[i] = 0.2$	$P^{[161,320]}[i] = 0.6$	$P^{[321,480]}[i] = 0$
Class 5	81%	$P^0[i] = 0.1$	$P^{[1,160]}[i] = 0.1$	$P^{[161,320]}[i] = 0.8$	$P^{[321,480]}[i] = 0$
Class 6	100%	$P^0[i] = 0$	$P^{[1,160]}[i] = 0.3$	$P^{[161,320]}[i] = 0.7$	$P^{[321,480]}[i] = 0$

Fig. 2. Summary of the test instances, grouped per degree of dynamism. $P^{[t,t']}[i]$ represents the probability that a request gets revealed during the time slice defined by interval $[t, t']$.

25 minutes of offline computation are allocated to MSA, in order to decide the first online action at time $t = 1$. During online execution, each time unit within the time horizon was executed during 7.5 seconds by the simulation framework. In order to compensate the technology difference, we decided in this study to allow only 10 minutes of offline computation and 4 seconds of online computation per time unit. Thereafter, in order to highlight the contribution of the offline computation in our approach, the amount of time allowed at pre-computation is increased to 60 minutes, while each time unit still lasts 4 seconds. According to preliminary experiments, both the size of the scenario pool and the resampling rate are set to $\alpha = \beta = 150$ for all our algorithms except GLS*df*.

Figure 3 gives a graphical representation of our algorithms results, through performance profiles. Performance profiles provide, for each algorithm, a cumulative distribution of its performance compared to other algorithms. For a given algorithm, a point (x, y) on its curve means that, in $(100 \cdot y)\%$ of the instances, this algorithm performed at most x times worse than the best algorithm on each instance taken separately. Instances are grouped by DOD and by offline computation time. Classes 1, 2 and 3 have a DOD of 44%, hence they are grouped together. An algorithm is strictly better than another one if its curve stays above the other algorithm's curve. For example on the 60min plot of Class 6, GLS*df* is the worst algorithm in 95% of Class 6 instances, outperforming GSA*df* in the remaining 5% (but not the other algorithms). On the other hand, provided these 60 minutes of offline computation, GSA*ro* obtains the best results in 55% of the instances, whereas only 30% for GSA*df* and GSA*dfr*. See [11] for a complete description of performance profiles. Detailed results are provided in the extended version [30].

Our algorithms compare fairly with MSA, especially on lately dynamic instances of Class 4. Given more offline computation, our algorithms get stronger, although that MSA benefits of the same offline time in every plots. Surprisingly, GLS*df* performs well compared to other algorithms on classes 1,2 and 3. The low DOD that characterizes these instances tends to lower the contribution of stochastic knowledge against the computational power of GLS*df*. Indeed, approximating the stochastic evaluation function over 150 scenarios is about 10^3 times more expensive than GLS*df* evaluation function. However, as the offline computation time and the DOD increase, stochastic algorithms tend to outperform their deterministic counterpart.

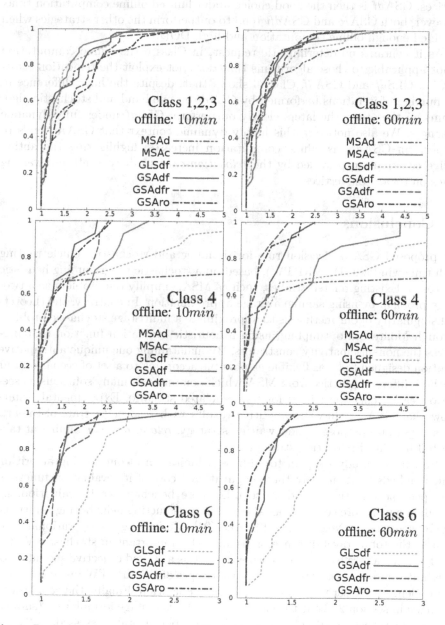

Fig. 3. Performance profiles on classes [1, 2 ,3], Class 4 and Class 6 problem instances

We notice that the relocation strategy gets stronger as the offline computation time increases. This is due to the computational overhead induced by relocation vertices. GSA*df* is then the good choice under limited offline computation time. However, both GSA*ro* and GSA*dfr* tend to outperform the other strategies when provided enough offline computation and high DOD.

As it contains no deterministic request, in Class 6 the offline computation is not applicable to those algorithms that does not exploit the relocation strategy, i.e. GLS*df* and GSA*df*. Class 6 shows that, despite the huge difference in the number of iterations performed by GLS*df* on one hand and stochastic algorithms on the other, the laters clearly outperform GLS*df* under fully dynamic instances. We also notice in this highly dynamic context that GSA*ro* tends to outperform GSA*dfr* as offline computation increases, highlighting the anticipative contribution provided by the *relocation-only* strategy, centering waiting times on relocation vertices.

7 Conclusions

We proposed GSA, a decision rule for dynamic and stochastic vehicle routing with time windows (DS-VRPTW), based on a stochastic programming heuristic approach. Existing related studies, such as MSA, simplify the problem as a two-stage problem by using sample average approximation. In contrary, the theoretical singularity of our method is to approximate a multistage stochastic problem through Monte Carlo sampling, using a heuristic evaluation function that preserves the nonanticipativity constraints. By maintaining one unique anticipative solution designed to be as flexible as possible according to a set of scenarios, our method differs in practice from MSA which computes as many solutions as scenarios, each being specialized for its associated scenario. Experimental results show that GSA produces competitive results with respect to state-of-the-art. This paper also proposes a new waiting strategy, *relocation-only*, aiming at taking full benefit of relocation strategy.

In a future study we plan to address a limitation of our solving algorithm which embeds GSA, namely the computational cost of its evaluation function. One possible direction would be to take more benefit of each evaluation, by spending much more computational effort in constructing neighboring solutions, e.g. by using Large Neighborhood Search [24]. Minimizing the operational cost, such as the total travel distance, is usually also important in stochastic VRPs. Studying the aftereffect when incorporating it as a second objective should be of worth. It is also necessary to consider other types of DS-VRPTW instances, such as problem sets closer to public or good transportation. Finally, the conclusions we made in section 2 about the shortcoming of a two-stage formulation (showed in Fig. 1) are theoretical only, and should be experimentally assessed.

Acknowledgments. Christine Solnon is supported by the LABEX IMU (ANR-10-LABX-0088) of Universit? de Lyon, within the program "Investissements d'Avenir" (ANR-11-IDEX-0007) operated by the French National Research Agency (ANR). This research is also partially supported by the UCLouvain Action de Recherche Concert?e ICTM22C1.

References

1. Ahmed, S., Shapiro, A.: The sample average approximation method for stochastic programs with integer recourse (Submitted for publication, 2002)
2. Asmussen, S., Glynn, P.W.: Stochastic Simulation: Algorithms and Analysis: Algorithms and Analysis, vol. 57. Springer (2007)
3. Bent, R., Van Hentenryck, P.: Regrets only! online stochastic optimization under time constraints. In: AAAI, pp. 501–506 (2004)
4. Bent, R., Van Hentenryck, P.: The value of consensus in online stochastic scheduling. In: ICAPS, (1), pp. 219–226 (2004)
5. Bent, R., Van Hentenryck, P.: Waiting and relocation strategies in online stochastic vehicle routing. In: IJCAI, pp. 1816–1821 (2007)
6. Bent, R., Katriel, I., Van Hentenryck, P.: Sub-optimality approximations. In: van Beek, P. (ed.) CP 2005. LNCS, vol. 3709, pp. 122–136. Springer, Heidelberg (2005)
7. Bent, R.W., Van Hentenryck, P.: Scenario-based planning for partially dynamic vehicle routing with stochastic customers. Operations Research 52(6), 977–987 (2004)
8. Bertsimas, D.J., Van Ryzin, G.: A stochastic and dynamic vehicle routing problem in the Euclidean plane. Operations Research (1991)
9. Bertsimas, D.J., Van Ryzin, G.: Stochastic and Dynamic Vehicle Routing in the Euclidean Plane with Multiple Capacitated Vehicles. Operations Research (1993)
10. Cordeau, J.-F., Laporte, G.: The dial-a-ride problem (DARP): Variants, modeling issues and algorithms. 4OR: A Quarterly Journal of Operations Research 1(2), 89–101 (2003)
11. Dolan, E.D., Moré, J.J.: Benchmarking optimization software with performance profiles. Mathematical Programming 91(2), 201–213 (2002)
12. Flatberg, T., Hasle, G., Kloster, O., Nilssen, E.J., Riise, A.: Dynamic and stochastic vehicle routing in practice. In: Dynamic Fleet Management, pp. 41–63. Springer (2007)
13. Ghiani, G., Manni, E., Quaranta, A., Triki, C.: Anticipatory algorithms for same-day courier dispatching. Transportation Research Part E: Logistics and Transportation Review 45(1), 96–106 (2009)
14. Hvattum, L.M., Løkketangen, A., Laporte, G.: Solving a Dynamic and Stochastic Vehicle Routing Problem with a Sample Scenario Hedging Heuristic. Transportation Science 40(4), 421–438 (2006)
15. Ichoua, S., Gendreau, M., Potvin, J.-Y.: Exploiting Knowledge About Future Demands for Real-Time Vehicle Dispatching. Transportation Science 40(2), 211–225 (2006)
16. Kindervater, G.A.P., Savelsbergh, M.W.P.: Vehicle routing: handling edge exchanges. In: Local Search in Combinatorial Optimization, pp. 337–360 (1997)
17. Mitrović-Minić, S., Laporte, G.: Waiting strategies for the dynamic pickup and delivery problem with time windows. Transportation Research Part B: Methodological 38(7), 635–655 (2004)
18. Pillac, V., Gendreau, M., Guéret, C., Medaglia, A.L.: A review of dynamic vehicle routing problems. European Journal of Operational Research 225(1), 1–11 (2013)
19. Pillac, V., Guéret, C., Medaglia, A.L.: An event-driven optimization framework for dynamic vehicle routing. Decision Support Systems 54(1), 414–423 (2012)

20. Psaraftis, H.N.: A dynamic programming solution to the single vehicle many-to-many immediate request dial-a-ride problem. Transportation Science **14**(2), 130–154 (1980)
21. Psaraftis, H.N.: Dynamic vehicle routing: Status and prospects. Annals of Operations Research **61**(1), 143–164 (1995)
22. Schilde, M., Doerner, K.F., Hartl, R.F.: Metaheuristics for the dynamic stochastic dial-a-ride problem with expected return transports. Computers & Operations Research **38**(12), 1719–1730 (2011)
23. Shapiro, A., Dentcheva, D., Ruszczyński, A.P.: Lectures on stochastic programming: modeling and theory, vol. 9. SIAM (2009)
24. Shaw, P.: Using constraint programming and local search methods to solve vehicle routing problems. In: Maher, M.J., Puget, J.-F. (eds.) CP 1998. LNCS, vol. 1520, pp. 417–431. Springer, Heidelberg (1998)
25. Solomon, M.M.: Algorithms for the vehicle routing and scheduling problems with time window constraints. Operations Research **35**(2) (1987)
26. Taillard, É., Badeau, P.: A tabu search heuristic for the vehicle routing problem with soft time windows. Transportation..., pp. 1–36 (1997)
27. Van Hentenryck, P., Bent, R., Upfal, E.: Online stochastic optimization under time constraints, vol. 177 (September 2009)
28. Verweij, B., Ahmed, S., Kleywegt, A.J., Nemhauser, G., Shapiro, A.: The sample average approximation method applied to stochastic routing problems: a computational study. Computational Optimization and Applications **24**(2–3), 289–333 (2003)
29. Wilson, N.H.M., Colvin, N.J.: Computer control of the Rochester dial-a-ride system. Massachusetts Institute of Technology, Center for Transportation Studies (1977)
30. Saint-Guillain, M., Deville, Y., Solnon, C.: A Multistage Stochastic Programming Approach to the Dynamic and Stochastic VRPTW (2015). Extended version. arXiv:1502.01972 [cs.AI]

Constraint Solving on Bounded String Variables

Joseph D. Scott$^{(\boxtimes)}$, Pierre Flener, and Justin Pearson

Uppsala University, Uppsala, Sweden
{joseph.scott,pierre.flener,justin.pearson}@it.uu.se

Abstract. Constraints on strings of unknown length occur in a wide variety of real-world problems, such as test case generation, program analysis, model checking, and web security. We describe a set of constraints sufficient to model many standard benchmark problems from these fields. For strings of an unknown length bounded by an integer, we describe propagators for these constraints. Finally, we provide an experimental comparison between a state-of-the-art dedicated string solver, CP approaches utilising fixed-length string solving, and our implementation extending an off-the-shelf CP solver.

1 Introduction

Constraints on strings occur in a wide variety of problems, such as test case generation [8], program analysis [6], model checking [11], and web security [5].

As a motivating example, we consider the *symbolic execution* [7,20] of string-manipulating programs. Symbolic execution is a semantics for a programming language, wherein program variables are represented by symbols, and language operators are redefined to accept symbolic inputs and produce symbolic expressions. In symbolic execution, a program P is represented by a *control flow graph* [2]: a directed graph with nodes representing the basic blocks in P, and arcs representing possible branchings. A *path* on a control flow graph is a finite sequence of arc-connected nodes. A *symbolic state* for a path π on a program P consists of a mapping, μ, from the program variables of P to symbolic expressions, and a *path constraint*, PC, which is associated with the path π, over the symbols used in μ. Solving the path constraint results in either a set of concrete inputs that yields an execution following the path π, or, when PC is unsatisfiable, a proof that π is an infeasible path.

Example 1. In Fig. 1 is some JavaScript-like code (it is uninteresting, but small enough to illustrate our points), with a corresponding control flow graph. For the path $\pi = 1$-2-4-5-6, a corresponding path constraint may be as follows:

$$PC_\pi : y = |s| \land y \bmod 2 = 0 \land s \in \mathrm{L}((\mathrm{a}^*\mathrm{b})\mathrm{c}\backslash 1) \land x = s_{1:\,y/2} \tag{1}$$

(Notation will be introduced in Section 2 but should not be an obstacle here.) Note that PC_π is unsatisfiable: only a string of odd length can match the expression on line 4, due to the required symbol 'c' in the middle, so if the condition on that line is true, then the condition on line 2 must also be true, and therefore π is infeasible, as node 3 should be visited at least once between nodes 2 and 4.

© Springer International Publishing Switzerland 2015
L. Michel (Ed.): CPAIOR 2015, LNCS 9075, pp. 375–392, 2015.
DOI: 10.1007/978-3-319-18008-3_26

```
0  function doSomething(s) {
1     y = s.len();
2     while (y mod 2 != 0)
3        { x += s[y-1]; y = y/2; }
4     if (s.match(/^(a*b)c\1$/))
5        x = s.substr(1,s.len()/2);
6     return x;
7  }
```

Fig. 1. JavaScript-like code with a corresponding control flow graph

Example 1 helps to illuminate the type of string constraints needed for software verification purposes: the constraint language should be rich enough to model the kinds of string operators typically found in programming languages. For example, the `substr()` operation on line 5 of Fig. 1 suggests that a constraint stating "string y is a substring of x, starting from index i" would be useful. Similarly, the `match()` operation on line 4 suggests the utility of a constraint stating "string x is a member of the regular language $\mathcal{L}$". However, this second constraint is somewhat misleading, as the '\1' in the pattern on line 4 is a *back-reference*: the parentheses delineate a subexpression (in this case, 'a*b'), and the '\1' indicates that the *value* matched by that subexpression is repeated in the same string. In the absence of a bound on string length, the languages defined by expressions with back-references are not regular, but rather context sensitive.[1] Nevertheless, back-references are a common feature of regular expressions as implemented in modern programming languages, and hence are a feature we would like to model. To avoid confusion, we will write *regular expression* to refer *only* to the formal language concept, while we will write *regex* to refer to a possibly non-regular pattern allowed by a programming language.

String constraint solving has been the focus of a large amount of research in recent years. Current string constraint solving methods may be broadly classified by their treatment of string length. At one extreme are solvers for string variables of *unbounded* length, such as [1,9,11,15,17,18,23,33,36]. These solvers define the set of all satisfying strings intensionally, typically by formal languages. Constraint reasoning in these solvers generally reduces to a question of language intersection; research centres on the question of how to avoid the exponential blowup of these intersection operations. At the opposite end of the spectrum, *fixed*-length string solvers, such as [19,24], are extensional in the sense that they generate solutions individually. Fixed-length solvers are generally superior to

---

[1] In fact, for strings of *bounded* length, expressions with back-references *do* correspond to regular languages; however the size of a finite-automaton encoding grows exponentially in the size of the bound, so even in the bounded case back-references tend to result in inefficient encodings.

unbounded-length solvers for producing a single solution, but suffer comparatively when producing the set of all solutions.

In this paper, we address a problem between these two extremes: we consider string variables of *bounded* length. While bounded-length solvers exist in other fields, such as [6,29], to this point *constraint programming* (CP) models have handled bounded-length strings either by iterating over a series of fixed string lengths, or by representing a string variable as an array of variables long enough to accommodate the maximum considered string length, while allowing occurrences of a padding symbol at the end of the string (e. g., [16]). We focus instead on encoding the string length and contents directly; nevertheless, our implementation (Sect. 6) can be seen as an encapsulation of padding into a variable type. Using fixed-length strings with a padding character is appealingly simple in theory, but in practice it leads to complicated and error-prone models. Our approach allows simpler modelling, without requiring an extension to the solver's modelling language, by introducing a new structured variable type for strings. In this framework, the choice to use a padding character and the consequences of that choice are implementation details, which may be ignored during modelling.[2] We also note that bounded-length string variables ease the design of string-specific branching heuristics.

The contributions and organisation of this paper are as follows, after defining notation and terminology (Sect. 2) and outlining related work (Sect. 3):

- a formalisation of string variables and a specification of several interesting string constraints, all applicable to strings of fixed, bounded, or unbounded length (Sect. 4);
- a definition of a bounded-length string variable representation, called the *open-sequence representation*, which is directly implementable for any existing finite-domain CP solver, and propagator descriptions for the specified string constraints (Sect. 5);
- an implementation of our bounded-length string variable representation and a principled derivation of actual propagators for the specified constraints, all for the CP solver GECODE [12] (Sect. 6);
- an experimental evaluation of our implementation: despite being only a prototype, it already outperforms not only off-the-shelf fixed-length CP approaches [16], but also, by orders of magnitude, the state-of-the-art dedicated string solvers SUSHI [9] and KALUZA [29], on their benchmarks (Sect. 7).

Finally, we conclude in Sect. 8.

2   Notation and Terminology for Strings and Languages

An *alphabet* Σ is a finite set of *symbols*. A *string* s of length $|s| = n$ over an alphabet Σ is a finite sequence of n symbols of Σ, denoted $s_1 s_2 \cdots s_n$, where

---

[2] As noted elsewhere (e. g., [13]), the case for structured variable types is similar to that for global constraints: both capture commonly recurring combinatorial substructure.

$s_i \in \Sigma$ for all $1 \leq i \leq n$. For a given string s, we denote its ith symbol by s_i. We denote the *empty string*, of length 0, by ϵ. We denote the *concatenation* of strings x and y by $x \cdot y$. We say that a string y is a *substring* of a string s if there exist strings x and z such that $s = x \cdot y \cdot z$. For $1 \leq i, j \leq |s|$, we define $s_{i:\,j}$ as the substring $s_i \cdots s_j$ from the ith to the jth symbol of s; if $i > j$, then $s_{i:\,j} = \epsilon$. The *reverse* of a string $s = s_1 \cdots s_n$ is the string $s^{\mathrm{rev}} = s_n \cdots s_1$.

We denote by Σ^n the set of strings over Σ of length n. The infinite set of all strings over Σ, including ϵ, is denoted by Σ^*. A *language* over Σ is a possibly infinite subset of Σ^*. The language of a regex r is denoted by $L(r)$.

A *constraint* C of *arity* k is a pair $\langle R, S \rangle$ where R is the underlying relation on ground instances of the variable tuple $S = \langle X_1, \ldots, X_k \rangle$, called the *scope* of C. We denote the *domain* of a variable X by $\mathcal{D}(X)$.

We denote scalar variables in uppercase (e.g., N, N_1, etc. for integers, and A for a symbol of a finite alphabet) and string variables (to be introduced in Sect. 4) in boldface uppercase (e.g., $\mathbf{S}$). We denote sets in script (e.g., $\mathcal{A}$, $\mathcal{B}$, etc.), and write $|\mathcal{A}|$ for the cardinality of a set $\mathcal{A}$. We refer to the set of integers $\{\ell, \ell+1, \ldots, u-1, u\}$, which is the empty set $\emptyset$ if $\ell > u$, using the notation $[\ell, u]$. We use angled brackets to denote an ordered sequence, or tuple, $\langle a_1, \ldots, a_n \rangle$.

3   Related Work

We distinguish between string variables of fixed, unbounded, and bounded length.

Fixed-Length String Variables. In CP, a fixed-length string variable has a natural representation as an array of scalar variables that may be acted upon by a wide variety of constraints. Of particular interest for solving string problems are constraints for membership in regular [4,26] and context-free [27,32] languages. For example, the propagator in [26] for regular language membership works by maintaining a *layered graph*: each layer replicates the nodes of a finite automaton representing the regular language, but with each transition connecting to a node in the next layer. The labels of the arcs between nodes in two consecutive layers determine the feasible values for the corresponding variable in the string, and propagation works by removing arcs not on a path between the start node in the first layer and any accepting node in the last layer. Fixed-length bit-vector variables have also been explored [24].

HAMPI [19] provides a theory of fixed-length strings for *satisfaction modulo theories* (SMT) solvers, using the bit-vector solver STP [10]. HAMPI handles constraints of membership in both regular and context-free languages. For a set of such constraints, on a *single* fixed-length string variable, HAMPI either returns one satisfying string, or reports that the constraints are unsatisfiable.

Unbounded-Length String Variables. At the other extreme are solvers for string variables of unbounded length. An example of this approach in CP is [15], in which the *regular domain* of a string variable is defined by a regular language.

A regular domain is represented as a finite automaton accepting that language, and propagation of a constraint over string variables is achieved by computing a series of automaton operations, such as intersection or negation. The expressivity of regular domains is balanced out by relatively expensive propagation: several of the presented propagators take time quadratic in the size of the automata, and the size of the automata themselves may grow exponentially with the number of constraints. It is not surprising that performing propagation on string variables of unbounded length by computing on a set of strings is expensive. The equivalent for integer domains would be propagation over multiple integer variables through computation on value tuples, which is not generally reasonable. Constraints of regular language membership are, of course, trivially enforceable on the regular domain, although extension to context-free languages is impractical.

A decision procedure for Boolean combinations of equalities on unbounded-length string variables, called word equations, is provided in [23]. A *word equation* [21] is a constraint such as $x \oplus y = z$, where $\oplus$ is a string operator and x, y, z are string variables. Word equations are not decidable in the general case, and their decidability in conjunction with other constraints, including length constraints, remains open [6]. Nonetheless, for fragments of the logic of word equations with constraints on length or regular or context-free language membership, there exist several decision procedures. For example, SUSHI [9] handles a restricted fragment of word equations called *simple linear string equations* (SLSE); in essence, these are word equations in which no string variable appears more than once, and string variables occur only on the left-hand side. SUSHI allows concatenation, substring, regular membership, and regular replacement. Other solvers handling weak fragments include the stand-alone solver DPRLE [17], which handles only language subset and language concatenation constraints for regular languages, and Z3-STR [36], an extension for the SMT solver Z3 that provides a theory of word equations with length constraints, but does not include language membership. The algorithm in NORN [1] is sound for the complete logic of word equations with both length and regular language membership constraints, and is a decision procedure for a restricted fragment (strictly stronger than that of SUSHI). Also, S3 [33] improves the Z3-STR solver and adds a procedure for unfolding unbounded repetitions in regular expressions.

Another line of work has focused on avoiding the exponential blowup encountered in language intersection. Both REVENANT [11] and NORN utilise interpolation, albeit in different contexts, while STRSOLVE [18] handles automaton intersection operations by lazily constructing cross-products.

Also worth noting is that automaton-based approaches do not allow a natural handling of length constraints. The latter may be directly encoded as automata (e.g., [35]); however, this results in only a weak connection between string lengths and other numerical constraints. Solvers that combine automata and numerical reasoning [1,14,28] strengthen this connection to varying degrees.

Bounded-Length String Variables. Less work has been done on bounded-length string solvers. Probably the best known solver in this category is KALUZA [29], which solves constraints in two stages. First, an SMT solver is used to find possible lengths for strings that satisfy explicit length constraints, length constraints implied by the string constraints of the problem, and any other integer constraints present in the model. Second, these lengths are applied to create a fixed-length bit-vector problem, solved with STP [10] in the same manner as by HAMPI. If the problem in the second stage is unsatisfiable, then the first stage is repeated, with the addition of new constraints to avoid previously tried lengths. Further, a stand-alone bit-vector solver for bounded-length strings is described in [6]; however, it does not handle regular language membership, and no information is propagated from numerical constraints to the string variables.

In CP, propagation of constraints for bounded-length sequences of variables is described in [22], which treats *open* global constraints. A constraint is *global* if the cardinality of its scope is not determined a priori. In a *closed* global constraint, the cardinality of the scope is determined by the model, and remains constant throughout the solution process; however, in an *open* global constraint, the cardinality of the scope is determined as the solution process progresses [3]. In [22], the scope of an open global constraint is a sequence of scalar variables with a length that that has an upper *bound* that is an integer variable. During the solving process, scalar variables are *added* to the end (never the beginning) of the sequence, in connection with changes to the bounds of the length. We here take inspiration from that work, particularly in regard to propagation for constraints of regular and context-free language membership; nevertheless, the two approaches are essentially orthogonal.

In [31], we introduced a representation for bounded-length string variables by prefix-suffix pairs, and we designed propagators for this representation in an *ad hoc* way, testing them only on a *home-made* benchmark. We here introduce a *much simpler* representation, leading to propagators that are *different* and achieve an *incomparable* level of consistency, show how to derive such propagators in a *principled* way, and test them on *third-party* standard benchmarks.

4   String Variables and String Constraints

In a model of a constraint problem, we refer to unknown strings over a finite alphabet Σ as *string variables*. The most precise representation of the domain of a string variable is a subset of Σ^*; in other words, such a domain of a string variable is the language of all strings that are not (yet) known to be infeasible. Operations on this representation are expensive [15], making the representation unsuitable for propagation. We use this representation as an ideal *starting* point, suitable for strings of fixed, bounded, or unbounded length. We introduce in Sect. 5 a representation more suited to propagation.

We divide constraints involving string variables into three groups: pure string constraints, mixed string constraints, and language membership constraints.

Pure String Constraints. We refer to constraints involving only string variables as *pure string constraints*.

The constraint $\text{EQ}(\mathbf{X}, \mathbf{Y})$ holds if string variables $\mathbf{X}$ and $\mathbf{Y}$ are equal, where equality for strings means that they have equal length and the same symbol at each index. The underlying relation $\mathcal{E}$ is $\{\langle x, y \rangle \,|\, |x| = |y| \wedge \forall i \in [1, |x|] : x_i = y_i\}$.

The constraint $\text{NEQ}(\mathbf{X}, \mathbf{Y})$ holds if string variables $\mathbf{X}$ and $\mathbf{Y}$ are not equal, where inequality for two strings holds if the strings have different lengths, or if there exists an index for which the two strings have a different symbol. The underlying relation is $\{\langle x, y \rangle \,|\, |x| \neq |y| \vee \exists i \in [1, |x|] : x_i \neq y_i\}$.

The constraint $\text{REV}(\mathbf{X}, \mathbf{Y})$, for string variables $\mathbf{X}$ and $\mathbf{Y}$, holds if $\mathbf{X}$ is equal to the reverse of $\mathbf{Y}$. The underlying relation is $\{\langle x, y \rangle \,|\, x = y^{\text{rev}}\}$.

The constraint $\text{CAT}(\mathbf{X}, \mathbf{Y}, \mathbf{Z})$ holds if string variable $\mathbf{Z}$ is the concatenation of string variables $\mathbf{X}$ and $\mathbf{Y}$. The underlying relation is $\{\langle x, y, z \rangle \,|\, z = x \cdot y\}$.

Mixed String Constraints. We refer to constraints involving at least one string variable and at least one non-string variable as *mixed string constraints*.

The constraint $\text{SUB}(\mathbf{X}, \mathbf{Y}, N)$ holds if string variable $\mathbf{Y}$ is a contiguous substring of string variable $\mathbf{X}$, starting at the index given by the integer variable N. The underlying relation is $\{\langle x, y, n \rangle \,|\, y = x_{n:\, n+|y|-1}\}$.

For the special case of SUB in which $\mathbf{Y}$ has a fixed length of one (i.e., where $\mathbf{Y}$ can be replaced by a scalar variable), we instead propose $\text{CHAR}(\mathbf{X}, A, N)$, whose underlying relation is $\{\langle x, a, n \rangle \,|\, x_n = a\}$.

The constraint $\text{LEN}(\mathbf{X}, N)$ holds if the string variable $\mathbf{X}$ has a length equal to the integer variable N. The underlying relation is $\{\langle x, n \rangle \,|\, n = |x|\}$.

Language Membership Constraints. Conceptually, a constraint that holds if a string variable $\mathbf{X}$ is a member of a given formal language $\mathcal{L}$ may be viewed as a unary constraint on $\mathbf{X}$, parameterised by $\mathcal{L}$, irrespective of the class of $\mathcal{L}$. In practice, propagators for such a *language membership constraint* are specific to the language class, so it is common to name such constraints by the language class. For a language $\mathcal{L}$, we have the constraints $\text{REGULAR}(\mathbf{X}, \mathcal{L})$, if $\mathcal{L}$ is regular, and $\text{CONTEXTFREE}(\mathbf{X}, \mathcal{L})$, if $\mathcal{L}$ is context-free, with $\mathcal{L}$ as the underlying relation.

Example 2. Consider once again the path $\pi = 1\text{-}2\text{-}4\text{-}5\text{-}6$ of Example 1. We can now express its path constraint PC_π in (1) using the string constraints defined in this section, along with some primitive numerical constraints:

$$\text{LEN}(\mathbf{S}, Y) \wedge \text{MOD}(Y, 2, 0) \wedge \text{REGULAR}(\mathbf{S}_1, \text{L}(a^*b)) \wedge$$
$$\text{CAT}(\text{``c''}, \mathbf{S}_1, \mathbf{S}_2) \wedge \text{CAT}(\mathbf{S}_1, \mathbf{S}_2, \mathbf{S}) \wedge \text{DIV}(Y, 2, Z) \wedge \text{LEN}(\mathbf{X}, Z) \wedge \text{SUB}(\mathbf{S}, \mathbf{X}, 1)$$

Note the use of CAT in eliminating the back-reference in the regex `/^(a*b)c\1$/`.

5 Open-Sequence Representation and Propagation

As previously noted, a language is a natural representation of the domain of a string variable, but the complexity of computation over languages makes this representation unsuitable for propagation. As a more practical representation, we consider an over-approximation of a finite set of strings, upon which we describe propagators for the constraints defined in Sect. 4.

Open-Sequence Representation. Inspired by [22], we now introduce a string variable representation, called the *open-sequence representation*: $\langle \mathcal{A}, \mathcal{N} \rangle$ consists of a sequence $\mathcal{A} = \langle \mathcal{A}_1, \ldots, \mathcal{A}_m \rangle$ of sets over the same alphabet, and a set $\mathcal{N}$ of natural numbers, representing the possible lengths of the string, with $\max(\mathcal{N}) \leq m$. An $\langle \mathcal{A}, \mathcal{N} \rangle$ pair corresponds to the set of all strings that have a length $\ell \in \mathcal{N}$ and are constructed by selecting a symbol from $\mathcal{A}_i$ at each index $i \in [1, \ell]$; in other words, the domain of a string variable represented by $\langle \mathcal{A}, \mathcal{N} \rangle$ is given by:

$$\mathcal{D}(\langle \mathcal{A}, \mathcal{N} \rangle) = \bigcup_{\ell \in \mathcal{N}} \{ s \in \Sigma^\ell \mid \forall i \in [1, \ell] : s_i \in \mathcal{A}_i \} \tag{2}$$

Intuitively, (2) shows that if any $\mathcal{A}_i$ is empty, then $\mathcal{D}(\langle \mathcal{A}, \mathcal{N} \rangle)$ contains no strings of length at least i. This insight leads to the following representation invariant:

$$\forall i \in [1, m] : \mathcal{A}_i = \emptyset \iff \max(\mathcal{N}) < i \tag{3}$$

Note that we are purposefully general in this section: in Sect. 6, we discuss *a* possible implementation of the open-sequence representation, namely by treating the value sets as the domains of scalar variables. However, the open-sequence representation could *also* be implemented for a CP solver as a new first-class string variable type. In this section, we consider an $\langle \mathcal{A}, \mathcal{N} \rangle$ pair as the representation of a string variable, without regard to the choice of implementation.

Open-Sequence Propagators. We now describe, for some representative constraints specified in Sect. 4, the strongest pruning that may be achieved by a propagator implementing that constraint for string variables that are *all* in the open-sequence representation. In Sect. 6, we will use these propagator descriptions as the basis for automatically generating an implementation of the pure and mixed string constraints.

We give the following propagation descriptions, which specify exactly what pruning can be achieved in the open sequence representation. It is possible to derive the propagator descriptions in a principled manner, using a methodology that has been omitted from this paper for reasons of space and orthogonality.

The propagator for EQ performs set intersections between $\mathcal{A}^x$ and $\mathcal{A}^y$:

$$\text{EQ}^{\text{P}} \left(\langle \mathcal{A}^x, \mathcal{N}^x \rangle, \langle \mathcal{A}^y, \mathcal{N}^y \rangle \right)$$
$$= \left\langle \begin{array}{l} \left\langle \left\langle \mathcal{A}_1^x \cap \mathcal{A}_1^y, \ldots, \mathcal{A}_{\max(\mathcal{N}^x \cap \mathcal{N}^y)}^x \cap \mathcal{A}_{\max(\mathcal{N}^x \cap \mathcal{N}^y)}^y, \emptyset, \ldots, \emptyset \right\rangle, \mathcal{N}^x \cap \mathcal{N}^y \right\rangle, \\ \left\langle \left\langle \mathcal{A}_1^x \cap \mathcal{A}_1^y, \ldots, \mathcal{A}_{\max(\mathcal{N}^x \cap \mathcal{N}^y)}^x \cap \mathcal{A}_{\max(\mathcal{N}^x \cap \mathcal{N}^y)}^y, \emptyset, \ldots, \emptyset \right\rangle, \mathcal{N}^x \cap \mathcal{N}^y \right\rangle \end{array} \right\rangle$$

Note that EQ^{P} does not enforce the representation invariant (3): a separation between the invariant and the propagators presented in this section significantly simplifies design, at the level of both theory and implementation.

Example 3. If $\mathbf{X}$ and $\mathbf{Y}$ are string variables with open-sequence representations $\mathbf{X} = \langle \langle [1, 3], \{3\}, [1, 3], \emptyset, \ldots \rangle, [2, 3] \rangle$ and $\mathbf{Y} = \langle \langle [1, 2], [1, 2], [1, 2], \emptyset, \ldots \rangle, [1, 3] \rangle$, then propagation by EQ^{P} yields $\mathbf{X}' = \mathbf{Y}' = \langle \langle [1, 2], \emptyset, [1, 2], \emptyset, \ldots \rangle, [2, 3] \rangle$. The invariant (3) reveals that neither string has a feasible length, resulting in failure.

The propagator for CAT is similar to Eq^{P} in regards to the relationship between $\mathbf{\mathcal{A}}^x$ and $\mathbf{\mathcal{A}}^z$, but the relationship between $\mathbf{\mathcal{A}}^y$ and $\mathbf{\mathcal{A}}^z$ is complicated by a dependency on $\mathcal{N}^x$:

$$\mathrm{CAT}^{\mathrm{P}}(\langle\mathbf{\mathcal{A}}^x,\mathcal{N}^x\rangle,\langle\mathbf{\mathcal{A}}^y,\mathcal{N}^y\rangle,\langle\mathbf{\mathcal{A}}^z,\mathcal{N}^z\rangle)$$
$$=\left\langle\langle\langle\mathcal{A}_1^{x\prime},\ldots,\mathcal{A}_m^{x\prime}\rangle,\mathcal{N}^{x\prime}\rangle,\langle\langle\mathcal{A}_1^{y\prime},\ldots,\mathcal{A}_m^{y\prime}\rangle,\mathcal{N}^{y\prime}\rangle,\langle\langle\mathcal{A}_1^{z\prime},\ldots,\mathcal{A}_m^{z\prime}\rangle,\mathcal{N}^{z\prime}\rangle\right\rangle$$

where

$$\mathcal{N}^{x\prime}=\mathcal{N}^x\cap[\min(\mathcal{N}^z)-\max(\mathcal{N}^y),\max(\mathcal{N}^z)-\min(\mathcal{N}^y)]$$
$$\mathcal{N}^{y\prime}=\mathcal{N}^y\cap[\min(\mathcal{N}^z)-\max(\mathcal{N}^x),\max(\mathcal{N}^z)-\min(\mathcal{N}^y)]$$
$$\mathcal{N}^{z\prime}=\mathcal{N}^z\cap[\min(\mathcal{N}^x)+\min(\mathcal{N}^y),\max(\mathcal{N}^x)+\max(\mathcal{N}^y)]$$

and

$$\mathcal{A}_i^{x\prime}=\begin{cases}\mathcal{A}_i^x\cap\mathcal{A}_i^z & \text{if }i<\max(\mathcal{N}^{x\prime})\\\emptyset & \text{otherwise}\end{cases}\qquad\mathcal{A}_i^{y\prime}=\begin{cases}\mathcal{A}_i^y\cap\bigcup_{j\in\mathcal{N}^x\cap[1,\max(\mathcal{N}^{z\prime})-i]}\mathcal{A}_{i+j}^z & \text{if }i<\max(\mathcal{N}^{y\prime})\\\emptyset & \text{otherwise}\end{cases}$$

$$\mathcal{A}_i^{z\prime}=\begin{cases}\mathcal{A}_i^z\cap\mathcal{A}_i^x & \text{if }i<\min(\mathcal{N}^{x\prime})\\[2mm]\mathcal{A}_i^z\cap\mathcal{A}_i^x\cap\bigcup_{j\in\mathcal{N}^{x\prime}\cap[i-\max(\mathcal{N}^{y\prime}),i]}\mathcal{A}_{i-j}^y & \text{if }\min(\mathcal{N}^{x\prime})\le i<\max(\mathcal{N}^{x\prime})\\[2mm]\mathcal{A}_i^z\cap\bigcup_{j\in\mathcal{N}^{x\prime}\cap[i-\max(\mathcal{N}^{y\prime}),i]}\mathcal{A}_{i-j}^y & \text{if }\max(\mathcal{N}^{x\prime})\le i<\max(\mathcal{N}^{z\prime})\\[2mm]\emptyset & \text{if }\max(\mathcal{N}^{z\prime})\le i\end{cases}$$

Propagation of LEN on the open-sequence representation is trivial:

$$\mathrm{LEN}^{\mathrm{P}}(\langle\mathbf{\mathcal{A}},\mathcal{N}\rangle,\mathcal{S})=\left\langle\langle\langle\mathcal{A}_1,\ldots,\mathcal{A}_{\max(\mathcal{N}\cap\mathcal{S})},\emptyset,\ldots,\emptyset\rangle,\mathcal{N}\cap\mathcal{S}\rangle,\mathcal{N}\cap\mathcal{S}\right\rangle$$

It is also easy to express the desired propagation for the REGULAR constraint, although the description in this case is of little help in regards to efficient implementation (see Sect. 6).

$$\mathrm{REGULAR}^{\mathrm{P}}(\langle\mathbf{\mathcal{A}},\mathcal{N}\rangle,\mathcal{L})=\left\langle\langle\{s_1'\in\mathcal{A}_1\mid s'\in\mathcal{L}\},\ldots,\{s_{\max(\mathcal{N})}'\in\mathcal{A}_{\max(\mathcal{N})}\mid s'\in\mathcal{L}\},\emptyset,\ldots,\emptyset\rangle,\{\ell\in\mathcal{N}\}\right\rangle$$

Propagators for the remaining constraints from Sect. 4 are omitted for reasons of space; all may be described similarly to the propagators detailed above.

6   Implementation

While the open-sequence representation described in the previous section could be implemented as a new variable type for a CP solver, the correspondence between sets of feasible values and the domains of scalar variables suggests another method of implementing the open-sequence representation, namely as an aggregation of two components: an array of scalar variables over the alphabet of the string, and an integer variable for the length of the string. Without loss of generality, we focus on strings of integers.

This implementation is similar to [22], which also involves a sequence of scalar variables that may be extended at the end but not at the beginning, and an integer variable that determines the length of that sequence. Beyond that similarity, our treatment diverges significantly. The open constraints described in [22] rely on the existence of a meta-programming framework to dynamically add variables to the model during search. In contrast, we extend a (closed) CP solver by adding a variable type representing a bounded-length sequence, eliminating the need for meta-programming and maintaining the declarative nature of CP solving. Unlike [22], we have no concept of *adding* a variable to the sequence: our implementation uses a fixed sequence of scalar variables, each of which may or may not participate in a solution as determined by the length variable. Additionally, we choose to treat each sequence-length pair as a single bounded-length sequence variable; whereas in [22] OPENREGULAR is defined as a global constraint of bounded arity, in our treatment REGULAR is a unary (non-global) constraint on a string variable of bounded length. This choice allows us to define constraints conventionally as relations over tuples (constraint semantics in [22] are described using formal languages), and eases the presentation of n-ary constraints on sequence variables (constraints in [22] involve only one sequence).

After discussing the technical challenges to such an aggregate implementation, we show how to derive actual propagators in a principled way, both from the underlying relations of the constraints in Sect. 4 and from the propagator descriptions in Sect. 5.

Aggregate Implementation. The open-sequence representation $\langle \mathcal{A}, \mathcal{N} \rangle$ for a string of integers is here implemented as an array of integer variables $\boldsymbol{N} = \langle N_1, \ldots, N_m \rangle$ representing $\boldsymbol{\mathcal{A}} = \langle \mathcal{A}_1, \ldots, \mathcal{A}_m \rangle$, and an integer variable N representing $\mathcal{N}$.

In regards to consistency level, the length variable and the sequence variables seem to have different requirements. For the length the most interesting values are the bounds (i.e., the lengths of the shortest and longest feasible strings). It seems unlikely, however, that maintaining bounds consistency on the variables of $\boldsymbol{N}$ is useful, as the set of feasible symbols at any index will rarely form a meaningful interval. We therefore choose to maintain a mixed consistency level, which considers the upper and lower bounds of N, and all domain values of the variables of $\boldsymbol{N}$; other choices are certainly possible.

For correctness, the representation invariant (3) must be enforced for each $\langle \boldsymbol{N}, N \rangle$ pair. Some care needs to be taken in the interpretation of (3), however: while $\mathcal{A}_i = \emptyset$ merely means $\max(\mathcal{N}) < i$, we have that $\mathcal{D}(N_i) = \emptyset$ leads to a failed search node. One way to avoid this is to include a reserved character, NULL, in the domains of all variables of $\boldsymbol{N}$. The representation invariant may then be enforced by propagating the following conjunction of reified constraints:

$$\forall i \in [1, m] : N_i = \text{NULL} \iff N < i \tag{4}$$

```
1 def EQ(vint[] X, vint LenX, vint[] Y, vint LenY){
2   checker{
3     (val(LenX) == val(LenY)) and
4     and(i in (min(rng(X)) .. ((min(rng(X)) + val(LenX)) + -1)))
5       (val(X[i]) == val(Y[i]))
6   }
```

Fig. 2. Checker for the string equality constraint $\mathrm{EQ}(\mathbf{X}, \mathbf{Y})$

Pure and Mixed String Constraints. An interesting feature of the propagator descriptions in Sect. 5 for pure and mixed string constraints is that they consist solely of a conjunction of range restriction operations. When applied to a variable domain, such an operation is called an *indexical* [34]: it is of the form $X \in \sigma$ and restricts the domain of the variable X to its intersection with the interval σ. An *indexical language* is a high-level solver-independent language for writing a propagator description with indexicals. The extended indexical language of [25] includes arrays and n-ary operations, and its system includes the following two automated transformations:

- A solver-independent *synthesis* of an indexical description of a propagator from a ground checker of its constraint.[3]
- A solver-specific *code generation* of an actual propagator from an indexical description thereof.

Following our ideas in [30], applied there to our more complex representation of bounded-length string variables in [31], we use this system to generate automatically a prototype implementation of the pure and mixed string constraints of Sect. 4 for GECODE [12].

We illustrate this process using the string equality constraint $\mathrm{EQ}(\mathbf{X}, \mathbf{Y})$. Its underlying relation $\mathcal{E}$ from Sect. 4

$$\mathcal{E} = \{\langle x, y \rangle \mid x = y\} = \{\langle x, y \rangle \mid |x| = |y| \wedge \forall i \in [1, |x|] : x_i = y_i\} \qquad (5)$$

can be seen as a ground checker for the constraint. We first replace the string variables $\mathbf{X}$ and $\mathbf{Y}$ with the pairs $\langle \mathcal{A}^x, \mathcal{N}^x \rangle$ and $\langle \mathcal{A}^y, \mathcal{N}^y \rangle$, respectively, as in Sect. 5. We then manually translate $\mathcal{E}$ into the checker sub-language of the extended indexical language, yielding Fig. 2. For our purposes, it suffices to illuminate a few less obvious features of the syntax. The aggregate variable $\langle \mathcal{A}^x, \mathcal{N}^x \rangle$ is represented by two variables: an integer variable for the length (vint LenX) and an array of integer variables for the string (vint[] X). One constraint is on the lengths of the two strings (line 3). Another constraint is on the contents of the arrays (lines 4 and 5): it is expressed as an n-ary conjunction of equality constraints, corresponding to the universal quantification in (5).

From this checker, automatic synthesis yields an indexical description of a propagator for EQ, given in Fig. 3. Compared with the hand-derived propagator description EQ^{P} in Sect. 5, we note that while the synthesised propagator

---

[3] This synthesiser is not mentioned in [25], but described in a paper under preparation.

```
1 def EQ(vint[] X, vint LenX, vint[] Y, vint LenY){
2   propagator(gen)::DR{
3     LenX in dom(LenY);
4     LenY in dom(LenX);
5     forall(i in (min(rng(X)) .. ((min(rng(X)) + min(LenX)) + -1))){
6       X[i] in dom(Y[i]);
7       Y[i] in dom(X[i]);
8 }}}
```

Fig. 3. Synthesised indexical description of a propagator for EQ($\mathbf{X}, \mathbf{Y}$)

correctly filters values in the arrays X and Y at indices below the current minimum length, it misses some propagation on LenX and LenY. Intuitively, if the intersection of the domains of X[i] and Y[i] is empty, then all feasible strings in $\mathbf{X}$ and $\mathbf{Y}$ must be shorter than i. This additional reasoning is expressed with the following forall construct that can be added to Fig. 3:

```
forall(i in ((min(rng(X)) + min(LenX)) .. (min(rng(X)) + max(LenX))))
    {(dom(X[i]) inter dom(Y[i])) == emptyset -> LenX in inf .. (i - 1);}
```

Automatic code generation from the extended version of Fig. 3 results in a C++ implementation of an EQ propagator for GECODE.

Alternatively, one can hand-code an EQ propagator for GECODE directly from the mathematical description that was derived in Sect. 5. This is much more labour-intensive and error-prone than the tool-assisted approach. Hence, the implementation we evaluate in Sect. 7 started as code generated by the indexical compiler; however, portions have been modified for efficiency reasons.

Language Membership Constraints. Indexicals are no help when it comes to language membership constraints, because propagators for those constraints rely on internal data structures. However, there are propagators for open constraints of language membership.

The REGULAR propagator of [26] is extended in [22] to handle bounded-length sequences. Propagation proceeds by dynamically increasing the number of layers in the layered automaton as the minimum feasible length of the string increases. We implemented this bounded-length extension of REGULAR in GECODE. Bounded-length propagators for the GCC and CONTEXTFREE constraints are also described in [22]; the addition of these constraints to our implementation has been left to future work.

7 Experimental Results

We compare our *bounded*-length CP implementation of the open representation,[4] called 'open' below, against *fixed*-length CP models and against state-of-the-art string solvers, on benchmarks provided by the latter. It outperforms the

---

[4] It is available at https://github.com/jossco/gecode-string

Table 1. Runtimes in seconds (fastest in **bold**) for SUSHI word equations

	$n = 37$			$n = 50$			$n = 100$		
	open	pad	SUSHI	open	pad	SUSHI	open	pad	SUSHI
Eq. 1	**0.02**	0.05	0.30	**0.02**	0.07	1.11	**0.09**	0.24	2.56
Eq. 2	**<0.01**	**<0.01**	0.37	**<0.01**	**<0.01**	0.88	**0.01**	0.02	19.24
Eq. 3	**0.01**	0.03	0.29	**0.01**	0.03	0.64	**0.02**	0.09	1.14
Eq. 4	**<0.01**	0.01	42.16	**<0.01**	0.03	>300	**0.06**	0.07	>300
Eq. 5	**<0.01**	**<0.01**	1.56	**<0.01**	**<0.01**	2.93	**<0.01**	0.02	6.37

implementation of our previous representation of bounded string variables [31]. In each experiment, all CP models used the same upper bound for string length.

Benchmark of SUSHI. SUSHI [9] is a word equation solver for *unbounded*-length string variables (see Sect. 3). Being automaton-based, it computes the entire solution set in one go, rather than seeking solutions one by one. Nevertheless, the applicability of SUSHI as a satisfiability solver for string constraints is considered in [9]: SUSHI is compared to the *bounded*-length string solver KALUZA [29] (see Sect. 3) on a benchmark of five satisfiable word equations, each parameterised by a natural number n. To solve an equation with KALUZA, a bounded-length version of the equation is created for each n.

Example 4. We can model SUSHI word equation 1, namely $x \cdot a^n = (a|b)^{2n}$, as follows: $\text{CAT}(\mathbf{X}, \mathbf{Y}, \mathbf{Z}) \wedge \text{REGULAR}(\mathbf{Y}, \text{L}(a^n)) \wedge \text{REGULAR}(\mathbf{Z}, \text{L}((a|b)^{2n}))$.

Using the CP models in [16] for the five word equations, we also test against two *fixed*-length CP approaches. In the first, the string lengths are fixed at a pessimistically large upper bound [16], and multiple occurrences of a padding symbol are allowed at the end of each string [16]. In the second (not tried in [16]), the string lengths are initially fixed to a lower bound and a set of satisfying strings is sought; upon unsatisfiability, the lengths are increased lexicographically and search is restarted. These models use the REGULAR constraint, and concatenation is modelled with reified channelling constraints [16]. The padding approach, called 'pad' below, is always faster, so we omit results on the iterative approach. Our *bounded*-length CP models use the OPENREGULAR propagator [22] and our indexical-based CAT propagator of Sect. 6. For all CP models, we use the same deterministic first-fail search heuristic, and stop at the first solution, with a time-out of 300 seconds. The tests were run on a 2.66 GHz Intel Core 2 Duo with 4 GB of RAM, on VirtualBox 4.3.10 (the recommended way to run SUSHI) running Ubuntu 10.04 on 1 GB of RAM, using SUSHI 2.0 and GECODE 4.3.2.

In Table 1 we give runtimes for all five word equations. We compare the CP approaches only against SUSHI; experimental results reported in [9] (and replicated in [16]) show that SUSHI typically outperforms KALUZA, often significantly, and we do not attempt to replicate those results here. Our smallest instance size, $n = 37$, is the largest size tried in [9,16]. Even for $n \in \{50, 100\}$ the benchmark

Table 2. Runtimes (in seconds) and backtracks (best in **bold**) for KALUZA instances

	GECODE(open)		GECODE(pad)		KALUZA
instance name	runtime	backtracks	runtime	backtracks	runtime
concat	**0.003**	**0**	0.008	0	0.088
indexof	**0.003**	**0**	0.010	0	1.560
bettermatch1	**0.002**	**0**	0.005	0	0.223
bettermatch2	**0.003**	**0**	**0.003**	0	0.192
streq	**0.003**	**0**	0.006	0	0.077
replace	**0.006**	**30**	0.019	30	0.364

turns out to be trivial for *all* CP approaches, outperforming the state-of-the-art SUSHI solver by up to three orders of magnitude, as already observed in [16] for $n = 37$. On all instances, our bounded-length prototype implementation results in the same search tree as the fixed-length padding CP approach of [16], but with a lower runtime.

It turns out that all instances run *without* backtracks in both CP approaches! The reason is that the underlying constraint graph (with variables as vertices and constraints as hyper-arcs) is Berge-acyclic, so that domain consistency on the entire model is achieved by maintaining domain consistency on each constraint: this follows from the definition of the SLSE fragment, as one can observe in Example 4. We argue that a CP model preserves problem structure that is lost by KALUZA when translating to a bit-vector representation, and that knowledge of the complexity results of CP applicable to high-level models could have prevented the creation of the SUSHI word equation benchmark in the first place.

We thus look now at another third-party benchmark (which we did not try in [31]), also in order to see if our bounded-length prototype implementation can outperform the fixed-length CP padding approach of [16] by a larger margin.

Benchmark of KALUZA. The bounded-length string solver KALUZA [29] (see Sect. 3) includes over 50,000 instances that were generated for the symbolic execution of JavaScript, based on real-world Ajax web applications. Unfortunately, they all turn out to be trivial for CP approaches, with runtimes below 0.01 seconds, and even the KALUZA runtimes are below half a second. Hence this extensive benchmark is also not particularly interesting. In order not to be biased by hand-picking among the 50,000 instances, we pick all the 14 instances that are in the KALUZA code. It turns out that KALUZA gives erroneous results or crashes on several of these instances, as reported also by [36]. The results on the remaining instances are in Table 2; note that KALUZA does not report backtracks (incomparable in any case to those of CP approaches), and that Z3-STR [36] can only be applied to REGULAR-free versions of the actual instances.

Once again, the state-of-the-art solver is beaten, but the difference between the CP models with bounded-length string variables (open) and padded fixed-length string variables (pad) is small: we address this issue in the conclusion. We are not aware of a hard third-party benchmark for string variables.

8   Conclusion

We have formalised string variables and specified several interesting string constraints, all applicable to strings of fixed, bounded, or unbounded length. We have defined a bounded-length string variable representation, called the *open-sequence representation*, which is directly implementable for any existing CP solver, and we have given propagator descriptions for the specified string constraints. We have implemented the open-sequence representation and derived in a principled way actual propagators for the specified constraints, for the CP solver GECODE. Despite being only a prototype, our implementation already outperforms not only off-the-shelf fixed-length CP approaches, but also, by orders of magnitude, state-of-the-art dedicated string solvers, on their own benchmarks.

The experimental *time* comparison of our advocated CP approach of bounded-length string variables against the existing CP approach of padded fixed-length string variables has shown only minor speed-ups on the third-party benchmarks. In retrospect, this is not so surprising, as propagation is similar, witness the *backtrack* counts in Table 2 and the zero backtracks behind Table 1, and as those benchmarks seem not to exercise the string length reasoning that could give an advantage to our approach. The invariant (4) connecting the length of strings N_i and the length variable N can be seen as an implementation, via reification, of padding, thus it is unlikely that the bounded-length representation will perform more propagation than using padding symbols, unless non-trivial reasoning is required on string lengths. Also, at the *modelling* level, we argue that it is much easier to model a bounded-length string problem without using padding symbols, since encoding such a problem as a fixed-length one is both labour-intensive and error-prone: by designing the required propagators once and for all, we allow modellers to save the encoding effort and risk. Indeed, in [16] the automaton representation had to be modified to include the padding symbol, adding an extra level of complexity to the modelling. Since our bounded-length approach subsumes the fixed-length one, it suffices to fix the length instead of bounding it when one has a fixed-length string variable. Future work consists of strengthening our length reasoning, implementing our open-sequence representation as a first-class string variable type, and adapting our propagators.

We argue that CP is well-suited for string variables and constraints: unlike for many non-CP solvers mentioned here, there is no difficulty in upgrading from a single string variable to multiple ones, possibly with shared element variables, in having both string and numeric variables in a model, or in handling numeric variables and constraints without unnatural encodings. Indeed, it suffices to extend any CP solver, coming with existing numeric variables and numeric or symbolic constraints, in plug-and-play fashion, by adding the new type of string variables and providing propagators for the new constraints. This may result in high-level models that preserve problem structure and are amenable to faster solving than by lower-level encodings in ad hoc solvers.

Acknowledgments. This work is supported by grants 2009-4384, 2011-6133, and 2012-4908 of VR, the Swedish Research Council. We thank J.-N. Monette.

References

1. Abdulla, P.A., Atig, M.F., Chen, Y.-F., Holík, L., Rezine, A., Rümmer, P., Stenman, J.: String constraints for verification. In: Biere, A., Bloem, R. (eds.) CAV 2014. LNCS, vol. 8559, pp. 150–166. Springer, Heidelberg (2014)
2. Allen, F.E.: Control flow analysis. ACM Sigplan Notices 5(7), 1–19 (1970)
3. Barták, R.: Dynamic global constraints in backtracking based environments. Annals of Operations Research 118(1), 101–119 (2003)
4. Beldiceanu, N., Carlsson, M., Petit, T.: Deriving filtering algorithms from constraint checkers. In: Wallace, M. (ed.) CP 2004. LNCS, vol. 3258, pp. 107–122. Springer, Heidelberg (2004)
5. Bisht, P., Hinrichs, T., Skrupsky, N., Venkatakrishnan, V.N.: WAPTEC: white-box analysis of web applications for parameter tampering exploit construction. In: Chen, Y., Danezis, G., Shmatikov, V. (eds.) Computer and Communications Security (CCS 2011), pp. 575–586. ACM (2011)
6. Bjørner, N., Tillmann, N., Voronkov, A.: Path feasibility analysis for string-manipulating programs. In: Kowalewski, S., Philippou, A. (eds.) TACAS 2009. LNCS, vol. 5505, pp. 307–321. Springer, Heidelberg (2009)
7. Clarke, L.A.: A system to generate test data and symbolically execute programs. IEEE Transactions on Software Engineering 2(3), 215–222 (1976)
8. Emmi, M., Majumdar, R., Sen, K.: Dynamic test input generation for database applications. In: Rosenblum, D.S., Elbaum, S.G. (eds.) Software Testing and Analysis (ISSTA 2007), pp. 151–162. ACM (2007)
9. Fu, X., Powell, M.C., Bantegui, M., Li, C.C.: Simple linear string constraints. Formal Aspects of Computing 25, 847–891 (2013). SUSHI is available from http://people.hofstra.edu/Xiang_Fu/XiangFu/projects/SAFELI/SUSHI.php
10. Ganesh, V., Dill, D.L.: A decision procedure for bit-vectors and arrays. In: Damm, W., Hermanns, H. (eds.) CAV 2007. LNCS, vol. 4590, pp. 519–531. Springer, Heidelberg (2007). STP is available from https://sites.google.com/site/stpfastprover/
11. Gange, G., Navas, J.A., Stuckey, P.J., Søndergaard, H., Schachte, P.: Unbounded model-checking with interpolation for regular language constraints. In: Piterman, N., Smolka, S.A. (eds.) TACAS 2013. LNCS, vol. 7795, pp. 277–291. Springer, Heidelberg (2013)
12. Gecode Team: Gecode: A generic constraint development environment (2006). http://www.gecode.org
13. Gervet, C.: Constraints over structured domains. In: Rossi, F., van Beek, P., Walsh, T. (eds.) Handbook of Constraint Programming, chap. 17, pp. 605–638. Elsevier (2006)
14. Ghosh, I., Shafiei, N., Li, G., Chiang, W.F.: JST: an automatic test generation tool for industrial Java applications with strings. In: Notkin, D., Cheng, B.H.C., Pohl, K. (eds.) International Conference on Software Engineering (ICSE 2013), pp. 992–1001. IEEE / ACM (2013)
15. Golden, K., Pang, W.: Constraint reasoning over strings. In: Rossi, F. (ed.) CP 2003. LNCS, vol. 2833, pp. 377–391. Springer, Heidelberg (2003)
16. He, J., Flener, P., Pearson, J., Zhang, W.M.: Solving string constraints: the case for constraint programming. In: Schulte, C. (ed.) CP 2013. LNCS, vol. 8124, pp. 381–397. Springer, Heidelberg (2013)

17. Hooimeijer, P., Weimer, W.: A decision procedure for subset constraints over regular languages. In: Hind, M., Diwan, A. (eds.) Programming Language Design and Implementation (PLDI 2009), pp. 188–198. ACM (2009)
18. Hooimeijer, P., Weimer, W.: StrSolve: solving string constraints lazily. Automated Software Engineering 19(4), 531–559 (2012)
19. Kieżun, A., Ganesh, V., Guo, P.J., Hooimeijer, P., Ernst, M.D.: HAMPI: a solver for string constraints. In: Rothermel, G., Dillon, L.K. (eds.) International Symposium on Software Testing and Analysis (ISSTA 2009), pp. 105–116. ACM (2009). HAMPI is available from http://people.csail.mit.edu/akiezun/hampi/
20. King, J.C.: Symbolic execution and program testing. Commun. ACM 19(7), 385–394 (1976)
21. Lothaire, M.: Combinatorics on words. Cambridge Mathematical Library, Cambridge University Press (1997)
22. Maher, M.J.: Open constraints in a boundable world. In: van Hoeve, W.-J., Hooker, J.N. (eds.) CPAIOR 2009. LNCS, vol. 5547, pp. 163–177. Springer, Heidelberg (2009)
23. Makanin, G.: The problem of solvability of equations in a free semigroup. Sbornik: Mathematics 32(2), 129–198 (1977)
24. Michel, L.D., Van Hentenryck, P.: Constraint satisfaction over bit-vectors. In: Milano, M. (ed.) CP 2012. LNCS, vol. 7514, pp. 527–543. Springer, Heidelberg (2012)
25. Monette, J.-N., Flener, P., Pearson, J.: Towards solver-independent propagators. In: Milano, M. (ed.) CP 2012. LNCS, vol. 7514, pp. 544–560. Springer, Heidelberg (2012). the indexical compiler is available from http://www.it.uu.se/research/group/astra/software#indexicals
26. Pesant, G.: A regular language membership constraint for finite sequences of variables. In: Wallace, M. (ed.) CP 2004. LNCS, vol. 3258, pp. 482–495. Springer, Heidelberg (2004)
27. Quimper, C.-G., Walsh, T.: Global grammar constraints. In: Benhamou, F. (ed.) CP 2006. LNCS, vol. 4204, pp. 751–755. Springer, Heidelberg (2006)
28. Redelinghuys, G., Visser, W., Geldenhuys, J.: Symbolic execution of programs with strings. In: Kroeze, J.H., de Villiers, R. (eds.) South African Institute of Computer Scientists and Information Technologists Conference (SAICSIT 2012), pp. 139–148. ACM (2012)
29. Saxena, P., Akhawe, D., Hanna, S., Mao, F., McCamant, S., Song, D.: A symbolic execution framework for JavaScript. In: Security and Privacy (S&P 2010), pp. 513–528. IEEE Computer Society (2010). KALUZA is available from http://webblaze.cs.berkeley.edu/2010/kaluza/
30. Scott, J.D.: Rapid prototyping of a structured domain through indexical compilation. In: Schaus, P., Monette, J.N. (eds.) Domain Specific Languages in Combinatorial Optimization (CoSpeL workshop at CP 2013) (2013). http://cp2013.a4cp.org/workshops/cospel
31. Scott, J.D., Flener, P., Pearson, J.: Bounded strings for constraint programming. In: Tools with Artificial Intelligence (ICTAI 2013), pp. 1036–1043. IEEE Computer Society (2013)
32. Sellmann, M.: The theory of grammar constraints. In: Benhamou, F. (ed.) CP 2006. LNCS, vol. 4204, pp. 530–544. Springer, Heidelberg (2006)

33. Trinh, M.T., Chu, D.H., Jaffar, J.: S3: A symbolic string solver for vulnerability detection in web applications. In: Computer and Communications Security (CCS 2014) (2014)
34. Van Hentenryck, P., Saraswat, V.A., Deville, Y.: Design, implementation, and evaluation of the constraint language cc(FD). techreport CS-93-02, Brown University, Providence, USA (January 1993), revised version: Journal of Logic Programming **37**(1–3), 293–316 (1998)
35. Yu, F., Bultan, T., Ibarra, O.H.: Symbolic string verification: combining string analysis and size analysis. In: Kowalewski, S., Philippou, A. (eds.) TACAS 2009. LNCS, vol. 5505, pp. 322–336. Springer, Heidelberg (2009)
36. Zheng, Y., Zhang, X., Ganesh, V.: Z3-str: A Z3-based string solver for web application analysis. In: Meyer, B., Baresi, L., Mezini, M. (eds.) Foundations of Software Engineering (FSE 2013), pp. 114–124. ACM (2013)

Freight Train Threading with Different Algorithms

Ilankaikone Senthooran[1], Mark Wallace[1(✉)], and Leslie De Koninck[2]

[1] Faculty of Information Technology, Monash University, Wellinghton Road,
Clayton, VIC 3800, Australia
{ilankaikone.senthooran,mark.wallace}@monash.edu
[2] Opturion pty ltd, Melbourne, VIC 3006, Australia
ldekoninck@opturion.com

Abstract. The problem addressed in this paper, of routing and scheduling freight trains within a scheduled passenger rail network, prevails in many large countries. The actual departure and arrival times of freight trains are not important, and nor are their routes. Only a starting station and destination station are specified on the day of travel. The current paper addresses the problem of how to thread the maximum number of freight trains through the passenger network, minimising the latest arrival time of the last freight train. This problem contrasts with the more traditional rail scheduling requirement of matching as closely as possible an ideal schedule. The rail network is modelled topologically, so the size of the network does not grow as the temporal granularity is made finer. Our use of the modelling language MiniZinc enables us to compare several different solvers and solving approaches applied to the model. In particular we investigate constraint programming, using global constraints and constraint propagation; mathematical programming naively, without using any of the decomposition techniques; and finally a hybrid of constraint propagation, learning, and dynamic search control called lazy clause generation. The paper describes two kinds of experiments. Firstly it gives results for a series of benchmark tests to investigate the flexibility and scalability of the algorithm, and secondly it is applied to a freight train scheduling problem on the Victorian rail network in Australia.

Keywords: MiniZinc · Algorithms · Freight train scheduling · Lazy clause generation

1 Introduction

The demand for rail transport is increasing in many countries, both for passengers and freight. In Melbourne, passenger demand has been increasing at 10% per year. According to the European commission, a single freight train on a track can replace 280 trucks on a road, reducing fuel use, congestion, and emissions, and passenger travel by rail produces three to ten times less $CO2$ than cars or airplanes [1].

---

[1] http://ec.europa.eu/digital-agenda/futurium/en/content/trends-rail-transport

© Springer International Publishing Switzerland 2015
L. Michel (Ed.): CPAIOR 2015, LNCS 9075, pp. 393–409, 2015.
DOI: 10.1007/978-3-319-18008-3_27

Passenger trains run to a specific published timetable, planned months in advance, while freight trains are scheduled on-demand at shorter timescales. Passenger and freight trains often share the same rail network. The problem addressed in this paper, of routing and scheduling freight trains within a scheduled passenger rail network, prevails in many large countries. The actual departure and arrival times of freight trains are not important, and nor are their routes. Only a starting station and destination station are specified on the day of travel. The objective is to minimize the time required to complete the rail movements. This problem contrasts with the more traditional rail scheduling requirement of matching an ideal schedule as closely as possible.

In a highly connected rail network, such as that in India, there are a large number of possible routes a freight train can take from its specified origin to its destination. The problem therefore is to thread as many freight trains as possible through the rail network and the passenger rail schedule in such a way that passenger trains are not disrupted and the freight trains can avoid stopping repeatedly en route to their destinations.

The train scheduling problem benefits from a rich research literature, and many variants of the problem have been addressed. The passenger rail scheduling problem is surveyed in [3]. Periodic passenger schedules are addressed in [8]. Train routing is tackled in [1]. The underlying models for all these varieties is a network of potential events - potential arrival and departure times at stations on the rail network.

Two papers address similar problems to this paper -threading freight trains through a passenger schedule. The 2010 paper [2] minimises the deviation of freight train schedules from an ideal schedule, and is modelled using an time-expanded network. The paper [6] on the other hand, does not start from an ideal schedule. They consider the decision problem of routing and scheduling freight trains in a passenger rail network in a manner that minimizes Makespan (time to complete all freight movements) or Flow Times (sum of completion times for each freight movement). The model used to tackle this problem is a topological model of the rail network, and the algorithm seeks a (temporally feasible) shortest path through this network for each freight train.

The number of freight trains threaded through the network is broadly similar in the benchmark tests in both papers [2] and [6]. The smallest benchmarks have around 20 freight trains, and the largest benchmark has 96 trains in [2] and 160 freight trains in [6]. A major difference is that [2] use a complete search method and prove their solutions are within a given percentage of the optimum (around 30% for the largest instances). [6] use a heuristic method for which there is no quality measure.

The current paper addresses the problem of how to thread the maximum number of freight trains through the passenger network, minimising the latest arrival time of the last freight train (termed the Makespan in [6]). The rail network is modelled topologically, so the size of the network does not grow as the temporal granularity is made finer. Different from [6], the algorithm used here is able to prove optimality. Of particular interest is the ability to directly model the number of freight trains that could be routed along a network link in a given time period. This gives a bound on the number of train paths available, which is essential for detecting and pruning infeasible parts of the search space.

Our use of the modelling language MiniZinc [10] enables us to compare several different solvers and solving approaches applied to the model. In particular we investigate constraint programming, using global constraints, constraint propagation; mathematical programming naively, without using any of the decomposition techniques; and finally a hybrid of constraint propagation, learning, and dynamic search control called lazy clause generation [14].

The next section presents the train threading problem informally and describes the algorithmic approaches used in this paper. In section 3 we contrast the constraint programming and mathematical programming models for the problem, and then formally present our model. The following section describes a number of experiments we have carried out on the model, mapping it down to various solvers and solving techniques. Our experiments use a small network, but the number of passenger trains varies from 68 to 136, and the number of freight trains from 4 up to 100. In the section 5 we model a rail network and passenger timetable taken from the shared rail network in Melbourne and its suburbs. While based on a real timetable, with nearly 150 passenger trains, our model of the network is still highly simplified. The final section presents our conclusions and plans for extensions of this work.

2   The Train Threading Problem

The problem of routing and scheduling a number of freight trains through a rail network combines both a routing subproblem and a timetabling subproblem. The requirement is to find a route and a timetable for each freight train from its specified origin to its specified destination, so that no two trains are scheduled to run too close together on any segment of rail track.

By expanding the network to represent each time point at each station, the problem can be mapped onto a single routing problem, but at the expense of dramatically increasing the size of the network. The number of nodes in such a network is the number of stations multiplied by the number of possible event times. For a periodic timetable of one hour, a discretisation of time into one-minute intervals, ensures the number of nodes is at most only 60 times the number of stations. However, scheduling a 24-hour period at a time-discretisation of one minute leads to a number of nodes 1440 times the number of stations. The huge increase in network size brings a huge increase in the number of variables, with the result that a large problem instance cannot be loaded into a mathematical solver and some problem decomposition becomes essential.

Instead we can model the problem as finding a path for each train through the rail network, and scheduling the train on that path so that it avoids all other passenger and freight trains. Unfortunately it is not possible to choose a path for each train and then try and schedule the trains on their paths. The risk is that too many of the train paths share a rail segment and that there is not sufficient time available to schedule all the trains through the segment. A natural problem decomposition is to plan paths for the trains and if a rail segment becomes a bottleneck to add a constraint limiting the set of train paths that can share the segment. This can be done using Lagrangian relaxation (as described in [2]) or Benders' decomposition.

The number of paths from origin to destination can grow exponentially in case there are multiple opportunities to swap between two lines (such as the 10 junctions that allow trains to pass between the fast and slow lines between Florence and Rome [2]).

Rather than train paths, an algorithm can select which rail links to traverse. The number of links traversed by a freight train can be limited to a given maximum, and the number of decision variables therefore limited to possible links traversed at each successive stage of the route. Naturally the links possible for a train at successive stages are restricted to pairs of links that meet at a station of junction. In fact all junctions can be modelled as stations.

Novel in this paper, we include rail link and temporal constraints in the same model, and the lazy clause generation algorithm automatically learns which combinations of decisions are infeasible.

Each link is traversed by a number of passenger trains during the day. This leaves a set of gaps during which freight trains can traverse the link. Due to the necessary headway between trains, the maximum number of freight trains that can traverse the link can be computed a priori. However, during search for a solution, the time period during which a freight train could possibly traverse an link becomes more constrained. The cumulative constraint in constraint programming, when extended to allow optional tasks, provides a powerful way to model these constraints.

The different speeds of freight trains give rise to more nodes and more links in a time-expanded network. The conflict between two links is dependent on the actual trains that are traversing the links. In our model this difference in speeds can be represented by two cumulative constraints, the first maintaining the headways between trains starting to traverse an link, and the second maintaining the headways at the end of the link. This model makes it possible to associate a different speed with each train on each station-station link. It even allows train speeds to be variable, by constraining the time at the end of the link to be within a certain minimum and maximum time after the time at the beginning of the link.

3 Problem Formulation

3.1 Mathematical Programming Model

The two subproblems of the train threading problem are

1. routing the freight trains through the network from their start locations to their destinations
2. scheduling the trains so as to avoid all conflicts.

For some earlier work such as [1], routing is the key problem. In other work such as [2], the choice of routes is more constrained and the heart of the problem is the scheduling.

In the above mathematical programming approaches, the two subproblems are merged by constructing a network that represents both the routes and the schedules. Each network node represents a potential event - an arrival or departure of a train at or from a station. Each network edge represents a potential train journey leg: starting at a particular time from a particular station and ending at the next station at a specific later time. The time gap represents the journey leg duration. If a new train is added to the problem with a different speed from previous trains, then new edges must be added to represent every possible journey leg that the train could take.

The left part of Figure 1 represents two stations, and 8 time-points at each station. The blue edges represent potential journey legs between the two stations for one train, and the red edges represent potential journey legs for a slower train. The middle part of the figure shows a set of compatible edges, which can be chosen in a schedule without violating any headways between pairs of trains. The right hand part of the figure shows a dual of the network, where the original blue and red edges are now shown as nodes, and the edges in the dual network show edges that are incompatible (i.e. pairs of potential journey legs that would violate headway constraints).

Notice that if different trains with the same speed need different headways,then edges in the original network must be duplicated in order to capture the conflicts. If every potential train on a leg between two stations had a different speed or headway, then the original graph would need different edges for every potential journey leg of every train. The size of the resulting network would be $\#Legs \times \#Periods \times \#Trains$. Such a network model can remain computationally tractable as long as the number of edges is limited. This limitation on the number of edges is achieved by limiting the divergence from an ideal, or target, schedule.

Even with a single standard headway and a small number of train types, the size of the networks is too large to be solved as a single MIP. In the case of [1], the problem is tackled using column generation, while in [2] Lagrangian relaxation is used for train conflict constraints, added only when the constraints are violated as the algorithm progresses. For the larger instances (96 freight trains) the optimality gap grows to a few percent, in the uni-directional case which is most similar to the benchmarks tackled in this paper.

3.2   Constraint Programming Model

The constraint programming model represents the problem as a set of tasks. A task is, of course, a train traversing the link, and in the case of freight trains the tasks are optional because the train may take a different route avoiding the link. However each train must perform a sequence of tasks corresponding to a path from its starting point to its destination.

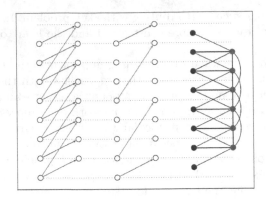

Fig. 1. Network representation of possible journey legs between two stations (from [1])

Task scheduling is modelled in constraint programming using the cumulative constraint. We use the constraint `cumulative_optional` in which each task has a given duration, and requires a given amount of resource. Its parameters are:

- duration of each task
- amount of resource used by each task while it is active
- total available capacity of the resource

The decision variables involved in the constraint are

- start time of each task
- boolean variable for each task indicating whether or not the task is performed

To represent potential journey legs, each task represents a different train. Just two constraints are needed to represent the set of potential journey legs between two stations, and to enforce the headway between all trains going from the first station to the other.

The duration of the task is the headway needed by that train (the delay before the following train can start on the same leg). The resource consumption of the task is 1. The available resource is the number of tracks (in the given direction). One cumulative constraint is imposed for each end of the leg. The link between the two cumulative constraints is as follows

- the start times associated with the corresponding tasks at the beginning and end of the leg (i.e. the same train) are constrained to differ by the time needed for the train to cover the leg.
- the same boolean is associated with the corresponding tasks at the beginning and end of the leg

As an example of the constraint, suppose there are three trains, t1, t2, t3 going from station **a** to station **b**. Each train needs 5 minutes to complete the leg and a headway of 3 minutes. The first cumulative constraint associates a time

Table 1. Tasks at Station **a**

Task name	Duration	Resource Needed	Start Time	Boolean
$t1_{astart}$	3	1	ta1	ba1
$t2_{astart}$	3	1	ta2	ba2
$t3_{astart}$	3	1	ta3	ba3

at station **a** with each train: $ta1, ta2, ta3$, and "optionality" boolean variable $ba1, ba2, ba3$. Thus the "tasks" can be represented in a simple table 1.

The second cumulative constraint also has three tasks, with modifed start times, but the same booleans. The associated tasks are shown in table 2.

Table 2. Tasks at Station **b**

Task name	Duration	Resource Needed	Start Time	Boolean
$t1_{aend}$	3	1	ta1+5	ba1
$t2_{aend}$	3	1	ta2+5	ba2
$t3_{aend}$	3	1	ta3+5	ba3

If we additionally constraint the task starts times all to lie within the range 1..10, then the third boolean variable $ba3$ must be `false`, and `t2` must go before `t1`. A feasible solution is therefore $ta1 = 1, ta2 = 4, ba1 = $ `true`$, ba2 = $ `true`$, ba3 = $ `false`. Note that the time $ta3$ is now meaningless, since it represents the start time of a task that is never active.

It is a limitation of our model using the cumulative constraint that we must assume dual (or multiple) tracks, so each track is only used in one direction. On the other hand the model is very flexible in that it allows each train to have a different speed and a different headway. It can even allow the train to have the speed as a decision variable optimised as part of the solution (in case, for example, we seek a solution where freight trains avoid stopping between journey legs).

The cumulative constraint detects early when a link is becoming congested during a certain time period, because too many freight trains have been routed across the link within a certain time period. This enables infeasible combinations of freight train paths to be detected before any attempt to schedule times for their journey legs. One of our solution algorithms uses lazy clause generation [4], and our implementation of `cumulative_optional` generates explanations of any infeasibilities it detects. It uses, and explains, both timetabling and edge-findings, as described in [12,14].

The optionality of tasks can be modelled using optional types in MiniZinc [9]. In fact our implementation instead uses a pair of variables to represent the upper and lower bounds on the start times [13]. It sets the boolean variable to `false` as soon as the lower bound exceeds the upper bound. Naturally the upper and lower bound are constrained to be equal if the boolean is `true`.

3.3   Formal Model of the Freight Train Threading Problem

For general case we assume that we have S stations, T^p passenger trains and their daily schedule. Our objective is to find earliest feasible schedule for T^f freight trains that does not affect the usual passenger train operation.

We assume that trains flow in one direction. In short, an origin for one train is not a destination for another.

3.4   Assumptions

We consider this problem with following assumptions:

- For all freight trains, their origin and destination are known.
- No train can overtake any preceding train.
- The passenger train schedule is known. That is, occupancy of each and every railway link in the network is known.
- Restricted operational times of passenger and freight trains. The passenger train and freight trains operate between 05:00 - 22:00 and 03:00 - 23:00, respectively.
- All freight trains traverse at same average speed, which is different from the passenger train average speed.
- A minimum headway between consecutive trains (freight-freight and freight-passenger) that share the same link is always maintained.

3.5   Sets and Parameters

Sets

T^p	passenger trains	S	stations
T^f	freight trains	R	railway links
T	all trains		

Parameters

$Factor$	speed factor between passenger and freight trains
Gap	minimum headway between consecutive trains
$MaxLinks$	maximum number of legs in a freight train path
$d_r : r \in R$	time taken by a passenger train to travel the railway link r
$s_r : r \in R$	start station of the directional railway link r
$e_r : r \in R$	end station of the directional railway link r
$P_i : i \in T^p$	number of links in the path of passenger train i
$l_{i,j} : i \in T^p, j \in 1..P_i$	j^{th} railway link of passenger train i
$st_{i,j} : i \in T^p, j \in 1..P_i$	start time of passenger train i on link $l_{i,j}$
$orig_i : i \in T^f$	starting location of freight train i
$dest_i : i \in T^f$	destination of freight train i
σ_l^f	lower bound of freight train operation time in a day
σ_u^f	upper bound of freight train operation time in a day

3.6   Decision Variables

Find a feasible schedule for freight trains on an existing passenger railway network that satisfies some given objective function. The decisions are represented by the following variables.

$P_i : i \in T^f$ The number of links in the path of freight train i

$l_{i,j} : i \in T^f, j \in 1..P_i$ j^{th} railway link of freight train i

$st_{i,j} : i \in T^f, j \in 1..P_i$ start time on $l_{i,j}$ of freight train i

$prec_{i,i'} : i, i' \in T^f$ Train precedence order is true when train i precedes i' on any link.

3.7   Constraints

1. Length of path
$$0 \leq P_i \leq MaxLinks \qquad : i \in T^f$$

2. Start location of freight trains
$$s_{l_{i,1}} = orig_i \qquad : i \in T^f$$

3. Destination of freight trains
$$e_{l_{i,P_i}} = dest_i \qquad : i \in T^f$$

4. Preceding railway links
$$s_{l_{i,j+1}} = e_{l_{i,j}} \qquad : i \in T^f, j \in 1..(P_i - 1)$$

5. Each freight train travels at different average speed than a passenger train by *Factor*. This constraint allows a freight train to stop at a station before its next journey leg.
$$st_{i,j+1} \geq st_{i,j} + (d_{l_{i,j}} \times Factor) \qquad : i \in T^f, j \in 1..(P_i - 1)$$

6. Freight trains that use the same railway link have a precedence between them. The precedence order between two trains is the same throughout the journey, so no train can overtake any preceding train.
$$l_{i,j} = l_{i',j'} \Rightarrow prec_{i,i'} \lor prec_{i',i} \qquad : i \neq i' \in T^f, j \in 1..P_i, j' \in 1..P_{i'}$$

7. Minimum headway time between trains is *Gap*
$$prec_{i,i'} \land l_{i,j} = l_{i',j'} \Rightarrow st_{i',j'} \geq st_{i,j} + Gap$$
$$i, i' \in T^f, j \in 1..P_i, j' \in 1..P_{i'}$$
$$l_{i,j} = l_{i',j'} \Rightarrow$$
$$(st_{i',j'} \geq st_{i,j} + Gap) \lor (st_{i,j} \geq st_{i',j'} + Gap + d_{l_{i,j}} \times (Factor - 1))$$
$$i \in T^p, i' \in T^f, j \in 1..P_i, j' \in 1..P_{i'}$$

8. Daily operational times of freight trains
$$\sigma_l^f < t_{i,j} < \sigma_u^f \qquad : i \in T^f, j \in 1..P_i$$

3.8   Model with the Cumulative Optional Constraint

We can model the problem using the cumulative constraint, and make the precedence decision variables and constraints redundant. Instead we introduce a boolean variable for each train and rail link. A cumulative constraint enforces the required headway between trains at the beginning and end of each rail link

Decision Variables

$bb_{i,r} : i \in T, r \in R$ A boolean variable for each train and rail link

$tt_{i,r} : i \in T, r \in R$ A start time for each train on each rail link

1. Durations
   - $dd_{i,r} = d_r$ $: i \in T^p$
   - $dd_{i,r} = d_r * Factor$ $: i \in T^f$
2. Linking constraints
   - $tt_{i,r} = st_{i,l_{i,j}}$ $: i \in T, j \in 1..P$
   - $bb_{i,l_{i,j}} = \texttt{true}$ $: i \in T, j \in 1..P$
3. Cumulative constraint: (start-time-list, duration-list, resource-list, boolean-list,total-resource)
   - $\texttt{cumulative\_optional}([tt_{i,r} : i \in T],[Gap : i \in T],[1 : i \in T],[bb_{i,r} : i \in T], 1)$ $: r \in R$
   - $\texttt{cumulative\_optional}([tt_{i,r} + dd_{i,r} : i \in T],[Gap : i \in T],[1 : i \in T],[bb_{i,r} : i \in T], 1)$ $: r \in R$

3.9 Objective

The objective is to find earliest or a feasible schedule for freight trains on an existing railway network that doesn't affect the passenger train schedule. In other words, to find a schedule which satisfies all above constraints from (1) to (8) and minimises a new variable *Latest*.

$Latest \geq st_{i,P} + (d_{l_{i,P}}) \times Factor$ $: i \in T^f$

minimize(Latest)

3.10 Problem Variants

The freight train threading problem can involve a variety of side constraints. We explored several of these in combination to determine the impact of side-constraints on the performance of the solving methods. The side-constraints we investigated were:

1. lines crossing constraint - if two train paths cross, then they must be separated by a headway
2. no stopping constraint - a freight train cannot stop en route
3. cyclic network constraint - a constraint precluding freight trains from taking paths with cycles.

Each side-constraint requires its own parameters and variables.

Lines Crossing.

- Parameters
  $< r1, r2, r3, r4 >$ $: r1, r2, r3, r4 \in R \land e_{r1} = e_{r2} = s_{r3} = s_{r4}$ Each crossing is listed as an array of 4 links.
- Constraints $l_{i,j} = r1 \land l_{i,j+1} = r3 \land l_{i',j} = r2 \land l_{i',j'+1} = r4 \Rightarrow$
  $st_{i,j+1} \geq st_{i',j'+1} + Gap \lor st_{i',j'+1} \geq st_{i,j+1} + Gap$
  $i, i' \in T, j \in 1..P_i, j' \in 1..P_{i'}$

No Stopping.

- This constraint prevents a freight train from stopping at a station before its next journey leg.

$$st_{i,j+1} = st_{i,j} + (d_{l_{i,j}} \times Factor) \qquad : i \in T^f, \, j \in 1..(P_i - 1)$$

No cycles.

- Constraints: all the stations visited on a freight train path must be distinct (This alldifferent is implemented using MiniZinc's optional types [9]).

$$\texttt{all\_different}([e(l_{i,j}) : j \in 1..MaxLinks]) \qquad : i \in T^f$$

4 Experiments

4.1 Benchmark Rail Network

For benchmarking the rail freight threading solvers, we used the rail network shown in Figure 2, which comprises 15 stations and 16 rail links. This hypothetical network was constructed to connect four possible origins to four possible destinations allowing 16 possible freight train routes, with 8 rail links where conflicts could occur, and 7 stations where lines cross. The dotted links were used in the experiments on side-constraints described below. The node i represents a main station in the network. A main station is either an end station ($O_{1..4}/D_{1..4}$) of a line or a junction station. A junction station is a station that has more than one incoming edges (railway links/tracks) and/or more than one outgoing edges. Each edge $r = \{s,e\}$ represents a directional rail link from station s to station e. Intermediate stations ($I_{1..7}$) between two main stations which have one incoming track and one outgoing track each, are not modelled, as it will not affect the solution to our problem.

4.2 Benchmark Tests

We conducted freight train scheduling for ten different scenarios. These scenarios were formulated combining different passenger train schedules, freight

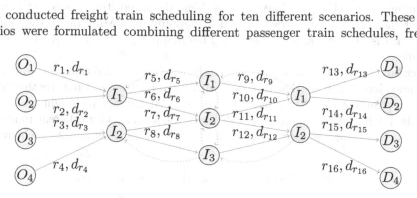

Fig. 2. Benchmark Rail Network

train speeds and minimum headway. The scenarios are described in Appendix 1. For each scenario, the benchmark test involved threading a given number of freight trains, with given origins and destinations, through the given passenger timetable. To test the scalability of the solvers, we tried different numbers of freight trains: 4, 8, 16, 32, 64 and 100.

We then extended this rail network to have some cycles by adding rail links as shown in dotted arrows (Figure 2), and repeated the experiments with additional combinations of side constraints, namely, cross-links and no-stopping. In the next section, we discuss the results obtained.

4.3 Testing Environment

We implemented our model using **MiniZinc** [10]. The Minizinc software supports the mapping of MiniZinc models to an underlying (Flatzinc) form suitable for different kinds of solvers. In particular we used mappings to an integer-linear form, and to constraint-programming forms to suit different solvers. The solvers we tested were Cplex [7], Gecode [5] and Opturion CPX [11]. All results reported here were obtained on an Intel(R) Core(TM) i5-3320M CPU @ 2.60Hz and 8GB RAM. We used MiniZinc version 2.0, Cplex version 12.5.1, Opturion CPX version 1.0.2 and Gecode version 4.3.2. Each of the instances (for a certain test scenario and a certain number of freight trains) was allowed to run a maximum of 10 minutes.

4.4 Results

We tested our model on the three solvers, Cplex, Opturion CPX and Gecode, each with varying numbers of freight trains, 4, 8, 16, 32, 64 and 100. As shown in Table 3, both Opturion CPX and Gecode solvers were able to arrive at feasible solutions for all cases.

The poor scalability of Cplex on these tests is partly due to the highly disjunctive nature of the problem, which is the reason previous mathematical programming approaches [1,2] have used problem decompositions. However it may be that the linearisation carried out when mapping from MiniZinc to the underlying form is not be the best possible. This mapping is the subject of ongoing research.

Scalability Test Table 4 summarises the results obtained using the Opturion CPX and Gecode solvers. The execution times given on the last row are the times taken for running all 10 test scenarios for a given number of freight trains, in seconds. Each of the remaining rows displays the solutions for different numbers of freight trains for a given test scenario. The solutions shown are the scheduled

Table 3. Maximum size solved on different solvers

	Cplex	CPX	Gecode
No. of Freight Trains	8	100	100

Table 4. Scalability Test Results of CPX and Gecode Solvers

	Opturion-CPX						Gecode					
	No. of Freight Trains						No. of Freight Trains					
Test Scenario	4	8	16	32	64	100	4	8	16	32	64	100
1	150	155	165	185	225	275	150	155	165	185	225	275
2	300	305	315	335	375	425	300	305	315	335	375	530
3	150	155	165	185	225	275	150	155	165	185	225	275
4	300	305	315	335	375	425	300	305	315	335	430	550
5	150	155	165	185	225	275	150	155	165	185	225	275
6	300	305	315	335	375	425	300	305	315	335	375	530
7	150	155	165	185	225	275	150	155	165	185	225	275
8	300	305	315	335	375	425	300	305	315	335	425	560
9	150	160	180	220	300	400	150	160	180	220	300	400
10	150	160	180	220	300	400	150	160	180	220	300	400
Execution Time (s)	0.3	1.0	7.1	45	519	4474	4	7	19	79	1700	5224

departure times for the last legs of the last freight train, given in minutes. For example, for test scenario 1 with 4 freight trains, the last train is scheduled to depart on its last leg at 2:30AM. The results shaded in red show the instances where the optimal solution could not be reached within the given time limit of 10 minutes.

In the cases of 4, 8, 16 and 32 freight trains, both the solvers arrived at optimal solutions for all test scenarios. In the case of 64 and 100 freight trains, however, only the Opturion CPX solver succeeded in obtaining the optimal solutions across all scenarios. For all problem sizes Opturion CPX produced solutions faster than Gecode.

Test on side-constraints To investigate the impact of the side-constraints: lines-crossing, no-stopping and cyclic network on the solving method, we performed a test using the Opturion CPX solver. In order to test the cyclic constraint, we introduced some cyclic routes into our hypothetical network.

We performed tests with instances sized 32 and 64 freights trains. For all tests in all combinations CPX found and proved optimality for the size 32 instances. Optimally could not be proven for size 64 instances.

5   A Real World Example

5.1   Melbourne Network

Finally we applied our model to a network based on real world data. We created a simplified version of the Melbourne V-Line network shown in Figure 3. In this network, freight trains start from any of the four destinations: Sunbury, Werribee, Pakenham and Craigieburn, and travel to any other. Inbound and outbound trains are on different tracks, and freight trains use different lines from the passenger services between Noth Melbourne and Flinders Street.

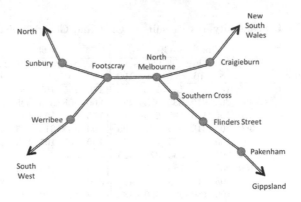

Fig. 3. V-Line Simplified Rail Network

The input timetables for the model were constructed based on the actual incoming and outgoing timetables for each passenger line between 7 AM and 12 Noon. Inbound passenger trains run from Werribee to North Melbourne, from Sunbury to North Melbourne, from Craigieburn to North Melbourne, and from Pakenham to Southern Cross. All outbound passenger trains going to Pakenham depart from Flinders Street, whereas trains going to all other destinations depart from North Melbourne.

5.2   Experiments

We tested the Melbourne rail network with varying number of freight trains, 12, 24, 48, 60, 72, 84, 96 and 120. The reason for choosing multiple of twelve was to have equal number of all possible combinations of origin and destination. Table 5 shows the maximum number of freight trains for which the Opturion CPX solver gave at least one feasible solution within the allowed runtime (10 minutes for each instance). We restricted the freight train operational time to match the passenger train timetable, which is from 7 AM till 12 Noon. This is to simulate an actual situation where freight train threading is required during usual operational time. For example, when minimum headway (*Gap*) is 3 minutes and the speed factor (*Factor*) is 1, a maximum of 84 freight trains can be scheduled.

Table 5. Scheduling varying number of freight trains

	Gap (*mins*)		
Factor	3	4	5
1	84	72	36
2	24	NP	NP

NP - Not possible to schedule

6   Conclusion

The freight train threading problem is both an important practical problem and an interesting scheduling problem which has been addressed using mathematical programming approaches as well as metaheuristics. In this paper we explored a novel approach to the problem using cumulative scheduling and optional tasks. We compared the available complete methods, and we were pleased to find that our new approach is able to prove optimality for problem instances with 100 freight trains, which is comparable to the numbers previously tackled using mathematical programming.

We have also implemented an approach where freight trains can slow down between stations. This is a variant that cannot be represented in current mathematical modelling approach with a space-time network.

We compared an approach using lazy clause generation with a traditional tree-search implemented in Gecode. The scalability of the traditional tree-search approach surprised us, though learning made an increasing improvement in the larger instances. We also investigated the impact of side-constraints on solving performance. While side-constraints did not prevent the solver from finding solutions, the number of instances where optimality could be proven was dramatically reduced when different side-constraints were combined.

We recognise that the rail networks investigated in this paper are small and over-simplified. We plan next to augment our tests on scalability with the number of trains on the network, to investigate scalability with the network itself, employing a more detailed model of a real network.

Acknowledgments. We are grateful to Sudhir Sinha and Narayan Rangaraj of Tata Consulting Services for introducing us to the problem Guido Tack took time and effort to help us construct the Gecode model, when he was under great time pressure!

This research was supported under Australian Research Council's Discovery Projects funding scheme (project number DP140100058).

Appendix: Data

Benchmark Scenarios

Passenger Train Timetables

The passenger train timetable for the hypothetical network can be downloaded from https://sakai-vre.its.monash.edu.au/access/content/user/wallace/FreightTrainThreading

Table 6. Passenger train schedules (termed set in table 7) used for freight rail threading benchmarks)

Set	Description
1	Frequency of trains is one per hour per origin
2	Frequency of trains is two per hour per origin
3	Train frequency is two per origin during peak hour and one per origin during off-peak hour. Peak hours: 6am - 9am, 4pm - 7pm
4	In set 1,2 and 3, passenger trains that start from origin (say O_1) always takes a particular path to reach its destination (D_1). In this set, passenger train's origin and destination remains the same but it takes two different paths. Therefore, within an hour two trains depart from same origin (O_1). Each will take different paths, but will end up at the same destination (D_1).

Table 7. Ten scenarios used for freight rail threading benchmarks

Test No.	Set	Total T^p	Factor	Gap (mins)	σ_l^p - σ_u^p
1	1	68	1	5	05:00 - 23:59
2	1	68	2	5	05:00 - 23:59
3	2	132	1	5	05:00 - 23:59
4	2	132	2	5	05:00 - 23:59
5	3	92	1	5	05:00 - 23:59
6	3	92	2	5	05:00 - 23:59
7	4	136	1	5	05:00 - 23:59
8	4	136	2	5	05:00 - 23:59
9	2	132	1	10	05:00 - 23:59
10	3	92	1	10	05:00 - 23:59

References

1. Borndörfer, R., Schlechte, T.: Models for railway track allocation. In: ATMOS 2007–7th Workshop on Algorithmic Approaches for Transportation Modeling, Optimization, and Systems. Internationales Begegnungs- und Forschungszentrum für Informatik (IBFI), Schloss Dagstuhl, Germany, Dagstuhl, Germany (2007)
2. Cacchiani, V., Caprara, A., Toth, P.: Scheduling extra freight trains on railway networks. Transportation Research Part B **44**, 215–231 (2010)
3. Caprara, A., Kroon, L., Monaci, M., Peeters, M., Toth, P.: Chapter 3 passengerrailway optimization. In: Barnhart, C., Laporte, G. (eds.) Transportation, Handbooks in Operations Research and Management Science, vol. 14, pp. 129–187. Elsevier (2007)
4. Feydy, T., Stuckey, P.J.: Lazy clause generation reengineered. In: Gent, I.P. (ed.) CP 2009. LNCS, vol. 5732, pp. 352–366. Springer, Heidelberg (2009)
5. Gecode Team: Gecode: Generic constraint development environment (2014). http://www.gecode.org
6. Godwin, T., Gopalan, R., Narendran, T.: Freight train routing and scheduling in a passenger rail network: computational complexity and the stepwise dispatching heuristic. Asia-Pacific Journal of Operational Research **24**(4), 499–533 (2007)
7. IBM Inc: CPLEX optimizer (2014). http://www-01.ibm.com/software/commerce/optimization/cplex-optimizer/
8. Liebchen, C., Proksch, M., Wagner, F.: Performance of algorithms for periodic timetable optimization. In: Hickman, M., Mirchandani, P., Voss, S. (eds.) Computer-aided Systems in Public Transport. Lecture Notes in Economics and Mathematical Systems, vol. 600, pp. 151–180. Springer, Berlin Heidelberg (2008)
9. Mears, C., Schutt, A., Stuckey, P.J., Tack, G., Marriott, K., Wallace, M.: Modelling with option types in MiniZinc. In: Simonis, H. (ed.) CPAIOR 2014. LNCS, vol. 8451, pp. 88–103. Springer, Heidelberg (2014)
10. Nethercote, N., Stuckey, P.J., Becket, R., Brand, S., Duck, G.J., Tack, G.R.: MiniZinc: towards a standard CP modelling language. In: Bessière, C. (ed.) CP 2007. LNCS, vol. 4741, pp. 529–543. Springer, Heidelberg (2007)
11. Opturion: Opturion CPX (2014). http://www.opturion.com/cpx.html
12. Schutt, A., Feydy, T., Stuckey, P.J.: Explaining time-table-edge-finding propagation for the cumulative resource constraint. In: Gomes, C., Sellmann, M. (eds.) CPAIOR 2013. LNCS, vol. 7874, pp. 234–250. Springer, Heidelberg (2013)
13. Schutt, A., Feydy, T., Stuckey, P.J.: Scheduling optional tasks with explanation. In: Schulte, C. (ed.) CP 2013. LNCS, vol. 8124, pp. 628–644. Springer, Heidelberg (2013)
14. Schutt, A., Feydy, T., Stuckey, P., Wallace, M.: Explaining the cumulative propagator. Constraints **16**(3), 173–194 (2011)

Learning General Constraints in CSP

Michael Veksler and Ofer Strichman$^{(\boxtimes)}$

Information Systems Engineering, Technion, Haifa, Israel
mveksler@tx.technion.ac.il, ofers@ie.technion.ac.il

Abstract. We present a new learning scheme for CSP solvers, which is based on learning (general) constraints rather than generalized no-goods or signed-clauses that were used in the past. The new scheme is integrated in a conflict-analysis algorithm reminiscent of a modern systematic SAT solver: it traverses backwards the conflict graph and gradually builds an asserting conflict constraint. This construction is based on new inference rules that are tailored for various pairs of constraints types, e.g., $x \leq y_1 + k_1$ and $x \geq y_2 + k_2$, or $y_1 \leq x$ and $[x, y_2] \not\subseteq [a, b]$. The learned constraint is stronger than what can be learned via signed resolution. Our experiments show that our solver HCSP backtracks orders of magnitude less than other state-of-the-art solvers, and is overall on par with the winner of this year's MiniZinc challenge.

1 Introduction

The ability of CSP solvers to learn new constraints during the solving process possibly shortens run-time by an exponential factor (see, e.g., [20]). Despite this fact, and in contrast to SAT solvers, only few CSP solvers use learning, owing to the difficulty of making it cost-effective. Learning in a limited form was present in early CSP solvers, where it was called *nogood learning* [10]. Nogoods are defined as partial assignments that cannot be extended to a full solution. Later *generalized nogoods* [20] (g-nogoods for short) were proposed, which allow *non-assignments* as well, e.g., a g-nogood $(x \leftarrow\!\!\!\!/\; 1, y \leftarrow 1)$ means that an assignment in which x is assigned anything but 1 and y is assigned 1 cannot be extended to a solution. This formalism is convenient for representing knowledge obtained by propagators. The g-nogood above, for example, can result from removing 1 from the domain of x, which leads by propagation to removing 1 from the domain of y. G-nogoods may be exponentially stronger than nogoods, as shown in [20].

A more general and succinct representation of learned knowledge is in the form of *signed clauses*. Such clauses are disjunctions of *signed literals*, where a signed literal has the form $v \in D$ or $v \notin D$ (called positive and negative signed literals, respectively), where v is a variable and D is a domain of values. Beckert et al. [5] studied the satisfiability problem of signed CNF, i.e., satisfiability of a conjunction of signed clauses. They proposed an inference system, based on simplification rules and a rule for binary resolution of signed clauses:

© Springer International Publishing Switzerland 2015
L. Michel (Ed.): CPAIOR 2015, LNCS 9075, pp. 410–426, 2015.
DOI: 10.1007/978-3-319-18008-3_28

$$\frac{((v \in A) \vee X) \quad ((v \in B) \vee Y)}{(v \in (A \cap B) \vee X \vee Y)} \quad \text{[Signed Resolution}(v)\text{]} \tag{1}$$

where X and Y consist of a disjunction of zero or more literals, A and B are sets of values, and v is called the *pivot* variable. Note that in case v is Boolean and A, B are complementary Boolean domains (e.g., $A = \{0\}, B = \{1\}$) then this rule simplifies to the standard resolution rule for propositional clauses that is used in SAT, namely the consequent becomes $(X \vee Y)$.

As we showed in an earlier publication [27], we used this rule in our CSP solver HCSP (short for HaifaCSP)[1], as part of a general learning scheme based on signed clauses. Using a special inference rule for each type of non-clausal constraint, HCSP inferred a signed clause e that *explains* a propagation by that constraint. This means that e is implied by the constraint, but at the same time is strong enough to make the same propagation as the constraint, at the same state. Using such explanations in combination with rule (1) for resolving signed clauses, HCSP can generate a signed *conflict clause* via *conflict analysis*. By construction this clause is *asserting* (i.e., it necessarily leads to additional propagation after backtracking). In contrast to the CSP solver EFC [20], which generates a g-nogood *eagerly* for each removed *value*, HCSP generates a signed explanation clause *lazily*, only as part of conflict analysis. Lazy learning of g-nogoods was also implemented on top of MINION [14]. There has also been work on extending explanations with new Boolean variables, which encode equalities and inequalities [19,25], and more recently constraint-specific inference [15], such as partial sums in the case of linear constraints. In all these works there is no direct inference between general constraints.

In this article we study a different learning scheme, which is based on inference rules with non-clausal consequents. Non-clausal learning has been studied before in the context of several first-order quantifier-free theories: Pseudo-Boolean constraints (see, e.g., Sect. 22.6.4 in [6] and [11]), difference constraints [9], and integer linear constraints, e.g., [18,24]. The congruence-closure algorithm for equality logic with uninterpreted functions, which is implemented in most SMT solvers, can also be seen as inferring non-clausal constraints, since it infers new equalities. In all of these cases such learning was shown to improve the search, which motivated us to develop such a scheme for CSP, that is strongly tied to the conflict-analysis procedure. What we suggest here is very general, as it can be used with most of the constraints that are supported by modern CSP solvers, and allows non-clausal inference between different types of constraints.

Our main goal in introducing this scheme is to learn a conflict constraint that is logically stronger and easier to compute than its clausal counterpart. The emphasis is on the first of these goals as it may improve the search itself. To that end, we propose a generic inference rule called *Combine* that for many popular (pairs of) constraints indeed fulfills these two goals. For example, suppose that in a state in which the domains of three variables are defined by $x \in \{2, 6, 10, 14, \ldots, 30\}, y_1 \in \{8, 12, 16, 20\}, y_2 \in \{1, 2, 3, \ldots, 9\}$, the constraint $c_1 \doteq y_1 \leq x$ propagates $x \in \{10, 14, \ldots, 30\}$, which leads to a contradiction with

---

[1] In [27] it was still called PCS, for Proof-producing Constraint Solver.

a constraint $c_2 \doteq x \leq y_2$. During conflict-analysis, HCSP now infers from this propagation the constraint $[y_1, y_2] \subseteq [8, 9] \implies x \in [8, 9]$, which is clearly implied by c_1, c_2 (square brackets denote a range). Rewriting this constraint as $[y_1, y_2] \not\subseteq [8, 9] \lor x \in [8, 9]$, we see that each of the disjuncts has less variables than the set of input variables, which potentially makes it easier to solve. Instantiations of *Combine* always have this property. For some combinations of rules we do not use *Combine* since the result is too complicated to derive or too computationally expensive to support. In such cases we revert to clausal explanations.

Our experimental results indicates that indeed the new scheme is better than clausal explanation. For reference, we also compared HCSP to MISTRAL [16], CPX [4] and IZPLUS [13], where the last two won the second and first places, respectively, in the 'free-search, single-core' track of the 2014 'MiniZinc Challenge' (a CSP competition). HCSP performs better than these tools in terms of average run-time and the number of runs it is able to complete within the given time limit, although in optimization problems IZPLUS typically finds better solutions. HCSP performs an order of magnitude less backtracks than MISTRAL and three orders of magnitudes less backtracks than CPX, which proves that the constraints it learns are far more effective in pruning the search.

The rest of the article is structured as follows. The next section covers background material, including the learning framework that we use and clausal explanations [27]. Sections 3 and 4 describe the new set of inference rules, the requirements from them and the proofs that they fulfill these requirements. In Sec. 3 we also explain how we use clausal explanations as a fallback solution when we are unable to infer a general constraint that satisfies the required properties. We conclude in Sec. 5 with an empirical evaluation and some proposals for future research.

2 Background

Our solver HCSP supports all the constraint types specified in the FlatZinc format [23]. The engine of HCSP adopts classical ideas from the CSP and SAT literature. We assume the reader is mostly familiar with those, and only mention several highlights briefly for lack of space. It makes a *decision* (variable ordering) by selecting a variable with the highest ratio of *score* to domain-size, where *score* is calculated similarly to Chaff's VSIDS technique [22]. This can be seen as a variant of the *dom/wdeg* strategy [7]. The value is initially chosen to be the minimal value in the domain, and after that according to the last assigned value, a technique that is typically referred to by the name *phase saving* in SAT [26]. It includes *restarts*, *learning*, and *deletion* of learnt-constraints with low activity. The rest of this section is focused on the learning mechanism.

2.1 Conflict Analysis

Conflict-analysis and learning in HCSP is based on the familiar pattern of traversing backward the conflict graph and computing an asserting constraint.

Conflict-analysis was used in CSP before, but only while assuming that the constraints are signed clauses, as in MVS [21], or made into signed clauses via explanations (to be described in Sect. 2.2), as in [12,28]. The conflict-analysis in HCSP is not restricted to clausal inference, and includes various adaptations and optimizations as we describe now. Alg. 1 shows pseudo-code of ANALYZE-CONFLICT as implemented in HCSP. It maintains a set of nodes F, which is initialized to the set of nodes that contradict the input constraint cc. In line 4 it performs a *relaxation* of F. Relaxation appeared first in our technical report [28]; A similar idea appeared also in [17]. Relaxation means that each node in F is 'pushed' to the left as long as the constraint $Cons$ remains conflicting. Generally this is possible when domain reductions are redundant, as demonstrated in the following example.

Example 1. Consider the constraints

$$c_1 \doteq y \geq x \quad c_2 \doteq x \geq y \quad c_3 \doteq x > y + u .\tag{2}$$

and the conflict graph in Fig. 1 (left).

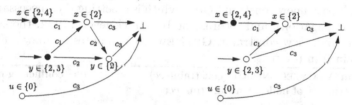

Fig. 1. Part of a conflict graph, based on the constraints in (2). Empty circles represent nodes in the set F. The left and right drawings are before and after relaxation, respectively. Relaxation discovers that the domain reduction by c_2 is not necessary for conflicting the constraint $Cons$ (c_3 in this case).

In Alg. 1, initially $Cons = c_3$, and hence after line 2 $F = \{x \in \{2\}, y \in \{2\}, u \in \{0\}\}$ (those are marked with empty circles). Relaxation in line 4 replaces in F the node $y \in \{2\}$ with the node $y \in \{2,3\}$, because the new F also contradicts the current constraint $Cons$. Fig. 1(right) shows this. The reason that this is possible is that the domain reduction by c_2 is redundant in the current state, because when $u = 0$, c_3 is capable of removing this value by itself. Such cases appear frequently, because the order in which constraints are processed is not optimal. □

Relaxation is necessary for several reasons: a) preventing a situation in which the learned clause is still conflicting immediately after backtracking, instead of being asserting, b) in Sec. 4.3 we rely on relaxation in the development of some of the inference rules, and c) our experiments show that without it many more cases fall back to clausal explanations, because relaxation enables to circumvent them.

Let us return to the description of Alg. 1. In lines 5–9 ANALYZECONFLICT gradually updates the constraint $Cons$. It does so by traversing the conflict graph backwards (i.e., going left, from the conflict node towards the decision node) while updating F and the constraint $Cons$ such that the following loop invariants are maintained:

1. ***Invar1.*** $Cons$ contradicts the domains defined by F, and is able to detect it via propagation.[2]
2. ***Invar2.*** No two nodes in F refer to the same variable.

It should be clear that these invariants are maintained at the entry to the loop, because of the definition of F, $Cons$, and relaxation. INFER and GETNEWSET are targeted towards maintaining it as will be evident later. The traversal stops in line 5 once the function STOP detects that $Cons$ is asserting, or that it conflicts the domains at decision level 0. In the latter case the function ASSERTINGLEVEL returns -1 to the solver, which accordingly declares the CSP to be unsatisfiable. In line 8 the current constraint $Cons$ is replaced with a constraint that is inferred from $Cons$ itself and the antecedent constraint of a node in F. The function INFER is the main contribution of this article and will be discussed in later sections.

Algorithm 1. ANALYZECONFLICT receives as input the currently conflicting constraint, learns a new constraint $Cons$ which is asserting (i.e., necessarily leads to further propagation), and returns the backtrack level. INFER, the subject of Sect. 3–4, infers a new constraint. GETNEWSET computes the new set of nodes F, as explained in the text.

```
 1: function ANALYZECONFLICT (constraint cc)          ▷ cc = conflicting constraint
 2:      F ← the set of nodes contradicting cc;
 3:      Cons ← cc;
 4:      F ← RELAX (F, Cons);
 5:      while !STOP (F, Cons) do          ▷ stop if Cons is asserting or UNSAT detected
 6:          pivot ← node of F that was propagated last;
 7:          antecedent ← incoming constraint of pivot;
 8:          Cons ← INFER (Cons, antecedent, pivot, F);
 9:          F ← GetNewSet(F, Cons, pivot);
10:          Remove from F nodes referring to variables not in Cons.
11:          F ← RELAX (F, Cons);          ▷ Go left as long as F contradicts Cons
12:      Add Cons to the constraints database;
13:      return ASSERTINGLEVEL (Cons, F);  ▷ the backtracking level, or -1 if UNSAT

14: function GETNEWSET(node-set F, node pivot)
15:      F ← (F \ {pivot}) ∪ parents of pivot;
16:      F ← DISTINCT (F);          ▷ Chooses right-most node of each variable in F
17:      Return F;
```

Let us now shift our focus to GETNEWSET, which updates the set F. Initially it replaces $pivot$ with its parents. In case there is more than one node in F representing the same variable, in line 16 the function DISTINCT leaves only the

---

[2] Detection is not a given, because not all constraints have a precise propagator, i.e., they are all sound but not all are complete. Bounds consistency is an example of such imprecise propagation.

right-most one. The reason that there may be multiple entries of a variable in F is that a parent of *pivot* may represent a variable that already labels a different node in F because of relaxation (line 11) in a previous iteration.

2.2   Clausal Explanations

Generic explanations were used in the past (e.g., [14,20]) for learning of g-nogoods. The scheme we describe here uses inference rules specialized for each constraint type, resulting in signed clauses. Such clausal explanations are important in our context both for understanding the alternative mechanism that we used in [27] (we use it as one of the points of reference for comparing the results), and because we still use it as a fallback solution when, e.g., we reach pairs of constraints for which we do not have an inference rule. Let us begin by formally defining the notion of explanation.

Definition 1 (Clausal explanation). *Let $l_1, \ldots, l_n$ be signed literals at the current state (each literal represents the current domain of a variable), and let c be a constraint that propagates the new signed literal l, i.e., $(l_1 \wedge \ldots \wedge l_n \wedge c) \to l$. Then a clause e is an* explanation *of this propagation if the following two conditions hold:*

$$c \to e \tag{3}$$

$$(l_1 \wedge \cdots \wedge l_n \wedge e) \to l . \tag{4}$$

Eq. (3) guarantees that the new clause e is logically implied by an existing constraint, hence we do not lose soundness. Eq. (4) guarantees that it is still strong enough to imply the same literal. It is always possible to derive an explanation from a constraint, regardless of the constraint type [27].

Example 2. The following rule from [27] provides a clausal explanation for an inequality constraint:

$$\frac{x \leq y}{x \in (-\infty, m] \vee y \in [m + 1, \infty)} \text{ (LE(m))} \tag{5}$$

where m is a parameter instantiating it (the rule is sound for any m). Note that the consequent is a signed clause. Now consider two literals:

$$l_1 \doteq (x \in [1, 3]), l_2 \doteq (y \in [0, 2])$$

and the constraint

$$c \doteq x \leq y ,$$

which implies in the context of l_1, l_2 the literal $l \doteq x \in [1, 2]$. Using (5) with $m = \max(y) = 2$ we obtain the explanation

$$e \doteq (x \in (-\infty, 2] \vee y \in [3, \infty)) ,$$

and indeed (3) and (4) hold, since $c \to e$ and $(l_1 \wedge l_2 \wedge e) \to l$. In [27] alternatives to choosing $m = \max(y)$ are discussed.   □

In [27] we showed how HCSP generates a signed conflict clause with an inference system based on signed resolution (1), that is reminiscent of how SAT solvers use binary resolution. Explanations are used for bridging between non-clausal constraints and a signed clause (as in the example above), and (1) is used for resolving signed clauses.

Example 3. The following demonstrates conflict analysis with clausal explanations. In addition to (5), we will use a variant of this rule for strict inequality:

$$\frac{x < y}{x \in (-\infty, m-1] \lor y \in [m+1, \infty)} \quad (\text{L(m)}) \tag{6}$$

We will also use the observation that if $c \to e$, then $(l \lor c) \to (l \lor e)$, to handle constraints with disjunctions. Let $D_x = \{0,1\}$, $D_y = \{0,1\}$, $D_z = \{0..100\}$, and

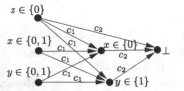

$c_1 \doteq (z = 9 \lor x < y) \quad c_2 \doteq (z = 10 \lor x \geq y)$. The conflict graph on the left shows the decision $(D_z = \{0\})$, and then that c_1 propagates $D_x = \{0\}$, $D_y = \{1\}$ in this order, and finally that c_2 detects a conflict. Now $F = \{z \in \{0\}, x \in \{0\}, y \in \{1\}\}$ and $pivot = \{y \in \{1\}\}$.

Then c_2 generates the explanation

$$(z \in \{10\} \lor x \in [1, \infty) \lor y \in (-\infty, 0])$$

based on LE(0) (see (5)), and c_1 generates the explanation

$$(z \in \{9\} \lor y \in [1, \infty) \lor x \in (-\infty, -1])$$

based on L(0) (see (6)). Resolving the two explanations on y yields

$$(z \in \{9, 10\} \lor x \notin \{0\}). \tag{7}$$

Now $pivot = x \in \{0\}$. c_1 explains the propagation of x with the clause

$$(z \in \{9\} \lor y \in [2, \infty) \lor x \in (-\infty, 0])$$

based on L(1). Resolving it with (7) on x yields

$$(z \in \{9, 10\} \lor x \in (\infty, -1] \lor y \in [2, \infty)) . \tag{8}$$

Now F is equal to the three nodes on the left. (8) is now asserting, since e.g., at the previous decision level $z \in \{9, 10\}$ is implied. □

3   Non-clausal Inference: Requirements

In Alg. 1 INFER is given the constraints $Cons(x, \vec{y})$ and $antecedent(x, \vec{y})$ with a joint variable x that appears at the node $pivot$, and some set of variables $\vec{y}$, which may or may not be common to both[3]. It outputs a new constraint over $x, \vec{y}$ that is assigned back into $Cons$. In the presentation that follows we will use $c_1(x, \vec{y})$ to denote $Cons(x, \vec{y})$, $c_2(x, \vec{y})$ to denote $antecedent(x, \vec{y})$, and $c^*(x, \vec{y})$ to denote the output constraint. We also define

$$c_{12}(x, \vec{y}) \doteq c_1(x, \vec{y}) \wedge c_2(x, \vec{y}).$$

Typically we will discard the parameters and write c_1, c_2, c^*, c_{12} instead.

Our first requirement from c^* is that it preserves soundness:

$$c_{12} \rightarrow c^* . \tag{9}$$

This guarantees that the constraint eventually learned in line 12 is inferred via sound derivations, and hence is guaranteed to be implied by the original CSP.

Let $D'_x, D'_{\vec{y}}$ denote the domains of $x, \vec{y}$ right before the propagation of c_1. Also, let $\vdash_{cp}$ denote the *provability relation* by constraints propagation, i.e., $\phi \vdash_{cp} \psi$ denotes that starting with a set of constraints and domains ϕ, the set of literals ψ is derivable through constraint propagation. Then to preserve *Invar1* (see Sec. 2), our second requirement from c^* is:

$$c^*, D'_x, D'_{\vec{y}} \vdash_{cp} \bot . \tag{10}$$

Finally, we aspire to find the strongest c^* that satisfies the above requirements, and which is easy to propagate.

4   Non-clausal Inference: Rules and Their Proofs

Rules R1–R7 in Table 1 are triples $\langle c_1, c_2, c^* \rangle$ that satisfy the two requirements (9) and (10). Rules R8 and R9 satisfy (9) but not necessarily (10). We use them to infer constraints, and then test whether they happen to satisfy (10). In addition, we use the following meta-rule for handling disjunctions:

$$\frac{(A \vee c_1) \quad (B \vee c_2)}{A \vee B \vee c^*} \tag{11}$$

If $\langle c_1, c_2, c^* \rangle$ satisfies (9) and (10), then so does (11). Detailed proofs for all of these rules can be found in a technical report extending this article [29].

Example 4. We now show two examples in which the rules lead to stronger learning than explanation-based learning

- Recall example 3, which yielded the conflict clause (8). Given the same conflict graph but using the meta rule (11) with pivot y, we learn instead $z \in \{9, 10\}$, which is clearly stronger.

---

[3] It is of course not necessarily the case that they share all the variables, but the description is simplified if we do not consider the shared and unshared variables separately, without sacrificing correctness.

Table 1. Triples $\langle c_1, c_2, c^* \rangle$ that we use for deriving conflict constraints. R1 – R7 are rules that satisfy both (9) and (10), whereas R8 – R9 are only guaranteed to satisfy (9). When using them we *test* if they also satisfy (10). The various min, max operators refer to the domain values at the point in time in which the rule is activated.

	c_1	c_2	c^*
R1	$x \in X_1 \vee A_1(\vec{y})$	$x \in X_2 \vee A_2(\vec{y})$	$x \in (X_1 \cap X_2) \vee A_1(\vec{y}) \vee A_2(\vec{y})$
R2	$y_1 \leq x - k_1$	$x \leq y_2 - k_2$	$(x \in [k_1 + \min(D'_{y_1}), \max(D'_{y_2}) - k_2]) \vee ([y_1, y_2 - k_2 - k_1] \not\subseteq [\min(D'_{y_1}), \max(D'_{y_2}) - k_2 - k_1])$
R3	$y_1 \leq x$	$[x, y_2] \not\subseteq [a, b]$	$(a > x \geq \min(D'_{y_1})) \vee ([y_1, y_2] \not\subseteq [\min(D'_{y_1}), b])$
R4	$x \leq y_1 - k_1$	$[y_2, x - k_2] \not\subseteq [a, b]$	$(\max(D'_{y_1}) - k_1 \geq x > b + k_2) \vee [y_2, y_1 - k_1 - k_2] \not\subseteq [a, \max(b, \max(D'_{y_1} - k_1 - k_2))]$
R5	$[y_1, x] \not\subseteq [a_1, b_1]$	$[x, y_2] \not\subseteq [a_2, b_2]$	$(a_2 > x > b_1) \vee ([y_1, y_2] \not\subseteq [a_1, b_2])$
R6	$[x, y] \not\subseteq [a_1, b_1]$	$[y, x] \not\subseteq [a_2, b_2]$	$(x \in (D'_y \setminus ([a_1, b_1] \cup [a_2, b_2])) \vee y \notin (D'_y \cup [a_1, b_1] \cup [a_2, b_2]))$
R7	$xor(x, \vec{y})$	$xor(x, \vec{z})$	$xor(\vec{y}, \vec{z}, 1)$
R8	$y \leq x + k_1$	$x \leq y + k_2$	$\begin{cases} -k_1 \leq x - y \leq k_2 & \text{if } k_1 + k_2 \geq 0 \\ \bot & \text{otherwise} \end{cases}$
R9	$ax + \sum_{i=1}^{n} a_i y_i \geq k_1$	$-ax + \sum_{i=1}^{n} b_i y_i \geq k_2$	$\sum_{i=1}^{n}(a_i + b_i)x_i \geq k_1 + k_2$

- Consider a variant of the example that was described in the introduction: $x \in \{2, 6, 10, 14, \ldots, 30\}, y_1 \in \{8, 12, 16, 20\}, y_2 \in \{1, 2, 3, \ldots, 9\}$, and constraints

$$c_1 \doteq (z \in \{1\} \vee y_1 \leq x) \quad c_2 \doteq (z \in \{1\} \vee x \leq y_2) \,.$$

Suppose we make a decision $z \in \{0\}$. Then c_1 propagates $x \in \{10, 14, \ldots, 30\}$ and c_2 detects a conflict. Using rule R2 with $k1 = k2 = 0$, and the meta rule (11) we obtain:

$$(z \in \{1\} \vee x \in [8, 9] \vee [y_1, y_2] \not\subseteq [8, 9]) \,. \tag{12}$$

On the other hand if we use explanations, c_2's explanation via LE(9) is $(z \in \{1\} \vee x \in (-\infty, 9] \vee y_2 \in [10, \infty))$, c_1's explanation via LE(7) is

$$(z \in \{1\} \vee y_1 \in (-\infty, 7] \vee x \in (8, \infty]) \,,$$

and resolving these explanations on the pivot x yields

$$(z \in \{1\} \vee x \in [8, 9] \vee y_1 \in (-\infty, 7] \vee y_2 \in [10, \infty)) \,.$$

This constraint is strictly weaker than (12) because the right disjunct of (12) implies $y_1 \leq y_2$. $\qquad \square$

Most of the entries in the table were developed by instantiating a general inference rule called *Combine* (see Sec. 4.1 below), which satisfies these requirements. In some other cases instantiating it turned out to be too complicated and we found c^* without it. Sec. 4.3 includes proofs for some of these other rules.

Since not all combinations of rule types are supported, not all propagators are precise (i.e., logically complete) and not all rules are precise (see R8, R9 in the table), then INFER uses explanation-based inference (see Sec. 2.2) as a fallback solution. Pseudocode of INFER, which is rather self-explanatory, appears in Alg. 2.

Algorithm 2. INFER infers a new constraint c^* from c_1, c_2, which satisfies (9) and (10), the requirements listed in Sec. 3.

---

function INFER(constraint c_1, constraint c_2, node *pivot*, node-set F)
   $F' = $ GETNEWSET $(F, pivot)$;
   if the combination of c_1, c_2 is supported **then**
      $con = $ Combine $(c_1, c_2, pivot)$;   ▷ One of the rules in Table 1.
      if $F', con \vdash_{cp} \bot$ **then return** con;   ▷ con satisfies *Invar1*
   $e_1 \leftarrow explain(c_1, parents(pivot), pivot)$;   ▷ Fallback: use explanations.
   $e_2 \leftarrow explain(c_2, F, \bot)$;
   return resolve$(e_1, e_2, pivot)$;   ▷ Signed resolution

4.1 A Generic Inference Rule: *Combine*

Let S be some set of values. Then it is not hard to see that the following is a contradiction for any constraint $c(x, \vec{y})$:

$$c(x, \vec{y}) \wedge x \in S \wedge \forall x' \in S. \neg c(x', \vec{y}), \tag{13}$$

or, equivalently, that the following implication is valid:

$$c(x, \vec{y}) \rightarrow (x \notin S \vee \exists x' \in S. c(x', \vec{y})). \tag{14}$$

Let $\mathcal{X}$ denote the set of values of x which have no support in $D'_{\vec{y}}$:

$$\mathcal{X} = \{x' \mid \forall \vec{y'} \in D'_{\vec{y}}. \neg c_{12}(x', \vec{y'})\}. \tag{15}$$

Instantiating (14) with c_{12} for c and with $\mathcal{X}$ for S yields the inference rule that we call *Combine*:

$$\frac{c_{12}(x, \vec{y})}{(x \notin \mathcal{X} \vee \exists x' \in \mathcal{X}. c_{12}(x', \vec{y}))} \quad (Combine) \tag{16}$$

Since (16) is just an instantiation of (14), then (16) is clearly sound, and hence (9) is satisfied. To satisfy (10) we first prove logical entailment ($\models$), which is weaker than the requirement of (10) for provability ($\vdash_{cp}$).

Lemma 1. $c^*, D'_x, D'_{\bar{y}} \models \perp$.

Proof. In our case $c^* \doteq (x \notin \mathcal{X} \vee \exists x' \in \mathcal{X}.c_{12}(x', \vec{y}))$. Falsely assume that c^* is satisfied for an assignment of values $a \in D'_x, \vec{b} \in D'_{\bar{y}}$ to $x, \vec{y}$, respectively. Consider the two disjuncts of c^*:

- Suppose $x \notin \mathcal{X}$ is satisfied. Considering the definition of $\mathcal{X}$ in (15), this implies that a is supported in c_{12}, or formally

$$\exists \vec{y'} \in D'_{\bar{y}}. \, c_{12}(a, \vec{y'}) \, . \tag{17}$$

  Based on *Invar1* we know that $c_{12}(x, \vec{y}), D'_x, D'_{\bar{y}} \models \perp$, and hence $\forall x \in D'_x. \neg \exists \vec{y} \in D'_{\bar{y}}. \, c_{12}(x, \vec{y})$, and particularly for $x = a$, $\neg \exists \vec{y} \in D'_{\bar{y}}. \, c_{12}(a, \vec{y})$, which contradicts (17).
- Now suppose $\exists x' \in \mathcal{X}. \, c_{12}(x', \vec{y})$ is satisfied. Expanding $\mathcal{X}$ and substituting $\vec{y}$ with its assignment $\vec{b}$ yields

$$\exists x'. \, \forall \vec{y'} \in D'_{\bar{y}}. \, \neg c_{12}(x', \vec{y'}) \wedge c_{12}(x', \vec{b}) \, .$$

  Since $\vec{b} \in D'_{\bar{y}}$ and $\neg c_{12}(x', \vec{y'})$ is satisfied for all $\vec{y'} \in D'_{\bar{y}}$, then it is satisfied for $\vec{y'} = \vec{b}$. This implies a contradiction: $\exists x'. \, \neg c_{12}(x', \vec{b}) \wedge c_{12}(x', \vec{b}) \, .$

Hence, $x \in D'_x, \vec{y} \in D'_{\bar{y}}$ falsifies c^*, which completes our proof.   □

It is trivial to see that this lemma implies (10) when $\vdash_{cp}$ is precise constraint propagation. When imprecise propagation is involved, e.g., $\vdash_{cp}$ is defined by bounds consistency [8], HCSP checks whether the constraint happens to be conflicting, and if not it falls back to clausal explanation.

The Relative Strength of *Combine*. Two observations about the strength of *Combine* that we prove in [29] are:

- There is no alternative to $\mathcal{X}$ for replacing S in (14) that makes the resulting constraint stronger, and
- The signed clause that we obtain through the explanation mechanism—see Sec. 2.2—cannot yield a stronger consequent.

4.2   Selected Rules Based on Instantiating *Combine*

We now instantiate *Combine* (16) with several specific constraints of interest.

Rule R2: $c_1 \doteq y_2 - x \geq k_2$ $c_2 \doteq x - y_1 \geq k_1$
Expanding c_{12} in (15) yields

$$\begin{aligned}
\mathcal{X} &= \left\{ x \mid \forall \vec{y'} \in D'_{\bar{y}}. \, [y_2 - x < k_2 \vee x - y_1 < k_1] \right\} \\
&= \left\{ x \mid \max(D'_{y_2}) - x < k_2 \vee x - \min(D'_{y_1}) < k_1 \right\} \\
&= \left\{ x \mid \max(D'_{y_2}) - k_2 < x \vee x < k_1 + \min(D'_{y_1}) \right\} \, .
\end{aligned}$$

The complement of $\mathcal{X}$ can be written as

$$\mathcal{X}^c = \left[k_1 + \min(D'_{y_1}),\ \max(D'_{y_2}) - k_2\right] . \tag{18}$$

Recall (15): $x \notin \mathcal{X} \vee \exists x' \in \mathcal{X}.\ c_{12}(x', \vec{y})$. The right disjunct is equal to:

$$\exists x'.\ x' \in \mathcal{X} \wedge [y_2 - x' \geq k_2 \wedge x' - y_1 \geq k_1]$$
$$= \exists x'.\ x' \in \mathcal{X} \wedge [y_2 - k_2 \geq x' \geq y_1 + k_1]$$
$$= \exists x'.\ x' \in \mathcal{X} \wedge x' \in [y_1 + k_1, y_2 - k_2] . \tag{19}$$

We use (18) to rewrite (19):

$$\exists x'.\ x' \notin \left[k_1 + \min(D'_{y_1}),\ \max(D'_{y_2}) - k_2\right] \wedge x' \in [y_1 + k_1, y_2 - k_2] ,$$

which implies

$$[y_1 + k_1, y_2 - k_2] \not\subseteq \left[k_1 + \min(D'_{y_1}),\ \max(D'_{y_2}) - k_2\right]$$
$$= [y_1, y_2 - k_2 - k_1] \not\subseteq \left[\min(D'_{y_1}),\ \max(D'_{y_2}) - k_2 - k_1\right] .$$

Hence, the rule is

$$\frac{y_2 - x \geq k_2 \qquad x - y_1 \geq k_1}{\left(x \in \left[k_1 + \min(D'_{y_1}),\ \max(D'_{y_2}) - k_2\right] \vee \right.} \tag{20}$$
$$\left. [y_1,\ y_2 - k_2 - k_1] \not\subseteq \left[\min(D'_{y_1}),\ \max(D'_{y_2}) - k_2 - k_1\right]\right)$$

Rule R7: $c_1 \doteq xor(x, \vec{y}) \qquad c_2 \doteq xor(x, \vec{z})$
Here $\vec{y} \doteq y_1, \ldots, y_n$ and $\vec{z} \doteq z_1, \ldots, z_m$. We assume that $\vec{y}$ and $\vec{z}$ are fully assigned and x is not; we further assume that, w.l.o.g, c_1 propagates the value of x and c_2 detects a conflict.

Under these assumptions it is clear that c_1 and c_2 cannot be simultaneously satisfied by either $x = 0$ or $x = 1$, and hence by definition $\mathcal{X} = \{0, 1\}$, and Combine amounts to:

$$\frac{xor(x, \vec{y}) \qquad xor(x, \vec{z})}{x \notin \{0, 1\} \vee \exists x' \in \{0, 1\}.\ [xor(x', \vec{y}) \wedge xor(x', \vec{z})]} .$$

We can replace $x \notin \{0, 1\}$ with **false** and obtain:

$$\frac{xor(x, \vec{y}) \qquad xor(x, \vec{z})}{\exists x'.\ [xor(x', \vec{y}) \wedge xor(x', \vec{z})]} .$$

It is not hard to see that the consequent is equivalent to $xor(\vec{y}) = xor(\vec{z})$, and hence also to $xor(\vec{y}, \vec{z}, 1)$, which brings us to the desired rule:

$$\frac{xor(x, \vec{y}) \qquad xor(x, \vec{z})}{xor(\vec{y}, \vec{z}, 1)} .$$

Note that variables that are shared by $\vec{y}$ and $\vec{z}$ can be removed from the xor predicate without changing its value.

4.3   Selected Rules Not Based on *Combine*

Rule R3: $c_1 \doteq y_1 \le x$ $c_2 \doteq [x, y_2] \not\subseteq [a, b]$

We assume that at the point of conflict, replacing c_2 with $x \le y_2$ makes c_{12} too weak to detect the conflict. Otherwise we simply use rule R2. Based on this assumption, which we denote by ψ, in [29] we show that here

$$\mathcal{X} = \{x' \mid x' < \min(D'_{y_1}) \vee a \le x'\} . \tag{21}$$

We propose the following consequent:

$$c^* \doteq x \notin \mathcal{X} \vee [y_1, y_2] \not\subseteq [\min(D'_{y_1}), b] \tag{22}$$
$$= a > x \ge \min(D'_{y_1}) \vee [y_1, y_2] \not\subseteq [\min(D'_{y_1}), b] . \tag{23}$$

Note that c^* still follows our general pattern, by which the pivot is separated and not referred-to by the other disjunct. Since we cannot rely on the correctness of the general rule, we now prove that (23) satisfies (9) and (10):

- Eq. (9): Falsely assume the contrary, i.e., there are x, y_1, y_2 such that

$$a \le \max(D'_{y_2}) \le b \wedge \min(D'_{y_1}) \le \max(D'_x) \wedge a \le \max(D'_x) \wedge y_1 \le x$$
$$\wedge [x, y_2] \not\subseteq [a, b] \wedge x \in \mathcal{X} \wedge [y_1, y_2] \subseteq [\min(D'_{y_1}), b] .$$

  Expanding $\mathcal{X}$ yields

$$a \le \max(D'_{y_2}) \le b \wedge \min(D'_{y_1}) \le \max(D'_x) \wedge a \le \max(D'_x) \wedge y_1 \le x$$
$$\wedge [x, y_2] \not\subseteq [a, b] \wedge (x < \min(D'_{y_1}) \vee a \le x) \wedge [y_1, y_2] \subseteq [\min(D'_{y_1}), b] .$$

  If $x < \min(D'_{y_1})$ then $y_1 \le x$ implies $y_1 < \min(D'_{y_1})$ which conflicts $[y_1, y_2] \subseteq [\min(D'_{y_1}), b]$. Otherwise, $x \ge a$ and $[x, y_2] \not\subseteq [a, b]$ implies $y_2 > b$ which conflicts $[y_1, y_2] \subseteq [\min(D'_{y_1}), b]$.
- Eq. (10): Proved in [29].

To summarize, the rule is

$$\frac{y_1 \le x \qquad [x, y_2] \not\subseteq [a, b] \qquad \psi}{a > x \ge \min(D'_{y_1}) \vee [y_1, y_2] \not\subseteq [\min(D'_{y_1}), b]} . \tag{24}$$

Rule R8: $c_1 \doteq (y \le x + k_1)$ $c_2 = (x \le y + k_2)$

Isolating $x - y$ on both sides yields $c_{12}(x, y) = -k_1 \le x - y \le k_2$, which is false if $k_1 + k_2 < 0$. Since it is simply a conjunction of the input constraints, then (9) and (10) are satisfied trivially.

5   Experimental Results

We performed two sets of experiments as described below. All experiments were ran on a 4 core Intel® Xeon® 2.5GHz.

The 2009 CSP solver competition benchmarks. Here we used a subset of benchmarks of the Fourth International CSP Solver competition [2] (this was the last CSP competition, before the MiniZinc challenge started). Specifically out of over 7000 in the competition's satisfiability benchmark-set, we focused on the 2162 benchmarks that have at least one comparison operator from $\{<, \leq, \geq, >\}$ (the reason being that the rules in Table 1 refer to combinations of constraints based on these operators and constraints that are consequents of these rules). The CPU time limit was set to 1200 seconds. Out of memory and time-outs are called 'fails' in the discussion below.

We compared three different settings: (1) HCSP with general constraints learning based on *Combine* (from hereon—HCSP), (2) HCSP using only clause-based learning with explanations, as described in Sec. 2 (from hereon—EXPLAIN)[4], and (3) MISTRAL [16] latest version (1.550). Fig. 2 compares these three engines.

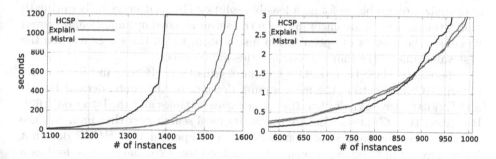

Fig. 2. Number of instances solved within the given time limit comparing HCSP, EXPLAIN, and MISTRAL. (left) Shows the time in linear scale; (right) A zoom-in of the left figure showing the cross-over between MISTRAL and HCSP occurring after 1-2 seconds.

Memory was limited to 1 GiB. Number of fails in HCSP was 25% less than MISTRAL. Number of fails ofHCSP was 4.9% less than EXPLAIN. The average number of backtracks in HCSP is 2045, in EXPLAIN 4389, and in MISTRAL 49562. This drastic difference in the average backtrack-count indicates that the cost of learning is compensated-for by a better search.

MiniZinc benchmarks. Given the recent results of the MiniZinc challenge, we compared HCSP and EXPLAIN to CPX and IZPLUS, which won the second and

---

[4] We emphasize that this is a far-improved engine in comparison to [27], owing to numerous optimizations that are beyond the scope of the current article.

first places, respectively, of the 'free-search/single-core' track of the MiniZinc challenge [3].[5] IZPLUS is based on the iZ-C constraint programming library, and includes stochastic local search for optimization problems. CPX is based on a lazy clause generation (i.e., lazy reduction to SAT). We used all the 100 benchmarks of the competition, 75 of which are optimization problems. The time-limit was set to 1800 sec., and the memory limit to 3GB. All benchmarks were converted to the FlatZinc format prior to benchmarking. The following table summarizes the results:

	time (avg.)	*time* (med.)	*backtracks*	*success*	*opt.*	*wins*
HCSP	897.8	686.3	35136.2	95	25	35
EXPLAIN	965.0	1232.1	37206.2	93	25	39
CPX	1055.0	1786.5	18451225.5	85	20	30
IZPLUS	972.7	1475.5		88	25	59

Detailed results can be found in [1]. The columns should be interpreted as follows: *time* (avg. and median) – the number of seconds in all benchmarks, including time-out and memout cases; *backtracks* – the number of backtracks in benchmarks in which all engines finished successfully before the time-out[6]; *success* – the number of instances solved within time and memory limits, and in the case of optimization problems found a feasible solution (but not necessarily optimal); *Opt.* – the number of instances in which the solver reached optimality and proved it; *wins* – the number of optimization instances in which the solver reached the best value among the four contenders (ties are counted).

The results show that IZPLUS has more wins than HCSP, and in all other criteria HCSP is better (the *wins* column depends on the contenders. If HCSP and EXPLAIN are on their own, the former wins 65 times and the latter only 62). It is likely that IZPLUS's wins are due to its local search part that improves the objective function once a solution is found, a component that HCSP does not have. Overall in these experiments 40% of the cases explanation was used as a fall-back solution.

Conclusion and Future Work. We have presented a new learning scheme based on inference of general constraints. We presented the development of various inference rules that are necessary for this scheme, but it is clear that there is still a lot of work in deriving such rules for additional popular pairs of constraints which are currently not supported and force HCSP into a fallback solution. In addition, currently learning general constraints is incompatible with producing machine-checkable proofs in case the formula is unsatisfiable, in contrast to our earlier explanation-based method [27]. HCSP is written in C++, contains 23k lines of non-comment code, and its architecture enables the addition of new constraints and new rules without changing the core solver. It is free software available from [1] under the GPL license.

---

[5] An early version of HCSP also participated in that competition and reached the 5th place. Since then we improved HCSP in multiple ways, including better data-structures and specialized code (instead of generic) to generate explanations for 'element' global constraints, e.g., var_vector[var_i] = var_0.

[6] IZPLUS, a closed-source program, does not print this information.

References

1. The HCSP Constraint Solver web page. http://tx.technion.ac.il/mveksler/HCSP/
2. Fourth international CSP solver competition (2009). http://cpai.ucc.ie/09/index.html
3. Minizinc challenge (2014). http://www.minizinc.org/challenge2014/
4. Opturion CPX - tool description (2014). http://www.minizinc.org/challenge2014/description_opturion_cpx.txt
5. Beckert, B., Hähnle, R., Manyá, F.: The SAT problem of signed CNF formulas, pp. 59–80 (2000)
6. Biere, A.: Bounded model checking. In: Biere, A., Heule, M., van Maaren, H., Walsh, T. (eds.) Handbook of Satisfiability. Frontiers in Artificial Intelligence and Applications, vol. 185, pp. 457–481. IOS Press (2009)
7. Boussemart, F., Hemery, F., Lecoutre, C., Sais, L.: Boosting systematic search by weighting constraints. In: López de Mántaras, R., Saitta, L. (eds.) ECAI, pp. 146–150. IOS Press (2004)
8. Choi, C.W., Harvey, W., Lee, J.H.M., Stuckey, P.J.: Finite domain bounds consistency revisited. In: Sattar, A., Kang, B.-H. (eds.) AI 2006. LNCS (LNAI), vol. 4304, pp. 49–58. Springer, Heidelberg (2006)
9. Cotton, S., Maler, O.: Fast and flexible difference constraint propagation for DPLL(T). In: Biere, A., Gomes, C.P. (eds.) SAT 2006. LNCS, vol. 4121, pp. 170–183. Springer, Heidelberg (2006)
10. Dechter, R.: Enhancement schemes for constraint processing: backjumping, learning, and cutset decomposition. Artif. Intell. **41**(3), 273–312 (1990)
11. Dixon, H.E., Ginsberg, M.L.: Inference methods for a pseudo-boolean satisfiability solver. In: Dechter, R., Sutton, R.S. (eds.) Proceedings of the Eighteenth National Conference on Artificial Intelligence and Fourteenth Conference on Innovative Applications of Artificial Intelligence, July 28 - August 1, 2002, Edmonton, Alberta, Canada, pp. 635–640. AAAI Press / The MIT Press (2002)
12. Feydy, T., Stuckey, P.J.: Lazy clause generation reengineered. In: Gent, I.P. (ed.) CP 2009. LNCS, vol. 5732, pp. 352–366. Springer, Heidelberg (2009)
13. Fujiwara, T.: iZplus - tool description (2014). http://www.minizinc.org/challenge2014/description_izplus.txt
14. Gent, I.P., Miguel, I., Moore, N.C.A.: Lazy explanations for constraint propagators. In: Carro, M., Peña, R. (eds.) PADL 2010. LNCS, vol. 5937, pp. 217–233. Springer, Heidelberg (2010)
15. Stuckey, P.J., Chu, G.: Structure based extended resolution for constraint programming. http://arxiv.org/abs/1306.4418
16. Hebrard, E.: Mistral, a constraints satisfiaction library. In: Third international CSP solver competition, pp. 31–40 (2008)
17. Jain, S., Sabharwal, A., Sellmann, M.: A general nogoodlearning framework for pseudo-boolean multi-valued SAT. In: Burgard, W., Roth, D. (eds.) Proceedings of the Twenty-Fifth AAAI Conference on Articial Intelligence, AAAI 2011, San Francisco, California, USA, August 7–11, 2011. AAAI Press (2011)
18. Jovanovic, D., de Moura, L.M.: Cutting to the chase - solving linear integer arithmetic. J. Autom. Reasoning **51**(1), 79–108 (2013)
19. Katsirelos, G., Bacchus, F.: Unrestricted nogood recording in CSP search. In: Rossi, F. (ed.) CP 2003. LNCS, vol. 2833, pp. 873–877. Springer, Heidelberg (2003)
20. Katsirelos, G., Bacchus, F.: Generalized nogoods in CSPs. In: Veloso, M.M., Kambhampati, S. (eds.) AAAI, pp. 390–396. AAAI Press/The MIT Press (2005)

21. Liu, C., Kuehlmann, A., Moskewicz, M.W.: CAMA: a multi-valued satisfiability solver. In: ICCAD, pp. 326–333. IEEE Computer Society/ACM (2003)
22. Moskewicz, M., Madigan, C., Zhao, Y., Zhang, L., Malik, S.: Chaff: engineering an efficient SAT solver. In: Proc. Design Automation Conference (DAC 2001) (2001)
23. Nethercote, N., Stuckey, P.J., Becket, R., Brand, S., Duck, G.J., Tack, G.R.: MiniZinc: towards a Standard CP Modelling Language. In: Bessière, C. (ed.) CP 2007. LNCS, vol. 4741, pp. 529–543. Springer, Heidelberg (2007)
24. Nieuwenhuis, R.: The IntSat method for integer linear programming. In: O'Sullivan, B. (ed.) CP 2014. LNCS, vol. 8656, pp. 574–589. Springer, Heidelberg (2014)
25. Ohrimenko, O., Stuckey, P.J., Codish, M.: Propagation via lazy clause generation. Constraints **14**(3), 357–391 (2009)
26. Strichman, O.: Tuning SAT checkers for bounded model checking. In: Emerson, E.A., Sistla, A.P. (eds.) CAV 2000. LNCS, vol. 1855, pp. 480–494. Springer, Heidelberg (2000)
27. Veksler, M., Strichman, O.: A proof-producing CSP solver. In: Proceedings of the Twenty-Fourth AAAI Conference on Artificial Intelligence (2010)
28. Veksler, M., Strichman, O.: A proof-producing CSP solver (a proof supplement). Technical Report IE/IS-2010-02, Industrial Engineering, Technion, Haifa, Israel, January 2010. Available also from [1]
29. Veksler, M., Strichman, O.: Learning non-clausal constraints in csp (long version). Technical report, Technion, Industrial Engineering, IE/IS-2014-05 (2014). Available also from [1]

Understanding the Potential of Propagators

Sascha Van Cauwelaert[1]([✉]), Michele Lombardi[2], and Pierre Schaus[1]

[1] UCLouvain, Louvain, Belgium
{sascha.vancauwelaert,pierre.schaus}@uclouvain.be
[2] University of Bologna, Bologna, Italy
michele.lombardi2@unibo.it

Abstract. Propagation is at the very core of Constraint Programming (CP): it can provide significant performance boosts as long as the search space reduction is not outweighed by the cost for running the propagators. A lot of research effort in the CP community is directed toward improving this trade-off, which for a given type of filtering amounts to reducing the computation cost. This is done chiefly by 1) devising more efficient algorithms or by 2) using on-line control policies to limit the propagator activations. In both cases, obtaining improvements is a long and demanding process with uncertain outcome. We propose a method to assess the potential gain of both approaches before actually starting the endeavor, providing the community with a tool to best direct the research efforts. Our approach is based on instrumenting the constraint solver to collect statistics, and we rely on *replaying* search trees to obtain more realistic assessments. The overall approach is easy to setup and is showcased on the Energetic Reasoning (ER) and the Revisited Cardinality Reasoning for BinPacking (RCRB) propagators.

Keywords: Constraint programming · Propagator · Analysis · Energetic Reasoning · BinPacking

1 Introduction

Propagation is undoubtedly one of the signature features of Constraint Programming (CP): it makes a constraint solver capable of skipping large portions of the search space, possibly achieving significant performance boosts. In practice, the effectiveness of the approach depends on the balance between the time saved by filtering values and the time spent in running the propagators. Improving this trade-off is the objective of huge research efforts in the CP community.

Here, *we consider the specific case where the goal is to optimize the performance of a given propagation technique, without changing its input-output behavior.* For example, we may be interested in finding a more efficient way to enforce Generalized Arc Consistency (GAC) for a specific constraint. In general, this goal can be achieved by either 1) devising more efficient algorithms that achieve the same filtering, or by 2) guarding the activation of the propagator

© Springer International Publishing Switzerland 2015
L. Michel (Ed.): CPAIOR 2015, LNCS 9075, pp. 427–436, 2015.
DOI: 10.1007/978-3-319-18008-3_29

with a necessary condition to reduce fruitless activations[1]. In both cases, obtaining improvements is a long and demanding process with uncertain outcome.

As an example, the SEQUENCE constraint was introduced in 1994 [5], but no poly-time GAC algorithm was available until 2006 [21]. Then, the original GAC run time of $O(n^3)$ was not low enough to consistently beat weaker (but cheaper) propagators. This motivated improvement efforts that are still ongoing [6,9,11]. The trade-off between computation time and pruning power is even more critical for NP-hard constraints. For example, Energetic Reasoning (ER) was proposed as a (powerful) filtering technique for CUMULATIVE in the nineties (see [3,14]): however, the approach has never been widely employed due to its large run time. Improving the original $O(n^3)$ complexity took in this case around 20 years [12], while an approach to reduce the overhead by guarding the ER activation with a necessary condition was presented only in 2011 [7].

In general, *this line of research would greatly benefit from tools and methods to probe the potential of propagation techniques and to assess the likely impact of specific improvement measures.* Such tools would allow the researchers to focus their efforts in the most promising directions (notice that for preliminary analysis, profiling tools already allow to reason about potential linear speedups).

A typical approach for evaluating propagators consists in measuring time and fails w.r.t. a baseline propagator, on a set of benchmark instances that are solved to completeness. This allows to asses the propagator performance, but provides little or no information on how to improve it. It is also common to use static search strategies to make the evaluation fair and rigorous, with the risk to reduce the analysis significance, since dynamic strategies are often preferred in practice. Finally, the need to solve the problems to completeness may bias the analysis toward relatively small instances.

We propose to extend this basic evaluation approach by: 1) instrumenting the solver to collect information about the constraint; 2) storing and *replaying* search trees to enable fair comparisons with arbitrary search strategies and instance sizes. Our approach is simple and allows to assess the amount of improvement that could be obtained by reducing the propagator run-time or by controlling its activation. We use the Energetic Reasoning (ER) and the Revisited Cardinality Reasoning for BinPacking (RCRB) propagators as case studies.

2   The Proposed Approach

Formally, we consider the problem of evaluating a filtering function ϕ that maps a set of domains $D_0, \ldots D_{n-1}$ to a second set of domains $D'_0, \ldots D'_{n-1}$ such that $D'_i \subseteq D_i$. In practice, ϕ may represent a propagator for enforcing GAC or a domain-specific consistency level (e.g. Energetic Reasoning), or it can be some kind of meta-propagation scheme such as Singleton Arc Consistency [8].

---

[1] A more general approach consists in trying to *predict* when the propagator should be triggered: we plan to develop tools to analyze this scenario as part of future research.

We assume we are interested in reducing the time for computing ϕ, without changing the function definition. In particular, our goal is to assess the potential of two improvement directions: 1) increasing the efficiency of the current implementation and 2) guarding the activation of ϕ with a necessary condition.

Measuring the Performance: In order to make such an assessment, we must first be able to measure the performance of the current implementation of ϕ. Like many other approaches, we do this by comparing the time needed to solve a target CSP with and without ϕ. Let M and $M \cup \phi$ denote the two CSPs, with their variables, domains, and constraints. For the comparison to be meaningful, two well known conditions must be respected:

1. The two runs must explore the same search space;
2. All search nodes that are visited by both runs are visited in the same order.

The first requirement is always met as long as M and $M \cup \phi$ are semantically equivalent (i.e. they have the same solutions) and the problem is solved to completeness (feasibility or optimality).

Without the second requirement, one of the approaches could get an unfair advantage if the search strategy quickly hits a feasible solution (and stops, for feasibility problems), or a high-quality solution (and gets a good bound, for optimality problems). Moreover, if the second requirement is satisfied, then the nodes visited when solving $M \cup \phi$ will always be a subset of (or the same as) those visited when solving M. Typically, this is all guaranteed by using static search strategies. As an alternative, we propose an approach based on *replaying* search trees, which does not suffer from most of the drawbacks discussed in Section 1.

The Replay Technique: For the sake of precision, it is useful to introduce some notation at this point. As it is quite common in CP, we view tree search as a recursive process, where the search space is iteratively decomposed by opening choice points and posting constraints on each branch. Formally, we can define a search strategy as a function b that given the current state of the search and of a problem M returns a sequence of constraints $c_0, c_1, \ldots$ to be posted each on a different branch. By "search state" we refer to search parameters that are not part of the model (i.e. time markers for the *SetTimes* strategy). The whole search process can be seen as the evaluation of a recursive function $traverse(b, M)$ having as parameters the search strategy b and the target problem M.

We guarantee the satisfaction of both requirements for measuring the performance by storing in a tree-like structure, during one run: 1) the branching constraints and 2) the search state. We then force the following run to post exactly the same constraints at the same search nodes. This is done by introducing two wrapper search strategies called $store(b)$ and $replay(b)$ that respectively memorize and re-post the constraints returned by the strategy b. Then, in order to evaluate a propagator ϕ, we simply run in sequence:

$$traverse(store(b), M) \tag{1}$$

$$traverse(replay(b), M) \tag{2}$$

$$traverse(replay(b), M \cup \phi) \tag{3}$$

and we compare the results of the two latter runs, which both use $replay(b)$ and hence incur the same search overhead. It is important that the first run is done with the baseline problem M, because, thanks to the additional propagation performed by ϕ, the run with $M \cup \phi$ may skip some parts of the stored tree. However, all of the runs will always explore the same space and visit the shared nodes in the same order.

This approach offers two significant advantages: 1) it allows to tackle arbitrarily large instances, since a time limit can be enforced on the first run and the second run will still be guaranteed to explore the same search space. 2) It allows to use any search strategy, including dynamic ones, making the evaluation more realistic. The comparison remains artificial to some degree, because an actual dynamic strategy may behave differently on the two runs. Still, the ability to make fair comparisons using an arbitrary strategy is a very valuable contribution. Our replay technique is easy to implement on mosts solvers that allow the user to write custom search strategies.

Assessing the Propagator Potential: In order to assess the potential of improving the efficiency of ϕ or controlling its activation, we instrument the solver to collect detailed information about the propagator. Specifically, we store the total time for running ϕ, making a distinction between activations that actually lead to some pruning and fruitless activations. The two time statistics are respectively referred to as t_ϕ^+ and t_ϕ^-. We collect the information by introducing a wrapper function $stats(\phi)$ that checks the domain sizes, then runs ϕ, and finally checks the domains again and stores the elapsed time. The overhead for the collection process is properly subtracted. Once again, this approach is easy to implement on most solvers that allow the user to write new propagators.

It is now easy to get a rough, but valuable, estimate of the impact of specific measures on the solution time. In particular, let $t(b, M)$ be the time required to solve the problem M with the strategy b (i.e. to run $traverse(b, M)$). Then we can estimate the impact of reducing the run time of ϕ by a factor $\mu \in [0,1]$ by computing:

$$t(replay(b), M \cup stats(\phi)) - \mu \cdot (t_\phi^+ + t_\phi^-) \tag{4}$$

i.e. by subtracting a fraction of the total computation time of ϕ. Similarly, we can assess the impact of guarding ϕ with a necessary condition that stops a fraction $\mu \in [0,1]$ of the fruitless propagator activations. This is done by computing:

$$t(replay(b), M \cup stats(\phi)) - \mu \cdot (t_\phi^-) \tag{5}$$

This simple, linear, approach allows us to compare *fictional* implementations of ϕ with real ones. By doing so, we get a chance to explore which values of μ would be necessary for beating the baseline, and we get a better understanding of the effort required to achieve such goal. In particular, we can approximately evaluate the impact of having an hypothetical time complexity for a fictional propagator. For instance, if the current implementation for ϕ is in $O(n^3)$ (where n is the number of variables), then we can estimate roughly what would be its cost for an $O(n^2)$ algorithm by choosing $\mu = (n - 1)/n$ in equation 4.

Deeper and more general insights can be obtained by comparing fictional and real propagators on full benchmarks. To this purpose, we rely on *performance profiles* [13]. A performance profile is a cumulative distribution function $F(\tau)$ of a given performance metric τ. In our case, the τ value is the ratio between the solution time of a target approach and that of the baseline. For the sake of clarity, if $F(2) = 0.75$ for an approach, it means that its performance is within a factor of 2 from the baseline in 75% of the benchmark problems. Assuming the benchmark is representative enough, the value of $F(\tau)$ can be interpreted as a probability.

Formally, let $\phi_0, \phi_1, \ldots$ be the set of all considered implementations of ϕ (real and fictional alike), and let $\mathcal{M}$ be the set for all problems (instances) in the benchmark. Then the performance profile of ϕ_i is given by:

$$F_{\phi_i}(\tau) = \frac{1}{|\mathcal{M}|} \left| \left\{ M \in \mathcal{M} : \frac{t(replay(b), M \cup \phi_i)}{t(replay(b), M)} \leq \tau \right\} \right| \qquad (6)$$

where $t(replay(b), M \cup \phi_i)$ for fictional implementations of ϕ is computed using Equation (4) or (5).

Reading of Performance Profiles: An important value of a given performance profile $F_{\phi_i}(\tau)$ is in $\tau = 1$. For a given ϕ_i, $F_{\phi_i}(\tau = 1)$ gives the percentage of instances that can be solved using $M \cup \phi_i$ in a time less (or the same) time as the baseline model M. Although F_M is not represented, it would actually be a step function $F_M(\tau < 1) = 0$ and $F_M(\tau \geq 1) = 100\%$. The space of τ is therefore divided in two important regions, $\tau < 1$ and $\tau \geq 1$. If $F_{\phi_i}(\tau) = 100\%$ for some $\tau < 1$, then using the model $M \cup \phi_i$ is always better than using the baseline, i.e. $M \cup \phi_i$ provides a speed-up for every instance. Unfortunately, this situation rarely happens in practice and it is thus interesting to read more carefully the performance profile. For a given pair ϕ_i, ϕ_j it is interesting to observe $F_{\phi_i}(\tau)$ - $F_{\phi_j}(\tau)$. It indicates the *gain* of ϕ_i over ϕ_j. That is, $F_{\phi_i}(\tau)$ - $F_{\phi_j}(\tau)$ reflects how many more (or less) instances can be solved by using $M \cup \phi_i$ instead of $M \cup \phi_j$ within a factor τ of the baseline time. Finally, the region above $F_\phi(\tau)$ for $\tau < 1$ is very informative, as it exhibits the gain of a given ϕ_i compared to the baseline M **and** to $M \cup \phi$, i.e., the two non-fictional models. Finally, instances with similar performance give rise to step-like changes in $F(\tau)$, while a linearly growing $F(\tau)$ is symptomatic of a diversified performance across the benchmark.

Limitation of the approach: A bottleneck of our approach is the need to store the search tree in memory. After an experimentation on toy problems with only a few constraints (such as the n-queens) we found it reasonable that no more than $\sim 5 \times 10^6$ nodes are created per minute on a standard laptop. Our data structure to store the branching decisions does not use more than 40 bytes per node. Hence, assuming that 16 GB of memory are available, we can record search attempts up to 40 minutes long. We believe this time limit should be large enough to collect valuable statistics in practice.

3   Experimentation

We applied our approach to two propagators, namely *Energetic Reasoning* (ER, see [3,14]) and *Revisited Cardinality Reasoning for BinPacking* (RCRB, see [18]). Both the approaches provide powerful filtering, but are expensive to run, so that the design of more efficient implementations has a strong appeal. In order to assess the potential for improvements, we considered the following classes of fictional implementation:

- ϕ_μ^{cost}, i.e. an implementation for which the time is reduced by a factor μ.
- $\phi_{\mathcal{O}(f(n))}^{cost}$, i.e. an implementation for which the time complexity is $\mathcal{O}(f(n))$.
- ϕ_p^{oracle}, i.e. an implementation that guards ϕ with a necessary condition causing useless activations with a probability p.

The profile of ϕ_0^{oracle} (perfect necessary condition) bounds the gain that can be obtained by any necessary condition. The profile of $\phi_1^{cost}(\tau)$ (zero-cost implementation), or $\phi_{\mathcal{O}(1)}^{cost}(\tau)$, bounds the performance of any possible implementation. Against common intuition, ϕ_1^{cost} is not guaranteed to beat the baseline, since a weak filtering done by ϕ may trigger other (possibly expensive) propagators.

Experimental Set-up: We used the constraint solver *OscaR* [17] and ran instances on AMD Opteron processors (2.7 GHz). For each instance, we limited the run-time of $traverse(store(b), M)$ to 600 seconds and the run-time of $traverse(replay(b), M)$ and $traverse(replay(b), M \cup \phi)$ to 1200 seconds. Instances for which either $traverse(replay(b), M \cup \phi)$ timed out or $traverse(replay(b), M)$ took less than 1 second were filtered out. The target propagator ϕ was executed with low priority by the constraint scheduler.

Energetic Reasoning: We analyzed the *ER* propagator for the CUMULATIVE constraint[1,2] on *Resource Constrained Project Scheduling Problems* (RCPSP). The baseline model M employs the Timetabling algorithm from [4] and the ER Checker [3], which both run in $\mathcal{O}(n^2)$ [3,12]. We did not use the improvements proposed in [12]. We use a dynamic search strategy, i.e. the classical *SetTimes* approach from [16]. We consider two benchmarks: the BL instances [2] (20-25 activities) and the PSPLIB (j30 and j90, with 30 and 90 activities) [15]. We focus on investigating, for the chosen benchmarks: 1) the potential benefit of having an ER algorithm running in $O(n^2)$ rather than in $O(n^3)$; 2) the potential benefit of a perfect necessary condition (see [10] and [7] for related works).

Figure 1 and 2 report profiles respectively for the BL and j90 instances. The real ER propagator beats the baseline in $\sim 50\%$ of the cases for BL, but only in $\sim 10\%$ of the cases for j90. The larger problem size is a likely reason for the performance drop, so it is interesting to analyze the fictional, reduced-cost implementations (left-most figures). In the BL benchmark a cost reduction translates to roughly proportional benefits. On j90, an $O(n^2)$ ER would lead to dramatic performance improvement, but it would beat the baseline on only 40% of the cases. More interestingly, there is a 30% portion of instances where the baseline would win *no matter what the efficiency of ER is*, i.e. where the

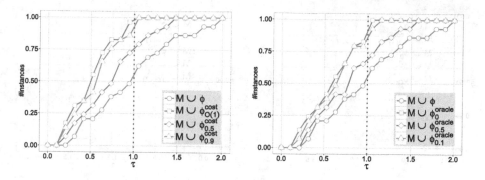

Fig. 1. Performance profiles for real and fictional ER propagators on the BL instances

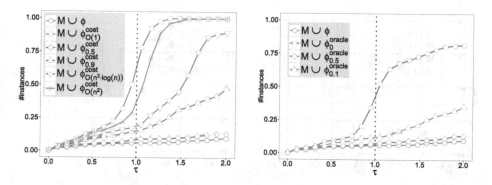

Fig. 2. Performance profiles for real and fictional ER propagators on the j90 instances

additional pruning of ER is sometimes detrimental rather than beneficial. On such instances, ER cannot lead to benefits unless we find a way to activate it only when it provides an actual advantage. As for using a necessary condition, a perfect approach would enable the same performance of a $O(n^2)$ ER, but even a small mistake probability would cancel most of the benefits.

Figure 3 compares profiles for different search strategies on j30 (*SetTimes* and a binary static approach): the potential gain of reducing the cost is very different for the two strategies, even if the performance of the real propagator is roughly identical. This points out the importance of having an approach for the rigorous comparison of propagators using practical search strategies.

Revisited Cardinality Reasoning for BinPacking: In our analysis of the RCRB propagator, we use as a benchmark the instances of the Balanced Academic Curriculum Problem (BACP) from [18,19]. The baseline model M employs the BinPacking propagator from [20] and a GCC constraint (model **A** in [18]). The search heuristic is *binary first-fail*, i.e. we choose for branching the variable with the smallest domain and we assign on the left branch the minimum value.

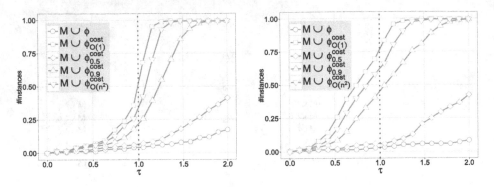

Fig. 3. Performance profiles for the SetTimes (left) and binary static (right) strategies

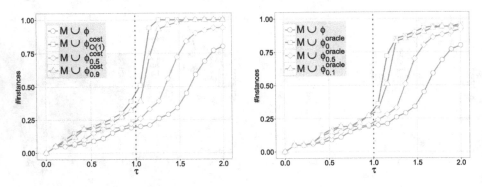

Fig. 4. Performance profiles with fictionally cost-reduced RCRB propagators

Figure 4 (left) is very informative about the cost of RCRB. We can see that less than 25% of the instances are solved faster than the baseline model. Moreover, reducing its cost down to 0 provides a small gain before $\tau = 1.1$. From then, reducing the cost by a factor 0.9 is enough to solve a lot more of the instances. Hence, reducing the cost would improve considerably the RCRB, but not that much compared to the baseline model as the benefits come "too late" in terms of τ. A similar analysis can be done for figure 4 (right).

4   Conclusion

Evaluating the potential advantages of reducing the cost of a given filtering procedure is of great importance to make our research efforts as fruitful as possible. In addition, being able to measure exactly the time gain provided by a filtering algorithm permits to reduce the bias in empirical evaluations. As a first step in this direction, we proposed a systematic methodology to simulate the performance of *fictional implementations of a propagator having reduced activation cost.* This is done *before* starting time-consuming research activities to actually reduce the cost. The approach was illustrated for Energetic Reasoning and Revisited Cardinality Reasoning for BinPacking over popular sets of instances. We found that reducing the propagator costs, *even to the point of making it negligible,* might actually be beneficial only on a small subset of a given instance set. Furthermore, this outcome can differ substantially depending on the considered benchmark and on the search strategy.

Acknowledgments. We thank the anonymous reviewers for their valuable comments. This research is partially supported by the UCLouvain Action de Recherche Concerté ICTM22C1.

References

1. Aggoun, A., Beldiceanu, N.: Extending chip in order to solve complex scheduling and placement problems. Mathematical and Computer Modelling **17**(7), 57–73 (1993)
2. Baptiste, P., Le Pape, C.: Constraint propagation and decomposition techniques for highly disjunctive and highly cumulative project scheduling problems. Constraints **5**(1–2), 119–139 (2000)
3. Baptiste, P., Le Pape, C., Nuijten, W.: Constraint-Based Scheduling: Applying Constraint Programming to Scheduling Problems, vol. 39. Springer (2001)
4. Beldiceanu, N., Carlsson, M.: A new multi-resource *cumulatives* constraint with negative heights. In: Van Hentenryck, P. (ed.) CP 2002. LNCS, vol. 2470, pp. 63–79. Springer, Heidelberg (2002)
5. Beldiceanu, N., Contejean, E.: Introducing global constraints in chip. Mathematical and computer Modelling **20**(12), 97–123 (1994)
6. Bergman, D., Ciré, A.A., van Hoeve, W.J.: MDD propagation for sequence constraints. J. Artif. Intell. Res. (JAIR) **50**, 697–722 (2014)
7. Berthold, T., Heinz, S., Schulz, J.: An approximative criterion for the potential of energetic reasoning. In: Marchetti-Spaccamela, A., Segal, M. (eds.) TAPAS 2011. LNCS, vol. 6595, pp. 229–239. Springer, Heidelberg (2011)
8. Bessière, C., Debruyne, R.: Optimal and suboptimal singleton arc consistency algorithms. In: IJCAI-2005, Proceedings of the Nineteenth International Joint Conference on Artificial Intelligence, Edinburgh, Scotland, UK, July 30-August 5, 2005, pp. 54–59 (2005)
9. Brand, S., Narodytska, N., Quimper, C.-G., Stuckey, P.J., Walsh, T.: Encodings of the SEQUENCE constraint. In: Bessière, C. (ed.) CP 2007. LNCS, vol. 4741, pp. 210–224. Springer, Heidelberg (2007)
10. Van Cauwelaert, S., Lombardi, M., Schaus, P.: Supervised learning to control energetic reasoning: feasibility study. In: Proceedings of the Doctoral Program CP2014 (2014)
11. Cheng, K.C.K., Yap, R.H.C.: An mdd-based generalized arc consistency algorithm for positive and negative table constraints and some global constraints. Constraints **15**(2), 265–304 (2010)
12. Derrien, A., Petit, T.: A new characterization of relevant intervals for energetic reasoning. In: O'Sullivan, B. (ed.) CP 2014. LNCS, vol. 8656, pp. 289–297. Springer, Heidelberg (2014)

13. Dolan, E.D., Moré, J.J.: Benchmarking optimization software with performance profiles. Mathematical programming **91**(2), 201–213 (2002)
14. Erschler, J., Lopez, P.: Energy-based approach for task scheduling under time and resources constraints. In: 2nd international workshop on project management and scheduling, pp. 115–121 (1990)
15. Kolisch, R., Schwindt, C., Sprecher, A.: Benchmark instances for project scheduling problems. In: Project Scheduling, pp. 197–212. Springer (1999)
16. Le Pape, C., Couronné, P., Vergamini, D., Gosselin, V.: Time-Versus-Capacity Compromises in Project Scheduling. (1994)
17. OscaR Team. OscaR: Scala in OR (2012). https://bitbucket.org/oscarlib/oscar
18. Pelsser, F., Schaus, P., Régin, J.-C.: Revisiting the cardinality reasoning for bin-packing constraint. In: Schulte, C. (ed.) CP 2013. LNCS, vol. 8124, pp. 578–586. Springer, Heidelberg (2013)
19. Schaus, P. et al.: Solving balancing and bin-packing problems with constraint programming. PhD thesis, Université catholique de Louvain, Louvain-la-Neuve (2009)
20. Shaw, P.: A constraint for bin packing. In: Wallace, M. (ed.) CP 2004. LNCS, vol. 3258, pp. 648–662. Springer, Heidelberg (2004)
21. van Hoeve, W.-J., Pesant, G., Rousseau, L.-M., Sabharwal, A.: Revisiting the sequence constraint. In: Benhamou, F. (ed.) CP 2006. LNCS, vol. 4204, pp. 620–634. Springer, Heidelberg (2006)

Failure-Directed Search for Constraint-Based Scheduling

Petr Vilím[1](✉), Philippe Laborie[2], and Paul Shaw[3]

[1] IBM, V Parku 2294/4, 148 00 Praha 4 - Chodov, Czech Republic
petr_vilim@cz.ibm.com
[2] IBM, 9 rue de Verdun, 94253 Gentilly Cedex, France
laborie@fr.ibm.com
[3] IBM, Les Taissounieres HB2, 2681 Route des Dolines, 06560 Valbonne, France
paul.shaw@fr.ibm.com

Abstract. This paper presents a new constraint programming search algorithm that is designed for a broad class of scheduling problems. Failure-directed Search (FDS) assumes that there is no (better) solution or that such a solution is very hard to find. Therefore, instead of looking for solution(s), it focuses on a systematic exploration of the search space, first eliminating assignments that are most likely to fail. It is a "plan B" strategy that is used once a less systematic "plan A" strategy – here, Large Neighborhood Search (LNS) – is not able to improve current solution any more. LNS and FDS form the basis of the automatic search for scheduling problems in CP Optimizer, part of IBM ILOG CPLEX Optimization Studio.

FDS and LNS+FDS (the default search in CP Optimizer) are tested on a range of scheduling benchmarks: Job Shop, Job Shop with Operators, Flexible Job Shop, RCPSP, RCPSP/max, Multi-mode RCPSP and Multi-mode RCPSP/max. Results show that the proposed search algorithm often improves best-known lower and upper bounds and closes many open instances.

Keywords: Constraint programming · Scheduling · Search · Job shop · Job shop with Operators · Flexible job shop · RCPSP · RCPSP/max · Multi-mode RCPSP · Multi-mode RCPSP/max · CPLEX · CP optimizer

1 Introduction

Generic search algorithms have become quite successful in constraint programming solvers in recent years, see for example impact-based search [22], weighted-degree heuristics [6] and activity-based search [18]. However, the authors are aware of only one attempt to use such a generic search (in particular impact-based search) for scheduling problems [37].

One of the obstacles for using the search algorithms mentioned above for scheduling is that they make branching decisions of the form $x = n \lor x \neq n$, where

© Springer International Publishing Switzerland 2015
L. Michel (Ed.): CPAIOR 2015, LNCS 9075, pp. 437–453, 2015.
DOI: 10.1007/978-3-319-18008-3_30

x is a variable and n is a value from its domain. In case of scheduling the decision $x \neq n$ usually does not propagate at all because the majority of propagation algorithms take into account only minimum and maximum values of domains (see propagation by temporal constraint networks [7], timetable [4, chapters 3.3.1 and 2.1.1] or family of edge-finding algorithms for unary and discrete cumulative resources [35,36]). A possible solution is to branch on disjunctions as proposed in [37]. However, this approach is hard to generalize from disjunctive resources to other scheduling constraints. Failure-directed search overcomes this problem in a different way: it branches by splitting a domain into two disjoint intervals so that one of the bounds of the domain is always changed.

2 Scheduling Using Constraint Programming

In this paper we consider a broad class of scheduling problems with activities, precedences between activities, unary or discrete cumulative resources and also alternatives and optional activities. For a more detailed description on modeling those problems using Constraint Programming (and in particular using IBM ILOG CP Optimizer that implements FDS) please refer to [12,14]. Here we briefly introduce the main concepts used in this paper.

Interval variable is a decision variable that represents a task/activity with unknown start and end times. It is possible to express the fact that the task is optional and may be left unperformed (*e.g.* because an alternative task was used instead). More formally domain D_v of an interval variable v is a subset of:

$$\{\bot\} \cup \{[s,e)|s,e \in Z, s \leq e\}$$

Where $\bot$ represents the case that the task is left unperformed. When $D_v = \bot$ then we say that v is *absent*, when $\bot \notin D_v$ then v is *present* and otherwise then v is *optional*.

Precedence models the fact that end/start of some interval variable must be before/after the start/end of another interval possibly with minimum/maximum delay (the delay can be negative). Precedences are propagated by a dedicated global constraint called the Temporal Network as described in [12] (inspired by [7]).

Unary resource (noOverlap) forbids any pair of intervals variables from a given set to overlap (*e.g.* because all the interval variables require the same machine). There has been a lot of work on propagation of unary resources— we are using the methods described in [35].

Discrete cumulative resource models a machine or any other resource that can process multiple tasks at once but which has a limited capacity for processing tasks simultaneously. In CP Optimizer, cumulative resources are modeled by *cumulative function expressions*. Again, there are many algorithms for propagating this constraint: we are using Timetable Edge Finding [36] and Timetabling [4, chapters3.3.1and2.1.1].

Alternative models an alternative between several optional intervals. Alternatives are used to model, for example, different modes of a task. Propagation of alternatives is described in [12].

3   Choices

Failure-directed search does not operate on decision variables directly, instead it works on a set of binary *choices*. A choice is an abstraction of anything that needs to be decided in order to obtain a solution. An obvious example of a choice is assigning a value to a variable (*e.g.* choice between $x = 5$ and $x \neq 5$). However failure-directed search is using *domain splitting* instead (see *e.g.* [9]) and the following kinds of choices:

Presence choice: for an interval variable v whether v is present ($\perp \notin D_v$) or absent ($D_v = \{\perp\}$). This kind of choice is used only for optional interval variables.

Start time choice: for an interval variable v and a time t whether startOf(v) $\leq$ t or startOf(v) $> t$. Function startOf(v) returns start time of interval variable v in a solution. Start time choice can be used only on a present interval variable v, if v is still optional then the presence choice must be applied first (see later).

At the highest level, failure-directed search knows only a set of choices that needs to be decided: it is ignorant of what the choices are doing.

For now we assume that all possible presence choices and start time choices are generated before the search starts. The topic of actual set of choices is further discussed in Section 6.1.

FDS search operates under the assumption that the current problem is infeasible, or alternatively, if there is a solution then it is hard to find (heuristic methods already failed to find it). Therefore it supposes that it will explore the whole search space (to prove infeasibility or optimality) or at least a significant part of it (before a solution is found).

With this assumption in mind, failure-directed search gives up on the idea of guiding the search towards possible solutions. It does exactly the opposite: it drives the search into conflicts in order to prove that the current branch is infeasible. Choices that fail the most are preferred. From two branches of a choice the one that fails the most is preferred. It is the well-known first-fail principle but applied also on the branch ordering.

Let's assume for a while that it is we, not the search algorithm, who decide how the search space is explored. We are given an infeasible problem, a set of predefined choices and our task is to build a complete but small search tree. We can imagine it as a game: there is a box of bricks (the choices) and the task is to build from them a search tree in a depth-first way. Our task is to repeatedly pick a choice from the box and add it into the tree. When we pick a choice, it is possible that the choice is already decided, in this case we continue picking. Otherwise the choice is added into the tree and it produces two new branches. Thanks to constraint propagation branches can fail and therefore one of the following three possibilities will happen (see Figure 1):

0-fails: Neither branch fails. From our tree-builder point of view it is a disappointment because we ended up increasing the number of open branches.

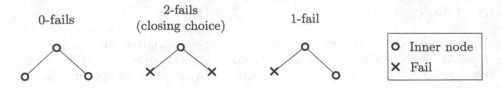

Fig. 1. Types of internal nodes

Instead of making our tree smaller we ended up increasing it. However at some points, especially near the root node, there is no other way.

2-fails: Both branches fail. In this case lets calls this choice a *closing choice* because it closes the current branch. As we are looking for a small search tree, closing choice is the best that can happen. Search tree cannot be fully explored without closing choices.

1-fail: Only one branch fails. We did not close the current branch, but at least we did not open a new one. Constraint propagation tightened the bounds, so we have better chances to close the branch next time.

Of course we do not know in advance which of the three possibilities above will happen. Instead FDS uses a system of ratings that reflects recent behavior of a choice.

4 Ratings

Ratings are the measure that failure-directed search uses in order to pick the next choice to explore. Smaller ratings are preferred. The algorithm simply picks an available choice with the best rating.

For every available choice c, the system maintains separate ratings for its positive and negative branches[1]: rating$^+[c]$ and rating$^-[c]$. Both rating$^+[c]$ and rating$^-[c]$ are initially set to 1.0. Rating of choice c is defined as:

$$\text{rating}[c] = \text{rating}^+[c] + \text{rating}^-[c] \tag{1}$$

Additionally, for every search depth, d there is average rating of choices on the given depth: avgRating$[d]$. Its initial value is also 1.0.

Like impact-based search [22], FDS computes an estimate $0 \leq R \leq 1$ of the *reduction in effort* to search the rest of the problem, given a particular assignment. For example, [22] uses the ratios of the search space sizes (using variable domains only) before and after propagation of each decision:

$$R = \frac{|D'_{x_1}| \times \cdots \times |D'_{x_n}|}{|D_{x_1}| \times \cdots \times |D_{x_n}|}$$

---

[1] Note the difference between positive/negative branch and left/right branch. When a choice is generated one of the branches is called positive and the second negative and this assignment does not change. It is up to the search algorithm to decide which of the two branches will be explored first and become the left branch of a node.

where D_v and D'_v are the domains of variable v before and after the decision, respectively.

Each time a branch of a choice c is explored its rating $\text{rating}^+[c]$ or $\text{rating}^-[c]$ is updated using the estimation of search effort reduction. The computation starts with localRating:

$$\text{localRating} := \begin{cases} 0 & \text{if the branch fails immediately} \\ 1 + R & \text{otherwise} \end{cases} \tag{2}$$

Notice that this measure puts a much greater emphasis on failures than traditional impact-based search, making FDS much more aggressive in seeking out immediate failures during search.

The local rating of a decision depends a lot on the current subproblem. In particular the same decision usually has a higher local rating near the root node than in the depths of the search tree. To compensate for this effect, localRating is normalized using the average rating on the current depth d. With this in mind, the rating of a branch (positive or negative) is updated to:

$$\text{rating}^{+/-}[c] := \alpha \cdot \text{rating}^{+/-}[c] + (1 - \alpha) \cdot \frac{\text{localRating}}{\text{avgRating}[d]} \tag{3}$$

Where α is a constant controlling the speed of decay (typical values of α range from 0.9 to 0.99). Note that update of the rating of the branch by (3) has immediate effect on rating[c] according to (1).

As ratings are decaying by factor α, they reflect the recent behavior of the choice. Ratings can change quite quickly, especially when closing decisions are encountered.

5   Search Algorithm

The search algorithm is using several data structures to store choices according to their current state, see Figure 2. The state of a choice can be:

Unchecked: The choice was not picked for branching in the current branch. Initially all choices are unchecked. Unchecked choices are stored in a heap that allows fast access to the choice with the best rating.

Decided: The choice was picked for branching, it was found applicable (*i.e.* the choice is not *resolved* or *waiting*, see below) and one of the branches was applied. The choice remains decided until the search backtracks from the decision about the choice. Decided decisions are kept on a stack in order to facilitate fast backtracking.

Resolved: Again the choice was already picked for branching but it was found to be already resolved. *e.g.* consider a choice is between startOf(v) ≤ 5 and startOf(v) > 5 where v is a present interval variable. If in the current node startOf(v) is known to be in interval $[7, 12]$ then there is no point in branching on the choice. The choice remain resolved until the search backtracks above the point where the choice was found to be resolved. Therefore resolved choices are also kept on the stack.

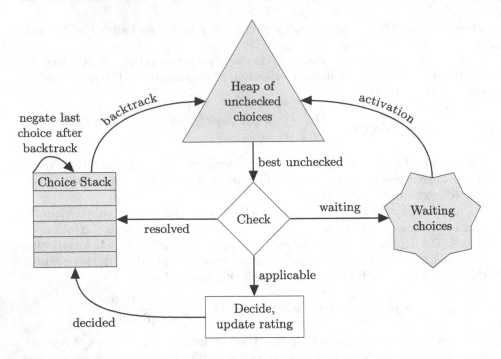

Fig. 2. Choice states and data structures

Waiting: Let's consider once more the choice between startOf(v) $\leq$ 5 and startOf(v) $>$ 5 but this time consider that v is an optional interval variable. The choice does not split the domain of v into two disjunctive subsets: the case v is absent is possible in both branches. For this reason choices like this one can be applied only when v is already present. The choice remains waiting as long as v is optional. In order to be activated at the right time the choice must monitor the status of v. Once v becomes present the choice automatically returns into the heap and becomes unchecked again.

The search proceeds as follows, see Algorithm 1. A choice with the best rating is taken from the heap and its state is checked (line 7). If it is *waiting* then another choice is taken from the heap. Similarly if the choice is *resolved* then it is put on the stack and also another choice is drawn. This process continues until an applicable choice is found (line 15). The applicable choice is decided (the branch with the better rating first), put on stack and constraint propagation is run until the fixed point. The process continues this way until constraint propagation finds the current subproblem infeasible (dead end, line 21). In this case the search backtracks: choices are removed from the stack and put back into the heap until the last choice with open branch is found. The choice is switched (the right branch is applied), constraint propagation is run and branch rating is updated.

```
 1  pick:
 2     if heap is empty then begin
 3        solution found;
 4        add improving objective cut;
 5        goto backtrack;
 6     end;
 7     remove choice c with the best rating from heap;
 8     if c is waiting then begin
 9        let c monitor the underlying interval variable;
10        goto pick;
11     end;
12     add c to stack;
13     if c is resolved then
14        goto pick;
15     b := branch of c with the better rating
16  propagate:
17     apply branch b;
18     propagate until a fixed point;
19     update rating of branch b;
20     if not infeasible then goto pick;
21  backtrack:
22     let c is the last choice with open branch on the stack;
23     if there is no such c then
24        terminate;  // whole search space was explored
25     put back into the heap all choices from the stack until c;
26     b := the unexplored branch of c;
27     goto propagate;
```

Algorithm 1. Search algorithm

6 Other Components

The previous section describes the basic failure-directed search. However there are more components that contribute to the performance of failure-directed search.

6.1 Initial Set of Choices

FDS as described in so far requires that all possible choices to be generated before the search starts. However it may be more efficient to start with only a subset of choices and generate additional ones when needed. In particular, inspired by search techniques for square packing described in [5, 26], the initial set of choices only makes sure that if they are all decided then every interval variable is either absent or has mandatory part (see Figure 3). Then, if the search gets to a point when all choices have been decided but some decision variables remain unfixed, then either a more traditional depth-first approach can be used to complete the solution, or alternatively more choices can be generated at that point, allowing the search to continue.

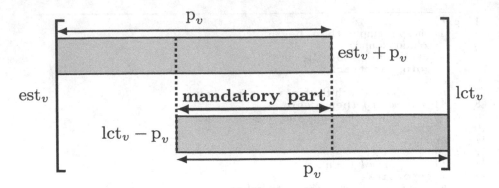

Fig. 3. Consider present interval variable v with duration p_v, earliest starting time est_v and latest completion time lct_v. If $lct_v - est_v < 2\,p_v$ then v always occupies interval $[lct_v - p_v, est_v + p_v]$. This interval is called *mandatory part*.

6.2   Restarts and Nogoods

Restarting the search is a widely used technique to improve performance by breaking out of heavy-tailed behavior typical of depth-first search [8, 19]. Similar generic search algorithms also use restarts [18, 22, 24].

The search is restarted for the first time after 100 backtracks. The restart limit is increased by 15% after each restart. These values correspond to the default parameterization of CP Optimizer (parameters `RestartFailLimit` and `RestartGrowthFactor`).

Nogoods from restarts are recorded and propagated as described in [15], which is also the same manner in which nogoods are propagated for integer search inside CP Optimizer.

FDS assumes that it will be restarted many times before it fully explores the search space. The only result that remains after each restart is a set of nogoods. The shorter they are, the easier they are to apply. That is the reason why FDS always explores first the branch that is more likely to fail.

6.3   Strong Branching and Shaving

Ratings try to estimate the behavior of choices, but they are still only estimations. At the top of the search tree, where it is most important to pick the right choices, the ratings are most imprecise.

Therefore at the root node of each restart it pays off to pre-evaluate a limited number of best choices to find out their "actual" behavior. FDS tries both branches of a number of best applicable choices from the heap and updates their ratings. After that it picks for branching a choice with a branch with the best localRating as defined by (2).

The process of pre-evaluation of different choices before committing to one of them is not new, see for example *strong branching* in MIP solvers [1] and *shaving* in CP (e.g. [29]). Shaving in particular has a different goal: to find a choice that has an infeasible branch and improve the filtering by applying the opposite branch. FDS does a similar thing: if one of the choices evaluated during strong branching has an infeasible branch then the opposite branch is applied and the pre-evaluation process continues with the reduced set of choices.

Finally while evaluating one of the choices, it can happen that some variable x is updated by constraint propagation in similar way in both branches. For example, minimum start time of x is increased from 0 to 7 in one branch and to 10 in the second branch. In this case the minimum start time of x can be increased from 0 to 7 immediately.

6.4   Coupling with LNS

As explained earlier, failure-directed search is designed for the case when the problem is infeasible or a solution is very hard to find. As FDS heads first into conflicts, it finds solutions just by a happy accident. If there are many easy-to-find solutions then FDS may not work well.

Therefore FDS is a good "plan B": when other approaches fail or are not able to improve any more then FDS can explore the whole search space. In another words, it pays off to couple FDS with another "plan A" strategy that is able to find near-optimal solutions and this way limit the search space for FDS.

In CP Optimizer the "plan A" strategy is a self-adapting Large Neighborhood Search (LNS) [11]. It consists of a process of continual relaxation and re-optimization: a first solution is computed and iteratively improved. Each iteration consists of a relaxation step followed by a re-optimization of the relaxed solution. This process continues until some condition is satisfied, typically, when the solution can be proved to be optimal or when a time limit is reached. In CP Optimizer this approach is made more robust by using portfolios of large neighborhoods and completion strategies in combination with Machine Learning techniques to converge on the most efficient neighborhoods and completion strategies for the problem being solved. Furthermore, in case of non-regular objective function (like earliness costs), some completion strategies are guided by a linear relaxation of the problem solved with LP techniques [13].

7   Experimental Results

FDS together with LNS are tested on a number of classical scheduling benchmarks listed below. Only instances that are still open are considered. The purpose of the experiments is solely to show that FDS is powerful enough to close number of open problems. A detailed study of individual features of FDS is out of scope of the paper.

Experiments are performed on a machine with Intel Core i7 2.60GHz processor (4 cores and hyperthreading) and 16GB RAM using slightly modified[2] IBM ILOG CP Optimizer version 12.6.1. The instances are solved by two different methods:

---

[2] With minor performance improvements.

```
1      LB := best known lower bound − 1;
2   checkLB:
3      solve with upper bound set to LB and specified time limit;
4      if solution found then begin
5         terminate; // Optimal solution found
6      end;
7      if infeasible then begin
8         LB := LB+1;
9         goto checkLB;
10     end;
11     terminate; // Time limit was hit. LB is new valid lower bound
```

Algorithm 2. Destructive lower bounds

LNS+FDS: This configuration is just the standard automatic search of CP
Optimizer with a parameterization to use two CPU threads (`Workers=2`)
and more aggressive FDS (`FailureDirectedSearchEmphasis=0.99`). At the
beginning, both threads use LNS, but once LNS is not able to improve the
current solution for some time (determined automatically by auto-tuning in
CP Optimizer) one of the two threads switches from LNS to FDS.

DestructLB: This approach tries to improve best known lower bounds by proving them wrong iteratively, see Algorithm 2. This time only one CPU thread
is used and FDS is started immediately[3]. In order to make a fair comparison
with the state of the art, the algorithm first tries to confirm the current best
known lower bound and only then tries to improve it. The time limit for
DestructLB specified in the benchmark description is used for each *iteration*
of the algorithm, the total running time of the algorithm is not limited (this
way the result is not biased by the initial value of the lower bound).

Results are summarized in Table 1. Column *Instances* gives the number of
open instances of the benchmark, the *Lower bound* and *Upper bound improvements* columns give the number of lower and upper bounds improved by CP
Optimizer respectively. The last column represents the number of instances that
were closed by CP Optimizer. Detailed lists of improved lower and upper bounds
can be found at http://ibm.biz/FDSearch.

7.1 Job Shop ($J||C_{max}$)

For job shop scheduling problems, we focus on the open classical instances of [28]
(`tail*`, 32 open instances), [2] (`abz*`, 3 open instances), [27] (`swv*`, 9 open
instances) and [38] (`yam*`, 4 open instances). The current lower and upper bounds
for these instances were gathered from [30] and [31].

In the case of job shop, computation of current best known lower and upper
bounds usually took a very long time. *e.g.* computation of upper bounds in [21]
used a time limit of 30000 seconds (8 hours 20 minutes) using a dedicated local

---

[3] In version 12.6.1 IBM ILOG CP Optimizer does not offer yet a public API to run
FDS directly and replicate the reported results.

Table 1. Results summary

Benchmark set	Number of instances	Lower bound improvements	Upper bound improvements	Closed instances
JobShop	48	40	3	15
JobShopOperators	222	107	215	208
FlexibleJobShop	107	67	39	74
RCPSP	472	52	1	0
RCPSPMax	58	51	23	1
MultiModeRCPSP (j30)	552	No reference	3	535
MultiModeRCPSPMax	85	84	77	85

search algorithm. FDS is not a local search, it explores the whole search space, so an even bigger time limit would make sense. We decided to use the same time limit of 30000s but two threads in LNS+FDS approach and 10 minutes per iteration for DestructLB.

The DestructLB approach, despite the small time limit, was able to improve lower bounds for 40 of the 48 instances and close 4 instances (improving the upper bound for 2 of them). The LNS+FDS approach closed 15 instances, including the 4 instances already closed by DestructLB. Solve times ranged from 50 minutes (`tail12`) to 7.5 hours (`tail21`).

This benchmark illustrates the benefits of the automatic search of CP Optimizer that couples LNS and FDS together using two threads (LNS+FDS). Let's take a closer look at instance `tail19` by Taillard. After 388s LNS finds a solution with makespan 1352 which is only 1.5% from the optimum value of 1332. Such a tight upper bound limits the search space for FDS and at time 1061.2s FDS finds a solution with makespan 1351. This solution is passed to LNS and LNS improves it immediately (in 0.32s) to 1350. LNS continues improving the solution, reaching the optimal value of 1332 at 8518s. In parallel, FDS is systematically exploring the search space while taking advantage of the new upper bounds as they come from LNS. Finally after 12853s in total, FDS proves that there is no better solution and the search stops.

In general, FDS is able to help LNS to escape local minima by providing a new (possibly totally different) solution. LNS can use this solution as a new starting point and further improve it. And in the opposite direction, LNS provides tight upper bounds to FDS and removes from FDS the burden to guide the search towards possible solutions. This way FDS can concentrate only on the fastest way to explore the search space.

7.2 Job Shop with Operators

The job shop scheduling problem with operators is an extension of the classical job shop scheduling problem proposed in [3] where each operation also requires an operator to aid in the processing of the operation (beside the machine). An operator can process only one operation at a time and the total number of operators in the shop is limited. The whole set of operators is modelled by a single discrete cumulative resource. Results are compared with the current best known lower and upper bounds provided by the approach described in [17] on the 222 open problems. We used a time limit of 600s for the LNS+FDS approach and 300s per iteration of the DestructLB algorithm. Both LNS+FDS and DestructLB were able to close many instances. In all, 208 instances were closed, with 107 lower bounds and 215 upper bounds being improved.

7.3   Flexible Job Shop ($FJ||C_{\max}$)

Flexible job shop scheduling problems are an extension of classical job shop scheduling problems for production environments where it is possible to run an operation on more than one machine. Current lower and upper bounds were taken from [32][4]. Out of the 107 open instances, the LNS+FDS approach closes 74 instances (resulting in an indirect improvement of 61 lower bounds among these instances) and improves 39 upper bounds. Those results with LNS+FDS were obtained using a time-limit of up to 8h. The DestructLB approach with a time limit of 3600s per iteration was able to additionally improve 10 lower bounds.

7.4   RCPSP ($PS|prec|C_{\max}$)

For Resource Constrained Project Scheduling Problems (RCPSP), we focus on the 472 open instances of the PSPLib [10]. Current lower and upper bounds were taken from [24].

   This benchmark allows direct comparison with the approach of [24] as they also compute destructive lower bounds in exactly the same way on a machine with the same speed. Therefore we used the same time limits as [24]: 10 minutes for LNS+FDS and 10 minutes for one iteration of DestructLB. The DestructLB approach improves 52 lower bounds (by 1, 2 or 3), proves the same lower bound as [24] for 330 instances and is not able to prove the same lower bound within the time limit for 90 instances. We conclude that in terms of lower bounds FDS achieve similar results as [24] despite the fact that FDS does not use use explanations as Schutt *et al.* does.

   In terms of upper bounds, LNS+FDS is clearly worse despite using two threads instead of one. Only one upper bound is improved and only in 78 cases the upper bound is the same. No open instance of RCPSP was closed.

7.5   RCPSP/max ($PS|temp|C_{\max}$)

For Resource Constrained Project Scheduling Problems with minimal and maximal time lags (RCPSP/max), we use the best known lower and upper bounds reported in [25]. We used again a time limit of 10 minutes for both LNS+FDS and DestructLB iteration.

---

[4] Note that this page already includes most of the results reported in the present article under the reference [CPO].

The DestructLB approach improves lower bound for 56 out of 57 open instances and proves optimality for psp_j30_73. However as [25] did not compute destructive lower bounds, a direct comparison is not possible.

The LNS+DFS approach improves best upper bounds for 23 instances, for 10 instances it reaches the same upper bound (again proving optimality for psp_j30_73) but produces a poorer upper bound for 25 instances. Direct comparison with [25] is again not possible because we used 2 threads on a faster machine.

To summarize, one instance is closed by CP Optimizer and the average gap (defined as $(UB - LB)/UB$) is reduced from 23.37% to 13.62%.

7.6 Multi-mode RCPSP ($MPS|prec|C_{\max}$)

Multi-Mode Resource Constrained Project Scheduling Problems are extensions of classical RCPSP allowing for alternative execution modes of the tasks. We worked with the j30* instances of the PSPLib [33] (all other instances are closed). The bounds reported in [33] include some recent improvements described in [20]. As [33] reports only upper bounds, we worked with all the 552 feasible instances of the problem. We used a time limit of 3600s for LNS+FDS and 600s per iteration for DestructLB. Both LNS+FDS and DestructLB were able to close many instances. In all, 535 instances were closed (almost 97%) and 3 upper bounds are improved.

7.7 Multi-mode RCPSP/max ($MPS|temp|C_{\max}$)

Multi-Mode Resource Constrained Project Scheduling Problems with minimal and maximal time lags combine the two extensions of the classical RCPSP. We focused on the 85 open instances of the PSPLib (7 in the mm50 group, 79 in the mm100 group) given the lower and upper bounds reported in [34]. These bounds include some recent improvements described in [23].

DestructLB with a time limit of 300s was able to improve 53 lower bounds. LNS+FDS using a time limit of 1800s was able to improve 73 upper bounds. The combination of the two approaches closes 10 instances.

In fact, this benchmark turns out to be very peculiar because the renewable (cumulative) resources are not the hardest part of the problem. We exploited this remark to implement an alternative approach that first solves a MIP relaxation of the problem that exactly handles all constraints except the renewable resources. The MIP model has numerical variables s_i for the start time of each activity i and boolean variables m_{ij} for selecting the mode of activity i. Renewable resources are relaxed using a basic energy reasoning over the schedule horizon. The MIP is solved using CPLEX 12.6.1. The optimal makespan of the MIP clearly is a lower bound on the makespan of the original problem. In a second step, we find with CP Optimizer the optimal solution to the RCPSP/max problem that uses the optimal mode allocation of the MIP. It turns out that for 83 instances out of 85, the optimal makespan of this problem is equal to the MIP lower bound and thus is an optimal solution to the original problem. For the 2 remaining instances, we re-injected the optimal solution of the RCPSP/max as a warm start into the original Multi-Mode RCPSP/max model using CP Optimizer starting point functionality. The LNS+FDS approach then improves on the starting point and produces a solution with a makespan equal to the MIP lower bound. In conclusion, all 85 open instances were closed.

8   Analysis of FDS Behavior

It is still not completely clear why FDS is so successful on some problem instances. Therefore we tried to analyze the behavior of FDS in order to get better understanding. This section summarizes our observations.

One (perhaps not surprising) observation is that FDS produces unbalanced search trees. Left branches often fail immediately or at least they are explored much faster than right branches. Common are long chains of 1-fail nodes ended by a closing node as demonstrated in Figure 4.

An important feature of failure-directed search is that once a closing choice is found, it is likely to be reused again immediately after the backtrack. It is common that the same closing choice is reused several times before it is no longer closing. This way the search can quickly escape even from deep search depths. Similarly 1-fail choices are also usually quickly reused. The behavior of FDS is in this sense very similar to the one of *quick shaving* [16].

A choice chosen for branching in a root node after a restart is most likely to be a choice that was recently closing before the restart. As usual, the choice is probably unbalanced, *e.g.* it could be a choice between start time in interval $[1, 4]$ versus $[5, 100]$. Therefore when the left branch of the root node is proved infeasible, FDS improves the domain only a little (from $[1, 100]$ to $[5, 100]$). However as the choice used to be closing, even such a small improvement is probably important. As FDS accumulates those small improvements (and nogoods in general), search space is reduced and constraint propagation becomes stronger.

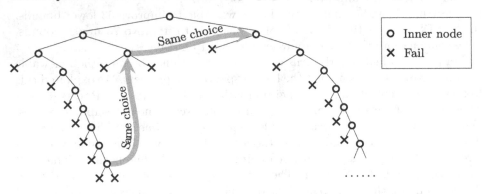

Fig. 4. Reusing closing choices

Heading First into Conflict. The importance of heading first into conflict can be demonstrated for example on job shop instance `tail50`. It takes 465 seconds for FDS to prove that there is no solution with makespan 1832 or lower. Lets compare that with reverse branching order: pick the choices as usual (low rating first) but switch the branching order to worse rating first. With this change the same proof takes 1023 seconds.

The reason why branching order is important seems to be the fact that the search is periodically restarted. When low rating branches are explored first then the generated nogoods from restarts are shorter and easier to apply.

Preferring conflicts. We perform one more experiment with the same job shop instance, this time to demonstrate the importance of preferring immediate failures in computation of ratings. Lets replace $1 + R$ by R in formula (2):

$$\text{localRating} := \begin{cases} 0 & \text{if the branch fails immediately} \\ R & \text{otherwise} \end{cases}$$

This new version of localRating resembles much more impact-based search. With this change the proof that used to take 465 seconds does not finish within 24 hours.

9 Conclusions

Using failure-directed search we were able to improve the state-of-the-art results for a number of scheduling benchmarks covering disjunctive and cumulative resources, minimum and maximum lags and multiple modes. Results demonstrate that FDS and CP Optimizer's automatic search (LNS+FDS) can compete with specialized algorithms and even outperform them.

Failure-directed search has been an integral part of the CP Optimizer automatic search algorithm since version 12.6.0.

References

1. Achterberg, T., Koch, T., Martin, A.: Branching rules revisited. Operations Research Letters **33**, 42–54 (2004)
2. Adams, J., Balas, E., Zawack, D.: The shifting bottleneck procedure for job shop scheduling. Management Science **34**(3), 391–401 (1988)
3. Agnetis, A., Flamini, M., Nicosia, G., Pacifici, A.: A job-shop problem with one additional resource type. Journal of Scheduling **14**, 225–237 (2011)
4. Baptiste, P., Pape, C.L., Nuijten, W.: Constraint-Based Scheduling: Applying Constraint Programming to Scheduling Problems. Kluwer Academic Publishers (2001)
5. Beldiceanu, N., Carlsson, M., Demassey, S., Poder, E.: New filtering for the cumulative constraint in the context of non-overlapping rectangles. Annals of Operations Research **184**(1), 27–50 (2011)

6. Boussemart, F., Hemery, F., Lecoutre, C., Saïs, L.: Boosting systematic search by weighting constraints. In: de Mántaras, R.L., Saitta, L. (eds.) ECAI. pp. 146–150. IOS Press (2004)
7. Dechter, R., Meiri, I., Pearl, J.: Temporal constraint networks. Artificial Intelligence **49**(1–3), 61–95 (1991)
8. Gomes, C.P., Selman, B., Crato, N., Kautz, H.: Heavy-tailed phenomena in satisfiability and constraint satisfaction problems. J. Autom. Reason. **24**(1–2), 67–100 (2000)
9. Jussien, N., Lhomme, O.: Dynamic domain splitting for numeric CSPs. In: Proc. European Conference on Artificial Intelligence, pp. 224–228 (1998)
10. Kolisch, R., Sprecher, A.: PSPLIB - a project scheduling problem library. European Journal of Operational Research **96**, 205–216 (1996). http://www.om-db.wi.tum.de/psplib/main.html
11. Laborie, P., Godard, D.: Self-adapting large neighborhood search: application to single-mode scheduling problems. In: Proceedings of the 3rd Multidisciplinary International Conference on Scheduling: Theory and Applications (MISTA), pp. 276–284 (2007)
12. Laborie, P., Rogerie, J.: Reasoning with conditional time-intervals. In: Wilson, D., Lane, H.C. (eds.) Proceedings of the 21st International Florida Artificial Intelligence Research Society Conference, pp. 555–560. AAAI Press (2008)
13. Laborie, P., Rogerie, J.: Temporal Linear Relaxation in IBM ILOG CP Optimizer. Journal of Scheduling (2014)
14. Laborie, P., Rogerie, J., Shaw, P., Vilím, P.: Reasoning with conditional time-intervals. part II: an algebraical model for resources. In: Lane, H.C., Guesgen, H.W. (eds.) Proceedings of the 22nd International Florida Artificial Intelligence Research Society Conference. AAAI Press, Sanibel Island, Florida, USA, 19–21 May 2009
15. Lecoutre, C., Saïs, L., Tabary, S., Vidal, V.: Nogood recording from restarts. In: 20th International Joint Conference on Artificial Intelligence (IJCAI 2007), pp. 131–136 (2007)
16. Lhomme, O.: Quick shaving. In: Veloso, M.M., Kambhampati, S. (eds.) AAAI, pp. 411–415. AAAI Press/The MIT Press (2005)
17. Mencía, R., Sierra, M.R., Mencía, C., Varela, R.: A genetic algorithm for job-shop scheduling with operators enhanced by weak lamarckian evolution and search space narrowing. Natural Computing **13**, 179–192 (2014)
18. Michel, L., Van Hentenryck, P.: Activity-based search for black-box constraint programming solvers. In: Beldiceanu, N., Jussien, N., Pinson, É. (eds.) CPAIOR 2012. LNCS, vol. 7298, pp. 228–243. Springer, Heidelberg (2012)
19. Moskewicz, M.W., Madigan, C.F., Zhao, Y., Zhang, L., Malik, S.: Chaff: engineering an efficient SAT solver. In: Annual ACM IEEE Design Automation Conference, pp. 530–535. ACM (2001)
20. Muller, L.F.: An adaptive large neighborhood search algorithm for the multi-mode RCPSP. Tech. Rep. Report 3.2011, Department of Management Engineering, Technical University of Denmark (2011)
21. Pardalos, P.M., Shylo, O.V.: An algorithm for the job shop scheduling problem based on global equilibrium search techniques. Computational Management Science **3**(4), 331–348 (2006)
22. Refalo, P.: Impact-based search strategies for constraint programming. In: Wallace, M. (ed.) CP 2004. LNCS, vol. 3258, pp. 557–571. Springer, Heidelberg (2004)

23. Schnell, A., Hartl, R.F.: Optimizing the multi-mode resource-constrained project scheduling problem with standard and generalized precedence relations by constraint programming and boolean satisfiability solving techniques (working Paper) (2014)
24. Schutt, A., Feydy, T., Stuckey, P.J.: Explaining time-table-edge-finding propagation for the cumulative resource constraint. In: Gomes, C., Sellmann, M. (eds.) CPAIOR 2013. LNCS, vol. 7874, pp. 234–250. Springer, Heidelberg (2013)
25. Schutt, A., Feydy, T., Stuckey, P.J., Wallace, M.G.: Solving RCPSP/max by lazy clause generation. Journal of Scheduling 16(3), 273–289 (2013). http://ww2.cs.mu.oz.au/pjs/rcpsp/rcpspmax_all.html (Accessed 1 November 2014)
26. Simonis, H., O'Sullivan, B.: Search strategies for rectangle packing. In: Stuckey, P. (ed.) CP 2008. LNCS, vol. 5202, pp. 52–66. Springer, Heidelberg (2008)
27. Storer, R., Wu, S., Vaccari, R.: New search spaces for sequencing problems with application to job shop scheduling. Management Science 38(10), 1495–1509 (1992)
28. Taillard, E.: Benchmarks for basic scheduling problems. European Journal of Operations Research 64, 278–285 (1993)
29. Torres, P., Lopez, P.: Overview and possible extensions of shaving techniques for job-shop problems. In: Junker, U., Karisch, S., Tschöke, S. (eds.) Proceedings of 2nd International Workshop on the Integration of AI and OR Techniques in Constraint Programming for Combinatorial Optimization Problems, pp. 181–186 (2000)
30. http://optimizizer.com/jobshop.php (Accessed 1 November 2014)
31. http://tinyurl.com/nl85fhy (Accessed 1 November 2014)
32. http://tinyurl.com/kvm8nuk (Accessed 1 November 2014)
33. http://tinyurl.com/nn2j599 (Accessed 1 November 2014)
34. http://tinyurl.com/n8oahua (Accessed 1 November 2014)
35. Vilím, P.: Global Constraints in Scheduling. Ph.D. thesis, Charles University in Prague, Faculty of Mathematics and Physics, Department of Theoretical Computer Science and Mathematical Logic (2007)
36. Vilím, P.: Timetable edge finding filtering algorithm for discrete cumulative resources. In: Achterberg, T., Beck, J. (eds.) CPAIOR 2011. LNCS, vol. 6697, pp. 230–245. Springer, Heidelberg (2011)
37. Wolf, A.: Impact-based search in constraint-based scheduling. In: Hegering, H., Lehmann, A., Ohlbach, H.J., Scheideler, C. (eds.) Informatik 2008, Beherrschbare Systeme - dank Informatik, Band 2, Beiträge der 38. Jahrestagung der Gesellschaft für Informatik e.V. (GI), in München, 8–13 September, LNI, vol. 134, pp. 523–528. GI (2008)
38. Yamada, T., Nakano, R.: A genetic algorithm applicable to large-scale job-shop problems. In: Männer, R., Manderick, B. (eds.) Proc. 2nd International Workshop on Parallel Problem Solving from Nature, pp. 281–290 (1992)

Author Index

Printed in the United States
By Bookmasters